SIXTH EDITION

Fundamentals of Physics

PART 1

ENHANCED PROBLEMS VERSION

SIXTH EDITION

Fundamentals of Physics

PART 1

ENHANCED PROBLEMS VERSION

David Halliday

University of Pittsburgh

Robert Resnick

Rensselaer Polytechnic Institute

Jearl Walker

Cleveland State University

John Wiley & Sons, Inc.

ACQUISITIONS EDITOR   Stuart Johnson
DEVELOPMENTAL EDITOR   Ellen Ford
MARKETING MANAGER   Bob Smith
SENIOR PRODUCTION MANAGER   Lucille Buonocore
SENIOR PRODUCTION EDITORS   Monique Calello, Elizabeth Swain
TEXT/COVER DESIGNER   Madelyn Lesure
COVER PHOTO   Tsuyoshi Nishiinoue/Orion Press
PHOTO MANAGER   Hilary Newman
DUMMY DESIGNER   Lee Goldstein
ILLUSTRATION EDITORS   Edward Starr and Anna Melhorn
ILLUSTRATION   Radiant/Precision Graphics

This book was set in 10/12 Times Roman by Progressive Information Technologies and printed and bound by Von Hoffmann Press, Inc. The cover was printed by Von Hoffman Press.

This book is printed on acid-free paper.

The paper in this book was manufactured by a mill whose forest management programs include sustained yield harvesting of its timberlands. Sustained yield harvesting principles ensure that the number of trees cut each year does not exceed the amount of new growth.

To order books or for customer service please, call 1(800)-CALL-WILEY (225-5945).

ISBN 0-471-22860-5

Printed in the United States of America

10 9 8 7 6 5 4 3

BRIEF CONTENTS

TABLES

CONTENTS

CHAPTER 12

Rolling, Torque, and Angular Momentum   245

Why is a quadruple somersault so difficult in trapeze acts?

APPENDICES

ANSWERS TO CHECKPOINTS AND ODD-NUMBERED QUESTIONS, EXERCISES AND PROBLEMS   AN-1

INDEX   I-1

PREFACE

The Goal

The principle goal of this book is to provide instructors with a tool by which they can teach students how to effectively read scientific material and successfully reason through scientific questions. To sharpen this tool, this enhanced version of the sixth edition of *Fundamentals of Physics* contains over 1000 new, high-quality problems that require thought and reasoning rather than simplistic plugging of data into formulas.

The new problems are printed at the end of each chapter. They are not presorted in any way for students but, pedagogically, most of them can be grouped into three types:

➤ *Graphs as puzzles.* These are problems that give a graph and ask for a result that requires much more than just reading off a data point from the graph. Rather, the solution requires an understanding of the physical arrangement in a problem and the principles behind the associated equations. These problems are more like puzzles in that a student must decide what data are important, like Sherlock Holmes must decide what evidence is important. Here are three examples from over 130 similar problems:
- Chapter 9, Problem N14
- Chapter 22, Problem N3
- Chapter 30, Problem N6

➤ *Novel situations.* Here is one example:
- Chapter 5, problem N1: A true story of how Air Canada flight 143 ran out of fuel at an altitude of 7.9 km because the crew and airport personnel did not consider the units for the fuel. An important lesson for students who tend to "blow off" units.

➤ *Sorting-out problems.* These are problems that focus a student on a derivation, a commonly confusing point, or the limitation of an approximation. Here are three examples:
- Chapter 11, Problem N9
- Chapter 18, Problem N14
- Chapter 26, Problem N12

Sixth Edition

The sixth edition of *Fundamentals of Physics* contains a redesign and major rewrites of the widely used fifth edition, while maintaining many elements of the classic text first

written by David Halliday and Robert Resnick. Nearly all the changes are based on suggestions from instructors and students using the fifth edition, from reviewers of the manuscripts for the sixth edition, and from research done on the process of learning. You can send suggestions, corrections, and positive or negative comments to John Wiley & Sons (http://www.wiley.com/college/halliday) or Jearl Walker (mail address: Physics Department, Cleveland State University, Cleveland OH 44115 USA; fax number: (USA) (216) 687-2424; or email address: physics@wiley.com). We may not be able to respond to all suggestions, but we keep and study each of them.

Design Changes

➤ *More open format.* Previous editions have been printed in a double-column format, which many students and instructors have found cluttered and distracting. In this edition, the narrative is presented in a single-column format with a wide margin for note-taking.

➤ *Streamlined presentation.* It is a common complaint of all texts that they cover too much material. As a response to this criticism, the sixth edition has been shortened in two ways.

1. Material regarding special relativity and quantum physics has been moved from the early chapters to the later chapters devoted to those subjects.

2. Only the essential sample problems have been retained in the book.

➤ *Vector notation.* Vectors are now presented with an overhead arrow (such as $\vec{F}$) instead of as a bold symbol (such as **F**).

➤ *Emphasis on metric units.* Except in Chapter 1 (in which various systems of units are employed) and certain problems involving baseball (in which English units are traditional), metric units are used almost exclusively.

➤ *Structured versus unstructured order of problems.* The homework problems in this book are still ordered approximately according to their difficulty and grouped under section titles corresponding to the narrative of the chapter. However, the new problems that are now at the end of each chapter are not ordered or grouped in any way.

➤ *Icons for additional help.* When worked-out solutions are provided either in print or electronically for certain of the odd-numbered homework problems, the statements for those problems include a trailing icon to alert both student and instructor as to where the solutions are located. An icon guide is provided here and at the beginning of each set of homework problems:

> ssm   Solution is in the Student Solutions Manual.
> www   Solution is available on the World Wide Web at:
>   http://www.wiley.com/college/halliday
> ilw   Solution is available on the Interactive LearningWare

These resources are described later in this preface.

Pedagogy Changes

➤ *Reasoning versus plug-and-chug.* A primary goal of this book is to teach students to reason through challenging situations, from basic principles to a solution. Although some plug-and-chug homework problems remain in this book, most homework problems emphasize reasoning.

➤ *Key Ideas in the sample problems.* The solutions to all the sample problems have been rewritten to begin with one or more Key Ideas based on basic principles.

➤ *Lengthened solutions to sample problems.* Most of the solutions to the sample problems are now longer because they build step by step from the beginning Key Ideas to an answer, often repeating some of the important reasoning of the narrative preceding the sample problems. For example, see Sample Problem 8-3 on page 148 and Sample Problem 10-2 on pages 200–201.

➤ *Use of vector-capable calculators.* When vector calculations in a sample problem can be performed directly on-screen with a vector-capable calculator, the solution of the sample problem indicates that fact but still carries through the traditional component analysis. When vector calculations cannot be performed directly on-screen, the solution explains why.

➤ *Problems with applied physics,* based on published research, have been added in many places, either as sample problems or homework problems. For example, see Sample Problem 11-6 on page 229, homework problem 64 on page 71, and homework problem 56 on page 214. For an example of homework problems that build with a continuing story, see problems 4, 32, and 48 (on pages 112, 114, and 115, respectively) in Chapter 6.

Content Changes

➤ *Chapter 5 on force and motion* contains clearer explanations of the gravitational force, weight, and normal force (pages 80–82).

➤ *Chapter 7 on kinetic energy and work* begins with a rough definition of energy. It then defines kinetic energy, work, and the work–kinetic energy theorem in ways that are more closely tied to Newton's second law than in the fifth edition, while keeping those definitions consistent with thermodynamics (pages 117–120).

➤ *Chapter 8 on the conservation of energy* avoids the much criticized definition of work done by a nonconservative force by explaining, instead, the energy transfers that occur due to a nonconservative force (page 153). (The wording still allows an instructor to superimpose a definition of work done by a nonconservative force.)

➤ *Chapter 10 on collisions* now presents the general situation of inelastic one-dimensional collisions (pages 198–200) before the special situation of elastic one-dimensional collisions (pages 202–204).

➤ *Chapters 16, 17, and 18 on SHM and waves* have been rewritten to better ease a student into these difficult subjects.

➤ *Chapter 21 on entropy* now presents a Carnot engine as the ideal heat engine with the greatest efficiency.

Chapter Features

➤ *Opening puzzlers.* A curious puzzling situation opens each chapter and is explained somewhere within the chapter, to entice a student to read the chapter.

➤ *Checkpoints* are stopping points that effectively ask the student, "Can you answer this question with some reasoning based on the narrative or sample problem that you just read?" If not, then the student should go back over that previous material before traveling deeper into the chapter. For example, see Checkpoint 3 on page 78 and Checkpoint 1 on page 101. **Answers to all checkpoints are in the back of the book.**

➤ *Sample problems* have been chosen to help the student organize the basic concepts of the narrative and to develop problem-solving skills. Each sample problem builds step by step from one or more Key Ideas to a solution.

➤ *Problem-solving tactics* contain helpful instructions to guide the beginning physics student as to how to solve problems and to avoid common errors.

➤ *Review & Summary* is a brief outline of the chapter contents that contains the essential concepts but which is not a substitute for reading the chapter.

➤ *Questions* are like the checkpoints and require reasoning and understanding rather than calculations. **Answers to the odd-number questions are in the back of the book.**

➤ *Exercises & Problems* are ordered approximately according to difficulty and grouped under section titles. **The odd-numbered ones are answered in the back of the book.** Worked-out solutions to the odd-numbered problems with trailing icons are available either in print or electronically. (See the icon guide at the beginning of the Exercises & Problems.) A problem number with a star indicates an especially challenging problem.

➤ *Additional Problems* appear at the end of the Exercises & Problems in certain chapters. They are not sorted according to section titles and many involve applied physics.

➤ *New Problems* appear at the ends of Chapters 1 through 38. They are not ordered or sorted in any way. Many use graphs where students must decide what data are important.

Versions of the Text

The Enhanced Problems Version of the sixth edition of *Fundamentals of Physics* is available in a number of different versions, to accommodate the individual needs of instructors and students. The Regular Edition consists of Chapters 1 through 38 (ISBN 0-471-22863-X). The Extended Edition contains seven additional chapters on quantum physics and cosmology (Chapters 1–45) (ISBN 0-471-22858-3). Both editions are available as single, hardcover books, or in the following alternative versions:

➤ *Volume 1—Chapters 1–21 (Mechanics/Thermodynamics), hardcover, 0-471-22861-3*

➤ *Volume 2—Chapters 22–45 (E&M and Modern Physics), hardcover, 0-471-22858-3*

➤ *Part 1—Chapters 1–12, paperback, 0-471-22860-5*

➤ *Part 2—Chapters 13–21, paperback, 0-471-22859-1*

➤ *Part 3—Chapters 22–33, paperback, 0-471-22857-5*

➤ *Part 4—Chapters 34–38, paperback, 0-471-22856-7*

➤ *Part 5—Chapters 39–45, paperback, 0-471-36038-4*

Supplements

The sixth edition of *Fundamentals of Physics* is supplemented by a comprehensive ancillary package carefully developed to help teachers teach and students learn.

Instructor's Supplements

➤ *Instructor's Manual* by J. RICHARD CHRISTMAN, U.S. Coast Guard Academy. This manual contains lecture notes outlining the most important topics of each chapter, demonstration experiments, laboratory and computer projects, film and video sources, answers to all Questions, Exercises & Problems, and Checkpoints, and a correlation guide to the Questions and Exercises & Problems in the previous edition.

➤ *Instructor's Solutions Manual* by JAMES WHITENTON, Southern Polytechnic University. This manual provides worked-out solutions for all the exercises and problems found at the end of each chapter within the text and in the Problem Supplement #1. *This supplement is available only to instructors.*

➤ *Test Bank* by J. RICHARD CHRISTMAN, U.S. Coast Guard Academy. More than 2200 multiple-choice questions are included in this manual. These items are also available in the Computerized Test Bank (see below).

➤ *Instructor's Resource CD.* This CD contains:
- All of the Instructor's Solutions Manual in both LaTex and PDF files.
- Computerized Test Bank in both IBM and Macintosh versions, with full editing features to help instructors customize tests.
- All text illustrations suitable for both classroom presentation and printing.

➤ *Transparencies.* More than 200 four-color illustrations from the text are provided in a form suitable for projection in the classroom.

➤ *On-line Course Management.*
- WebAssign, CAPA, and WebTest are on-line homework and quizzing programs that give instructors the ability to deliver and grade homework and quizzes over the Internet.
- Instructors will also have access to WebCT course materials. WebCT is a powerful Web site program that allows instructors to set up complete on-line courses with chat rooms, bulletin boards, quizzing, student tracking, etc. Please contact your local Wiley representative for more information.

Student's Supplements

➤ *A Student Companion* by J. RICHARD CHRISTMAN, U.S. Coast Guard Academy. This student study guide consists of a traditional print component and an accompanying Web site, which together provide a rich, interactive environment for review and study. The Student Companion Web site includes self-quizzes, simulation exercises, hints for solving end-of-chapter problems, the *Interactive LearningWare* program (see the next page), and links to other Web sites that offer physics tutorial help.

➤ *Student Solutions Manual* by J. RICHARD CHRISTMAN, U.S. Coast Guard Academy and EDWARD DERRINGH, Wentworth Institute. This manual provides students with complete worked-out solutions to 30 percent of the exercises and problems found at the end of each chapter within the text. These problems are indicated with an `ssm` icon in the text.

➤ *Interactive LearningWare.* This software guides students through solutions to 200 of the end-of-chapter problems. The solutions process is developed interactively, with appropriate feedback and access to error-specific help for the most common mistakes. These problems are indicated with an `ilw` icon in the text.

➤ *CD-Physics, 3.0.* This CD-ROM based version of *Fundamentals of Physics,* Sixth Edition, contains the complete, extended version of the text, *A Student's Companion,* the *Student's Solutions Manual,* the *Interactive LearningWare,* and numerous simulations all connected with extensive hyperlinking.

➤ *Take Note!* This bound notebook lets students take notes directly onto large, black-and-white versions of textbook illustrations. All of the illustrations from the transparency set are included. In-class time spent copying illustrations is substantially reduced by this supplement.

➤ *Physics Web Site.* This Web site, **http://www.wiley.com/ college/halliday**, was developed specifically for *Fundamentals of Physics,* Sixth Edition, and is designed to further assist students in the study of physics and offers additional physics resources. The site also includes solutions to selected end-of-chapter problems. These problems are identified with a `www` icon in the text.

ACKNOWLEDGMENTS

A textbook contains far more contributions to the elucidation of a subject than those made by the authors alone. J. Richard Christman, of the U.S. Coast Guard Academy, has once again created many fine supplements for us; his knowledge of our book and his recommendations to students and faculty are invaluable. James Tanner, of Georgia Institute of Technology, and Gary Lewis, of Kennesaw State College, have provided us with innovative software, closely tied to the text exercises and problems. James Whitenton, of Southern Polytechnic State University, and Jerry Shi, of Pasadena City College, performed the Herculean task of working out solutions for every one of the Exercises & Problems in the text. We thank John Merrill, of Brigham Young University, and Edward Derringh, of the Wentworth Institute of Technology, for their many contributions in the past. We also thank George W. Hukle of Oxnard, California, for his check of the "old problems" and Renee M. Goertzen of Cleveland State University for her check of the "new problems."

At John Wiley publishers, we have been fortunate to receive strong coordination and support from our former editor, Cliff Mills. Cliff guided our efforts and encouraged us along the way. When Cliff moved on to other responsibilities at Wiley, we were ably guided to completion by his successor, Stuart Johnson. Ellen Ford has coordinated the developmental editing and multilayered preproduction process. Sue Lyons, our marketing manager, has been tireless in her efforts on behalf of this edition. Joan Kalkut has built a fine supporting package of ancillary materials. Thomas Hempstead managed the reviews of manuscript and the multiple administrative duties admirably.

We thank Lucille Buonocore and Pam Kennedy, our production directors, and Monique Calello and Elizabeth Swain, our production editors, for pulling all the pieces together and guiding us through the complex production process. We also thank Maddy Lesure, for her design; Helen Walden for her copy editing; Edward Starr and Anna Melhorn, for managing the line art program; Georgia Mederer, Katrina Avery, and Lilian Brady, for their proofreading; and all other members of the production team.

Hilary Newman and her team of photo researchers were inspired in their search for unusual and interesting photographs that communicate physics principles beautifully. We also owe a debt of gratitude for the line art to the late John Balbalis, whose careful hand and understanding of physics can still be seen in every diagram.

We especially thank Edward Millman for his developmental work on the manuscript. With us, he has read every word, asking many questions from the point of view of a student. Many of his questions and suggested changes have added to the clarity of this volume.

We owe a particular debt of gratitude to the numerous students who used the previous editions of *Fundamentals of Physics* and took the time to fill out the response cards and return them to us. As the ultimate consumers of this text, students are extremely important to us. By sharing their opinions with us, your students help us ensure that we are providing the best possible product and the most value for their textbook dollars. We encourage the users of this book to contact us with their thoughts and concerns so that we can continue to improve this text in the years to come.

Finally, our external reviewers have been outstanding and we acknowledge here our debt to each member of that team:

Edward Adelson
Ohio State University

Mark Arnett
Kirkwood Community College

Arun Bansil
Northeastern University

J. Richard Christman
U.S. Coast Guard Academy

Robert N. Davie, Jr.
St. Petersburg Junior College

Cheryl K. Dellai
Glendale Community College

Eric R. Dietz
California State University at Chico

N. John DiNardo
Drexel University

Harold B. Hart
Western Illinois University

Rebecca Hartzler
Edmonds Community College

Joey Huston
Michigan State University

Shawn Jackson
University of Tulsa

Hector Jimenez
University of Puerto Rico

Sudhakar B. Joshi
York University

Leonard M. Kahn
University of Rhode Island

Yuichi Kubota
Cornell University

Priscilla Laws
Dickinson College

Edbertho Leal
Polytechnic University of Puerto Rico

Dale Long
Virginia Tech

Andreas Mandelis
University of Toronto

Paul Marquard
Caspar College

James Napolitano
Rensselaer Polytechnic Institute

Des Penny
Southern Utah University

Joe Redish
University of Maryland

Timothy M. Ritter
University of North Carolina at Pembroke

Gerardo A. Rodriguez
Skidmore College

John Rosendahl
University of California at Irvine

Michael Schatz
Georgia Institute of Technology

Michael G. Strauss
University of Oklahoma

Dan Styer
Oberlin College

Marshall Thomsen
Eastern Michigan University

Fred F. Tomblin
New Jersey Institute of Technology

B. R. Weinberger
Trinity College

William M. Whelan
Ryerson Polytechnic University

William Zimmerman, Jr.
University of Minnesota

Reviewers of the Fifth and Previous Editions

Maris A. Abolins
Michigan State University

Barbara Andereck
Ohio Wesleyan University

Albert Bartlett
University of Colorado

Michael E. Browne
University of Idaho

Timothy J. Burns
Leeward Community College

Joseph Buschi
Manhattan College

Philip A. Casabella
Rensselaer Polytechnic Institute

Randall Caton
Christopher Newport College

J. Richard Christman
U.S. Coast Guard Academy

Roger Clapp
University of South Florida

W. R. Conkie
Queen's University

Peter Crooker
University of Hawaii at Manoa

William P. Crummett
Montana College of Mineral Science and Technology

Eugene Dunnam
University of Florida

Robert Endorf
University of Cincinnati

F. Paul Esposito
University of Cincinnati

Jerry Finkelstein
San Jose State University

Alexander Firestone
Iowa State University

Alexander Gardner
Howard University

Andrew L. Gardner
Brigham Young University

John Gieniec
Central Missouri State University

John B. Gruber
San Jose State University

Ann Hanks
American River College

Samuel Harris
Purdue University

Emily Haught
Georgia Institute of Technology

Laurent Hodges
Iowa State University

John Hubisz
North Carolina State University

Joey Huston
Michigan State University

Darrell Huwe
Ohio University

Claude Kacser
University of Maryland

Leonard Kleinman
University of Texas at Austin

Earl Koller
Stevens Institute of Technology

Arthur Z. Kovacs
Rochester Institute of Technology

Kenneth Krane
Oregon State University

Sol Krasner
University of Illinois at Chicago

Peter Loly
University of Manitoba

Robert R. Marchini
Memphis State University

David Markowitz
University of Connecticut

Howard C. McAllister
University of Hawaii at Manoa

W. Scott McCullough
Oklahoma State University

James H. McGuire
Tulane University

David M. McKinstry
Eastern Washington University

Joe P. Meyer
Georgia Institute of Technology

Roy Middleton
University of Pennsylvania

Irvin A. Miller
Drexel University

Eugene Mosca
United States Naval Academy

Michael O'Shea
Kansas State University

Patrick Papin
San Diego State University

George Parker
North Carolina State University

Robert Pelcovits
Brown University

Oren P. Quist
South Dakota State University

Jonathan Reichart
SUNY—Buffalo

Manuel Schwartz
University of Louisville

Darrell Seeley
Milwaukee School of Engineering

Bruce Arne Sherwood
Carnegie Mellon University

John Spangler
St. Norbert College

Ross L. Spencer
Brigham Young University

Harold Stokes
Brigham Young University

Jay D. Strieb
Villanova University

David Toot
Alfred University

J. S. Turner
University of Texas at Austin

T. S. Venkataraman
Drexel University

Gianfranco Vidali
Syracuse University

Fred Wang
Prairie View A & M

Robert C. Webb
Texas A & M University

George Williams
University of Utah

David Wolfe
University of New Mexico

1 Measurement

You can watch the Sun set and disappear over a calm ocean once while you lie on a beach, and then once again if you stand up. Surprisingly, by measuring the time between the two sunsets, you can approximate Earth's radius.

How can such a simple observation be used to measure Earth?

The answer is in this chapter.

1-1 Measuring Things

Physics is based on measurement. We discover physics by learning how to measure the quantities that are involved in physics. Among these quantities are length, time, mass, temperature, pressure, and electric current.

We measure each physical quantity in its own units, by comparison with a **standard.** The **unit** is a unique name we assign to measures of that quantity—for example, meter (or m) for the quantity length. The standard corresponds to exactly 1.0 unit of the quantity. As you will see, the standard for length, which corresponds to exactly 1.0 m, is the distance traveled by light in a vacuum during a certain fraction of a second. We can define a unit and its standard in any way we care to. However, the important thing is to do so in such a way that scientists around the world will agree that our definitions are both sensible and practical.

Once we have set up a standard, say, for length, we must work out procedures by which any length whatever, be it the radius of a hydrogen atom, the wheelbase of a skateboard, or the distance to a star, can be expressed in terms of the standard. Rulers, which approximate our length standard, give us one such procedure for measuring length. However, many of our comparisons must be indirect. You cannot use a ruler, for example, to measure the radius of an atom or the distance to a star.

There are so many physical quantities that it is a problem to organize them. Fortunately, they are not all independent; for example, speed is the ratio of a length to a time. Thus, what we do is pick out—by international agreement—a small number of physical quantities, such as length and time, and assign standards to them alone. We then define all other physical quantities in terms of these *base quantities* and their standards (called *base standards*). Speed, for example, is defined in terms of the base quantities length and time and the associated base standards.

Base standards must be both accessible and invariable. If we define the length standard as the distance between one's nose and the index finger on an outstretched arm, we certainly have an accessible standard—but it will, of course, vary from person to person. The demand for precision in science and engineering pushes us to aim first for invariability. We then exert great effort to make duplicates of the base standards that are accessible to those who need them.

1-2 The International System of Units

In 1971, the 14th General Conference on Weights and Measures picked seven quantities as base quantities, thereby forming the basis of the International System of Units, abbreviated SI from its French name and popularly known as the *metric system.* Table 1-1 shows the units for the three base quantities—length, mass, and time—that we use in the early chapters of this book. These units were defined to be on a "human scale."

Many SI *derived units* are defined in terms of these base units. For example, the SI unit for power, called the **watt** (symbol: W), is defined in terms of the base units for mass, length, and time. Thus, as you will see in Chapter 7,

$$1 \text{ watt} = 1 \text{ W} = 1 \text{ kg} \cdot \text{m}^2/\text{s}^3, \tag{1-1}$$

where the last collection of unit symbols is read as kilogram–square meter per second–cubed.

To express the very large and very small quantities that we often run into in physics, we use *scientific notation*, which employs powers of 10. In this notation,

$$3\,560\,000\,000 \text{ m} = 3.56 \times 10^9 \text{ m} \tag{1-2}$$

and

$$0.000\,000\,492 \text{ s} = 4.92 \times 10^{-7} \text{ s.} \tag{1-3}$$

TABLE 1-1 Some SI Base Units

Quantity	Unit Name	Unit Symbol
Length	meter	m
Time	second	s
Mass	kilogram	kg

TABLE 1-2   Prefixes for SI Units

Factor	Prefix[a]	Symbol
10^{24}	yotta-	Y
10^{21}	zetta-	Z
10^{18}	exa-	E
10^{15}	peta-	P
10^{12}	tera-	T
10^{9}	**giga-**	**G**
10^{6}	**mega-**	**M**
10^{3}	**kilo-**	**k**
10^{2}	hecto-	h
10^{1}	deka-	da
10^{-1}	deci-	d
10^{-2}	**centi-**	**c**
10^{-3}	**milli-**	**m**
10^{-6}	**micro-**	**μ**
10^{-9}	**nano-**	**n**
10^{-12}	**pico-**	**p**
10^{-15}	femto-	f
10^{-18}	atto-	a
10^{-21}	zepto-	z
10^{-24}	yocto-	y

[a]The most commonly used prefixes are shown in bold type.

Scientific notation on computers sometimes takes on an even briefer look, as in 3.56 E9 and 4.92 E−7, where E stands for "exponent of ten." It is briefer still on some calculators, where E is replaced with an empty space.

As a further convenience when dealing with very large or very small measurements, we use the prefixes listed in Table 1-2. As you can see, each prefix represents a certain power of 10, as a factor. Attaching a prefix to an SI unit has the effect of multiplying by the associated factor. Thus, we can express a particular electric power as

$$1.27 \times 10^{9} \text{ watts} = 1.27 \text{ gigawatts} = 1.27 \text{ GW} \tag{1-4}$$

or a particular time interval as

$$2.35 \times 10^{-9} \text{ s} = 2.35 \text{ nanoseconds} = 2.35 \text{ ns}. \tag{1-5}$$

Some prefixes, as used in milliliter, centimeter, kilogram, and megabyte, are probably familiar to you.

1-3   Changing Units

We often need to change the units in which a physical quantity is expressed. We do so by a method called *chain-link conversion*. In this method, we multiply the original measurement by a **conversion factor** (a ratio of units that is equal to unity). For example, because 1 min and 60 s are identical time intervals, we have

$$\frac{1 \text{ min}}{60 \text{ s}} = 1 \quad \text{and} \quad \frac{60 \text{ s}}{1 \text{ min}} = 1.$$

Thus, the ratios (1 min)/(60 s) and (60 s)/(1 min) can be used as conversion factors. This is *not* the same as writing $\frac{1}{60} = 1$ or $60 = 1$; each *number* and its *unit* must be treated together.

Because multiplying any quantity by unity leaves it unchanged, we can introduce such conversion factors wherever we find them useful. In chain-link conversion, we use the factors to cancel unwanted units. For example, to convert 2 min to seconds, we have

$$2 \text{ min} = (2 \text{ min})(1) = (2 \text{ min}) \left(\frac{60 \text{ s}}{1 \text{ min}} \right) = 120 \text{ s}. \tag{1-6}$$

If you introduce a conversion factor in such a way that unwanted units do *not* cancel, invert the factor and try again. In conversions, the units obey the same algebraic rules as variables and numbers.

Appendix D and the inside back cover give conversion factors between SI and other systems of units, including non-SI units still used in the United States. However, the conversion factors are written in the style of "1 min = 60 s" rather than as a ratio as used previously. The following sample problem gives an example of how to set up such ratios.

Sample Problem 1-1

When Pheidippides ran from Marathon to Athens in 490 B.C. to bring word of the Greek victory over the Persians, he probably ran at a speed of about 23 rides per hour (rides/h). The ride is an ancient Greek unit for length, as are the stadium and the plethron: 1 ride was defined to be 4 stadia, 1 stadium was defined to be 6 plethra, and, in terms of a modern unit, 1 plethron is 30.8 m. How fast did Pheidippides run in kilometers per second (km/s)?

SOLUTION: The Key Idea in chain-link conversions is to write the conversion factors as ratios that will eliminate unwanted units. Here we write

$$23 \text{ rides/h} = \left(23 \frac{\text{rides}}{\text{h}} \right) \left(\frac{4 \text{ stadia}}{1 \text{ ride}} \right) \left(\frac{6 \text{ plethra}}{1 \text{ stadium}} \right)$$

$$\times \left(\frac{30.8 \text{ m}}{1 \text{ plethron}} \right) \left(\frac{1 \text{ km}}{1000 \text{ m}} \right) \left(\frac{1 \text{ h}}{3600 \text{ s}} \right)$$

$$= 4.7227 \times 10^{-3} \text{ km/s} \approx 4.7 \times 10^{-3} \text{ km/s}.$$

(Answer)

Sample Problem 1-2

The cran is a British volume unit for freshly caught herrings: 1 cran = 170.474 liters (L) of fish, about 750 herrings. Suppose that, to be cleared through customs in Saudia Arabia, a shipment of 1255 crans must be declared in terms of cubic covidos, where the covido is an Arabic unit of length: 1 covido = 48.26 cm. What is the required declaration?

SOLUTION: From Appendix D we see that 1 L is equivalent to 1000 cm³. A Key Idea then helps: To convert from *cubic* centimeters to *cubic* covidos, we must *cube* the conversion ratio between centimeters and covidos. Thus, we write the following chain-link conversion:

$$1255 \text{ crans}$$
$$= (1255 \text{ crans}) \left(\frac{170.474 \text{ L}}{1 \text{ cran}} \right) \left(\frac{1000 \text{ cm}^3}{1 \text{ L}} \right) \left(\frac{1 \text{ covido}}{48.26 \text{ cm}} \right)^3$$
$$= 1.903 \times 10^3 \text{ covidos}^3. \hspace{2cm} \text{(Answer)}$$

PROBLEM-SOLVING TACTICS

Tactic 1: *Significant Figures and Decimal Places*

If you calculated the answer to Sample Problem 1-1 without your calculator automatically rounding it off, the number 4.722 666 666 67 × 10⁻³ might have appeared in the display. The precision implied by this number is meaningless. We rounded the answer to 4.7 × 10⁻³ km/s so as not to imply that it is more precise than the given data. The given speed of 23 rides/h consists of two digits, called **significant figures.** Thus, we rounded the answer to two significant figures. In this book, final results of calculations are often rounded to match the least number of significant figures in the given data. (However, sometimes an extra significant figure is kept.) When the leftmost of the digits to be discarded is 5 or more, the last remaining digit is rounded up; otherwise it is retained as is. For example, 11.3516 is rounded to three significant figures as 11.4 and 11.3279 is rounded to three significant figures as 11.3. (The answers to sample problems in this book are usually presented with the symbol = instead of ≈ even if rounding off is involved.)

When a number such as 3.15 or 3.15 × 10³ is provided in a problem, the number of significant figures is apparent, but how about the number 3000? Is it known to only one significant figure (could it be written as 3 × 10³)? Or is it known to as many as four significant figures (could it be written as 3.000 × 10³)? In this book, we assume that all the zeros in such given numbers as 3000 are significant, but you had better not make that assumption elsewhere.

Don't confuse *significant figures* with *decimal places*. Consider the lengths 35.6 mm, 3.56 m, and 0.00356 m. They all have three significant figures but they have one, two, and five decimal places, respectively.

1-4 Length

In 1792, the newborn Republic of France established a new system of weights and measures. Its cornerstone was the meter, defined to be one ten-millionth of the distance from the north pole to the equator. Later, for practical reasons, this Earth standard was abandoned and the meter came to be defined as the distance between two fine lines engraved near the ends of a platinum–iridium bar, the **standard meter bar,** which was kept at the International Bureau of Weights and Measures near Paris. Accurate copies of the bar were sent to standardizing laboratories throughout the world. These **secondary standards** were used to produce other, still more accessible standards so that ultimately every measuring device derived its authority from the standard meter bar through a complicated chain of comparisons.

Eventually, modern science and technology required a more precise standard than the distance between two fine scratches on a metal bar. In 1960, a new standard for the meter, based on the wavelength of light, was adopted. Specifically, the standard for the meter was redefined to be 1 650 763.73 wavelengths of a particular orange-red light emitted by atoms of krypton-86 (a particular isotope, or type, of krypton) in a gas discharge tube. This awkward number of wavelengths was chosen so that the new standard would be close to the old meter-bar standard.

By 1983, however, the demand for higher precision had reached such a point that even the krypton-86 standard could not meet it, and in that year a bold step was taken. The meter was redefined as the distance traveled by light in a specified time interval. In the words of the 17th General Conference on Weights and Measures:

TABLE 1-3 Some Approximate Lengths

Measurement	Length in Meters
Distance to the first galaxies formed	2×10^{26}
Distance to the Andromeda galaxy	2×10^{22}
Distance to the nearest star (Proxima Centauri)	4×10^{16}
Distance to Pluto	6×10^{12}
Radius of Earth	6×10^{6}
Height of Mt. Everest	9×10^{3}
Thickness of this page	1×10^{-4}
Length of a typical virus	1×10^{-8}
Radius of a hydrogen atom	5×10^{-11}
Radius of a proton	1×10^{-15}

> The meter is the length of the path traveled by light in a vacuum during a time interval of 1/299 792 458 of a second.

This time interval was chosen so that the speed of light c is exactly

$$c = 299\ 792\ 458 \text{ m/s.}$$

Measurements of the speed of light had become extremely precise, so it made sense to adopt the speed of light as a defined quantity and to use it to redefine the meter.

Table 1-3 shows a wide range of lengths, from that of the universe to those of some very small objects.

PROBLEM-SOLVING TACTICS

Tactic 2: *Order of Magnitude*

The *order of magnitude* of a number is the power of ten when the number is expressed in scientific notation. For example, if $A = 2.3 \times 10^4$ and $B = 7.8 \times 10^4$, then the orders of magnitude of both A and B are 4.

Often, engineering and science professionals will estimate the result of a calculation to the *nearest* order of magnitude. For our example, the nearest order of magnitude is 4 for A and 5 for B. Such estimation is common when detailed or precise data required in the calculation are not known or easily found. Sample Problem 1-3 gives an example.

Sample Problem 1-3

The world's largest ball of string is about 2 m in radius. To the nearest order of magnitude, what is the total length L of the string in the ball?

SOLUTION: We could, of course, take the ball apart and measure the total length L, but that would take great effort and make the ball's builder most unhappy. A Key Idea here is that, because we want only the nearest order of magnitude, we can estimate any quantities required in the calculation.

Let us assume the ball is spherical with radius $R = 2$ m. The string in the ball is not closely packed (there are uncountable gaps between nearby sections of string). To allow for these gaps, let us somewhat overestimate the cross-sectional area of the string by assuming the cross section is square, with an edge length $d = $

4 mm. Then, with a cross-sectional area of d^2 and a length L, the string occupies a total volume of

$$V = \text{(cross-sectional area)(length)} = d^2 L.$$

This is approximately equal to the volume of the ball, given by $\frac{4}{3}\pi R^3$, which is about $4R^3$ because π is about 3. Thus, we have

$$d^2 L = 4R^3,$$

or

$$L = \frac{4R^3}{d^2} = \frac{4(2 \text{ m})^3}{(4 \times 10^{-3} \text{ m})^2}$$

$$= 2 \times 10^6 \text{ m} \approx 10^6 \text{ m} = 10^3 \text{ km.} \quad \text{(Answer)}$$

(Note that you do not need a calculator for such a simplified calculation.) Thus, to the nearest order of magnitude, the ball contains about 1000 km of string!

TABLE 1-4   Some Approximate Time Intervals

Measurement	Time Interval in Seconds
Lifetime of the proton (predicted)	1×10^{39}
Age of the universe	5×10^{17}
Age of the pyramid of Cheops	1×10^{11}
Human life expectancy	2×10^{9}
Length of a day	9×10^{4}
Time between human heartbeats	8×10^{-1}
Lifetime of the muon	2×10^{-6}
Shortest lab light pulse	6×10^{-15}
Lifetime of the most unstable particle	1×10^{-23}
The Planck time^a	1×10^{-43}

^{a}This is the earliest time after the big bang at which the laws of physics as we know them can be applied.

Fig. 1-1 When the metric system was proposed in 1792, the hour was redefined to provide a 10-hour day. The idea did not catch on. The maker of this 10-hour watch wisely provided a small dial that kept conventional 12-hour time. Do the two dials indicate the same time?

1-5 Time

Time has two aspects. For civil and some scientific purposes, we want to know the time of day so that we can order events in sequence. In much scientific work, we want to know how long an event lasts. Thus, any time standard must be able to answer two questions: *"When did it happen?"* and "What is its *duration*?" Table 1-4 shows some time intervals.

Any phenomenon that repeats itself is a possible time standard. Earth's rotation, which determines the length of the day, has been used in this way for centuries; Fig. 1-1 shows one novel example of a watch based on that rotation. A quartz clock, in which a quartz ring is made to vibrate continuously, can be calibrated against Earth's rotation via astronomical observations and used to measure time intervals in the laboratory. However, the calibration cannot be carried out with the accuracy called for by modern scientific and engineering technology.

To meet the need for a better time standard, atomic clocks have been developed. An atomic clock at the National Institute of Standards and Technology (NIST) in Boulder, Colorado, is the standard for Coordinated Universal Time (UTC) in the United States. Its time signals are available by shortwave radio (stations WWV and WWVH) and by telephone (303-499-7111). Time signals (and related information) are also available from the United States Naval Observatory at Web site http://tycho.usno.navy.mil/time.html. (To set a clock extremely accurately at your particular location, you would have to account for the travel time that is required for these signals to reach you.)

Figure 1-2 shows variations in the length of one day on Earth over a 4-year period, as determined by comparison with a cesium (atomic) clock. Because the variation displayed by Fig. 1-2 is seasonal and repetitious, we suspect the rotating Earth when there is a difference between Earth and atom as timekeepers. The variation is probably due to tidal effects caused by the Moon and to large-scale winds.

The 13th General Conference on Weights and Measures in 1967 adopted a standard second based on the cesium clock:

> One second is the time taken by 9 192 631 770 oscillations of the light (of a specified wavelength) emitted by a cesium-133 atom.

Atomic clocks are so consistent that, in principle, two cesium clocks would have to run for 6000 years before their readings would differ by more than 1 s. Even such accuracy pales in comparison to that of clocks currently being developed; their precision may be 1 part in 10^{18}—that is, 1 s in 1×10^{18} s (about 3×10^{10} y).

Fig. 1-2 Variations in the length of the day over a 4-year period. Note that the entire vertical scale amounts to only 3 ms (3 milliseconds = 0.003 s).

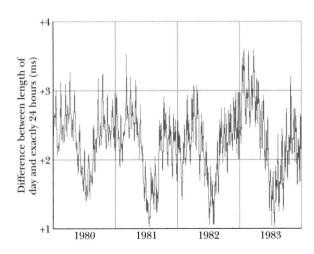

Sample Problem 1-4*

Suppose that while lying on a beach watching the Sun set over a calm ocean, you start a stopwatch just as the top of the Sun disappears. You then stand, elevating your eyes by a height $h = 1.70$ m, and stop the watch when the top of the Sun again disappears. If the elapsed time on the watch is $t = 11.1$ s, what is the radius r of Earth?

SOLUTION: A Key Idea here is that just as the Sun disappears, your line of sight to the top of the Sun is tangent to Earth's surface. Two such lines of sight are shown in Fig. 1-3. There your eyes are located at point A while you are lying, and at height h above point A while you are standing. For the latter situation, the line of sight is tangent to Earth's surface at point B. Let d represent the distance between point B and the location of your eyes when you are standing, and draw radii r as shown in Fig. 1-3. From the Pythagorean theorem, we then have

$$d^2 + r^2 = (r + h)^2 = r^2 + 2rh + h^2,$$

or
$$d^2 = 2rh + h^2. \qquad (1\text{-}7)$$

Because the height h is so much smaller than Earth's radius r, the term h^2 is negligible compared to the term $2rh$, and we can rewrite Eq. 1-7 as

$$d^2 = 2rh. \qquad (1\text{-}8)$$

In Fig. 1-3, the angle between the radii to the two tangent points A and B is θ, which is also the angle through which the Sun moves about Earth during the measured time $t = 11.1$ s. During a full day, which is approximately 24 h, the Sun moves through an angle of 360° about Earth. This allows us to write

$$\frac{\theta}{360°} = \frac{t}{24 \text{ h}},$$

*Adapted from "Doubling Your Sunsets, or How Anyone Can Measure the Earth's Size with a Wristwatch and Meter Stick," by Dennis Rawlins, *American Journal of Physics,* Feb. 1979, Vol. 47, pp. 126–128. This technique works best at the equator.

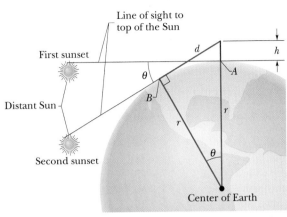

Fig. 1-3 Sample Problem 1-4. Your line of sight to the top of the setting Sun rotates through the angle θ when you stand up at point A and elevate your eyes by a distance h. (Angle θ and distance h are exaggerated here for clarity.)

which, with $t = 11.1$ s, gives us

$$\theta = \frac{(360°)(11.1 \text{ s})}{(24 \text{ h})(60 \text{ min/h})(60 \text{ s/min})} = 0.04625°.$$

Again in Fig. 1-3, we see that $d = r \tan \theta$. Substituting this for d in Eq. 1-8 gives us

$$r^2 \tan^2\theta = 2rh,$$

or
$$r = \frac{2h}{\tan^2\theta}.$$

Substituting $\theta = 0.04625°$ and $h = 1.70$ m, we find

$$r = \frac{(2)(1.70 \text{ m})}{\tan^2 0.04625°} = 5.22 \times 10^6 \text{ m}, \qquad \text{(Answer)}$$

which is within 20% of the accepted value (6.37×10^6 m) for the (mean) radius of Earth.

1-6 Mass

The Standard Kilogram

The SI standard of mass is a platinum–iridium cylinder (Fig. 1-4) kept at the International Bureau of Weights and Measures near Paris and assigned, by international agreement, a mass of 1 kilogram. Accurate copies have been sent to standardizing laboratories in other countries, and the masses of other bodies can be determined by balancing them against a copy. Table 1-5 shows some masses expressed in kilograms, ranging over about 83 orders of magnitude.

Fig. 1-4 The international 1 kg standard of mass, a platinum–iridium cylinder 3.9 cm in height and in diameter.

TABLE 1-5   Some Approximate Masses

Object	Mass in Kilograms
Known universe	1×10^{53}
Our galaxy	2×10^{41}
Sun	2×10^{30}
Moon	7×10^{22}
Asteroid Eros	5×10^{15}
Small mountain	1×10^{12}
Ocean liner	7×10^{7}
Elephant	5×10^{3}
Grape	3×10^{-3}
Speck of dust	7×10^{-10}
Penicillin molecule	5×10^{-17}
Uranium atom	4×10^{-25}
Proton	2×10^{-27}
Electron	9×10^{-31}

The U.S. copy of the standard kilogram is housed in a vault at NIST. It is removed, no more than once a year, for the purpose of checking duplicate copies that are used elsewhere. Since 1889, it has been taken to France twice for recomparison with the primary standard.

A Second Mass Standard

The masses of atoms can be compared with each other more precisely than they can be compared with the standard kilogram. For this reason, we have a second mass standard. It is the carbon-12 atom, which, by international agreement, has been assigned a mass of 12 **atomic mass units** (u). The relation between the two units is

$$1 \text{ u} = 1.6605402 \times 10^{-27} \text{ kg}, \tag{1-9}$$

with an uncertainty of ± 10 in the last two decimal places. Scientists can, with reasonable precision, experimentally determine the masses of other atoms relative to the mass of carbon-12. What we presently lack is a reliable means of extending that precision to more common units of mass, such as a kilogram.

REVIEW & SUMMARY

Measurement in Physics   Physics is based on measurement of physical quantities. Certain physical quantities have been chosen as **base quantities** (such as length, time, and mass); each has been defined in terms of a **standard** and given a **unit** of measure (such as meter, second, and kilogram). Other physical quantities are defined in terms of the base quantities and their standards and units.

SI Units   The unit system emphasized in this book is the International System of Units (SI). The three physical quantities displayed in Table 1-1 are used in the early chapters. Standards, which must be both accessible and invariable, have been established for these base quantities by international agreement. These standards are used in all physical measurement, for both the base quantities and the quantities derived from them. Scientific notation and the prefixes of Table 1-2 can be used to simplify measurement notation in many cases.

Changing Units   Conversion of units from one system to another (for example, from miles per hour to kilometers per second)

may be performed by using *chain-link conversions* in which the original data are multiplied successively by conversion factors written as unity and the units are manipulated like algebraic quantities until only the desired units remain.

Length   The unit of length—the meter—is defined as the distance traveled by light during a precisely specified time interval.

Time   The unit of time—the second—was formerly defined in terms of the rotation of Earth. It is now defined in terms of the oscillations of light emitted by an atomic (cesium-133) source. Accurate time signals are sent worldwide by radio signals keyed to atomic clocks in standardizing laboratories.

Mass   The unit of mass—the kilogram—is defined in terms of a particular platinum–iridium prototype kept near Paris, France. For measurements on an atomic scale, the atomic mass unit, defined in terms of the atom carbon-12, is usually used.

EXERCISES & PROBLEMS

ssm   Solution is in the Student Solutions Manual.
www   Solution is available on the World Wide Web at:
           http://www.wiley.com/college/hrw
ilw   Solution is available on the Interactive LearningWare.

SEC. 1-4   Length

1E. The micrometer (1 μm) is often called the *micron*. (a) How many microns make up 1.0 km? (b) What fraction of a centimeter equals 1.0 μm? (c) How many microns are in 1.0 yd?   ssm

2E. Two types of *barrel* units were in use in the 1920s in the United States. The apple barrel had a legally set volume of 7056 cubic inches; the cranberry barrel, 5826 cubic inches. If a merchant sells 20 cranberry barrels of goods to a customer who thinks he is receiving apple barrels, what is the discrepancy in the shipment volume in liters?

3E. Horses are to race over a certain English meadow for a distance of 4.0 furlongs. What is the race distance in units of (a) rods and (b) chains? (1 furlong = 201.168 m, 1 rod = 5.0292 m, and 1 chain = 20.117 m.)   ssm

4E. Spacing in this book was generally done in units of points and picas: 12 points = 1 pica, and 6 picas = 1 inch. If a figure was misplaced in the page proofs by 0.80 cm, what was the misplacement in (a) points and (b) picas?

5E. Earth is approximately a sphere of radius 6.37×10^6 m. What are (a) its circumference in kilometers, (b) its surface area in square kilometers, and (c) its volume in cubic kilometers? ssm

6E. An old manuscript reveals that a landowner in the time of King Arthur held 3.00 acres of plowed land plus a livestock area of 25.0 perches by 4.00 perches. What was the total area in (a) the old unit of roods and (b) the more modern unit of square meters? Here, 1 acre is an area of 40 perches by 4 perches, 1 rood is 40 perches by 1 perch, and 1 perch is 16.5 ft.

7P. Antarctica is roughly semicircular, with a radius of 2000 km (Fig. 1-5). The average thickness of its ice cover is 3000 m. How many cubic centimeters of ice does Antarctica contain? (Ignore the curvature of Earth.) ssm

Fig. 1-5 Problem 7.

8P. In the United States, a doll house has the scale of 1:12 of a real house (that is, each length of the doll house is $\frac{1}{12}$ that of the real house) and a miniature house (a doll house to fit within a doll house) has the scale of 1:144 of a real house. Suppose a real house (Fig. 1-6) has a front length of 20 m, a depth of 12 m, a height of 6.0 m, and a standard sloped roof (vertical triangular faces on the ends) of height 3.0 m. In cubic meters, what are the volumes of the corresponding (a) doll house and (b) miniature house?

Fig. 1-6 Problem 8.

9P. Hydraulic engineers in the United States often use, as a unit of volume of water, the *acre-foot*, defined as the volume of water that will cover 1 acre of land to a depth of 1 ft. A severe thunderstorm dumped 2.0 in. of rain in 30 min on a town of area 26 km². What volume of water, in acre-feet, fell on the town? ssm ilw www

SEC. 1-5 Time

10E. Physicist Enrico Fermi once pointed out that a standard lecture period (50 min) is close to 1 microcentury. (a) How long is a microcentury in minutes? (b) Using

$$\text{percentage difference} = \left(\frac{\text{actual} - \text{approximation}}{\text{actual}}\right)100,$$

find the percentage difference from Fermi's approximation.

11E. Express the speed of light, 3.0×10^8 m/s, in (a) feet per nanosecond and (b) millimeters per picosecond. ssm

12E. A unit of time sometimes used in microscopic physics is the *shake*. One shake equals 10^{-8} s. (a) Are there more shakes in a second than there are seconds in a year? (b) Humans have existed for about 10^6 years, whereas the universe is about 10^{10} years old. If the age of the universe now is taken to be 1 "universe day," for how many "universe seconds" have humans existed?

13P. Five clocks are being tested in a laboratory. Exactly at noon, as determined by the WWV time signal, on successive days of a week the clocks read as in the following table. Rank the five clocks according to their relative value as good timekeepers, best to worst. Justify your choice. ssm

Clock	Sun.	Mon.	Tues.	Wed.	Thurs.	Fri.	Sat.
A	12:36:40	12:36:56	12:37:12	12:37:27	12:37:44	12:37:59	12:38:14
B	11:59:59	12:00:02	11:59:57	12:00:07	12:00:02	11:59:56	12:00:03
C	15:50:45	15:51:43	15:52:41	15:53:39	15:54:37	15:55:35	15:56:33
D	12:03:59	12:02:52	12:01:45	12:00:38	11:59:31	11:58:24	11:57:17
E	12:03:59	12:02:49	12:01:54	12:01:52	12:01:32	12:01:22	12:01:12

14P. Three digital clocks A, B, and C run at different rates and do not have simultaneous readings of zero. Figure 1-7 shows simultaneous readings on pairs of the clocks for four occasions. (At the earliest occasion, for example, B reads 25.0 s and C reads 92.0 s.) If two events are 600 s apart on clock A, how far apart are they on (a) clock B and (b) clock C? (c) When clock A reads 400 s, what does clock B read? (d) When clock C reads 15.0 s, what does clock B read? (Assume negative readings for prezero times.)

Fig. 1-7 Problem 14.

15P. An astronomical unit (AU) is the average distance of Earth from the Sun, approximately 1.50×10^8 km. The speed of light is about 3.0×10^8 m/s. Express the speed of light in terms of astronomical units per minute. ssm

16P. Until 1883, every city and town in the United States kept its own local time. Today, travelers reset their watches only when the time change equals 1.0 h. How far, on the average, must you travel in degrees of longitude until your watch must be reset by 1.0 h? (*Hint:* Earth rotates 360° in about 24 h.)

17P. Assuming the length of the day uniformly increases by 0.0010 s per century, calculate the cumulative effect on the measure of time over 20 centuries. (Such slowing of Earth's rotation is indicated by observations of the occurrences of solar eclipses during this period.) ssm www

18P. Time standards are now based on atomic clocks. A promising second standard is based on *pulsars*, which are rotating neutron stars (highly compact stars consisting only of neutrons). Some rotate at a rate that is highly stable, sending out a radio beacon that sweeps briefly across Earth once with each rotation, like a lighthouse beacon. Pulsar PSR 1937+21 is an example; it rotates once every 1.557 806 448 872 75 ± 3 ms, where the trailing ±3 indicates the uncertainty in the last decimal place (it does *not* mean ±3 ms). (a) How many times does PSR 1937+21 rotate in 7.00 days? (b) How much time does the pulsar take to rotate 1.0×10^6 times and (c) what is the associated uncertainty?

SEC. 1-6   Mass

19E. Earth has a mass of 5.98×10^{24} kg. The average mass of the atoms that make up Earth is 40 u. How many atoms are there in Earth? ssm

20P. Gold, which has a mass of 19.32 g for each cubic centimeter of volume, is the most ductile metal and can be pressed into a thin leaf or drawn out into a long fiber. (a) If 1.000 oz of gold, with a mass of 27.63 g, is pressed into a leaf of 1.000 μm thickness, what is the area of the leaf? (b) If, instead, the gold is drawn out into a cylindrical fiber of radius 2.500 μm, what is the length of the fiber?

21P. (a) Assuming that each cubic centimeter of water has a mass of exactly 1 g, find the mass of one cubic meter of water in kilograms. (b) Suppose that it takes 10.0 h to drain a container of 5700 m³ of water. What is the "mass flow rate," in kilograms per second, of water from the container? ssm

22P. What mass of water fell on the town in Problem 9 during the thunderstorm? One cubic meter of water has a mass of 10^3 kg.

23P. Iron has a mass of 7.87 g per cubic centimeter of volume, and the mass of an iron atom is 9.27×10^{-26} kg. If the atoms are spherical and tightly packed, (a) what is the volume of an iron atom and (b) what is the distance between the centers of adjacent atoms? ssm

24P. Grains of fine California beach sand are approximately spheres with an average radius of 50 μm and are made of silicon dioxide. A solid cube of silicon dioxide with a volume of 1.00 m³ has a mass of 2600 kg. What mass of sand grains would have a total surface area (the total area of all the individual spheres) equal to the surface area of a cube 1 m on an edge?

Additional Problems

25. Harvard Bridge, which connects MIT with its fraternities across the Charles River, has a length of 364.4 Smoots plus one ear. The unit of one Smoot is based on the length of Oliver Reed Smoot, Jr., class of 1962, who was carried or dragged length by length across the bridge so that other pledge members of the Lambda Chi Alpha fraternity could mark off (with paint) 1-Smoot lengths along the bridge. The marks have been repainted biannually by fraternity pledges since the initial measurement, usually during times of traf-

fic congestion so that the police could not easily interfere. (Presumably, the police were originally upset because a Smoot is not an SI base unit, but these days they seem to have accepted the unit.) Figure 1-8 shows three parallel paths, measured in Smoots (S), Willies (W), and Zeldas (Z). What is the length of 50.0 Smoots in (a) Willies and (b) Zeldas?

26. An old English children's rhyme states, "Little Miss Muffet sat on her tuffet, eating her curds and whey, when along came a spider who sat down beside her. . . ." The spider sat down not because of the curds and whey but because Miss Muffet had a stash of 11 tuffets of dried flies. The volume measure of a tuffet is given by 1 tuffet = 2 pecks = 0.50 bushel, where 1 Imperial (British) bushel = 36.3687 liters (L). What was Miss Muffet's stash in (a) pecks, (b) bushels, and (c) liters?

27. During the summers at high latitudes, ghostly, silver-blue clouds occasionally appear after sunset when common clouds are in Earth's shadow and are no longer visible. The ghostly clouds have been called *noctilucent clouds* (NLC), which means "luminous night clouds," but now are often called *mesospheric clouds*, after the *mesosphere*, the name of the atmosphere at the altitude of the clouds.

These clouds were first seen in June 1885, after dust and water from the massive 1883 volcanic explosion of Krakatoa Island (near Java in the Southeast Pacific) reached the high altitudes in the Northern Hemisphere. In the low temperatures of the mesosphere, the water collected and froze on the volcanic dust (and perhaps on comet and meteor dust already present there) to form the particles that made up the first clouds. Since then, mesospheric clouds have generally increased in occurrence and brightness, probably because of the increased production of methane by industries, rice paddies, landfills, and livestock flatulence. The methane works its way into the upper atmosphere, undergoes chemical changes, and results in an increase of water molecules there, and also in bits of ice for the mesospheric clouds.

If mesospheric clouds are spotted 38 min after sunset and then quickly dim, what is their altitude if they are directly over the observer? (*Hint:* See Sample Problem 1-4.)

28. A standard interior staircase has steps each with a rise (height) of 19 cm and a run (horizontal depth) of 23 cm. Research suggests that the stairs would be safer for descent if the run were, instead, 28 cm. For a particular staircase of total height 4.57 m, how much farther would the staircase extend into the room at the foot of the stairs if this change in run were made?

29. As a contrast between the old and the modern and between the large and the small, consider the following: In old rural England 1 hide (between 100 and 120 acres) was the area of land needed to sustain one family with a single plough for one year. (An area of 1 acre is equal to 4047 m².) Also, 1 wapentake was the area of land needed by 100 such families. In quantum physics, the cross-sectional area of a nucleus (defined in terms of the chance of a particle hitting and being absorbed by it) is measured in units of barns, where 1 barn is 1×10^{-28} m². (In nuclear physics jargon, if a nucleus is "large," then shooting a particle at it is like shooting a bullet at a barn door, which can hardly be missed.) What is the ratio of 25 wapentakes to 11 barns?

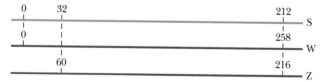

Fig. 1-8 Problem 25.

NEW PROBLEMS

N1. A person on a diet might lose 2.3 kg per week. Express the mass loss rate in milligrams per second, as if the dieter could sense the second-by-second loss.

N2. A mole of atoms is 6.02×10^{23} atoms. To the nearest order of magnitude, how many moles of atoms are in a large domestic cat? The masses of a hydrogen atom, an oxygen atom, and a carbon atom are 1.0 u, 16 u, and 12 u, respectively. (*Hint:* Cats are sometimes known to kill a mole.)

N3. A typical sugar cube has an edge length of 1 cm. If you had a cubical box that contained a mole of sugar cubes, what would its edge length be? (One mole = 6.02×10^{23} units.)

N4. A cubic centimeter in a typical cumulus cloud contains 50 to 500 water drops, which have a typical radius of 10 μm. (a) How many cubic meters of water are in a cylindrical cumulus cloud of height 3.0 km and radius 1.0 km? (b) How many 1-liter pop bottles would that water fill? (c) Water has a mass per unit volume (or density) of 1000 kg/m^3. How much mass does the water in the cloud have?

N5. Using conversions and data in the chapter, determine the number of hydrogen atoms required to obtain 1.0 kg of hydrogen. A hydrogen atom has a mass of 1.0 u.

N6. A tourist purchases a car in England and ships it home to the United States. The car sticker advertised that the car's fuel consumption was at the rate of 40 miles per gallon on the open road. The tourist does not realize that the U.K. gallon differs from the U.S. gallon:

$$1 \text{ U.K. gallon} = 4.545\ 963\ 1 \text{ liters}$$
$$1 \text{ U.S. gallon} = 3.785\ 306\ 0 \text{ liters.}$$

For a trip of 750 miles (in the United States), how many gallons of fuel does (a) the mistaken tourist believe she needs and (b) the car actually require?

N7. In purchasing food for a political rally, you erroneously order shucked medium-size Pacific oysters (which come 8 to 12 per U.S. pint) instead of shucked medium-size Atlantic oysters (which come 26 to 38 per U.S. pint). The filled oyster container delivered to you has the interior measure of 1.0 m $\times$ 12 cm $\times$ 20 cm, and a U.S. pint is equivalent to 0.4732 liter. By how many oysters is the order short of your anticipated count?

N8. A fortnight is a charming English measure of time equal to 2.0 weeks (the word is a contraction of "fourteen nights"). That is a nice amount of time in pleasant company but perhaps a painful string of microseconds in unpleasant company. How many microseconds are in a fortnight?

N9. A ton is a measure of volume frequently used in shipping, but that use requires some care because there are at least three types of tons: A *displacement ton* is equal to 7 barrels bulk, a *freight ton* is equal to 8 barrels bulk, and a *register ton* is equal to 20 barrels bulk. A *barrel bulk* is another measure of volume: 1 barrel bulk = 0.1415 m^3. Suppose you spot a shipping order for "73 tons" of M&M candies, and you are certain that the client who sent the order intended "ton" to refer to volume (instead of weight or mass, as discussed in Chapter 5). If the client actually meant displacement tons, how many extra U.S. bushels of the candies will you erroneously ship to the client if you interpret the order as (a) 73 freight tons and (b) 73 register tons? One cubic meter is equivalent to 28.378 U.S. bushels.

N10. A *gry* is an old English measure for length, defined as 1/10 of a line, where *line* is another old English measure for length, defined as 1/12 inch. A common measure for length in the publishing business is a *point*, defined as 1/72 inch. What is an area of 0.50 gry^2 in terms of points squared (points^2)?

N11. The *corn–hog ratio* is a financial term commonly used in the pig market and presumably is related to the cost of feeding a pig until it is large enough for market. It is defined as the ratio of the market price of a pig with a mass of 1460 slugs to the market price of a U.S. bushel of corn. The slug is the unit of mass in the English system. (The word "slug" is derived from an old German word that means "to hit"; we have the same meaning for "slug" as a verb in modern English.) A U.S. bushel is equal to 35.238 L. If the corn–hog ratio is listed as 5.7 on the market exchange, what is it in the metric units of

$$\frac{\text{price of 1 kilogram of pig}}{\text{price of 1 liter of corn}}?$$

(*Hint:* See the Mass table in Appendix D.)

N12. The wine for a large European wedding reception is to be served in a stunning cut-glass receptacle with the interior dimensions of 40 cm $\times$ 40 cm $\times$ 30 cm (height). The receptacle is to be initially filled to the top. The wine can be purchased in bottles of the sizes given in the following table, where the volumes of the larger bottles are given in terms of the volume of a standard wine bottle. Purchasing a larger bottle instead of multiple smaller bottles decreases the overall cost of the wine. To minimize that overall cost, (a) which bottle sizes should be purchased and how many of each should be purchased, and (b) how much wine is left over once the receptacle is filled?

1 standard

1 magnum = 2 standard

1 jeroboam = 4 standard

1 rehoboam = 6 standard

1 methuselah = 8 standard

1 salmanazar = 12 standard

1 balthazar = 16 standard = 11.356 L

1 nebuchadnezzar = 20 standard

N13. You receive orders to sail due east for 24.5 mi to put your salvage ship directly over a sunken pirate ship. However, when your divers probe the ocean floor at that location and find no evidence of a ship, you radio back to your source of information, only to discover that the sailing distance was supposed to be 24.5 *nautical miles*, not regular miles. Use the Length table in Appendix D to calculate how far horizontally you are from the pirate ship in kilometers.

N14. You can easily convert common units and measures electronically, but you still should be able to use a conversion table, such as those in Appendix D. Table 1-6 is part of a conversion table for a system of volume measures once common in Spain; a volume of 1 fanega is equivalent to 55.501 dm^3 (cubic decimeters). (a) Complete the table, using three significant figures. Then express 7.00 almude in terms of (b) medio, (c) cahiz, and (d) cubic centimeters (cm^3).

TABLE 1-6   Problem N14

	cahiz	fanega	cuartilla	almude	medio
1 cahiz =	1	12	48	144	288
1 fanega =		1	4	12	24
1 cuartilla =			1	3	6
1 almude =				1	2
1 medio =					1

N15. During heavy rain, a rectangular section of a mountainside measuring 2.5 km wide (horizontally), 0.80 km long (up along the slope), and 2.0 m deep suddenly slips into a valley in a mud slide. Assume that the mud ends up uniformly distributed over a valley section measuring 0.40 km × 0.40 km and that the mass of a cubic meter of mud is 1900 kg. What is the mass of the mud sitting above an area of 4.0 m^2 in that section?

N16. For about 10 years after the French revolution, the French government attempted to base measures of time on multiples of ten: One week consisted of 10 days, one day consisted of 10 hours, one hour consisted of 100 minutes, and one minute consisted of 100 seconds. What are the ratios of (a) the French decimal week to the standard week and (b) the French decimal second to the standard second?

N17. Traditional units of time have been based on astronomical measurements, such as the length of the day or year. However, one human-based measure of time can be found in Tibet, where the *dbug* is the average time between exhaled breaths. Estimate the number of dbugs in a day.

N18. Prior to adopting metric systems of measurement, the United Kingdom employed some challenging measures of liquid volume. A few are shown in Table 1-7. (a) Complete the table, using three significant figures. (b) The volume of 1 bag is equivalent to a volume of 0.1091 m^3. If an old British story has a witch cooking up some vile liquid in a cauldron with a volume of 1.5 chaldrons, what is the volume in terms of cubic meters?

TABLE 1-7   Problem N18

	wey	chaldron	bag	pottle	gill
1 wey =	1	10/9	40/3	640	120 240
1 chaldron =					
1 bag =					
1 pottle =					
1 gill =					

N19. You are to fix dinners for 400 people at a convention of Mexican food fans. Your recipe calls for 2 jalapeño peppers per serving (one serving per person). However, you have only habanero peppers on hand. The spiciness of peppers is measured in terms of the *scoville heat unit* (SHU). On average, one jalapeño pepper has a spiciness of 4000 SHU and one habanero pepper has a spiciness of 300 000 SHU. To salvage the situation, how many (total) habanero peppers should you substitute for the jalapeño peppers in the recipe for the convention?

2 Motion Along a Straight Line

On September 26, 1993, Dave Munday, a diesel mechanic by trade, went over the Canadian edge of Niagara Falls for the second time, freely falling 48 m to the water (and rocks) below. On this attempt, he rode in a steel ball with a hole for air. Munday, keen on surviving this plunge that had killed four other stuntmen, had done considerable research on the physics and engineering aspects of the plunge.

How long would he fall, and at what speed would he hit the turbulent water at the bottom?

The answer is in this chapter.

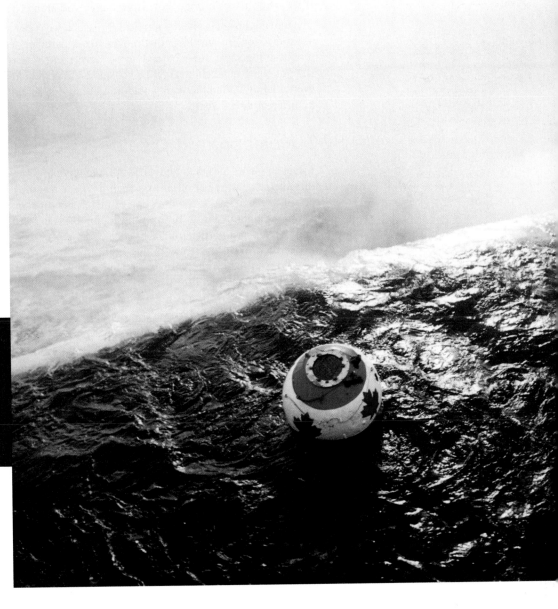

2-1 Motion

The world, and everything in it, moves. Even seemingly stationary things, such as a roadway, move with Earth's rotation, Earth's orbit around the Sun, the Sun's orbit around the center of the Milky Way galaxy, and that galaxy's migration relative to other galaxies. The classification and comparison of motions (called **kinematics**) is often challenging. What exactly do you measure, and how do you compare?

Before we attempt an answer, we shall examine some general properties of motion that is restricted in three ways.

1. The motion is along a straight line only. The line may be vertical (that of a falling stone), horizontal (that of a car on a level highway), or slanted, but it must be straight.

2. Forces (pushes and pulls) cause motion, but forces will not be discussed until Chapter 5. In this chapter you study only the motion itself, and changes in the motion. Does the moving object speed up, slow down, stop, or reverse direction? If the motion does change, how is time involved in the change?

3. The moving object is either a **particle** (by which we mean a pointlike object such as an electron) or an object that moves like a particle (such that every portion moves in the same direction and at the same rate). A stiff pig slipping down a straight playground slide might be considered to be moving like a particle; however, a tumbling tumbleweed would not, because different points inside it move in different directions.

2-2 Position and Displacement

To locate an object means to find its position relative to some reference point, often the **origin** (or zero point) of an axis such as the x axis in Fig. 2-1. The **positive direction** of the axis is in the direction of increasing numbers (coordinates), which is toward the right in Fig. 2-1. The opposite direction is the **negative direction.**

For example, a particle might be located at $x = 5$ m, which means that it is 5 m in the positive direction from the origin. If it were at $x = -5$ m, it would be just as far from the origin but in the opposite direction. On the axis, a coordinate of -5 m is less than one of -1 m, and both coordinates are less than a coordinate of $+5$ m. A plus sign for a coordinate need not be shown, but a minus sign must always be shown.

A change from one position x_1 to another position x_2 is called a **displacement** Δx, where

$$\Delta x = x_2 - x_1. \tag{2-1}$$

(The symbol Δ, the Greek uppercase delta, represents a change in a quantity, and it means the final value of that quantity minus the initial value.) When numbers are inserted for the position values x_1 and x_2, a displacement in the positive direction (toward the right in Fig. 2-1) always comes out positive, and one in the opposite direction (left in the figure), negative. For example, if the particle moves from $x_1 = 5$ m to $x_2 = 12$ m, then $\Delta x = (12\ \text{m}) - (5\ \text{m}) = +7$ m. The positive result indicates that the motion is in the positive direction. If the particle then returns to $x = 5$ m, the displacement for the full trip is zero. The actual number of meters covered for the full trip is irrelevant; displacement involves only the original and final positions.

A plus sign for a displacement need not be shown, but a minus sign must always be shown. If we ignore the sign (and thus the direction) of a displacement, we are

Fig. 2-1 Position is determined on an axis that is marked in units of length (here meters) and that extends indefinitely in opposite directions. The axis label, here x, is always on the positive side of the origin.

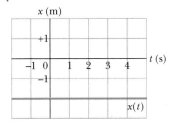

Fig. 2-2 The graph of $x(t)$ for an armadillo that is stationary at $x = -2$ m. The value of x is -2 m for all times t.

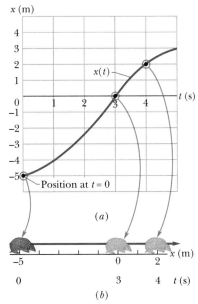

Fig. 2-3 (a) The graph of $x(t)$ for a moving armadillo. (b) The path associated with the graph. The scale below the x axis shows the times at which the armadillo reaches various x values.

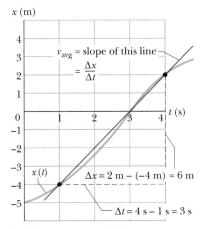

Fig. 2-4 Calculation of the average velocity between $t = 1$ s and $t = 4$ s as the slope of the line that connects the points (on the $x(t)$ curve) representing those times.

left with the **magnitude** (or absolute value) of the displacement. In the previous example, the magnitude of Δx is 7 m.

Displacement is an example of a **vector quantity**, which is a quantity that has both a direction and a magnitude. We explore vectors more fully in Chapter 3 (in fact, some of you may have already read that chapter), but here all we need is the idea that displacement has two features: (1) Its *magnitude* is the distance (such as the number of meters) between the original and final positions. (2) Its *direction*, from an original position to a final position, can be represented by a plus sign or a minus sign if the motion is along a single axis.

What follows is the first of many checkpoints that you will see in this book. Each consists of one or more questions whose answers require some reasoning or a mental calculation, and each gives you a quick check of your understanding. The answers are listed in the back of the book.

CHECKPOINT 1: Here are three pairs of initial and final positions, respectively, along an x axis. Which pairs give a negative displacement: (a) -3 m, $+5$ m; (b) -3 m, -7 m; (c) 7 m, -3 m?

2-3 Average Velocity and Average Speed

A compact way to describe position is with a graph of position x plotted as a function of time t—a graph of $x(t)$. (The notation $x(t)$ represents a function x of t, not the product x times t.) As a simple example, Fig. 2-2 shows the position function $x(t)$ for a stationary armadillo (which we treat as a particle) at $x = -2$ m.

Figure 2-3a, also for an armadillo, is more interesting, because it involves motion. The armadillo is apparently first noticed at $t = 0$ when it is at the position $x = -5$ m. It moves toward $x = 0$, passes through that point at $t = 3$ s, and then moves on to increasingly larger positive values of x.

Figure 2-3b depicts the actual straight-line motion of the armadillo and is something like what you would see. The graph in Fig. 2-3a is more abstract and quite unlike what you would see, but it is richer in information. It also reveals how fast the armadillo moves.

Actually, several quantities are associated with the phrase "how fast." One of them is the **average velocity** v_{avg}, which is the ratio of the displacement Δx that occurs during a particular time interval Δt to that interval:

$$v_{avg} = \frac{\Delta x}{\Delta t} = \frac{x_2 - x_1}{t_2 - t_1}. \tag{2-2}$$

The notation means that the position is x_1 at time t_1 and then x_2 at time t_2. A common unit for v_{avg} is the meter per second (m/s). You may see other units in the problems, but they are always in the form of length/time.

On a graph of x versus t, v_{avg} is the **slope** of the straight line that connects two particular points on the $x(t)$ curve: one is the point that corresponds to x_2 and t_2, and the other is the point that corresponds to x_1 and t_1. Like displacement, v_{avg} has both magnitude and direction (it is another vector quantity). Its magnitude is the magnitude of the line's slope. A positive v_{avg} (and slope) tells us that the line slants upward to the right; a negative v_{avg} (and slope), that the line slants downward to the right. The average velocity v_{avg} always has the same sign as the displacement Δx because Δt in Eq. 2-2 is always positive.

Figure 2-4 shows how to find v_{avg} for the armadillo of Fig. 2-3, for the time interval $t = 1$ s to $t = 4$ s. We draw the straight line that connects the point on the position curve at the beginning of the interval and the point on the curve at the end of the interval. Then we find the slope $\Delta x / \Delta t$ of the straight line. For the given time interval, the average velocity is

$$v_{avg} = \frac{6 \text{ m}}{3 \text{ s}} = 2 \text{ m/s}.$$

Average speed s_{avg} is a different way of describing "how fast" a particle moves. Whereas the average velocity involves the particle's displacement Δx, the average speed involves the total distance covered (for example, the number of meters moved), independent of direction; that is,

$$s_{avg} = \frac{\text{total distance}}{\Delta t}. \qquad (2\text{-}3)$$

Because average speed does *not* include direction, it lacks any algebraic sign. Sometimes s_{avg} is the same (except for the absence of a sign) as v_{avg}. However, as is demonstrated in Sample Problem 2-1, when an object doubles back on its path, the two can be quite different.

Sample Problem 2-1

You drive a beat-up pickup truck along a straight road for 8.4 km at 70 km/h, at which point the truck runs out of gasoline and stops. Over the next 30 min, you walk another 2.0 km farther along the road to a gasoline station.

(a) What is your overall displacement from the beginning of your drive to your arrival at the station?

SOLUTION: Assume, for convenience, that you move in the positive direction of an x axis, from a first position of $x_1 = 0$ to a second position of x_2 at the station. That second position must be at $x_2 = 8.4 \text{ km} + 2.0 \text{ km} = 10.4 \text{ km}$. Then the **Key Idea** here is that your displacement Δx along the x axis is the second position minus the first position. From Eq. 2-1, we have

$$\Delta x = x_2 - x_1 = 10.4 \text{ km} - 0 = 10.4 \text{ km}. \quad \text{(Answer)}$$

Thus, your overall displacement is 10.4 km in the positive direction of the x axis.

(b) What is the time interval Δt from the beginning of your drive to your arrival at the station?

SOLUTION: We already know the time interval Δt_{wlk} ($= 0.50$ h) for the walk, but we lack the time interval Δt_{dr} for the drive. However, we know that for the drive the displacement Δx_{dr} is 8.4 km and the average velocity $v_{avg,dr}$ is 70 km/h. A **Key Idea** to use here comes from Eq. 2-2: This average velocity is the ratio of the displacement for the drive to the time interval for the drive:

$$v_{avg,dr} = \frac{\Delta x_{dr}}{\Delta t_{dr}}.$$

Rearranging and substituting data then give us

$$\Delta t_{dr} = \frac{\Delta x_{dr}}{v_{avg,dr}} = \frac{8.4 \text{ km}}{70 \text{ km/h}} = 0.12 \text{ h}.$$

So,
$$\Delta t = \Delta t_{dr} + \Delta t_{wlk}$$
$$= 0.12 \text{ h} + 0.50 \text{ h} = 0.62 \text{ h}. \quad \text{(Answer)}$$

(c) What is your average velocity v_{avg} from the beginning of your drive to your arrival at the station? Find it both numerically and graphically.

SOLUTION: The **Key Idea** here again comes from Eq. 2-2: v_{avg} *for the entire trip* is the ratio of the displacement of 10.4 km *for the entire trip* to the time interval of 0.62 h *for the entire trip*. With Eq. 2-2, we find it is

$$v_{avg} = \frac{\Delta x}{\Delta t} = \frac{10.4 \text{ km}}{0.62 \text{ h}}$$
$$= 16.8 \text{ km/h} \approx 17 \text{ km/h}. \quad \text{(Answer)}$$

To find v_{avg} graphically, first we graph $x(t)$ as shown in Fig. 2-5, where the beginning and arrival points on the graph are the origin and the point labeled as "Station." The **Key Idea** here is that your average velocity is the slope of the straight line connecting those

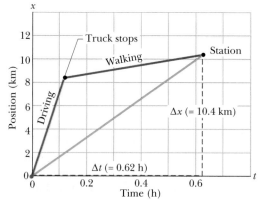

Fig. 2-5 Sample Problem 2-1. The lines marked "Driving" and "Walking" are the position–time plots for the driving and walking stages. (The plot for the walking stage assumes a constant rate of walking.) The slope of the straight line joining the origin and the point labeled "Station" is the average velocity for the trip, from the beginning to the station.

points; that is, it is the ratio of the *rise* ($\Delta x = 10.4$ km) to the *run* ($\Delta t = 0.62$ h), which gives us $v_{avg} = 16.8$ km/h.

(d) Suppose that to pump the gasoline, pay for it, and walk back to the truck takes you another 45 min. What is your average speed from the beginning of your drive to your return to the truck with the gasoline?

SOLUTION: The Key Idea here is that your average speed is the ratio of the total distance you move to the total time interval you take

to make that move. The total distance is 8.4 km + 2.0 km + 2.0 km = 12.4 km. The total time interval is 0.12 h + 0.50 h + 0.75 h = 1.37 h. Thus, Eq. 2-3 gives us

$$s_{avg} = \frac{12.4 \text{ km}}{1.37 \text{ h}} = 9.1 \text{ km/h}. \qquad \text{(Answer)}$$

✔CHECKPOINT 2:   In this sample problem, suppose that right after refueling the truck, you drive back to x_1 at 35 km/h. What is your average velocity for your entire trip?

PROBLEM-SOLVING TACTICS

Tactic 1: *Do You Understand the Problem?*
For beginning problem solvers, no difficulty is more common than simply not understanding the problem. The best test of understanding is this: Can you explain the problem in your own words?

Write down the given data, with units, using the symbols of the chapter. (In Sample Problem 2-1, the given data allow you to find your net displacement Δx in part (a) and the corresponding time interval Δt in part (b).) Identify the unknown and its symbol. (In the sample problem, the unknown in part (c) is your average velocity v_{avg}.) Then find the connection between the unknown and the data. (The connection is Eq. 2-2, the definition of average velocity.)

Tactic 2: *Are the Units OK?*
Be sure to use a consistent set of units when putting numbers into the equations. In Sample Problem 2-1, the logical units in terms of the given data are kilometers for distances, hours for time intervals, and kilometers per hour for velocities. You may sometimes need to make conversions.

Tactic 3: *Is Your Answer Reasonable?*
Does your answer make sense? Is it far too large or far too small? Is the sign correct? Are the units appropriate? In part (c) of Sample Problem 2-1, for example, the correct answer is 17 km/h. If you find 0.00017 km/h, −17 km/h, 17 km/s, or 17,000 km/h, you should realize at once that you have done something wrong. The error may lie in your method, in your algebra, or in your keystroking of numbers on a calculator.

Tactic 4: *Reading a Graph*
Figures 2-2, 2-3a, 2-4, and 2-5 are graphs that you should be able to read easily. In each graph, the variable on the horizontal axis is the time t, with the direction of increasing time to the right. In each, the variable on the vertical axis is the position x of the moving particle with respect to the origin, with the positive direction of x upward. Always note the units (seconds or minutes; meters or kilometers) in which the variables are expressed.

2-4 Instantaneous Velocity and Speed

You have now seen two ways to describe how fast something moves: average velocity and average speed, both of which are measured over a time interval Δt. However, the phrase "how fast" more commonly refers to how fast a particle is moving at a given instant—and that is its **instantaneous velocity** (or simply **velocity**) v.

The velocity at any instant is obtained from the average velocity by shrinking the time interval Δt closer and closer to 0. As Δt dwindles, the average velocity approaches a limiting value, which is the velocity at that instant:

$$v = \lim_{\Delta t \to 0} \frac{\Delta x}{\Delta t} = \frac{dx}{dt}. \qquad (2\text{-}4)$$

This equation displays two features of the instantaneous velocity v. First, v is the rate at which the particle's position x is changing with time at a given instant; that is, v is the derivative of x with respect to t. Second, v at any instant is the slope of the particle's position–time curve at the point representing that instant. Velocity is another vector quantity and thus has an associated direction.

Speed is the magnitude of velocity; that is, speed is velocity that has been stripped of any indication of direction, either in words or via an algebraic sign. (*Caution:* Speed and average speed can be quite different.) A velocity of +5 m/s and one of −5 m/s both have an associated speed of 5 m/s. The speedometer in a car measures the speed, not the velocity, because it cannot determine the direction.

Sample Problem 2-2

Figure 2-6a is an $x(t)$ plot for an elevator that is initially stationary, then moves upward (which we take to be the positive direction of x), and then stops. Plot v as a function of time.

SOLUTION: The Key Idea here is that we can find the velocity at any time from the slope of the curve of $x(t)$ at that time. The slope of $x(t)$, and so also the velocity, is zero in the intervals from 0 to 1 s and from 9 s on, so then the cab is stationary. During the interval bc the slope is constant and nonzero; so then the cab moves with constant velocity. We calculate the slope of $x(t)$ then as

$$\frac{\Delta x}{\Delta t} = v = \frac{24\text{ m} - 4.0\text{ m}}{8.0\text{ s} - 3.0\text{ s}} = +4.0\text{ m/s}.$$

The plus sign indicates that the cab is moving in the positive x direction. These intervals (where $v = 0$ and $v = 4$ m/s) are plotted in Fig. 2-6b. In addition, as the cab initially begins to move and then later slows to a stop, v varies as indicated in the intervals 1 s to 3 s and 8 s to 9 s. Thus, Fig. 2-6b is the required plot. (Figure 2-6c is considered in Section 2-5.)

Given a $v(t)$ graph such as Fig. 2-6b, we could "work backward" to produce the shape of the associated $x(t)$ graph (Fig. 2-6a). However, we would not know the actual values for x at various times, because the $v(t)$ graph indicates only *changes* in x. To find such a change in x during any interval, we must, in the language of calculus, calculate the area "under the curve" on the $v(t)$ graph for that interval. For example, during the interval 3 s to 8 s in which the cab has a velocity of 4.0 m/s, the change in x is

$$\Delta x = (4.0\text{ m/s})(8.0\text{ s} - 3.0\text{ s}) = +20\text{ m}.$$

(This area is positive because the $v(t)$ curve is above the t axis.) Figure 2-6a shows that x does indeed increase by 20 m in that interval. However, Fig. 2-6b does not tell us the *values* of x at the beginning and end of the interval. For that we need additional information such as the value of x for some given instant.

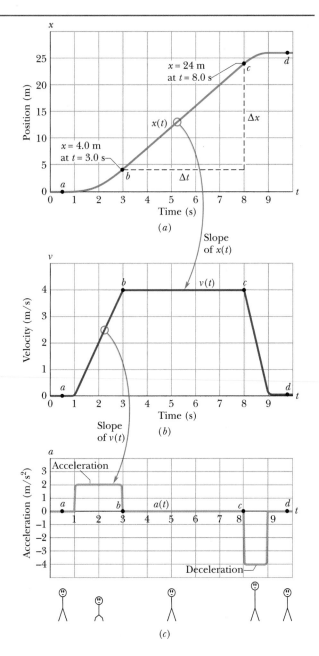

Fig. 2-6 Sample Problem 2-2. (a) The $x(t)$ curve for an elevator cab that moves upward along an x axis. (b) The $v(t)$ curve for the cab. Note that it is the derivative of the $x(t)$ curve ($v = dx/dt$). (c) The $a(t)$ curve for the cab. It is the derivative of the $v(t)$ curve ($a = dv/dt$). The stick figures along the bottom suggest how a passenger's body might feel during the accelerations.

Sample Problem 2-3

The position of a particle moving on an x axis is given by

$$x = 7.8 + 9.2t - 2.1t^3, \qquad (2\text{-}5)$$

with x in meters and t in seconds. What is its velocity at $t = 3.5$ s? Is the velocity constant, or is it continuously changing?

SOLUTION: For simplicity, the units have been omitted from Eq. 2-5, but you can insert them if you like by changing the coefficients to 7.8 m, 9.2 m/s, and -2.1 m/s³. The Key Idea here is that velocity is the first derivative (with respect to time) of the position function $x(t)$. Thus, we write

$$v = \frac{dx}{dt} = \frac{d}{dt}(7.8 + 9.2t - 2.1t^3),$$

which becomes

$$v = 0 + 9.2 - (3)(2.1)t^2 = 9.2 - 6.3t^2. \qquad (2\text{-}6)$$

At $t = 3.5$ s,

$$v = 9.2 - (6.3)(3.5)^2 = -68\text{ m/s}. \qquad \text{(Answer)}$$

At $t = 3.5$ s, the particle is moving in the negative direction of x (note the minus sign) with a speed of 68 m/s. Since the quantity t appears in Eq. 2-6, the velocity v depends on t and so is continuously changing.

✔CHECKPOINT 3: The following equations give the position $x(t)$ of a particle in four situations (in each equation, x is in meters, t is in seconds, and $t > 0$): (1) $x = 3t - 2$; (2) $x = -4t^2 - 2$; (3) $x = 2/t^2$; and (4) $x = -2$. (a) In which situation is the velocity v of the particle constant? (b) In which is v in the negative x direction?

2-5 Acceleration

When a particle's velocity changes, the particle is said to undergo **acceleration** (or to accelerate). For motion along an axis, the **average acceleration** a_{avg} over a time interval Δt is

$$a_{avg} = \frac{v_2 - v_1}{t_2 - t_1} = \frac{\Delta v}{\Delta t}, \tag{2-7}$$

where the particle has velocity v_1 at time t_1 and then velocity v_2 at time t_2. The **instantaneous acceleration** (or simply **acceleration**) is the derivative of the velocity with respect to time:

$$a = \frac{dv}{dt}. \tag{2-8}$$

In words, the acceleration of a particle at any instant is the rate at which its velocity is changing at that instant. Graphically, the acceleration at any point is the slope of the curve of $v(t)$ at that point.

We can combine Eq. 2-8 with Eq. 2-4 to write

$$a = \frac{dv}{dt} = \frac{d}{dt}\left(\frac{dx}{dt}\right) = \frac{d^2x}{dt^2}. \tag{2-9}$$

In words, the acceleration of a particle at any instant is the second derivative of its position $x(t)$ with respect to time.

A common unit of acceleration is the meter per second per second: m/(s · s) or m/s². You will see other units in the problems, but they will each be in the form of length/(time · time) or length/time². Acceleration has both magnitude and direction (it is yet another vector quantity). Its algebraic sign represents its direction on an axis just as for displacement and velocity; that is, acceleration with a positive value is in the positive direction of an axis, and acceleration with a negative value is in the negative direction.

Figure 2-6c is a plot of the acceleration of the elevator cab discussed in Sample Problem 2-2. Compare this $a(t)$ curve with the $v(t)$ curve—each point on the $a(t)$ curve shows the derivative (slope) of the $v(t)$ curve at the corresponding time. When v is constant (at either 0 or 4 m/s), the derivative is zero and so also is the acceleration. When the cab first begins to move, the $v(t)$ curve has a positive derivative (the slope is positive), which means that $a(t)$ is positive. When the cab slows to a stop, the derivative and slope of the $v(t)$ curve are negative; that is, $a(t)$ is negative.

Next compare the slopes of the $v(t)$ curve during the two acceleration periods. The slope associated with the cab's slowing down (commonly called "deceleration") is steeper, because the cab stops in half the time it took to get up to speed. The steeper slope means that the magnitude of the deceleration is larger than that of the acceleration, as indicated in Fig. 2-6c.

The sensations you would feel while riding in the cab of Fig. 2-6 are indicated by the sketched figures. When the cab first accelerates, you feel as though you are pressed downward; when later the cab is braked to a stop, you seem to be stretched upward. In between, you feel nothing special. Your body reacts to accelerations (it is an accelerometer) but not to velocities (it is not a speedometer). When you are in

Fig. 2-7 Colonel J. P. Stapp in a rocket sled as it is brought up to high speed (acceleration out of the page) and then very rapidly braked (acceleration into the page).

a car traveling at 90 km/h or an airplane traveling at 900 km/h, you have no bodily awareness of the motion. However, if the car or plane quickly changes velocity, you may become keenly aware of the change, perhaps even frightened by it. Part of the thrill of an amusement park ride is due to the quick changes of velocity that you undergo (you pay for the accelerations, not for the speed). A more extreme example is shown in the photographs of Fig. 2-7, which were taken while a rocket sled was rapidly accelerated along a track and then rapidly braked to a stop.

Large accelerations are sometimes expressed in terms of g units, with

$$1g = 9.8 \text{ m/s}^2 \qquad (g \text{ unit}). \qquad (2\text{-}10)$$

(As we shall discuss in Section 2-8, g is the magnitude of the acceleration of a falling object near Earth's surface.) On a roller coaster, you may experience brief accelerations up to $3g$, which is $(3)(9.8 \text{ m/s}^2)$ or about 29 m/s^2, more than enough to justify the cost of the ride.

PROBLEM-SOLVING TACTICS

Tactic 5: *An Acceleration's Sign*
In common language, the sign of an acceleration has a nonscientific meaning: positive acceleration means that the speed of an object is increasing, and negative acceleration means that the speed is decreasing (the object is decelerating). In this book, however, the sign of an acceleration indicates a direction, not whether an object's speed is increasing or decreasing.

For example, if a car with an initial velocity $v = -25$ m/s is braked to a stop in 5.0 s, then $a_{avg} = +5.0$ m/s^2. The acceleration is *positive*, but the car's speed has decreased. The reason is the difference in signs: the direction of the acceleration is opposite that of the velocity.

Here then is the proper way to interpret the signs:

> If the signs of the velocity and acceleration of a particle are the same, the speed of the particle increases. If the signs are opposite, the speed decreases.

✔**CHECKPOINT 4:** A wombat moves along an x axis. What is the sign of its acceleration if it is moving (a) in the positive direction with increasing speed, (b) in the positive direction with decreasing speed, (c) in the negative direction with increasing speed, and (d) in the negative direction with decreasing speed?

Sample Problem 2-4

A particle's position on the x axis of Fig. 2-1 is given by

$$x = 4 - 27t + t^3,$$

with x in meters and t in seconds.

(a) Find the particle's velocity function $v(t)$ and acceleration function $a(t)$.

SOLUTION: One Key Idea is that to get the velocity function $v(t)$, we differentiate the position function $x(t)$ with respect to time. Here we find

$$v = -27 + 3t^2, \qquad \text{(Answer)}$$

with v in meters per second.

Another Key Idea is that to get the acceleration function $a(t)$,

Fig. 2-7 *Continued*

we differentiate the velocity function $v(t)$ with respect to time. This gives us

$$a = +6t, \qquad \text{(Answer)}$$

with a in meters per second squared.

(b) Is there ever a time when $v = 0$?

SOLUTION: Setting $v(t) = 0$ yields

$$0 = -27 + 3t^2,$$

which has the solution

$$t = \pm 3 \text{ s.} \qquad \text{(Answer)}$$

Thus, the velocity is zero both 3 s before and 3 s after the clock reads 0.

(c) Describe the particle's motion for $t \geq 0$.

SOLUTION: The Key Idea is to examine the expressions for $x(t)$, $v(t)$, and $a(t)$.

At $t = 0$, the particle is at $x(0) = +4$ m and is moving with a velocity of $v(0) = -27$ m/s—that is, in the negative direction of the x axis. Its acceleration is $a(0) = 0$, because just then the particle's velocity is not changing.

For $0 < t < 3$ s, the particle still has a negative velocity, so it continues to move in the negative direction. However, its acceleration is no longer 0 but is increasing and positive. Because the signs of the velocity and the acceleration are opposite, the particle must be slowing.

Indeed, we already know that it stops momentarily at $t = 3$ s. Just then the particle is as far to the left of the origin in Fig. 2-1 as it will ever get. Substituting $t = 3$ s into the expression for $x(t)$, we find that the particle's position just then is $x = -50$ m. Its acceleration is still positive.

For $t > 3$ s, the particle moves to the right on the axis. Its acceleration remains positive and grows progressively larger in magnitude. The velocity is now positive, and it too grows progressively larger in magnitude.

2-6 Constant Acceleration: A Special Case

In many types of motion, the acceleration is either constant or approximately so. For example, you might accelerate a car at an approximately constant rate when a traffic light turns from red to green. Then graphs of your position, velocity, and acceleration would resemble those in Fig. 2-8. (Note that $a(t)$ in Fig. 2-8c is constant, which requires that $v(t)$ in Fig. 2-8b have a constant slope.) Later when you brake the car to a stop, the deceleration might also be approximately constant.

Such cases are so common that a special set of equations has been derived for dealing with them. One approach to the derivation of these equations is given in this section. A second approach is given in the next section. Throughout both sections and later when you work on the homework problems, keep in mind that *these equations are valid only for constant acceleration (or situations in which you can approximate the acceleration as being constant)*.

When the acceleration is constant, the average acceleration and instantaneous acceleration are equal and we can write Eq. 2-7, with some changes in notation, as

$$a = a_{\text{avg}} = \frac{v - v_0}{t - 0}.$$

Fig. 2-8 (*a*) The position $x(t)$ of a particle moving with constant acceleration. (*b*) Its velocity $v(t)$, given at each point by the slope of the curve in (*a*). (*c*) Its (constant) acceleration, equal to the (constant) slope of the curve of $v(t)$.

Here v_0 is the velocity at time $t = 0$, and v is the velocity at any later time t. We can recast this equation as

$$v = v_0 + at. \tag{2-11}$$

As a check, note that this equation reduces to $v = v_0$ for $t = 0$, as it must. As a further check, take the derivative of Eq. 2-11. Doing so yields $dv/dt = a$, which is the definition of a. Figure 2-8*b* shows a plot of Eq. 2-11, the $v(t)$ function; the function is linear and thus the plot is a straight line.

In a similar manner we can rewrite Eq. 2-2 (with a few changes in notation) as

$$v_{avg} = \frac{x - x_0}{t - 0}$$

and then as

$$x = x_0 + v_{avg}t, \tag{2-12}$$

in which x_0 is the position of the particle at $t = 0$, and v_{avg} is the average velocity between $t = 0$ and a later time t.

For the linear velocity function in Eq. 2-11, the *average* velocity over any time interval (say, from $t = 0$ to a later time t) is the average of the velocity at the beginning of the interval (= v_0) and the velocity at the end of the interval (= v). For the interval from $t = 0$ to the later time t then, the average velocity is

$$v_{avg} = \tfrac{1}{2}(v_0 + v). \tag{2-13}$$

Substituting the right side of Eq. 2-11 for v yields, after a little rearrangement,

$$v_{avg} = v_0 + \tfrac{1}{2}at. \tag{2-14}$$

Finally, substituting Eq. 2-14 into Eq. 2-12 yields

$$x - x_0 = v_0 t + \tfrac{1}{2}at^2. \tag{2-15}$$

As a check, note that putting $t = 0$ yields $x = x_0$, as it must. As a further check, taking the derivative of Eq. 2-15 yields Eq. 2-11, again as it must. Figure 2-8*a* shows a plot of Eq. 2-15; the function is quadratic and thus the plot is curved.

Equations 2-11 and 2-15 are the *basic equations for constant acceleration*; they can be used to solve any constant acceleration problem in this book. However, we can derive other equations that might prove useful in certain specific situations. First, note that five quantities can possibly be involved in any problem regarding constant acceleration, namely, $x - x_0$, v, t, a, and v_0. Usually, one of these quantities is *not* involved in the problem, *either as a given or as an unknown*. We are then presented with three of the remaining quantities and asked to find the fourth.

Equations 2-11 and 2-15 each contain four of these quantities, but not the same four. In Eq. 2-11, the "missing ingredient" is the displacement, $x - x_0$. In Eq. 2-15, it is the velocity v. These two equations can also be combined in three ways to yield three additional equations, each of which involves a different "missing variable." First, we can eliminate t to obtain

$$v^2 = v_0^2 + 2a(x - x_0). \tag{2-16}$$

This equation is useful if we do not know t and are not required to find it. Second, we can eliminate the acceleration a between Eqs. 2-11 and 2-15 to produce an equation in which a does not appear:

$$x - x_0 = \tfrac{1}{2}(v_0 + v)t. \tag{2-17}$$

TABLE 2-1   Equations for Motion with Constant Acceleration[a]

Equation Number	Equation	Missing Quantity
2-11	$v = v_0 + at$	$x - x_0$
2-15	$x - x_0 = v_0 t + \frac{1}{2}at^2$	v
2-16	$v^2 = v_0^2 + 2a(x - x_0)$	t
2-17	$x - x_0 = \frac{1}{2}(v_0 + v)t$	a
2-18	$x - x_0 = vt - \frac{1}{2}at^2$	v_0

[a] Make sure that the acceleration is indeed constant before using the equations in this table.

Finally, we can eliminate v_0, obtaining

$$x - x_0 = vt - \frac{1}{2}at^2. \tag{2-18}$$

Note the subtle difference between this equation and Eq. 2-15. One involves the initial velocity v_0; the other involves the velocity v at time t.

Table 2-1 lists the basic constant acceleration equations (Eqs. 2-11 and 2-15) as well as the specialized equations that we have derived. To solve a simple constant acceleration problem, you can usually use an equation from this list (*if* you have the list). Choose an equation for which the only unknown variable is the variable requested in the problem. A simpler plan is to remember only Eqs. 2-11 and 2-15, and then solve them as simultaneous equations whenever needed. An example is given in Sample Problem 2-5.

✓**CHECKPOINT 5:** The following equations give the position $x(t)$ of a particle in four situations: (1) $x = 3t - 4$; (2) $x = -5t^3 + 4t^2 + 6$; (3) $x = 2/t^2 - 4/t$; (4) $x = 5t^2 - 3$. To which of these situations do the equations of Table 2-1 apply?

Sample Problem 2-5

Spotting a police car, you brake a Porsche from a speed of 100 km/h to a speed of 80.0 km/h during a displacement of 88.0 m, at a constant acceleration.

(a) What is that acceleration?

SOLUTION: Assume that the motion is along the positive direction of an x axis. For simplicity, let us take the beginning of the braking to be at time $t = 0$, at position x_0. The Key Idea here is that, with the acceleration constant, we can relate the car's acceleration to its velocity and displacement via the basic constant acceleration equations (Eqs. 2-11 and 2-15). The initial velocity is $v_0 = 100$ km/h $= 27.78$ m/s, the displacement is $x - x_0 = 88.0$ m, and the velocity at the end of that displacement is $v = 80.0$ km/h $= 22.22$ m/s. However, we do not know the acceleration a and time t, which appear in both basic equations. So, we must solve those equations simultaneously.

To eliminate the unknown t, we use Eq. 2-11 to write

$$t = \frac{v - v_0}{a}, \tag{2-19}$$

and then we substitute this expression into Eq. 2-15 to write

$$x - x_0 = v_0\left(\frac{v - v_0}{a}\right) + \frac{1}{2}a\left(\frac{v - v_0}{a}\right)^2.$$

Solving for a and substituting known data then yield

$$a = \frac{v^2 - v_0^2}{2(x - x_0)} = \frac{(22.22 \text{ m/s})^2 - (27.78 \text{ m/s})^2}{2(88.0 \text{ m})}$$

$$= -1.58 \text{ m/s}^2. \tag{Answer}$$

Note that we could have used Eq. 2-16 instead to solve for a because the unknown t is the missing variable in that equation.

(b) How much time is required for the given decrease in speed?

SOLUTION: Now that we know a, we can use Eq. 2-19 to solve for t:

$$t = \frac{v - v_0}{a} = \frac{22.22 \text{ m/s} - 27.78 \text{ m/s}}{-1.58 \text{ m/s}^2}$$

$$= 3.519 \text{ s} \approx 3.52 \text{ s}. \tag{Answer}$$

If you are initially speeding and trying to slow to the speed limit, this is plenty of time for the police officer to measure your speed.

Tactic 6: *Check the Dimensions*

The dimension of a velocity is L/T—that is, length L divided by time T—and the dimension of an acceleration is L/T^2. In any equation, the dimensions of all terms must be the same. If you are in doubt about an equation, check its dimensions.

To check the dimensions of Eq. 2-15 ($x - x_0 = v_0 t + \frac{1}{2}at^2$), we note that every term must be a length, because that is the dimension of x and of x_0. The dimension of the term $v_0 t$ is $(L/T)(T)$, which is L. The dimension of $\frac{1}{2}at^2$ is $(L/T^2)(T^2)$, which is also L. Thus, this equation checks out.

2-7 Another Look at Constant Acceleration*

The first two equations in Table 2-1 are the basic equations from which the others are derived. Those two can be obtained by integration of the acceleration with the condition that a is constant. To find Eq. 2-11, we rewrite the definition of acceleration (Eq. 2-8) as

$$dv = a \, dt.$$

We next write the *indefinite integral* (or *antiderivative*) of both sides:

$$\int dv = \int a \, dt.$$

Since acceleration a is a constant, it can be taken outside the integration. Then we obtain

$$\int dv = a \int dt$$

or
$$v = at + C. \tag{2-20}$$

To evaluate the constant of integration C, we let $t = 0$, at which time $v = v_0$. Substituting these values into Eq. 2-20 (which must hold for all values of t, including $t = 0$) yields

$$v_0 = (a)(0) + C = C.$$

Substituting this into Eq. 2-20 gives us Eq. 2-11.

To derive Eq. 2-15, we rewrite the definition of velocity (Eq. 2-4) as

$$dx = v \, dt$$

and then take the indefinite integral of both sides to obtain

$$\int dx = \int v \, dt.$$

Generally v is not constant, so we cannot move it outside the integration. However, we can substitute for v with Eq. 2-11:

$$\int dx = \int (v_0 + at) \, dt.$$

Since v_0 is a constant, as is the acceleration a, this can be rewritten as

$$\int dx = v_0 \int dt + a \int t \, dt.$$

Integration now yields

$$x = v_0 t + \frac{1}{2}at^2 + C', \tag{2-21}$$

where C' is another constant of integration. At time $t = 0$, we have $x = x_0$. Substituting these values in Eq. 2-21 yields $x_0 = C'$. Replacing C' with x_0 in Eq. 2-21 gives us Eq. 2-15.

---

*This section is intended for students who have had integral calculus.

2-8 Free-Fall Acceleration

If you tossed an object either up or down and could somehow eliminate the effects of air on its flight, you would find that the object accelerates downward at a certain constant rate. That rate is called the **free-fall acceleration,** and its magnitude is represented by g. The acceleration is independent of the object's characteristics, such as mass, density, or shape; it is the same for all objects.

Two examples of free-fall acceleration are shown in Fig. 2-9, which is a series of stroboscopic photos of a feather and an apple. As these objects fall, they accelerate downward—both at the same rate g. Their speeds increase together.

The value of g varies slightly with latitude and with elevation. At sea level in Earth's midlatitudes the value is 9.8 m/s^2 (or 32 ft/s^2), which is what you should use for the problems in this chapter.

The equations of motion in Table 2-1 for constant acceleration also apply to free fall near Earth's surface; that is, they apply to an object in vertical flight, either up or down, when the effects of the air can be neglected. However, note that for free fall: (1) The directions of motion are now along a vertical y axis instead of the x axis, with the positive direction of y upward. (This is important for later chapters when combined horizontal and vertical motions are examined.) (2) The free-fall acceleration is now negative—that is, downward on the y axis, toward Earth's center—and so it has the value $-g$ in the equations.

Fig. 2-9 A feather and an apple, undergoing free fall in a vacuum, move downward with the same magnitude of acceleration g. The acceleration causes the increase in distance between successive images during the fall, but note that, in the absence of air, the feather and apple fall the same distance each time.

▶ The free-fall acceleration near Earth's surface is $a = -g = -9.8 \text{ m/s}^2$, and the *magnitude* of the acceleration is $g = 9.8 \text{ m/s}^2$. Do not substitute -9.8 m/s^2 for g.

Suppose that you toss a tomato directly upward with an initial (positive) velocity v_0 and then catch it when it returns to the release level. During its *free-fall flight* (just after its release and just before it is caught), the equations of Table 2-1 apply to its motion. The acceleration is always $a = -g = -9.8 \text{ m/s}^2$, negative and thus downward. The velocity, however, changes, as indicated by Eqs. 2-11 and 2-16: during the ascent, the magnitude of the positive velocity decreases, until it momentarily becomes zero. Because the tomato has then stopped, it is at its maximum height. During the descent, the magnitude of the (now negative) velocity increases.

Sample Problem 2-6

Let us return to the opening story about Dave Munday's Niagara free fall in a steel ball. He fell 48 m. Let us assume that his initial velocity was zero and neglect the effect of the air on the ball during the fall.

(a) How long did Munday fall to reach the water surface below the falls?

SOLUTION: The Key Idea here is that, because Munday's fall was a free fall, the equations of Table 2-1 apply. Let us place a y axis along the path of Munday's fall, with $y = 0$ at his starting point and the positive direction upward along the axis (Fig. 2-10). Then the acceleration is $a = -g$ along that axis and the water level is at $y = -48$ m (negative because it is below $y = 0$). Let the fall begin at time $t = 0$, with initial velocity $v_0 = 0$.

Then from Table 2-1 we choose Eq. 2-15 (but in y notation) because it contains the requested time t and all the other variables

Fig. 2-10 Sample Problem 2-6. The position, velocity, and acceleration of a freely falling object, here the steel ball ridden by Dave Munday over Niagara Falls.

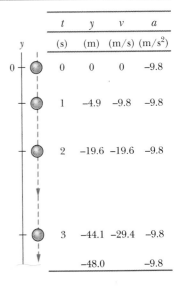

	t (s)	y (m)	v (m/s)	a (m/s^2)
	0	0	0	−9.8
	1	−4.9	−9.8	−9.8
	2	−19.6	−19.6	−9.8
	3	−44.1	−29.4	−9.8
		−48.0		−9.8

have known values. We find

$$y - y_0 = v_0 t - \tfrac{1}{2} g t^2,$$

$$-48 \text{ m} - 0 = 0t - \tfrac{1}{2}(9.8 \text{ m/s}^2) t^2,$$

$$t^2 = 48/4.9,$$

and $t = 3.1 \text{ s}.$ (Answer)

Note that Munday's displacement $y - y_0$ is a negative quantity—Munday fell down, in the *negative direction* of the y axis (he did not fall up!). Also note that 48/4.9 has two square roots: 3.1 and -3.1. Here we choose the positive root because Munday obviously reaches the water surface *after* he begins to fall at $t = 0$.

(b) Munday could count off the three seconds of free fall but could not see how far he had fallen with each count. Determine his position at each full second.

SOLUTION: We again use Eq. 2-15 but now we substitute, in turn, the values $t = 1.0$ s, 2.0 s, and 3.0 s, and solve for Munday's position y. The results are shown in Fig. 2-10.

(c) What was Munday's velocity as he reached the water surface?

SOLUTION: To find the velocity from the original data without using the time of fall from (a), we rewrite Eq. 2-16 in y notation and

then substitute known data:

$$v^2 = v_0^2 - 2g(y - y_0) = 0 - (2)(9.8 \text{ m/s}^2)(-48 \text{ m}),$$

so $v = -30.67 \text{ m/s} \approx -31 \text{ m/s} = -110 \text{ km/h}.$ (Answer)

We chose the negative root here, because the velocity was in the negative direction.

(d) What was Munday's velocity at each count of one full second? Was Munday aware of this increasing speed?

SOLUTION: To find the velocities from the original data without using the positions from (b), we let $a = -g$ in Eq. 2-11 and then substitute, in turn, the values $t = 1.0$ s, 2.0 s, and 3.0 s. Here is an example:

$$v = v_0 - gt$$
$$= 0 - (9.8 \text{ m/s}^2)(1.0 \text{ s}) = -9.8 \text{ m/s}.$$ (Answer)

The other results are shown in Fig. 2-10.

Once he was in free fall, Munday was unaware of the increasing speed because the acceleration during the fall was always -9.8 m/s², as noted in the last column of Fig. 2-10. He was, of course, sharply aware of hitting the water because then the acceleration abruptly changed. (Munday survived the fall but then faced stiff legal fines for his daredevil action.)

Sample Problem 2-7

In Fig. 2-11, a pitcher tosses a baseball up along a y axis, with an initial speed of 12 m/s.

(a) How long does the ball take to reach its maximum height?

SOLUTION: One Key Idea here is that once the ball leaves the pitcher and before it returns, its acceleration is the free-fall acceleration

Fig. 2-11 Sample Problem 2-7. A pitcher tosses a baseball straight up into the air. The equations of free fall apply for rising as well as for falling objects, provided any effects from the air can be neglected.

Ball
$v = 0$ at highest point

During ascent, $a = -g$, speed decreases, and velocity becomes less positive

During descent, $a = -g$, speed increases, and velocity becomes more negative

$y = 0$

$a = -g$. Because this is constant, Table 2-1 applies to the motion. A second Key Idea is that the velocity v at the maxium height must be 0. So, knowing v, a, and the initial velocity $v_0 = 12$ m/s, and seeking t, we solve Eq. 2-11, which contains those four variables. Rearranging yields

$$t = \frac{v - v_0}{a} = \frac{0 - 12 \text{ m/s}}{-9.8 \text{ m/s}^2} = 1.2 \text{ s}.$$ (Answer)

(b) What is the ball's maximum height above its release point?

SOLUTION: We can take the ball's release point to be $y_0 = 0$. We can then write Eq. 2-16 in y notation, set $y - y_0 = y$ and $v = 0$ (at the maximum height), and solve for y. We get

$$y = \frac{v^2 - v_0^2}{2a} = \frac{0 - (12 \text{ m/s})^2}{2(-9.8 \text{ m/s}^2)} = 7.3 \text{ m}.$$ (Answer)

(c) How long does the ball take to reach a point 5.0 m above its release point?

SOLUTION: We know v_0, $a = -g$, and the displacement $y - y_0 = 5.0$ m, and we want t, so we choose Eq. 2-15. Rewriting it for y and setting $y_0 = 0$ give us

$$y = v_0 t - \tfrac{1}{2} g t^2,$$

or $5.0 \text{ m} = (12 \text{ m/s})t - (\tfrac{1}{2})(9.8 \text{ m/s}^2) t^2.$

If we temporarily omit the units (having noted that they are consistent), we can rewrite this as

$$4.9 t^2 - 12t + 5.0 = 0.$$

Solving this quadratic equation for t yields

$$t = 0.53 \text{ s} \quad \text{and} \quad t = 1.9 \text{ s.} \qquad \text{(Answer)}$$

There are two such times! This is not really surprising because the ball passes twice through $y = 5.0$ m, once on the way up and once on the way down.

✔**CHECKPOINT 6:**   (a) In this sample problem, what is the sign of the ball's displacement for the ascent, from the release point to the highest point? (b) What is it for the descent, from the highest point back to the release point? (c) What is the ball's acceleration at its highest point?

PROBLEM-SOLVING TACTICS

Tactic 7: *Meanings of Minus Signs*
In Sample Problems 2-6 and 2-7, many answers emerged automatically with minus signs. It is important to know what these signs mean. For these two falling-body problems, we established a vertical axis (the y axis) and we chose—quite arbitrarily—its upward direction to be positive.

We then chose the origin of the y axis (that is, the $y = 0$ position) to suit the problem. In Sample Problem 2-6, the origin was at the top of the falls, and in Sample Problem 2-7 it was at the pitcher's hand. A negative value of y then means that the body is below the chosen origin. A negative velocity means that the body is moving in the negative direction of the y axis—that is, downward. This is true no matter where the body is located.

We take the acceleration to be negative (-9.8 m/s^2) in all problems dealing with falling bodies. A negative acceleration means that, as time goes on, the velocity of the body becomes either less positive or more negative. This is true no matter where the body is located and no matter how fast or in what direction it is moving. In Sample Problem 2-7, the acceleration of the ball is negative (downward) throughout its flight, whether the ball is rising or falling.

Tactic 8: *Unexpected Answers*
Mathematics often generates answers that you might not have thought of as possibilities, as in Sample Problem 2-7c. If you get more answers than you expect, do not automatically discard the ones that do not seem to fit. Examine them carefully for physical meaning. If time is your variable, even a negative value can mean something; negative time simply refers to time before $t = 0$, the (arbitrary) time at which you decided to start your stopwatch.

REVIEW & SUMMARY

Position   The *position* x of a particle on an x axis locates the particle with respect to the **origin,** or zero point, of the axis. The position is either positive or negative, according to which side of the origin the particle is on, or zero if the particle is at the origin. The **positive direction** on an axis is the direction of increasing positive numbers; the opposite direction is the **negative direction.**

Displacement   The *displacement* Δx of a particle is the change in its position:

$$\Delta x = x_2 - x_1. \qquad (2\text{-}1)$$

Displacement is a vector quantity. It is positive if the particle has moved in the positive direction of the x axis, and negative if the particle has moved in the negative direction.

Average Velocity   When a particle has moved from position x_1 to x_2 during a time interval $\Delta t = t_2 - t_1$, its *average velocity* during that interval is

$$v_{\text{avg}} = \frac{\Delta x}{\Delta t} = \frac{x_2 - x_1}{t_2 - t_1}. \qquad (2\text{-}2)$$

The algebraic sign of v_{avg} indicates the direction of motion (v_{avg} is a vector quantity). Average velocity does not depend on the actual distance a particle moves, but instead depends on its original and final positions.

On a graph of x versus t, the average velocity for a time interval Δt is the slope of the straight line connecting the points on the curve that represent the ends of the interval.

Average Speed   The *average speed* s_{avg} of a particle during a time interval Δt depends on the total distance the particle moves in that time interval:

$$s_{\text{avg}} = \frac{\text{total distance}}{\Delta t}. \qquad (2\text{-}3)$$

Instantaneous Velocity   The *instantaneous velocity* (or simply **velocity**) v of a moving particle is

$$v = \lim_{\Delta t \to 0} \frac{\Delta x}{\Delta t} = \frac{dx}{dt}, \qquad (2\text{-}4)$$

where Δx and Δt are defined by Eq. 2-2. The instantaneous velocity (at a particular time) may be found as the slope (at that particular time) of the graph of x versus t. **Speed** is the magnitude of instantaneous velocity.

Average Acceleration   *Average acceleration* is the ratio of a change in velocity Δv to the time interval Δt in which the change occurs:

$$a_{\text{avg}} = \frac{\Delta v}{\Delta t}. \qquad (2\text{-}7)$$

The algebraic sign indicates the direction of a_{avg}.

Instantaneous Acceleration   *Instantaneous acceleration* (or simply **acceleration**) a is the rate of change of velocity with time and the second derivative of position $x(t)$ with respect to time:

$$a = \frac{dv}{dt} = \frac{d^2x}{dt^2}. \qquad (2\text{-}8,\ 2\text{-}9)$$

On a graph of v versus t, the acceleration a at any time t is the slope of the curve at the point that represents t.

Constant Acceleration   The five equations in Table 2-1 describe the motion of a particle with constant acceleration:

$$v = v_0 + at, \qquad (2\text{-}11)$$

$$x - x_0 = v_0t + \tfrac{1}{2}at^2, \qquad (2\text{-}15)$$

$$v^2 = v_0^2 + 2a(x - x_0), \qquad (2\text{-}16)$$

$$x - x_0 = \tfrac{1}{2}(v_0 + v)t, \qquad (2\text{-}17)$$

$$x - x_0 = vt - \tfrac{1}{2}at^2. \qquad (2\text{-}18)$$

These are *not* valid when the acceleration is not constant.

Free-Fall Acceleration   An important example of straight-line motion with constant acceleration is that of an object rising or falling freely near Earth's surface. The constant acceleration equations describe this motion, but we make two changes in notation: (1) we refer the motion to the vertical y axis with $+y$ vertically *up*; (2) we replace a with $-g$, where g is the magnitude of the free-fall acceleration. Near Earth's surface, $g = 9.8$ m/s^2 ($= 32$ ft/s^2).

QUESTIONS

1. Figure 2-12 shows four paths along which objects move from a starting point to a final point, all in the same time. The paths pass over a grid of equally spaced straight lines. Rank the paths according to (a) the average velocity of the objects and (b) the average speed of the objects, greatest first.

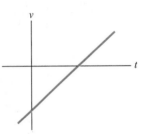

Fig. 2-12 Question 1.

2. Figure 2-13 gives the velocity of a particle moving on an x axis. What are (a) the initial and (b) the final directions of travel? (c) Does the particle stop momentarily? (d) Is the acceleration positive or negative? (e) Is it constant or varying?

Fig. 2-13 Question 2.

3. Figure 2-14 gives the acceleration $a(t)$ of a Chihuahua as it chases a German shepherd along an axis. In which of the time periods indicated does the Chihuahua move at constant speed?

Fig. 2-14 Question 3.

4. At $t = 0$, a particle moving along an x axis is at position $x_0 = -20$ m. The signs of the particle's initial velocity v_0 (at time t_0) and constant acceleration a are, respectively, for four situations: (1) +, +; (2) +, −; (3) −, +; (4) −, −. In which situation will the particle (a) undergo a momentary stop, (b) definitely pass through the origin (given enough time), and (c) definitely not pass through the origin?

5. The following equations give the velocity $v(t)$ of a particle in four situations: (a) $v = 3$; (b) $v = 4t^2 + 2t - 6$; (c) $v = 3t - 4$; (d) $v = 5t^2 - 3$. To which of these situations do the equations of Table 2-1 apply?

6. The driver of a blue car, moving at a speed of 80 km/h, suddenly realizes that she is about to rear-end a red car, moving at a speed of 60 km/h. To avoid a collision, what is the maximum speed the blue car can have just as it reaches the red car? (Warm-up for Problem 38)

7. At $t = 0$ and $x = 0$, an initially stationary blue car begins to accelerate at the constant rate of 2.0 m/s^2 in the positive direction of the x axis. At $t = 2$ s, a red car traveling in an adjacent lane and in the same direction, passes $x = 0$ with a speed of 8.0 m/s and a constant acceleration of 3.0 m/s^2. What pair of simultaneous equations should be solved to find when the red car passes the blue car? (Warm-up for Problem 36)

8. In Fig. 2-15, a cream tangerine is thrown directly upward past three evenly spaced windows of equal heights. Rank the windows according to (a) the average speed of the cream tangerine while passing them, (b) the time the cream tangerine takes to pass them, (c) the magnitude of the acceleration of the cream tangerine while passing them, and (d) the change Δv in the speed of the cream tangerine during the passage, greatest first.

9. You throw a ball straight up from the edge of a cliff, and it lands on the ground below the cliff. If you had, instead, thrown the ball down from the cliff edge with the same speed, would the ball's speed just before landing be larger than, smaller than, or the same as previously? (*Hint:* Consider Eq. 2-16.)

Fig. 2-15 Question 8.

EXERCISES & PROBLEMS

In several of the problems that follow you are asked to graph position, velocity, and acceleration versus time. Usually a sketch will suffice, appropriately labeled and with straight and curved portions apparent. If you have a computer or graphing calculator, you might use it to produce the graph.

SEC. 2-3 Average Velocity and Average Speed

1E. If a baseball pitcher throws a fastball at a horizontal speed of 160 km/h, how long does the ball take to reach home plate 18.4 m away? ssm

2E. A world speed record for bicycles was set in 1992 by Chris Huber riding Cheetah, a high-tech bicycle built by three mechanical engineering graduates. The record (average) speed was 110.6 km/h through a measured length of 200.0 m on a desert road. At the end of the run, Huber commented, "Cogito ergo zoom!" (I think, therefore I go fast!) What was Huber's elapsed time through the 200.0 m?

3E. An automobile travels on a straight road for 40 km at 30 km/h. It then continues in the same direction for another 40 km at 60 km/h. (a) What is the average velocity of the car during this 80 km trip? (Assume that it moves in the positive x direction.) (b) What is the average speed? (c) Graph x versus t and indicate how the average velocity is found on the graph. ssm

4P. A top-gun pilot, practicing radar avoidance maneuvers, is manually flying horizontally at 1300 km/h, just 35 m above the level ground. Suddenly, the plane encounters terrain that slopes gently upward at 4.3°, an amount difficult to detect visually (Fig. 2-16). How much times does the pilot have to make a correction to avoid flying into the ground?

Fig. 2-16 Problem 4.

5P. You drive on Interstate 10 from San Antonio to Houston, half the *time* at 55 km/h and the other half at 90 km/h. On the way back you travel half the *distance* at 55 km/h and the other half at 90 km/h. What is your average speed (a) from San Antonio to Houston, (b) from Houston back to San Antonio, and (c) for the entire trip? (d) What is your average velocity for the entire trip? (e) Sketch x versus t for (a), assuming the motion is all in the positive x direction. Indicate how the average velocity can be found on the sketch. ilw

6P. Compute your average velocity in the following two cases: (a) You walk 73.2 m at a speed of 1.22 m/s and then run 73.2 m at a speed of 3.05 m/s along a straight track. (b) You walk for 1.00 min at a speed of 1.22 m/s and then run for 1.00 min at 3.05 m/s along a straight track. (c) Graph x versus t for both cases and indicate how the average velocity is found on the graph.

7P. The position of an object moving along an x axis is given by $x = 3t - 4t^2 + t^3$, where x is in meters and t in seconds. (a) What is the position of the object at $t = 1, 2, 3$, and 4 s? (b) What is the object's displacement between $t = 0$ and $t = 4$ s? (c) What is its average velocity for the time interval from $t = 2$ s to $t = 4$ s? (d) Graph x versus t for $0 \leq t \leq 4$ s and indicate how the answer for (c) can be found on the graph. ssm www

8P. Two trains, each having a speed of 30 km/h, are headed at each other on the same straight track. A bird that can fly 60 km/h flies off the front of one train when they are 60 km apart and heads directly for the other train. On reaching the other train it flies directly back to the first train, and so forth. (We have no idea *why* a bird would behave in this way.) What is the total distance the bird travels?

9P. On two *different* tracks, the winners of the one-kilometer race ran their races in 2 min, 27.95 s and 2 min, 28.15 s. In order to conclude that the runner with the shorter time was indeed faster, how much longer can the other track be in *actual* length? ilw

SEC. 2-4 Instantaneous Velocity and Speed

10E. The graph in Fig. 2-17 is for an armadillo that scampers left (negative direction of x) and right along an x axis. (a) When, if ever, is the animal to the left of the origin on the axis? When, if ever, is its velocity (b) negative, (c) positive, or (d) zero?

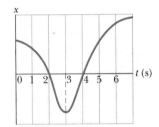

Fig. 2-17 Exercise 10.

11E. (a) If a particle's position is given by $x = 4 - 12t + 3t^2$ (where t is in seconds and x is in meters), what is its velocity at $t = 1$ s? (b) Is it moving in the positive or negative direction of x just then? (c) What is its speed just then? (d) Is the speed larger or smaller at later times? (Try answering the next two questions without further calculation.) (e) Is there ever an instant when the velocity is zero? (f) Is there a time after $t = 3$ s when the particle is moving in the negative direction of x?

12P. The position of a particle moving along the x axis is given in centimeters by $x = 9.75 + 1.50t^3$, where t is in seconds. Calculate (a) the average velocity during the time interval $t = 2.00$ s to $t = 3.00$ s; (b) the instantaneous velocity at $t = 2.00$ s; (c) the instantaneous velocity at $t = 3.00$ s; (d) the instantaneous velocity at $t = 2.50$ s; and (e) the instantaneous velocity when the particle is midway between its positions at $t = 2.00$ s and $t = 3.00$ s. (f) Graph x versus t and indicate your answers graphically.

13P. How far does the runner whose velocity–time graph is shown in Fig. 2-18 travel in 16 s? `ilw`

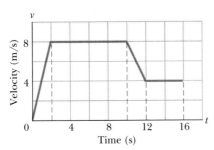

Fig. 2-18 Problem 13.

SEC. 2-5   Acceleration

14E. Sketch a graph that is a possible description of position as a function of time for a particle that moves along the x axis and, at $t = 1$ s, has (a) zero velocity and positive acceleration; (b) zero velocity and negative acceleration; (c) negative velocity and positive acceleration; (d) negative velocity and negative acceleration. (e) For which of these situations is the speed of the particle increasing at $t = 1$ s?

15E. What do the quantities (a) $(dx/dt)^2$ and (b) d^2x/dt^2 represent? (c) What are their SI units?

16E. A frightened ostrich moves in a straight line with velocity described by the velocity–time graph of Fig. 2-19. Sketch acceleration versus time.

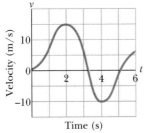

Fig. 2-19 Exercise 16.

17E. A particle had a speed of 18 m/s at a certain time, and 2.4 s later its speed was 30 m/s in the opposite direction. What were the magnitude and direction of the average acceleration of the particle during this 2.4 s interval? `ssm`

18P. From $t = 0$ to $t = 5.00$ min, a man stands still, and from $t = 5.00$ min to $t = 10.0$ min, he walks briskly in a straight line at a constant speed of 2.20 m/s. What are (a) his average velocity v_{avg} and (b) his average acceleration a_{avg} in the time interval 2.00 min to 8.00 min? What are (c) v_{avg} and (d) a_{avg} in the time interval 3.00 min to 9.00 min? (e) Sketch x versus t and v versus t, and indicate how the answers to (a) through (d) can be obtained from the graphs.

19P. A proton moves along the x axis according to the equation $x = 50t + 10t^2$, where x is in meters and t is in seconds. Calculate (a) the average velocity of the proton during the first 3.0 s of its motion, (b) the instantaneous velocity of the proton at $t = 3.0$ s, and (c) the instantaneous acceleration of the proton at $t = 3.0$ s. (d) Graph x versus t and indicate how the answer to (a) can be obtained from the plot. (e) Indicate the answer to (b) on the graph. (f) Plot v versus t and indicate on it the answer to (c). `ssm`

20P. A electron moving along the x axis has a position given by $x = 16te^{-t}$ m, where t is in seconds. How far is the electron from the origin when it momentarily stops?

21P. The position of a particle moving along the x axis depends on the time according to the equation $x = ct^2 - bt^3$, where x is in meters and t in seconds. (a) What units must c and b have? Let their numerical values be 3.0 and 2.0, respectively. (b) At what time does the particle reach its maximum positive x position? From $t = 0.0$ s to $t = 4.0$ s, (c) what distance does the particle move and (d) what is its displacement? At $t = 1.0$, 2.0, 3.0, and 4.0 s, what are (e) its velocities and (f) its accelerations? `ssm`

SEC. 2-6   Constant Acceleration: A Special Case

22E. An automobile driver increases the speed at a constant rate from 25 km/h to 55 km/h in 0.50 min. A bicycle rider speeds up at a constant rate from rest to 30 km/h in 0.50 min. Calculate their accelerations.

23E. A muon (an elementary particle) enters a region with a speed of 5.00×10^6 m/s and then is slowed at the rate of 1.25×10^{14} m/s². (a) How far does the muon take to stop? (b) Graph x versus t and v versus t for the muon. `ssm`

24E. The head of a rattlesnake can accelerate at 50 m/s² in striking a victim. If a car could do as well, how long would it take to reach a speed of 100 km/h from rest?

25E. An electron has a constant acceleration of $+3.2$ m/s². At a certain instant its velocity is $+9.6$ m/s. What is its velocity (a) 2.5 s earlier and (b) 2.5 s later? `ssm`

26E. The speed of a bullet is measured to be 640 m/s as the bullet emerges from a barrel of length 1.20 m. Assuming constant acceleration, find the time that the bullet spends in the barrel after it is fired.

27E. Suppose a rocket ship in deep space moves with constant acceleration equal to 9.8 m/s², which gives the illusion of normal gravity during the flight. (a) If it starts from rest, how long will it take to acquire a speed one-tenth that of light, which travels at 3.0×10^8 m/s? (b) How far will it travel in so doing? `ssm`

28E. A jumbo jet must reach a speed of 360 km/h on the runway for takeoff. What is the least constant acceleration needed for takeoff from a 1.80 km runway?

29E. An electron with initial velocity $v_0 = 1.50 \times 10^5$ m/s enters a region 1.0 cm long where it is electrically accelerated (Fig. 2-20). It emerges with velocity $v = 5.70 \times 10^6$ m/s. What is its acceleration, assumed constant? (Such a process occurs in conventional television sets.) `ssm`

Fig. 2-20 Exercise 29.

30E. A world's land speed record was set by Colonel John P. Stapp when in March 1954 he rode a rocket-propelled sled that moved along a track at 1020 km/h. He and the sled were brought to a stop in 1.4 s. (See Fig. 2-7.) In g units, what acceleration did he experience while stopping?

31E. The brakes on your automobile are capable of creating a deceleration of 5.2 m/s². (a) If you are going 137 km/h and suddenly see a state trooper, what is the minimum time in which you can get your car under the 90 km/h speed limit? (The answer reveals the futility of braking to keep your high speed from being detected with a radar or laser gun.) (b) Graph x versus t and v versus t for such a deceleration. ssm www

32E. Figure 2-21 depicts the motion of a particle moving along an x axis with a constant acceleration. What are the magnitude and direction of the particle's acceleration?

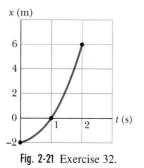
Fig. 2-21 Exercise 32.

33P. A car traveling 56.0 km/h is 24.0 m from a barrier when the driver slams on the brakes. The car hits the barrier 2.00 s later. (a) What is the car's constant deceleration before impact? (b) How fast is the car traveling at impact? ssm ilw

34P. A red train traveling at 72 km/h and a green train traveling at 144 km/h are headed toward one another along a straight, level track. When they are 950 m apart, each engineer sees the other's train and applies the brakes. The brakes decelerate each train at the rate of 1.0 m/s². Is there a collision? If so, what is the speed of each train at impact? If not, what is the separation between the trains when they stop?

35P. A car moving with constant acceleration covered the distance between two points 60.0 m apart in 6.00 s. Its speed as it passes the second point was 15.0 m/s. (a) What was the speed at the first point? (b) What was the acceleration? (c) At what prior distance from the first point was the car at rest? (d) Graph x versus t and v versus t for the car, from rest ($t = 0$). ssm www

36P. At the instant the traffic light turns green, an automobile starts with a constant acceleration a of 2.2 m/s². At the same instant a truck, traveling with a constant speed of 9.5 m/s, overtakes and passes the automobile. (a) How far beyond the traffic signal will the automobile overtake the truck? (b) How fast will the car be traveling at that instant?

37P. To stop a car, first you require a certain reaction time to begin braking; then the car slows under the constant braking deceleration. Suppose that the total distance moved by your car during these two phases is 56.7 m when its initial speed is 80.5 km/h, and 24.4 m when its initial speed is 48.3 km/h. What are (a) your reaction time and (b) the magnitude of the deceleration? ssm

38P. When a high-speed passenger train traveling at 161 km/h rounds a bend, the engineer is shocked to see that a locomotive has improperly entered onto the track from a siding and is a distance $D = 676$ m ahead (Fig. 2-22). The locomotive is moving at 29.0

km/h. The engineer of the high-speed train immediately applies the brakes. (a) What must be the magnitude of the resulting constant deceleration if a collision is to be just avoided? (b) Assume that the engineer is at $x = 0$ when, at $t = 0$, he first spots the locomotive. Sketch the $x(t)$ curves representing the locomotive and high-speed train for the situations in which a collision is just avoided and is not quite avoided.

39P. An elevator cab in the New York Marquis Marriott has a total run of 190 m. Its maximum speed is 305 m/min. Its acceleration and deceleration both have a magnitude of 1.22 m/s². (a) How far does the cab move while accelerating to full speed from rest? (b) How long does it take to make the nonstop 190 m run, starting and ending at rest? ilw

SEC. 2-8 Free-Fall Acceleration

40E. Raindrops fall 1700 m from a cloud to the ground. (a) If they were not slowed by air resistance, how fast would the drops be moving when they struck the ground? (b) Would it be safe to walk outside during a rainstorm?

41E. At a construction site a pipe wrench struck the ground with a speed of 24 m/s. (a) From what height was it inadvertently dropped? (b) How long was it falling? (c) Sketch graphs of y, v, and a versus t for the wrench. ssm

42E. A hoodlum throws a stone vertically downward with an initial speed of 12.0 m/s from the roof of a building, 30.0 m above the ground. (a) How long does it take the stone to reach the ground? (b) What is the speed of the stone at impact?

43E. (a) With what speed must a ball be thrown vertically from ground level to rise to a maximum height of 50 m? (b) How long will it be in the air? (c) Sketch graphs of y, v, and a versus t for the ball. On the first two graphs, indicate the time at which 50 m is reached. ssm

44E. The Zero Gravity Research Facility at the NASA Lewis Research Center includes a 145 m drop tower. This is an evacuated vertical tower through which, among other possibilities, a 1 m diameter sphere containing an experimental package can be dropped. (a) How long is the sphere in free fall? (b) What is its speed just as it reaches a catching device at the bottom of the tower? (c) When caught, the sphere experiences an average deceleration of $25g$ as its speed is reduced to zero. Through what distance does it travel during the deceleration?

45E. A rock is dropped from a 100-m-high cliff. How long does it take to fall (a) the first 50 m and (b) the second 50 m? ssm

46P. A ball is thrown *down* vertically with an initial *speed* of v_0 from a height of h. (a) What is its speed just before it strikes the ground? (b) How long does the ball take to reach the ground? What would be the answers to (c) part a and (d) part b if the ball were thrown *upward* from the same height and with the same initial speed? Before solving any equations, decide whether the answers to (c) and (d) should be greater than, less than, or the same as in (a) and (b).

47P. A startled armadillo leaps upward, rising 0.544 m in the first 0.200 s. (a) What is its initial speed as it leaves the ground? (b) What is its speed at the height of 0.544 m? (c) How much higher does it go? ssm www

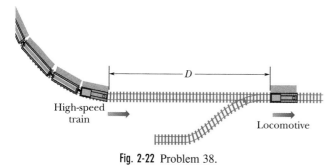
Fig. 2-22 Problem 38.

High-speed train

Locomotive

D

48P. A rock is dropped (from rest) from the top of a 60-m-tall building. How far above the ground is the rock 1.2 s before it reaches the ground?

49P. A key falls from a bridge that is 45 m above the water. It falls directly into a model boat, moving with constant velocity, that is 12 m from the point of impact when the key is released. What is the speed of the boat?   ssm   ilw

50P. A ball is thrown vertically downward from the top of a 36.6-m-tall building. The ball passes the top of a window that is 12.2 m above the ground 2.00 s after being thrown. What is the speed of the ball as it passes the top of the window?

51P. A ball of moist clay falls 15.0 m to the ground. It is in contact with the ground for 20.0 ms before stopping. What is the average acceleration of the ball during the time it is in contact with the ground? (Treat the ball as a particle.)   ssm

52P. A model rocket fired vertically from the ground ascends with a constant vertical acceleration of 4.00 m/s^2 for 6.00 s. Its fuel is then exhausted, so it continues upward as a free-fall particle and then falls back down. (a) What is the maximum altitude reached? (b) What is the total time elapsed from takeoff until the rocket strikes the ground?

53P. To test the quality of a tennis ball, you drop it onto the floor from a height of 4.00 m. It rebounds to a height of 2.00 m. If the ball is in contact with the floor for 12.0 ms, what is its average acceleration during that contact?   ssm

54P. A basketball player, standing near the basket to grab a rebound, jumps 76.0 cm vertically. How much (total) time does the player spend (a) in the top 15.0 cm of this jump and (b) in the bottom 15.0 cm? Does this help explain why such players seem to hang in the air at the tops of their jumps?

55P. Water drips from the nozzle of a shower onto the floor 200 cm below. The drops fall at regular (equal) intervals of time, the first drop striking the floor at the instant the fourth drop begins to fall. Find the locations of the second and third drops when the first strikes the floor.

56P. A ball is shot vertically upward from the surface of a planet in a distant solar system. A plot of y versus t for the ball is shown in Fig. 2-23, where y is the height of the ball above its starting point and $t = 0$ at the instant the ball is shot. What are the magnitudes of (a) the free-fall acceleration on the planet and (b) the initial velocity of the ball?

57P. Two diamonds begin a free fall from rest from the same height, 1.0 s apart. How long after the first diamond begins to fall will the two diamonds be 10 m apart?   ssm

58P. A certain juggler usually tosses balls vertically to a height H. To what height must they be tossed if they are to spend twice as much time in the air?

59P. A hot-air balloon is ascending at the rate of 12 m/s and is 80 m above the ground when a package is dropped over the side. (a) How long does the package take to reach the ground? (b) With what speed does it hit the ground?   ssm

60P. A stone is dropped into a river from a bridge 43.9 m above the water. Another stone is thrown vertically down 1.00 s after the first is dropped. Both stones strike the water at the same time. (a) What is the initial speed of the second stone? (b) Plot velocity versus time on a graph for each stone, taking zero time as the instant the first stone is released.

61P. An elevator without a ceiling is ascending with a constant speed of 10 m/s. A boy on the elevator shoots a ball directly upward, from a height of 2.0 m above the elevator floor, just as the elevator floor is 28 m above the ground. The initial speed of the ball with respect to the elevator is 20 m/s. (a) What maximum height above the ground does the ball reach? (b) How long does the ball take to return to the elevator floor?   ssm

62P. A stone is thrown vertically upward. On its way up it passes point A with speed v, and point B, 3.00 m higher than A, with speed $\frac{1}{2}v$. Calculate (a) the speed v and (b) the maximum height reached by the stone above point B.

63P. Figure 2-24 shows a simple device for measuring your reaction time. It consists of a cardboard strip marked with a scale and two large dots. A friend holds the strip *vertically*, with thumb and forefinger at the dot on the right in Fig. 2-24. You then position your thumb and forefinger at the other dot (on the left in Fig. 2-24), being careful not to touch the strip. Your friend releases the strip, and you try to pinch it as soon as possible after you see it begin to fall. The mark at the place where you pinch the strip gives your reaction time. (a) How far from the lower dot should you place the 50.0 ms mark? (b) How much higher should the marks for 100, 150, 200, and 250 ms be? (For example, should the 100 ms marker be two times as far from the dot as the 50 ms marker? Can you find any pattern in the answers?)

Reaction time (ms)

Fig. 2-24 Problem 63.

64P. A parachutist bails out and freely falls 50 m. Then the parachute opens, and thereafter she decelerates at 2.0 m/s^2. She reaches the ground with a speed of 3.0 m/s. (a) How long is the parachutist in the air? (b) At what height does the fall begin?

65P. A drowsy cat spots a flowerpot that sails first up and then down past an open window. The pot is in view for a total of 0.50 s, and the top-to-bottom height of the window is 2.00 m. How high above the window top does the flowerpot go?

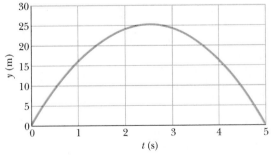

Fig. 2-23 Problem 56.

NEW PROBLEMS

N1. A car moves along an x axis through a distance of 900 m, starting at rest (at $x = 0$) and ending at rest (at $x = 900$ m). Through the first $\frac{1}{4}$ of that distance, its acceleration is $+2.25$ m/s^2. Through the next $\frac{3}{4}$ of that distance, its acceleration is -0.750 m/s^2. What are (a) its travel time through the 900 m and (b) its maximum speed? (c) Graph position x, velocity v, and acceleration a versus time t for the trip.

N2. At time $t = 0$, a rock climber accidentally allows a piton to fall freely from a high point on the rock wall to the valley below him. Then, after a short delay, his climbing partner, who is 10 m higher on the wall, throws a piton downward. The positions y of the pitons versus t during the falling are given in Fig. 2N-1. With what speed was the second piton thrown?

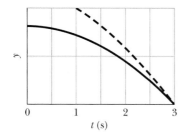

Fig. 2N-1 Problem N2.

N3. A bolt is dropped from a bridge under construction, falling 90 m to the valley below the bridge. (a) In how much time does it pass through the last 20% of its fall? What is its speed (b) when it begins that last 20% of its fall and (c) when it reaches the valley beneath the bridge?

N4. At time $t = 0$, apple 1 is dropped from a bridge onto a roadway beneath the bridge; somewhat later, apple 2 is thrown down from the same height. Figure 2N-2 gives the vertical positions y of the apples versus t during the falling, until both apples have hit the roadway. With approximately what speed is apple 2 thrown down?

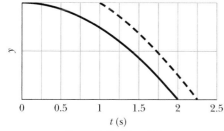

Fig. 2N-2 Problem N4.

N5. A mining cart is pulled up a hill at 20 km/h and then pulled back down the hill at 35 km/h through its original level. (The time required for the cart's reversal at the top of its climb is negligible.) What is the average speed of the cart for its round trip, from its original level back to its original level?

N6. As a runaway scientific balloon ascends at 19.6 m/s, one of its instrument packages breaks free of a harness and free-falls. Figure 2N-3 gives the vertical velocity of the package versus time, from before it breaks free to when it reaches the ground. (a) What maximum height above the break-free point does it rise? (b) How high was the break-free point above the ground?

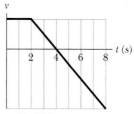

Fig. 2N-3 Problem N6.

N7. During a hard sneeze, your eyes might shut for 0.50 s. If you are driving a car at 90 km/h during such a sneeze, how far does the car move during that time?

N8. Figure 2N-4 shows the speed v versus height y of a ball tossed directly upward, along a y axis. The speed at height y_A is v_A. The speed at height y_B is $v_A/3$. What is speed v_A?

Fig. 2N-4 Problem N8.

N9. Figure 2N-5 is a plot of the age of ancient seafloor material, in millions of years, against the distance from a particular ocean ridge. Seafloor material extruded from this ridge moves away from it at approximately uniform speed. Find that speed in centimeters per year.

Fig. 2N-5 Problem N9.

N10. You are to drive to an interview in another town, at a distance of 300 km on an expressway. The interview is at 11:15 A.M. You plan to drive at 100 km/h, so you leave at 8:00 A.M. to allow some extra time. You drive at that speed for the first 100 km, but then construction work forces you to slow to 40 km/h for 40 km. What would be the least speed needed for the rest of the trip to arrive in time for the interview?

N11. In an arcade video game, a spot is programmed to move across the screen according to $x = 9.00t - 0.750t^3$, where x is distance in centimeters measured from the left edge of the screen and t is time in seconds. When the spot reaches a screen edge, at either $x = 0$ or $x = 15.0$ cm, t is reset to 0 and the spot starts moving again according to $x(t)$. (a) At what time after starting is the spot instantaneously at rest? (b) Where does this occur? (c) What is its acceleration when this occurs? (d) In what direction is it moving just prior to coming to rest? (e) Just after? (f) When does it first reach an edge of the screen after $t = 0$?

N12. As two trains move along a track, their conductors suddenly notice that they are headed toward each other. Figure 2N-6 gives their velocities v as functions of time t as the conductors slow the trains. The slowing processes begin when the trains are 200 m apart. What is their separation when both trains have stopped?

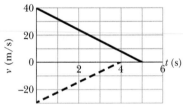

Fig. 2N-6 Problem N12.

N13. When a driver brings a car to a stop by braking as hard as possible, the stopping distance can be regarded as the sum of a "reaction distance," which is initial speed multiplied by the driver's reaction time, and a "braking distance," which is the distance traveled during braking. The following table gives typical values. (a) What reaction time is the driver assumed to have? (b) What is the car's stopping distance if the initial speed is 25 m/s?

Initial Speed (m/s)	Reaction Distance (m)	Braking Distance (m)	Stopping Distance (m)
10	7.5	5.0	12.5
20	15	20	35
30	22.5	45	67.5

N14. The acceleration of a particle along an x axis is $a = 5.0t$, with t in seconds and a in meters per second squared. At $t = 2.0$ s, its velocity is $+17$ m/s. What is its velocity at $t = 4.0$ s?

N15. An iceboat has a constant velocity toward the east when a sudden gust of wind causes the iceboat to have a constant acceleration toward the east for a period of 3.0 s. A plot of x versus t is shown in Fig. 2N-7, where $t = 0$ is taken to be the instant the wind

starts to blow and the positive x axis is toward the east. (a) What is the acceleration of the iceboat during the 3.0 s interval? (b) What is the velocity of the iceboat at the end of the 3.0 s interval? (c) If the acceleration remains constant for an additional 3.0 s, how far will the iceboat travel during this second 3.0 s interval?

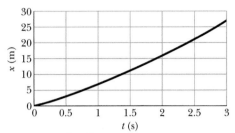

Fig. 2N-7 Problem N15.

N16. Cars A and B move in the same direction in adjacent lanes. The position x of car A is given in Fig. 2N-8, from time $t = 0$ to $t = 7.0$ s. At $t = 0$, car B is at $x = 0$, with a velocity of 12 m/s and a negative constant acceleration a_B. (a) What must a_B be such that the cars are (momentarily) side by side (momentarily at the same value of x) at $t = 4.0$ s? (b) For that value of a_B, how many times are the cars side by side? (c) Sketch the position x of car B versus time t on Fig. 2N-8. How many times will the cars be side by side if the magnitude of acceleration a_B is (d) more than and (e) less than the answer to part (a)?

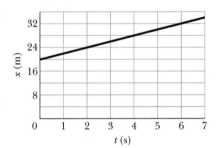

Fig. 2N-8 Problem N16.

N17. At the National Physical Laboratory in England, a measurement of the free-fall acceleration g was made by throwing a glass ball straight up in an evacuated tube and letting it return. Let ΔT_L in Fig. 2N-9 be the time interval between the two passages of the ball across a certain lower level, ΔT_U the time interval between the two passages across an upper level, and H the distance between the two levels. Show that

$$g = \frac{8H}{\Delta T_L^2 - \Delta T_U^2}.$$

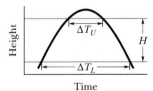

Time

Fig. 2N-9 Problem N17.

N18. Figure 2N-10 gives the acceleration a versus time t for a particle moving along an x axis. At $t = -2.0$ s, the particle's velocity is 7.0 m/s. What is its velocity at $t = 6.0$ s?

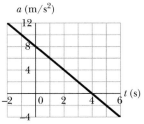

Fig. 2N-10 Problem N18.

N19. An electric vehicle starts from rest and accelerates at a rate of 2.0 m/s² in a straight line until it reaches a speed of 20 m/s. The vehicle then slows at a constant rate of 1.0 m/s² until it stops. (a) How much time elapses from start to stop? (b) How far does the vehicle travel from start to stop?

N20. You are arguing over a cell phone while trailing an unmarked police car by 25 m; both your car and the police car are traveling at 110 km/h. Your argument diverts your attention from the police car for 2.0 s (long enough for you to look at the phone and yell, "I won't do that!"). At the beginning of that 2.0 s, the police officer begins emergency braking at 5.0 m/s². (a) What is the separation between the two cars when your attention finally returns? Suppose that you take another 0.40 s to realize your danger and begin braking. (b) If you too brake at 5.0 m/s², what is your speed when you hit the police car?

N21. A rock is shot vertically upward from the edge of the top of a tall building. The rock reaches its maximum height above the top of the building 1.60 s after being shot. Then, after barely missing the edge of the building as it falls downward, the rock strikes the ground 6.00 s after it was launched. In SI units: (a) with what upward velocity was the rock shot, (b) what maximum height above the top of the building is reached by the rock, and (c) how tall is the building?

N22. Figure 2N-11 gives the velocity v (m/s) versus time t (s) for a particle moving along an x axis. The area between the time axis and the plotted curve is given for the two portions of the graph. At $t = t_A$ (at one of the crossing points in the plotted figure), the particle's position is $x = 14$ m. What is its position at (a) $t = 0$ and (b) $t = t_B$?

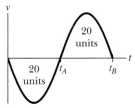

Fig. 2N-11 Problem N22.

N23. The position of a particle moving along an x axis is given by $x = 12t^2 - 2t^3$, where x is in meters and t is in seconds. (a) Determine the position, velocity, and acceleration of the particle at $t = 3.0$ s. (b) What is the maximum positive coordinate reached by the particle and at what time is it reached? (c) What is the maximum positive velocity reached by the particle and at what time is it reached? (d) What is the acceleration of the particle at the instant the particle is not moving (other than at $t = 0$)? (e) Determine the average velocity of the particle between $t = 0$ and $t = 3$ s.

N24. A shuffleboard disk is accelerated at a constant rate from rest to a speed of 6.0 m/s over a 1.8 m distance by a player using a cue. At this point the disk loses contact with the cue and slows at a constant rate of 2.5 m/s² until it stops. (a) How much time elapses from when the disk begins to accelerate until it stops? (b) What total distance does the disk travel?

N25. A model rocket, propelled by burning fuel, takes off vertically. Plot qualitatively (numbers not required) graphs of y, v, and a versus t for the rocket's flight. Indicate when the fuel is exhausted, when the rocket reaches maximum height, and when it returns to the ground.

N26. A particle moves along the x axis with position function $x(t)$ as shown in Fig. 2N-12. Make rough sketches of the particle's velocity versus time and its acceleration versus time for this motion.

Fig. 2N-12 Problem N26.

N27. The position of a particle as it moves along a y axis is given by

$$y = 2.0 \sin\left(\frac{\pi}{4}t\right),$$

where t is in seconds and y is in centimeters. (a) What is the average velocity of the particle between $t = 0$ and $t = 2.0$ s? (b) What is the instantaneous velocity of the particle at $t = 0$, 1.0, and 2.0 s? (c) What is the average acceleration of the particle between $t = 0$ and $t = 2.0$ s? (d) What is the instantaneous acceleration of the particle at $t = 0$, 1.0, and 2.0 s? (e) Plot v versus t for $0 \le t \le 2.0$ s, and estimate the instantaneous acceleration at $t = 1.0$ s from the graph.

N28. *Graphing.* Imagine that you throw an apple directly upward at time $t = 0$ and want it to return to the launch level (where it left your hand) at time $t = 1.5$ s. Let a y axis extend upward along the apple's path, with $y = 0$ at the launch point. The apple's position along the axis is then given by

$$y = v_0 t - 0.5(9.8 \text{ m/s}^2)t^2,$$

where y is in meters and t is in seconds. Graph the function for a launch speed of $v_0 = 6.5$ m/s, with a tick mark along the horizontal (time) axis at 1.5. Were the graphed curve to pass through the tick mark, then the apple would return to the launch level ($y = 0$) at $t = 1.5$ s. However, you will find that the curve misses the tick mark. (a) Do you increase or decrease the launch speed to get the curve closer to the tick mark? (b) More generally, in this type of graph, does, say, increasing the launch speed "pull" the curve rightward or leftward? (c) Try several launch speeds (or a list of launch speeds) to find the value needed to get the curve to pass through the tick mark. Approximately what is that value?

N29. A motorcyclist who is moving along an x axis directed toward the east has an acceleration given by $a = (6.1 - 1.2t)$ m/s^2 for $0 \leq t \leq 6.0$ s. At $t = 0$, the velocity and position of the cyclist are 2.7 m/s and 7.3 m. (a) What is the maximum speed achieved by the cyclist? (b) What total distance does the cyclist travel between $t = 0$ and 6.0 s?

N30. At time $t = 0$, a hockey puck is sent sliding over a frozen lake, directly into a strong wind. Figure 2N-13 gives the velocity v of the puck versus time, as the puck moves along a single axis. At $t = 14$ s, what is its position relative to its position at $t = 0$?

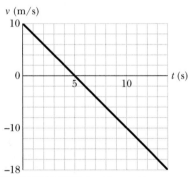

Fig. 2N-13 Problem N30.

N31. *Minimum trailing distance.* Your car trails a highway patrol car at separation d, with both cars at speed v_0, when suddenly the patrol car begins to decelerate to a stop at the constant negative rate a_1. After a time t_R to react to the patrol car's change (to perceive the change and then to begin applying your brake), you begin to decelerate at the constant negative rate a_2.

What is the stopping distance of (a) the patrol car and (b) your car? (c) What is the difference in those two stopping distances? This difference is the least separation d_{min} you should have originally if you are to avoid rear-ending the patrol car. (d) On a calculator, make a single graph of d_{min} versus v_0 for v_0 ranging from

0 to 50 m/s and for the following values of a_1 and a_2, used in lists:

(1) -5.00 and -5.00 m/s^2,

(2) -5.00 and -4.50 m/s^2,

(3) -4.50 and -4.00 m/s^2,

(4) -2.00 and -1.50 m/s^2.

Assume a response time $t_R = 1.00$ s. Set the separation of tick marks on the vertical axis to be the typical car length of 4.5 m. Using the calculator's evaluation option for the graph, evaluate d_{min} at $v_0 = 30$ m/s for the (e) second pair and (f) third pair of acceleration values.

(g) According to a conventional rule in the United States, d_{min} should be one car length for every 10 mi/h (= 4.5 m/s) in speed. Plot this dependence on the existing graph. (h) If the conventional rule is to be valid, what must be the relation between a_1 and a_2? (i) If this condition is not met, what is the error between the actual d_{min} and that predicted by the conventional rule? (j) Plot this error versus the given range of v_0 and for the given four pairs of accelerations. (k) Generally, does the error increase or decrease for weaker braking on the two cars? Evaluate the error at $v_0 = 30$ m/s for the (l) second pair and (m) third pair of acceleration values.

N32. In Fig. 2N-14, a red car and a green car, identical except for the color, move toward each other in adjacent lanes and parallel to an x axis. At time $t = 0$, the red car is at $x = 0$ and the green car is at $x = 220$ m. If the red car has a constant velocity of 20 km/h, the cars pass each other at $x = 44.5$ m, and if it has a constant velocity of 40 km/h, they pass each other at $x = 76.6$ m. What are (a) the initial velocity and (b) the acceleration of the green car?

Fig. 2N-14 Problem N32.

3 Vectors

For two decades spelunking teams crawled, climbed, and squirmed through 200 km of Mammoth Cave and the Flint Ridge cave system, seeking a connection. The photograph shows Richard Zopf pushing his pack through the Tight Tube, far inside the Flint Ridge system. After 12 hours of "caving" along a labyrinthine route, Zopf and six others waded through a stretch of chilling water and found themselves in Mammoth Cave. Their breakthrough established the Mammoth–Flint cave system as the longest cave in the world.

How does their final point relate to their initial point other than in terms of the actual route they covered?

The answer is in this chapter.

Fig. 3-1 (*a*) All three arrows have the same magnitude and direction and thus represent the same displacement. (*b*) All three paths connecting the two points correspond to the same displacement vector.

3-1 Vectors and Scalars

A particle moving along a straight line can move in only two directions. We can take its motion to be positive in one of these directions and negative in the other. For a particle moving in three dimensions, however, a plus sign or minus sign is no longer enough to indicate the direction of the motion. Instead, we must use a *vector*.

A **vector** has magnitude as well as direction, and vectors follow certain (vector) rules of combination, which we examine in this chapter. A **vector quantity** is a quantity that has both a magnitude and a direction and thus can be represented with a vector. Some physical quantities that are vector quantities are displacement, velocity, and acceleration. You will see many more throughout this book, so learning the rules of vector combination now will help you greatly in later chapters.

Not all physical quantities involve a direction. Temperature, pressure, energy, mass, and time, for example, do not "point" in the spatial sense. We call such quantities **scalars,** and we deal with them by the rules of ordinary algebra. A single value, with a sign (as in a temperature of $-40°F$), specifies a scalar.

The simplest vector quantity is displacement, or change of position. A vector that represents a displacement is called, reasonably, a **displacement vector.** (Similarly, we have velocity vectors and acceleration vectors.) If a particle changes its position by moving from *A* to *B* in Fig. 3-1*a*, we say that it undergoes a displacement from *A* to *B*, which we represent with an arrow pointing from *A* to *B*. The arrow specifies the vector graphically. To distinguish vector symbols from other kinds of arrows in this book, we use the outline of a triangle as the arrowhead.

In Fig. 3-1*a*, the arrows from *A* to *B*, from *A′* to *B′*, and from *A″* to *B″* have the same magnitude and direction. Thus, they specify identical displacement vectors and represent the same *change of position* for the particle. A vector can be shifted without changing its value, *if* its magnitude (length) and direction are not changed.

The displacement vector tells us nothing about the actual path that the particle takes. In Fig. 3-1*b*, for example, all three paths connecting points *A* and *B* correspond to the same displacement vector, that of Fig. 3-1*a*. Displacement vectors represent only the overall effect of the motion, not the motion itself.

3-2 Adding Vectors Geometrically

Suppose that, as in the vector diagram of Fig. 3-2*a*, a particle moves from *A* to *B* and then later from *B* to *C*. We can represent its overall displacement (no matter what its actual path) with two successive displacement vectors, *AB* and *BC*. The *net displacement* of these two displacements is a single displacement from *A* to *C*. We call *AC* the **vector sum** (or **resultant**) of the vectors *AB* and *BC*. This sum is not the usual algebraic sum.

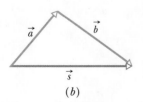

Fig. 3-2 (*a*) *AC* is the vector sum of the vectors *AB* and *BC*. (*b*) The same vectors relabeled.

In Fig. 3-2*b*, we redraw the vectors of Fig. 3-2*a* and relabel them in the way that we shall use from now on, namely, with an arrow over an italic symbol, as in $\vec{a}$. If we want to indicate only the magnitude of the vector (a quantity that lacks a sign or direction), we shall use the italic symbol, as in *a*, *b*, and *s*. (You can use just a handwritten symbol.) A symbol with an overhead arrow always implies both properties of a vector, magnitude and direction.

We can represent the relation among the three vectors in Fig. 3-2*b* with the *vector equation*

$$\vec{s} = \vec{a} + \vec{b}, \tag{3-1}$$

which says that the vector $\vec{s}$ is the vector sum of vectors $\vec{a}$ and $\vec{b}$. The symbol + in Eq. 3-1 and the words "sum" and "add" have different meanings for vectors than they do in the usual algebra because they involve both magnitude *and* direction.

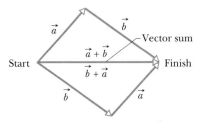

Fig. 3-3 The two vectors $\vec{a}$ and $\vec{b}$ can be added in either order; see Eq. 3-2.

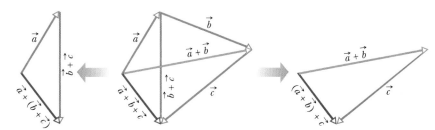

Fig. 3-4 The three vectors $\vec{a}$, $\vec{b}$, and $\vec{c}$ can be grouped in any way as they are added; see Eq. 3-3.

Figure 3-2 suggests a procedure for adding two-dimensional vectors $\vec{a}$ and $\vec{b}$ geometrically. (1) On paper, sketch vector $\vec{a}$ to some convenient scale and at the proper angle. (2) Sketch vector $\vec{b}$ to the same scale, with its tail at the head of vector $\vec{a}$, again at the proper angle. (3) The vector sum $\vec{s}$ is the vector that extends from the tail of $\vec{a}$ to the head of $\vec{b}$.

Vector addition, defined in this way, has two important properties. First, the order of addition does not matter. Adding $\vec{a}$ to $\vec{b}$ gives the same result as adding $\vec{b}$ to $\vec{a}$ (Fig. 3-3); that is,

$$\vec{a} + \vec{b} = \vec{b} + \vec{a} \quad \text{(commutative law)}. \tag{3-2}$$

Second, when there are more than two vectors, we can group them in any order as we add them. Thus, if we want to add vectors $\vec{a}$, $\vec{b}$, and $\vec{c}$, we can add $\vec{a}$ and $\vec{b}$ first and then add their vector sum to $\vec{c}$. We can also add $\vec{b}$ and $\vec{c}$ first and then add *that* sum to $\vec{a}$. We get the same result either way, as shown in Fig. 3-4. That is,

$$(\vec{a} + \vec{b}) + \vec{c} = \vec{a} + (\vec{b} + \vec{c}) \quad \text{(associative law)}. \tag{3-3}$$

Fig. 3-5 The vectors $\vec{b}$ and $-\vec{b}$ have the same magnitude and opposite directions.

The vector $-\vec{b}$ is a vector with the same magnitude as $\vec{b}$ but the opposite direction (see Fig. 3-5). Adding the two vectors in Fig. 3-5 would yield

$$\vec{b} + (-\vec{b}) = 0.$$

Thus, adding $-\vec{b}$ has the effect of subtracting $\vec{b}$. We use this property to define the difference between two vectors: let $\vec{d} = \vec{a} - \vec{b}$. Then

$$\vec{d} = \vec{a} - \vec{b} = \vec{a} + (-\vec{b}) \quad \text{(vector subtraction)}; \tag{3-4}$$

that is, we find the difference vector $\vec{d}$ by adding the vector $-\vec{b}$ to the vector $\vec{a}$. Figure 3-6 shows how this is done geometrically.

As in the usual algebra, we can move a term that includes a vector symbol from one side of a vector equation to the other, but we must change its sign. For example, if we are given Eq. 3-4 and need to solve for $\vec{a}$, we can rearrange the equation as

$$\vec{d} + \vec{b} = \vec{a} \quad \text{or} \quad \vec{a} = \vec{d} + \vec{b}.$$

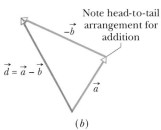

Fig. 3-6 (a) Vectors $\vec{a}$, $\vec{b}$, and $-\vec{b}$. (b) To subtract vector $\vec{b}$ from vector $\vec{a}$, add vector $-\vec{b}$ to vector $\vec{a}$.

Remember, although we have used displacement vectors here, the rules for addition and subtraction hold for vectors of all kinds, whether they represent velocities, accelerations, or any other vector quantity. However, we can add only vectors of the same kind. For example, we can add two displacements, or two velocities, but adding a displacement and a velocity makes no sense. In the arithmetic of scalars, that would be like trying to add 21 s and 12 m.

CHECKPOINT 1: The magnitudes of displacements $\vec{a}$ and $\vec{b}$ are 3 m and 4 m, respectively, and $\vec{c} = \vec{a} + \vec{b}$. Considering various orientations of $\vec{a}$ and $\vec{b}$, what is (a) the maximum possible magnitude for $\vec{c}$ and (b) the minimum possible magnitude?

Sample Problem 3-1

In an orienteering class, you have the goal of moving as far (straight-line distance) from base camp as possible by making three straight-line moves. You may use the following displacements in any order: (a) $\vec{a}$, 2.0 km due east (directly toward the east); (b) $\vec{b}$, 2.0 km 30° north of east (at an angle of 30° toward the north from due east); (c) $\vec{c}$, 1.0 km due west. Alternatively, you may substitute either $-\vec{b}$ for $\vec{b}$ or $-\vec{c}$ for $\vec{c}$. What is the greatest distance you can be from base camp at the end of the third displacement?

SOLUTION: Using a convenient scale, we draw vectors $\vec{a}$, $\vec{b}$, $\vec{c}$, $-\vec{b}$, and $-\vec{c}$ as in Fig. 3-7a. We then mentally slide the vectors over the page, connecting three of them at a time in head-to-tail arrangements to find their vector sum $\vec{d}$. The tail of the first vector represents base camp. The head of the third vector represents the point at which you stop. The vector sum $\vec{d}$ extends from the tail of the first vector to the head of the third vector. Its magnitude d is your distance from base camp.

We find that distance d is greatest for a head-to-tail arrangement of vectors $\vec{a}$, $\vec{b}$, and $-\vec{c}$. They can be in any order, because their vector sum is the same for any order. The order shown in Fig.

Scale of km

| | | |
|0|1|2|

(a)   (b)

Fig. 3-7 Sample Problem 3-1. (a) Displacement vectors; three are to be used. (b) Your distance from base camp is greatest if you undergo displacements $\vec{a}$, $\vec{b}$, and $-\vec{c}$, in any order. One choice of order is shown; it gives vector sum $\vec{d} = \vec{b} + \vec{a} - \vec{c}$.

3-7b is for the vector sum

$$\vec{d} = \vec{b} + \vec{a} + (-\vec{c}).$$

Using the scale given in Fig. 3-7a, we measure the length d of this vector sum, finding

$$d = 4.8 \text{ m}. \qquad \text{(Answer)}$$

3-3   Components of Vectors

Adding vectors geometrically can be tedious. A neater and easier technique involves algebra but requires that the vectors be placed on a rectangular coordinate system. The x and y axes are usually drawn in the plane of the page, as in Fig. 3-8a. The z axis comes directly out of the page at the origin; we ignore it for now and deal only with two-dimensional vectors.

A **component** of a vector is the projection of the vector on an axis. In Fig. 3-8a, for example, a_x is the component of vector $\vec{a}$ on (or along) the x axis and a_y is the component along the y axis. To find the projection of a vector along an axis, we draw perpendicular lines from the two ends of the vector to the axis, as shown. The projection of a vector on an x axis is its x component, and similarly the projection on the y axis is the y component. The process of finding the components of a vector is called **resolving the vector.**

A component of a vector has the same direction (along an axis) as the vector. In Fig. 3-8, a_x and a_y are both positive because $\vec{a}$ extends in the positive direction of both axes. (Note the small arrowheads on the components, to indicate their direction.) If we were to reverse vector $\vec{a}$, then both components would be negative and their arrowheads would point toward negative x and y. Resolving vector $\vec{b}$ in Fig. 3-9 yields a positive component b_x and a negative component b_y.

In general, a vector has three components, although for the case of Fig. 3-8a the component along the z axis is zero. As Figs. 3-8a and b show, if you shift a vector without changing its direction, its components do not change.

We can find the components of $\vec{a}$ in Fig. 3-8a geometrically from the right triangle there:

$$a_x = a \cos \theta \quad \text{and} \quad a_y = a \sin \theta, \qquad (3\text{-}5)$$

where θ is the angle that the vector $\vec{a}$ makes with the positive direction of the x axis, and a is the magnitude of $\vec{a}$. Figure 3-8c shows that $\vec{a}$ and its x and y components

(a)   (b)

(c)

Fig. 3-8 (a) The components a_x and a_y of vector $\vec{a}$. (b) The components are unchanged if the vector is shifted, as long as the magnitude and orientation are maintained. (c) The components form the legs of a right triangle whose hypotenuse is the magnitude of the vector.

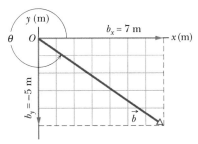

Fig. 3-9 The component of $\vec{b}$ on the x axis is positive, and that on the y axis is negative.

form a right triangle. It also shows how we can reconstruct a vector from its components: we arrange those components *head to tail*. Then we complete a right triangle with the vector forming the hypotenuse, from the tail of one component to the head of the other component.

Once a vector has been resolved into its components along a set of axes, the components themselves can be used in place of the vector. For example, $\vec{a}$ in Fig. 3-8a is given (completely determined) by a and θ. It can also be given by its components a_x and a_y. Both pairs of values contain the same information. If we know a vector in *component notation* (a_x and a_y) and want it in *magnitude-angle notation* (a and θ), we can use the equations

$$a = \sqrt{a_x^2 + a_y^2} \quad \text{and} \quad \tan \theta = \frac{a_y}{a_x} \tag{3-6}$$

to transform it.

In the more general three-dimensional case, we need a magnitude and two angles (say, a, θ, and ϕ) or three components (a_x, a_y, and a_z) to specify a vector.

✔**CHECKPOINT 2:** In the figure, which of the indicated methods for combining the x and y components of vector $\vec{a}$ are proper to determine that vector?

Sample Problem 3-2

A small airplane leaves an airport on an overcast day and is later sighted 215 km away, in a direction making an angle of 22° east of north. How far east and north is the airplane from the airport when sighted?

Fig. 3-10 Sample Problem 3-2. A plane takes off from an airport at the origin and is later sighted at P.

SOLUTION: The Key Idea here is that we are given the magnitude (215 km) and the angle (22° east of north) of a vector and need to find the components of the vector. We draw an xy coordinate system with the positive direction of x due east and that of y due north (Fig. 3-10). For convenience, the origin is placed at the airport. The airplane's displacement vector $\vec{d}$ points from the origin to where the airplane is sighted.

To find the components of $\vec{d}$, we use Eq. 3-5 with $\theta = 68°$ ($= 90° - 22°$):

$$d_x = d \cos \theta = (215 \text{ km})(\cos 68°)$$
$$= 81 \text{ km} \qquad \text{(Answer)}$$
$$d_y = d \sin \theta = (215 \text{ km})(\sin 68°)$$
$$= 199 \text{ km}. \qquad \text{(Answer)}$$

Thus, the airplane is 81 km east and 199 km north of the airport.

Sample Problem 3-3

The 1972 team that connected the Mammoth–Flint cave system went from Austin Entrance in the Flint Ridge system to Echo River in Mammoth Cave (Fig. 3-11a), traveling a net 2.6 km westward, 3.9 km southward, and 25 m upward. What was their displacement vector from start to finish?

SOLUTION: The Key Idea here is that we have the components of a three-dimensional vector, and we need to find the vector's magnitude and two angles to specify the vector's direction. We first draw the components as in Fig. 3-11b. The horizontal components (2.6 km west and 3.9 km south) form the legs of a horizontal right triangle. The team's horizontal displacement forms the hypotenuse of the triangle, and its magnitude d_h is given by the Pythagorean theorem:

$$d_h = \sqrt{(2.6 \text{ km})^2 + (3.9 \text{ km})^2} = 4.69 \text{ km}.$$

Also from the horizontal triangle in Fig. 3-11b, we see that this horizontal displacement is directed south of due west by an angle θ_h given by

$$\tan \theta_h = \frac{3.9 \text{ km}}{2.6 \text{ km}},$$

so

$$\theta_h = \tan^{-1} \frac{3.9 \text{ km}}{2.6 \text{ km}} = 56°, \text{(Answer)}$$

which is one of the two angles we need to specify the direction of the overall displacement.

To include the vertical component (25 m = 0.025 km), we now take a side view of Fig. 3-11b, looking northwest. We get Fig. 3-11c, where the vertical component and the horizontal displacement d_h form the legs of another right triangle. Now the team's overall displacement forms the hypotenuse of that triangle, with a magnitude d given by

$$d = \sqrt{(4.69 \text{ km})^2 + (0.025 \text{ km})^2} = 4.69 \text{ km}$$
$$\approx 4.7 \text{ km}. \text{(Answer)}$$

This displacement is directed upward from the horizontal displacement by the angle

$$\theta_v = \tan^{-1} \frac{0.025 \text{ km}}{4.69 \text{ km}} = 0.3°. \text{(Answer)}$$

Thus, the team's displacement vector had a magnitude of 4.7 km and was at an angle of 56° south of west and at an angle of 0.3° upward. The net vertical motion was, of course, insignificant compared to the horizontal motion. However, that fact would have been no comfort to the team, which had to climb up and down countless times to get through the cave. The route they actually covered was quite different from the displacement vector, which merely points in a straight line from start to finish.

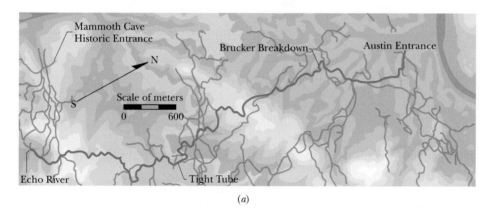

Fig. 3-11 Sample Problem 3-3. (a) Part of the Mammoth–Flint cave system, with the spelunking team's route from Austin Entrance to Echo River indicated in red. (b) The components of the team's overall displacement and their horizontal displacement d_h. (c) A side view showing d_h and the team's overall displacement vector $\vec{d}$. (Adapted from map by Cave Research Foundation.)

PROBLEM-SOLVING TACTICS

Tactic 1: *Angles—Degrees and Radians*
Angles that are measured relative to the positive direction of the x axis are positive if they are measured in the counterclockwise direction, and negative if measured clockwise. For example, 210° and −150° are the same angle.

Angles may be measured in degrees or radians (rad). You can relate the two measures by remembering that one full circle is equivalent to 360° and to 2π rad. If you needed to convert, say, 40° to radians, you would write

$$40° \frac{2\pi \text{ rad}}{360°} = 0.70 \text{ rad.}$$

Tactic 2: *Trig Functions*
You need to know the definitions of the common trigonometric functions—sine, cosine, and tangent—because they are part of the language of science and engineering. They are given in Fig. 3-12 in a form that does not depend on how the triangle is labeled.

You should also be able to sketch how the trig functions vary with angle, as in Fig. 3-13, in order to be able to judge whether a calculator result is reasonable. Even knowing the signs of the functions in the various quadrants can be of help.

Tactic 3: *Inverse Trig Functions*
When the inverse trig functions $\sin^{-1}$, $\cos^{-1}$, and $\tan^{-1}$ are taken on a calculator, you must consider the reasonableness of the answer you get, because there is usually another possible answer that the calculator does not give. The range of operation for a calculator in taking each inverse trig function is indicated in Fig. 3-13. As an example, $\sin^{-1} 0.5$ has associated angles of 30° (which is displayed by the calculator, since 30° falls within its range of operation) and 150°. To see both values, draw a horizontal line through 0.5 in Fig. 3-13a and note where it cuts the sine curve.

How do you distinguish a correct answer? It is the one that seems more reasonable for the given situation. As an example, reconsider the calculation of θ_h in Sample Problem 3-3, where $\tan \theta_h = 3.9/2.6 = 1.5$. Taking $\tan^{-1} 1.5$ on your calculator tells you that $\theta_h = 56°$, but $\theta_h = 236° (= 180° + 56°)$ also has a tangent

(a)

(b)

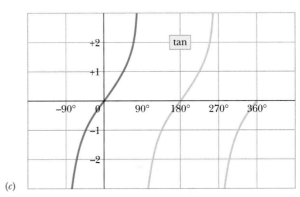

(c)

Fig. 3-13 Three useful curves to remember. A calculator's range of operation for taking *inverse* trig functions is indicated by the darker portions of the colored curves.

of 1.5. Which is correct? From the physical situation (Fig. 3-11b), 56° is reasonable and 236° is clearly not.

Tactic 4: *Measuring Vector Angles*
The equations for $\cos \theta$ and $\sin \theta$ in Eq. 3-5 and the equation for $\tan \theta$ in Eq. 3-6 are valid only if the angle is measured relative to the positive direction of the x axis. If it is measured relative to some other direction, then the trig functions in Eq. 3-5 may have to be interchanged, and the ratio in Eq. 3-6 may have to be inverted. A safer method is to convert the given angle into one that is measured from the positive direction of the x axis.

$$\sin \theta = \frac{\text{leg opposite } \theta}{\text{hypotenuse}}$$

$$\cos \theta = \frac{\text{leg adjacent to } \theta}{\text{hypotenuse}}$$

$$\tan \theta = \frac{\text{leg opposite } \theta}{\text{leg adjacent to } \theta}$$

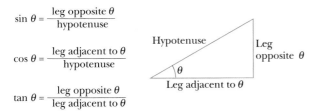

Fig. 3-12 A triangle used to define the trigonometric functions. See also Appendix E.

3-4 Unit Vectors

A **unit vector** is a vector that has a magnitude of exactly 1 and points in a particular direction. It lacks both dimension and unit. Its sole purpose is to point—that is, to specify a direction. The unit vectors in the positive directions of the x, y, and z axes are labeled $\hat{i}$, $\hat{j}$, and $\hat{k}$, where the hat ^ is used instead of an overhead arrow as for

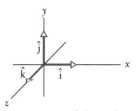

Fig. 3-14 Unit vectors $\hat{i}$, $\hat{j}$, and $\hat{k}$ define the directions of a right-handed coordinate system.

(a)

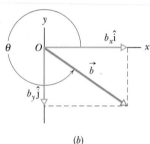

(b)

Fig. 3-15 (a) The vector components of vector $\vec{a}$. (b) The vector components of vector $\vec{b}$.

other vectors (Fig. 3-14). The arrangement of axes in Fig. 3-14 is said to be a **right-handed coordinate system.** The system remains right-handed if it is rotated rigidly to a new orientation. We use such coordinate systems exclusively in this book.

Unit vectors are very useful for expressing other vectors; for example, we can express $\vec{a}$ and $\vec{b}$ of Figs. 3-8 and 3-9 as

$$\vec{a} = a_x\hat{i} + a_y\hat{j} \tag{3-7}$$

and

$$\vec{b} = b_x\hat{i} + b_y\hat{j}. \tag{3-8}$$

These two equations are illustrated in Fig. 3-15. The quantities $a_x\hat{i}$ and $a_y\hat{j}$ are vectors and are called the **vector components** of $\vec{a}$. The quantities a_x and a_y are scalars and are called the **scalar components** of $\vec{a}$ (or, as before, simply its **components**).

As an example, let us write the displacement $\vec{d}$ of the spelunking team of Sample Problem 3-3 in terms of unit vectors. First, superimpose the coordinate system of Fig. 3-14 on the one shown in Fig. 3-11b. Then the direction of $\hat{i}$, $\hat{j}$, and $\hat{k}$ are toward the east, up, and toward the south, respectively. Thus, displacement $\vec{d}$ from start to finish is neatly expressed in unit-vector notation as

$$\vec{d} = -(2.6 \text{ km})\hat{i} + (0.025 \text{ km})\hat{j} + (3.9 \text{ km})\hat{k}. \tag{3-9}$$

Here $-(2.6 \text{ km})\hat{i}$ is the vector component $d_x\hat{i}$ along the x axis, and $-(2.6 \text{ km})$ is the x component d_x.

3-5 Adding Vectors by Components

Using a sketch, we can add vectors geometrically. On a vector-capable calculator, we can add them directly on the screen. A third way to add vectors is to combine their components, axis by axis.

To start, consider the statement

$$\vec{r} = \vec{a} + \vec{b}, \tag{3-10}$$

which says that the vector $\vec{r}$ is the same as the vector $(\vec{a} + \vec{b})$. If that is so, then each component of $\vec{r}$ must be the same as the corresponding component of $(\vec{a} + \vec{b})$:

$$r_x = a_x + b_x \tag{3-11}$$
$$r_y = a_y + b_y \tag{3-12}$$
$$r_z = a_z + b_z. \tag{3-13}$$

In other words, two vectors must be equal if their corresponding components are equal. Equations 3-10 to 3-13 tell us that to add vectors $\vec{a}$ and $\vec{b}$, we must (1) resolve the vectors into their scalar components; (2) combine these scalar components, axis by axis, to get the components of the sum $\vec{r}$; and (3) combine the components of $\vec{r}$ to get $\vec{r}$ itself. We have a choice in step 3. We can express $\vec{r}$ in unit-vector notation (as in Eq. 3-9) or in magnitude-angle notation (as in the answer to Sample Problem 3-3).

This procedure for adding vectors by components also applies to vector subtractions. Recall that a subtraction such as $\vec{d} = \vec{a} - \vec{b}$ can be rewritten as an addition $\vec{d} = \vec{a} + (-\vec{b})$. To subtract we simply add $\vec{a}$ and $-\vec{b}$ by components, to get

$$d_x = a_x - b_x, \quad d_y = a_y - b_y, \quad \text{and} \quad d_z = a_z - b_z,$$

where

$$\vec{d} = d_x\hat{i} + d_y\hat{j} + d_z\hat{k}.$$

CHECKPOINT 3: (a) In the figure here, what are the signs of the x components of $\vec{d}_1$ and $\vec{d}_2$? (b) What are the signs of the y components of $\vec{d}_1$ and $\vec{d}_2$? (c) What are the signs of the x and y components of $\vec{d}_1 + \vec{d}_2$?

Sample Problem 3-4

Figure 3-16a shows the following three vectors:

$$\vec{a} = (4.2 \text{ m})\hat{i} - (1.5 \text{ m})\hat{j},$$

$$\vec{b} = (-1.6 \text{ m})\hat{i} + (2.9 \text{ m})\hat{j},$$

and $$\vec{c} = (-3.7 \text{ m})\hat{j}.$$

What is their vector sum $\vec{r}$, which is also shown?

SOLUTION: The Key Idea here is that we can add the three vectors by components, axis by axis. For the x axis, we add the x components of $\vec{a}$, $\vec{b}$, and $\vec{c}$ to get the x component of $\vec{r}$:

$$r_x = a_x + b_x + c_x$$
$$= 4.2 \text{ m} - 1.6 \text{ m} + 0 = 2.6 \text{ m}.$$

Similarly, for the y axis,

$$r_y = a_y + b_y + c_y$$
$$= -1.5 \text{ m} + 2.9 \text{ m} - 3.7 \text{ m} = -2.3 \text{ m}.$$

Another Key Idea is that we can combine these components of $\vec{r}$ to write the vector in unit-vector notation:

$$\vec{r} = (2.6 \text{ m})\hat{i} - (2.3 \text{ m})\hat{j}, \qquad \text{(Answer)}$$

where $(2.6 \text{ m})\hat{i}$ is the vector component of $\vec{r}$ along the x axis and $-(2.3 \text{ m})\hat{j}$ is that along the y axis. Figure 3-16b shows one way to arrange these vector components to form $\vec{r}$. (Can you sketch the other way?)

A third Key Idea is that we can also answer the question by giving the magnitude and an angle for $\vec{r}$. From Eq. 3-6, the magnitude is

$$r = \sqrt{(2.6 \text{ m})^2 + (-2.3 \text{ m})^2} \approx 3.5 \text{ m} \qquad \text{(Answer)}$$

(a)

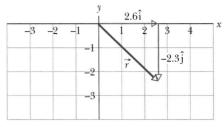

(b)

Fig. 3-16 Sample Problem 3-4. Vector $\vec{r}$ is the vector sum of the other three vectors.

and the angle (measured from the positive direction of x) is

$$\theta = \tan^{-1}\left(\frac{-2.3 \text{ m}}{2.6 \text{ m}}\right) = -41°, \qquad \text{(Answer)}$$

where the minus sign means that the angle is measured clockwise.

Sample Problem 3-5

Figure 3-17 gives an incomplete map of a road rally. From the starting point (at the origin), you must use available roads to go through the following displacements:
(1) $\vec{a}$ to checkpoint Able, magnitude 36 km, due east,
(2) $\vec{b}$ to checkpoint Baker, due north,
(3) $\vec{c}$ to checkpoint Charlie, magnitude 25 km, at the angle shown.
Your net displacement $\vec{d}$ from the starting point is 62.0 km. What is the magnitude b of $\vec{b}$?

SOLUTION: The Key Idea here is that the net displacement $\vec{d}$ is the vector sum of the three individual displacements, so we can write

$$\vec{d} = \vec{a} + \vec{b} + \vec{c},$$

which gives us

$$\vec{b} = \vec{d} - \vec{a} - \vec{c}. \qquad (3\text{-}14)$$

Although we know both magnitude and direction for $\vec{a}$ and $\vec{c}$, we do not know both for $\vec{d}$, so we cannot directly solve for $\vec{b}$

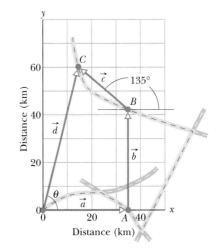

Fig. 3-17 Sample Problem 3-5. A rally route, showing the origin, checkpoints Able (A), Baker (B), and Charlie (C), and the roads.

on a vector-capable calculator. However, we can write Eq. 3-14 in terms of components along both the x axis and the y axis. Since $\vec{b}$ points parallel to the y axis, choosing that axis may give us the magnitude of $\vec{b}$. We thus write

$$b_y = d_y - a_y - c_y. \tag{3-15}$$

Following Eq. 3-5, inserting known data, and realizing that $b = b_y$, we have

$$b = (62 \text{ km}) \sin \theta - 0 - (25 \text{ km}) \sin 135°. \tag{3-16}$$

Unfortunately, we do not know θ. To find it, we write Eq. 3-14 for components along the x axis:

$$b_x = d_x - a_x - c_x, \tag{3-17}$$

which gives us

$$0 = (62 \text{ km}) \cos \theta - 36 \text{ km} - (25 \text{ km}) \cos 135°$$

and

$$\theta = \cos^{-1} \frac{36 + (25)(\cos 135°)}{62} = 72.81°.$$

Inserting this into Eq. 3-16, we find

$$b \approx 42 \text{ km}. \tag{Answer}$$

3-6   Vectors and the Laws of Physics

(a)

(b)

Fig. 3-18 (a) The vector $\vec{a}$ and its components. (b) The same vector, with the axes of the coordinate system rotated through an angle ϕ.

So far, in every figure that includes a coordinate system, the x and y axes are parallel to the edges of the book page. Thus, when a vector $\vec{a}$ is included, its components a_x and a_y are also parallel to the edges (as in Fig. 3-18a). The only reason for that orientation of the axes is that it looks "proper": there is no deeper reason. We could, instead, rotate the axes (but not the vector $\vec{a}$) through an angle ϕ as in Fig. 3-18b, in which case the components would have new values, call them a'_x and a'_y. Since there are an infinite number of choices of ϕ, there are an infinite number of different pairs of components for $\vec{a}$.

Which then is the "right" pair of components? The answer is that they are all equally valid because each pair (with its axes) just gives us a different way of describing the same vector $\vec{a}$; all produce the same magnitude and direction for the vector. In Fig. 3-18 we have

$$a = \sqrt{a_x^2 + a_y^2} = \sqrt{a'^2_x + a'^2_y} \tag{3-18}$$

and

$$\theta = \theta' + \phi. \tag{3-19}$$

The point is that we have great freedom in choosing a coordinate system, because the relations among vectors (including, for example, the vector addition of Eq. 3-1) do not depend on the location of the origin of the coordinate system or on the orientation of the axes. This is also true of the relations of physics; they are all independent of the choice of coordinate system. Add to that the simplicity and richness of the language of vectors and you can see why the laws of physics are almost always presented in that language: one equation, like Eq. 3-10, can represent three (or even more) relations, like Eqs. 3-11, 3-12, and 3-13.

3-7   Multiplying Vectors*

There are three ways in which vectors can be multiplied, but none is exactly like the usual algebraic multiplication. As you read this section, keep in mind that a vector-capable calculator will help you multiply vectors only if you understand the basic rules of that multiplication.

Multiplying a Vector by a Scalar

If we multiply a vector $\vec{a}$ by a scalar s, we get a new vector. Its magnitude is the product of the magnitude of $\vec{a}$ and the absolute value of s. Its direction is the direction

---

* This material will not be employed until later (Chapter 7 for scalar products and Chapter 12 for vector products), and so your instructor may wish to postpone assignment of the section.

of $\vec{a}$ if s is positive, but the opposite direction if s is negative. To divide $\vec{a}$ by s, we multiply $\vec{a}$ by $1/s$.

Multiplying a Vector by a Vector

There are two ways to multiply a vector by a vector: one way produces a scalar (called the *scalar product*), and the other produces a new vector (called the *vector product*). Students commonly confuse the two ways, and so starting now, you should carefully distinguish between them.

The Scalar Product

The **scalar product** of the vectors $\vec{a}$ and $\vec{b}$ in Fig. 3-19a is written as $\vec{a} \cdot \vec{b}$ and defined to be

$$\vec{a} \cdot \vec{b} = ab \cos \phi, \tag{3-20}$$

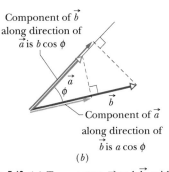

where a is the magnitude of $\vec{a}$, b is the magnitude of $\vec{b}$, and ϕ is the angle between $\vec{a}$ and $\vec{b}$ (or, more properly, between the directions of $\vec{a}$ and $\vec{b}$). There are actually two such angles: ϕ and $360° - \phi$. Either can be used in Eq. 3-20, because their cosines are the same.

Note that there are only scalars on the right side of Eq. 3-20 (including the value of $\cos \phi$). Thus $\vec{a} \cdot \vec{b}$ on the left side represents a *scalar* quantity. Because of the notation, $\vec{a} \cdot \vec{b}$ is also known as the **dot product** and is spoken as "a dot b."

A dot product can be regarded as the product of two quantities: (1) the magnitude of one of the vectors and (2) the scalar component of the second vector along the direction of the first vector. For example, in Fig. 3-19b, $\vec{a}$ has a scalar component $a \cos \phi$ along the direction of $\vec{b}$; note that a perpendicular dropped from the head of $\vec{a}$ to $\vec{b}$ determines that component. Similarly, $\vec{b}$ has a scalar component $b \cos \phi$ along the direction of $\vec{a}$.

Fig. 3-19 (a) Two vectors $\vec{a}$ and $\vec{b}$, with an angle ϕ between them. (b) Each vector has a component along the direction of the other vector.

If the angle ϕ between two vectors is 0°, the component of one vector along the other is maximum, and so also is the dot product of the vectors. If, instead, ϕ is 90°, the component of one vector along the other is zero, and so is the dot product.

Equation 3-20 can be rewritten as follows to emphasize the components:

$$\vec{a} \cdot \vec{b} = (a \cos \phi)(b) = (a)(b \cos \phi). \tag{3-21}$$

The commutative law applies to a scalar product, so we can write

$$\vec{a} \cdot \vec{b} = \vec{b} \cdot \vec{a}.$$

When two vectors are in unit-vector notation, we write their dot product as

$$\vec{a} \cdot \vec{b} = (a_x \hat{i} + a_y \hat{j} + a_z \hat{k}) \cdot (b_x \hat{i} + b_y \hat{j} + b_z \hat{k}), \tag{3-22}$$

which we can expand according to the distributive law: Each vector component of the first vector is to be dotted with each vector component of the second vector. By doing so, we can show that

$$\vec{a} \cdot \vec{b} = a_x b_x + a_y b_y + a_z b_z. \tag{3-23}$$

✓CHECKPOINT 4:   Vectors $\vec{C}$ and $\vec{D}$ have magnitudes of 3 units and 4 units, respectively. What is the angle between the directions of $\vec{C}$ and $\vec{D}$ if $\vec{C} \cdot \vec{D}$ equals (a) zero, (b) 12 units, and (c) -12 units?

Sample Problem 3-6

What is the angle ϕ between $\vec{a} = 3.0\hat{i} - 4.0\hat{j}$ and $\vec{b} = -2.0\hat{i} + 3.0\hat{k}$?

SOLUTION: First, a caution: Although many of the following steps can be bypassed with a vector-capable calculator, you will learn more about scalar products if, at least here, you use these steps.

One Key Idea here is that the angle between the directions of two vectors is included in the definition of their scalar product (Eq. 3-20):

$$\vec{a} \cdot \vec{b} = ab \cos \phi. \qquad (3\text{-}24)$$

In this equation, a is the magnitude of $\vec{a}$, or

$$a = \sqrt{3.0^2 + (-4.0)^2} = 5.00, \qquad (3\text{-}25)$$

and b is the magnitude of $\vec{b}$, or

$$b = \sqrt{(-2.0)^2 + 3.0^2} = 3.61. \qquad (3\text{-}26)$$

A second Key Idea is that we can separately evaluate the left side of

Eq. 3-24 by writing the vectors in unit-vector notation and using the distributive law:

$$\vec{a} \cdot \vec{b} = (3.0\hat{i} - 4.0\hat{j}) \cdot (-2.0\hat{i} + 3.0\hat{k})$$
$$= (3.0\hat{i}) \cdot (-2.0\hat{i}) + (3.0\hat{i}) \cdot (3.0\hat{k})$$
$$+ (-4.0\hat{j}) \cdot (-2.0\hat{i}) + (-4.0\hat{j}) \cdot (3.0\hat{k}).$$

We next apply Eq. 3-20 to each term in this last expression. The angle between the vectors in the first term ($3.0\hat{i}$ and $-2.0\hat{i}$) is $0°$, and in the other terms it is $90°$. We then have

$$\vec{a} \cdot \vec{b} = -(6.0)(1) + (9.0)(0) + (8.0)(0) - (12)(0) = -6.0.$$

Substituting this and the results of Eqs. 3-25 and 3-26 into Eq. 3-24 yields

$$-6.0 = (5.00)(3.61) \cos \phi,$$

so

$$\phi = \cos^{-1} \frac{-6.0}{(5.00)(3.61)} = 109° \approx 110°. \qquad \text{(Answer)}$$

The Vector Product

The **vector product** of $\vec{a}$ and $\vec{b}$, written $\vec{a} \times \vec{b}$, produces a third vector $\vec{c}$ whose magnitude is

$$c = ab \sin \phi, \qquad (3\text{-}27)$$

where ϕ is the *smaller* of the two angles between $\vec{a}$ and $\vec{b}$. (You must use the smaller of the two angles between the vectors because $\sin \phi$ and $\sin(360° - \phi)$ differ in algebraic sign.) Because of the notation, $\vec{a} \times \vec{b}$ is also known as the **cross product,** and in speech it is "a cross b."

▶ If $\vec{a}$ and $\vec{b}$ are parallel or antiparallel, $\vec{a} \times \vec{b} = 0$. The magnitude of $\vec{a} \times \vec{b}$, which can be written as $|\vec{a} \times \vec{b}|$, is maximum when $\vec{a}$ and $\vec{b}$ are perpendicular to each other.

The direction of $\vec{c}$ is perpendicular to the plane that contains $\vec{a}$ and $\vec{b}$. Figure 3-20a shows how to determine the direction of $\vec{c} = \vec{a} \times \vec{b}$ with what is known as the **right-hand rule.** Place the vectors $\vec{a}$ and $\vec{b}$ tail to tail without altering their orientations, and imagine a line that is perpendicular to their plane where they meet. Pretend to place your *right* hand around that line in such a way that your fingers would sweep $\vec{a}$ into $\vec{b}$ through the smaller angle between them. Your outstretched thumb points in the direction of $\vec{c}$.

The order of the vector multiplication is important. In Fig. 3-20b, we are determining the direction of $\vec{c}' = \vec{b} \times \vec{a}$, so the fingers are placed to sweep $\vec{b}$ into $\vec{a}$ through the smaller angle. The thumb ends up in the opposite direction from previously, and so it must be that $\vec{c}' = -\vec{c}$; that is,

$$\vec{b} \times \vec{a} = -(\vec{a} \times \vec{b}). \qquad (3\text{-}28)$$

In other words, the commutative law does not apply to a vector product.

In unit-vector notation, we write

$$\vec{a} \times \vec{b} = (a_x\hat{i} + a_y\hat{j} + a_z\hat{k}) \times (b_x\hat{i} + b_y\hat{j} + b_z\hat{k}), \qquad (3\text{-}29)$$

which can be expanded according to the distributive law; that is, each component of the first vector is to be crossed with each component of the second vector. The cross products of unit vectors are given in Appendix E (see "Products of Vectors").

Fig. 3-20 Illustration of the right-hand rule for vector products. (a) Sweep vector $\vec{a}$ into vector $\vec{b}$ with the fingers of your right hand. Your outstretched thumb shows the direction of vector $\vec{c} = \vec{a} \times \vec{b}$. (b) Showing that $\vec{a} \times \vec{b}$ is the reverse of $\vec{b} \times \vec{a}$.

For example, in the expansion of Eq. 3-29, we have

$$a_x\hat{i} \times b_x\hat{i} = a_xb_x(\hat{i} \times \hat{i}) = 0,$$

because the two unit vectors $\hat{i}$ and $\hat{i}$ are parallel and thus have a zero cross product. Similarly, we have

$$a_x\hat{i} \times b_y\hat{j} = a_xb_y(\hat{i} \times \hat{j}) = a_xb_y\hat{k}.$$

In the last step we used Eq. 3-27 to evaluate the magnitude of $\hat{i} \times \hat{j}$ as unity. (The vectors $\hat{i}$ and $\hat{j}$ each have a magnitude of unity, and the angle between them is 90°.) Also, we used the right-hand rule to get the direction of $\hat{i} \times \hat{j}$ as being in the positive direction of the z axis (thus in the direction of $\hat{k}$).

Continuing to expand Eq. 3-29, you can show that

$$\vec{a} \times \vec{b} = (a_yb_z - b_ya_z)\hat{i} + (a_zb_x - b_za_x)\hat{j} + (a_xb_y - b_xa_y)\hat{k}. \quad (3\text{-}30)$$

You can also evaluate a cross product by setting up and evaluating a determinant (as shown in Appendix E) or by using a vector-capable calculator.

To check whether any xyz coordinate system is a right-handed coordinate system, use the right-hand rule for the cross product $\hat{i} \times \hat{j} = \hat{k}$ with that system. If your fingers sweep $\hat{i}$ (positive direction of x) into $\hat{j}$ (positive direction of y) with the outstretched thumb pointing in the positive direction of z, then the system is right-handed.

✔**CHECKPOINT 5:**   Vectors $\vec{C}$ and $\vec{D}$ have magnitudes of 3 units and 4 units, respectively. What is the angle between the directions of $\vec{C}$ and $\vec{D}$ if the magnitude of the vector product $\vec{C} \times \vec{D}$ is (a) zero and (b) 12 units?

Sample Problem 3-7

In Fig. 3-21, vector $\vec{a}$ lies in the xy plane, has a magnitude of 18 units, and points in a direction 250° from the positive direction of x. Also, vector $\vec{b}$ has a magnitude of 12 units and points along the positive direction of z. What is the vector product $\vec{c} = \vec{a} \times \vec{b}$?

SOLUTION: One Key Idea is that when we have two vectors in magnitude-angle notation, we find the magnitude of their cross product (that is, the vector that results from taking their cross product) with Eq. 3-27. Here that means the magnitude of $\vec{c}$ is

$$c = ab \sin \phi = (18)(12)(\sin 90°) = 216. \quad \text{(Answer)}$$

A second Key Idea is that with two vectors in magnitude-angle notation, we find the direction of their cross product with the right-hand rule of Fig. 3-20. In Fig. 3-21, imagine placing the fingers of your right hand around a line perpendicular to the plane of $\vec{a}$ and $\vec{b}$ (the line on which $\vec{c}$ is shown) such that your fingers sweep $\vec{a}$

Fig. 3-21 Sample Problem 3-7. Vector $\vec{c}$ (in the xy plane) is the vector (or cross) product of vectors $\vec{a}$ and $\vec{b}$.

into $\vec{b}$. Your outstretched thumb then gives the direction of $\vec{c}$. Thus, as shown in Fig. 3-21, $\vec{c}$ lies in the xy plane. Because its direction is perpendicular to the direction of $\vec{a}$, it is at an angle of

$$250° - 90° = 160° \quad \text{(Answer)}$$

from the positive direction of x.

Sample Problem 3-8

If $\vec{a} = 3\hat{i} - 4\hat{j}$ and $\vec{b} = -2\hat{i} + 3\hat{k}$, what is $\vec{c} = \vec{a} \times \vec{b}$?

SOLUTION: The Key Idea is that when two vectors are in unit-vector notation, we can find their cross product by using the distributive law. Here that means we can write

$$\vec{c} = (3\hat{i} - 4\hat{j}) \times (-2\hat{i} + 3\hat{k})$$
$$= 3\hat{i} \times (-2\hat{i}) + 3\hat{i} \times 3\hat{k} + (-4\hat{j}) \times (-2\hat{i}) + (-4\hat{j}) \times 3\hat{k}.$$

We next evaluate each term with Eq. 3-27, determining the direction with the right-hand rule. For the first term here, the angle ϕ between the two vectors being crossed is 0. For the other terms, ϕ is 90°. We find

$$\vec{c} = 6(0) + 9(-\hat{j}) + 8(-\hat{k}) - 12\hat{i}$$
$$= -12\hat{i} - 9\hat{j} - 8\hat{k}. \quad \text{(Answer)}$$

This vector $\vec{c}$ is perpendicular to both $\vec{a}$ and $\vec{b}$, a fact you can check by showing that $\vec{c} \cdot \vec{a} = 0$ and $\vec{c} \cdot \vec{b} = 0$; that is, there is no component of $\vec{c}$ along the direction of either $\vec{a}$ or $\vec{b}$.

PROBLEM-SOLVING TACTICS

Tactic 5: *Common Errors with Cross Products*
Several errors are common in finding a cross product. (1) Failure to arrange vectors tail to tail is tempting when an illustration presents them head to tail: you must mentally shift (or better, redraw) one vector to the proper arrangement without changing its orientation. (2) Failing to use the right hand in applying the right-hand rule is easy when the right hand is occupied with a calculator or pencil. (3) Failure to sweep the first vector of the product into the second vector can occur when the orientations of the vectors require an awkward twisting of your hand to apply the right-hand rule. Sometimes that happens when you try to make the sweep mentally rather than actually using your hand. (4) Failure to work with a right-handed coordinate system results when you forget how to draw such a system (see Fig. 3-14).

REVIEW & SUMMARY

Scalars and Vectors   *Scalars,* such as temperature, have magnitude only. They are specified by a number with a unit (10°C) and obey the rules of arithmetic and ordinary algebra. *Vectors,* such as displacement, have both magnitude and direction (5 m, north) and obey the special rules of vector algebra.

Adding Vectors Geometrically   Two vectors $\vec{a}$ and $\vec{b}$ may be added geometrically by drawing them to a common scale and placing them head to tail. The vector connecting the tail of the first to the head of the second is the vector sum $\vec{s}$. To subtract $\vec{b}$ from $\vec{a}$, reverse the direction of $\vec{b}$ to get $-\vec{b}$; then add $-\vec{b}$ to $\vec{a}$. Vector addition is commutative and obeys the associative law.

Components of a Vector   The (scalar) *components* a_x and a_y of any two-dimensional vector $\vec{a}$ along the coordinate axes are found by dropping perpendicular lines from the ends of $\vec{a}$ onto the coordinate axes. The components are given by

$$a_x = a \cos \theta \quad \text{and} \quad a_y = a \sin \theta, \tag{3-5}$$

where θ is the angle between the positive direction of the x axis and the direction of $\vec{a}$. The algebraic sign of a component indicates its direction along the associated axis. Given its components, we can find the magnitude and orientation of the vector $\vec{a}$ with

$$a = \sqrt{a_x^2 + a_y^2} \quad \text{and} \quad \tan \theta = \frac{a_y}{a_x}. \tag{3-6}$$

Unit-Vector Notation   *Unit vectors* $\hat{i}$, $\hat{j}$, and $\hat{k}$ have magnitudes of unity and are directed in the positive directions of the x, y, and z axes, respectively, in a right-handed coordinate system. We can write a vector $\vec{a}$ in terms of unit vectors as

$$\vec{a} = a_x\hat{i} + a_y\hat{j} + a_z\hat{k}, \tag{3-7}$$

in which $a_x\hat{i}$, $a_y\hat{j}$, and $a_z\hat{k}$ are the **vector components** of $\vec{a}$ and a_x, a_y, and a_z are its **scalar components.**

Adding Vectors in Component Form   To add vectors in component form, we use the rules

$$r_x = a_x + b_x \quad r_y = a_y + b_y \quad r_z = a_z + b_z. \tag{3-11 to 3-13}$$

Here $\vec{a}$ and $\vec{b}$ are the vectors to be added, and $\vec{r}$ is the vector sum.

Vectors and Physical Laws   Any physical situation involving vectors can be described using many possible coordinate systems. We usually choose the one that most simplifies the situation. However, the relationship between the vector quantities does not depend on our choice of coordinates. The laws of physics are also independent of that choice.

Product of a Scalar and a Vector   The product of a scalar s and a vector $\vec{v}$ is a new vector whose magnitude is sv and whose direction is the same as that of $\vec{v}$ if s is positive, and opposite that of $\vec{v}$ if s is negative. To divide $\vec{v}$ by s, multiply $\vec{v}$ by $(1/s)$.

The Scalar Product   The **scalar** (or **dot**) **product** of two vectors $\vec{a}$ and $\vec{b}$ is written $\vec{a} \cdot \vec{b}$ and is the *scalar* quantity given by

$$\vec{a} \cdot \vec{b} = ab \cos \phi, \tag{3-20}$$

in which ϕ is the angle between the directions of $\vec{a}$ and $\vec{b}$. The scalar product may be positive, zero, or negative, depending on the value of ϕ. A scalar product is the product of the magnitude of one vector and the component of the second vector along the direction of the first vector.
In unit-vector notation,

$$\vec{a} \cdot \vec{b} = (a_x\hat{i} + a_y\hat{j} + a_z\hat{k}) \cdot (b_x\hat{i} + b_y\hat{j} + b_z\hat{k}), \tag{3-22}$$

which may be expanded according to the distributive law. Note that $\vec{a} \cdot \vec{b} = \vec{b} \cdot \vec{a}$.

The Vector Product   The **vector** (or **cross**) **product** of two vectors $\vec{a}$ and $\vec{b}$ is written $\vec{a} \times \vec{b}$ and is a *vector* $\vec{c}$ whose magnitude c is given by

$$c = ab \sin \phi, \tag{3-27}$$

in which ϕ is the smaller of the angles between the directions of $\vec{a}$ and $\vec{b}$. The direction of $\vec{c}$ is perpendicular to the plane defined by $\vec{a}$ and $\vec{b}$ and is given by a right-hand rule, as shown in Fig. 3-20. Note that $\vec{a} \times \vec{b} = -(\vec{b} \times \vec{a})$. In unit-vector notation,

$$\vec{a} \times \vec{b} = (a_x\hat{i} + a_y\hat{j} + a_z\hat{k}) \times (b_x\hat{i} + b_y\hat{j} + b_z\hat{k}), \tag{3-29}$$

which we may expand with the distributive law.

QUESTIONS

1. Displacement $\vec{D}$ points from coordinates (5 m, 3 m) to coordinates (7 m, 6 m) in the *xy* plane. Which of the following displacement vectors are equivalent to $\vec{D}$: vector $\vec{A}$, which points from (−6 m, −5 m) to (−4 m, −2 m); vector $\vec{B}$, which points from (−6 m, 1 m) to (−4 m, 4 m); and vector $\vec{C}$, which points from (−8 m, −6 m) to (−10 m, −9 m)?

2. Can the magnitude of the difference between two vectors ever be greater than (a) the magnitude of one of the vectors, (b) the magnitudes of both vectors, and (c) the magnitude of their sum?

3. Equation 3-2 shows that the addition of two vectors $\vec{a}$ and $\vec{b}$ is commutative. Does that mean subtraction is commutative, so that $\vec{a} - \vec{b} = \vec{b} - \vec{a}$?

4. If $\vec{d} = \vec{a} + \vec{b} + (-\vec{c})$, does (a) $\vec{a} + (-\vec{d}) = \vec{c} + (-\vec{b})$, (b) $\vec{a} = (-\vec{b}) + \vec{d} + \vec{c}$, and (c) $\vec{c} + (-\vec{d}) = \vec{a} + \vec{b}$?

5. Describe two vectors $\vec{a}$ and $\vec{b}$ such that
(a) $\vec{a} + \vec{b} = \vec{c}$ and $a + b = c$;
(b) $\vec{a} + \vec{b} = \vec{a} - \vec{b}$;
(c) $\vec{a} + \vec{b} = \vec{c}$ and $a^2 + b^2 = c^2$.

6. In Fig. 3-22, are (a) the *x* component and (b) the *y* component of vector $\vec{A}$ positive or negative? Are (c) the *x* component and (d) the *y* component of the vector combination $\vec{A} - \vec{B}$ positive or negative?

Fig. 3-22 Question 6.

7. Which of the arrangements of axes in Fig. 3-23 can be labeled "right-handed coordinate system"? As usual, each axis label indicates the positive side of the axis.

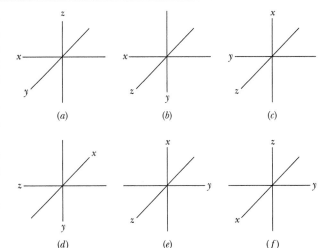

Fig. 3-23 Question 7.

8. If $\vec{a} \cdot \vec{b} = \vec{a} \cdot \vec{c}$, must $\vec{b}$ equal $\vec{c}$?

9. If $\vec{A} = 2\hat{i} + 4\hat{j}$, what is $\vec{A} \times \vec{B}$ when (a) $\vec{B} = 8\hat{i} + 16\hat{j}$ and (b) $\vec{B} = -8\hat{i} - 16\hat{j}$? (This question can be answered without computation.)

10. Figure 3-24 shows vector $\vec{A}$ and four other vectors that have the same magnitude but differ in orientation. (a) Which of those other four vectors have the same dot product with $\vec{A}$? (b) Which have a negative dot product with $\vec{A}$?

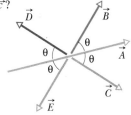

Fig. 3-24 Question 10.

EXERCISES & PROBLEMS

ssm Solution is in the Student Solutions Manual.
www Solution is available on the World Wide Web at:
          http://www.wiley.com/college/hrw
ilw Solution is available on the Interactive LearningWare.

SEC. 3-2 Adding Vectors Geometrically

1E. Consider two displacements, one of magnitude 3 m and another of magnitude 4 m. Show how the displacement vectors may be combined to get a resultant displacement of magnitude (a) 7 m, (b) 1 m, and (c) 5 m.

2P. A bank in downtown Boston is robbed (see the map in Fig. 3-25). To elude police, the robbers escape by helicopter, making three successive flights described by the following displacements: 32 km, 45° south of east; 53 km, 26° north of west; 26 km, 18° east of south. At the end of the third flight they are captured. In what town are they apprehended? (Use the geometrical method to add these displacements on the map.)

SEC. 3-3 Components of Vectors

3E. What are (a) the *x* component and (b) the *y* component of a vector $\vec{a}$ in the *xy* plane if its direction is 250° counterclockwise

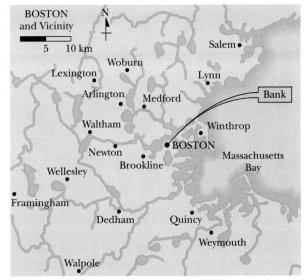

Fig. 3-25 Problem 2.

from the positive direction of the x axis and its magnitude is 7.3 m? ssm

4E. Express the following angles in radians: (a) 20.0°, (b) 50.0°, (c) 100°. Convert the following angles to degrees: (d) 0.330 rad, (e) 2.10 rad, (f) 7.70 rad.

5E. The x component of vector $\vec{A}$ is -25.0 m and the y component is $+40.0$ m. (a) What is the magnitude of $\vec{A}$? (b) What is the angle between the direction of $\vec{A}$ and the positive direction of x? ssm

6E. A displacement vector $\vec{r}$ in the xy plane is 15 m long and directed as shown in Fig. 3-26. Determine (a) the x component and (b) the y component of the vector.

Fig. 3-26 Exercise 6.

7P. A wheel with a radius of 45.0 cm rolls without slipping along a horizontal floor (Fig. 3-27). At time t_1, the dot P painted on the rim of the wheel is at the point of contact between the wheel and the floor. At a later time t_2, the wheel has rolled through one-half of a revolution. What are (a) the magnitude and (b) the angle (relative to the floor) of the displacement of P during this interval? ssm

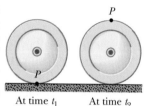

At time t_1 At time t_2

Fig. 3-27 Problem 7.

8P. Rock *faults* are ruptures along which opposite faces of rock have slid past each other. In Fig. 3-28, points A and B coincided before the rock in the foreground slid down to the right. The net displacement $\overrightarrow{AB}$ is along the plane of the fault. The horizontal component of $\overrightarrow{AB}$ is the *strike-slip* AC. The component of $\overrightarrow{AB}$ that is directly down the plane of the fault is the *dip-slip* AD. (a) What is the magnitude of the net displacement $\overrightarrow{AB}$ if the strike-slip is 22.0 m and the dip-slip is 17.0 m? (b) If the plane of the fault is inclined 52.0° to the horizontal, what is the vertical component of $\overrightarrow{AB}$?

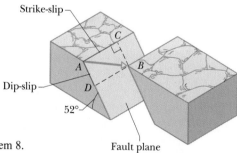

Strike-slip

Dip-slip

52°

Fault plane

Fig. 3-28 Problem 8.

9P. A room has dimensions 3.00 m (height) $\times$ 3.70 m $\times$ 4.30 m. A fly starting at one corner flies around, ending up at the diagonally opposite corner. (a) What is the magnitude of its displacement? (b) Could the length of its path be less than this magnitude? (c) Greater than this magnitude? (d) Equal to this magnitude? (e) Choose a suitable coordinate system and find the components of the displacement vector in that system. (f) If the fly walks rather than flies, what is the length of the shortest path it can take? (*Hint:* This can be answered without calculus. The room is like a box. Unfold its walls to flatten them into a plane.) ssm www

SEC. 3-5   Adding Vectors by Components

10E. A car is driven east for a distance of 50 km, then north for 30 km, and then in a direction 30° east of north for 25 km. Sketch the vector diagram and determine (a) the magnitude and (b) the angle of the car's total displacement from its starting point.

11E. A woman walks 250 m in the direction 30° east of north, then 175 m directly east. Find (a) the magnitude and (b) the angle of her final displacement from the starting point. (c) Find the distance she walks. (d) Which is greater, that distance or the magnitude of her displacement? ssm

12E. A person walks in the following pattern: 3.1 km north, then 2.4 km west, and finally 5.2 km south. (a) Sketch the vector diagram that represents this motion. (b) How far and (c) in what direction would a bird fly in a straight line from the same starting point to the same final point?

13E. (a) In unit-vector notation, what is the sum of

$$\vec{a} = (4.0 \text{ m})\hat{i} + (3.0 \text{ m})\hat{j} \quad \text{and} \quad \vec{b} = (-13.0 \text{ m})\hat{i} + (7.0 \text{ m})\hat{j}?$$

What are (b) the magnitude and (c) the direction of $\vec{a} + \vec{b}$ (relative to $\hat{i}$)? ssm

14E. Find the (a) x, (b) y, and (c) z components of the sum $\vec{r}$ of the displacements $\vec{c}$ and $\vec{d}$ whose components in meters along the three axes are $c_x = 7.4$, $c_y = -3.8$, $c_z = -6.1$; $d_x = 4.4$, $d_y = -2.0$, $d_z = 3.3$.

15E. Vector $\vec{a}$ has a magnitude of 5.0 m and is directed east. Vector $\vec{b}$ has a magnitude of 4.0 m and is directed 35° west of north. What are (a) the magnitude and (b) the direction of $\vec{a} + \vec{b}$? What are (c) the magnitude and (d) the direction of $\vec{b} - \vec{a}$? (e) Draw a vector diagram for each combination. ssm

16E. For the vectors

$$\vec{a} = (3.0 \text{ m})\hat{i} + (4.0 \text{ m})\hat{j} \quad \text{and} \quad \vec{b} = (5.0 \text{ m})\hat{i} + (-2.0 \text{ m})\hat{j},$$

give $\vec{a} + \vec{b}$ in (a) unit-vector notation, and as (b) a magnitude and (c) an angle (relative to $\hat{i}$). Now give $\vec{b} - \vec{a}$ in (d) unit-vector notation, and as (e) a magnitude and (f) an angle.

17E. Two vectors are given by

$$\vec{a} = (4.0 \text{ m})\hat{i} - (3.0 \text{ m})\hat{j} + (1.0 \text{ m})\hat{k}$$

and $$\vec{b} = (-1.0 \text{ m})\hat{i} + (1.0 \text{ m})\hat{j} + (4.0 \text{ m})\hat{k}.$$

In unit-vector notation, find (a) $\vec{a} + \vec{b}$, (b) $\vec{a} - \vec{b}$, and (c) a third vector $\vec{c}$ such that $\vec{a} - \vec{b} + \vec{c} = 0$. ssm

18P. Here are two vectors:

$$\vec{a} = (4.0 \text{ m})\hat{i} - (3.0 \text{ m})\hat{j} \quad \text{and} \quad \vec{b} = (6.0 \text{ m})\hat{i} + (8.0 \text{ m})\hat{j}.$$

What are (a) the magnitude and (b) the angle (relative to $\hat{i}$) of $\vec{a}$? What are (c) the magnitude and (d) the angle of $\vec{b}$? What are (e) the magnitude and (f) the angle of $\vec{a} + \vec{b}$; (g) the magnitude and (h) the angle of $\vec{b} - \vec{a}$; and (i) the magnitude and (j) the angle of $\vec{a} - \vec{b}$? (k) What is the angle between the directions of $\vec{b} - \vec{a}$ and $\vec{a} - \vec{b}$?

19P. Three vectors $\vec{a}$, $\vec{b}$, and $\vec{c}$ each have a magnitude of 50 m and lie in an xy plane. Their directions relative to the positive direction of the x axis are 30°, 195°, and 315°, respectively. What are (a) the magnitude and (b) the angle of the vector $\vec{a} + \vec{b} + \vec{c}$, and (c) the magnitude and (d) the angle of $\vec{a} - \vec{b} + \vec{c}$? What are (e) the magnitude and (f) the angle of a fourth vector $\vec{d}$ such that $(\vec{a} + \vec{b}) - (\vec{c} + \vec{d}) = 0$? ilw

20P. What is the sum of the following four vectors in (a) unit-vector notation and (b) magnitude-angle notation? For the latter, give the angle in both degrees and radians. Positive angles are counterclockwise from the positive direction of the x axis; negative angles are clockwise.

$\vec{E}$: 6.00 m at $+0.900$ rad $\vec{F}$: 5.00 m at $-75.0°$

$\vec{G}$: 4.00 m at $+1.20$ rad $\vec{H}$: 6.00 m at $-210°$

21P. The two vectors $\vec{a}$ and $\vec{b}$ in Fig. 3-29 have equal magnitudes of 10.0 m. Find (a) the x component and (b) the y component of their vector sum $\vec{r}$, (c) the magnitude of $\vec{r}$, and (d) the angle $\vec{r}$ makes with the positive direction of the x axis. ssm ilw www

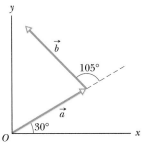

Fig. 3-29 Problem 21.

22P. In the sum $\vec{A} + \vec{B} = \vec{C}$, vector $\vec{A}$ has a magnitude of 12.0 m and is angled 40.0° counterclockwise from the $+x$ direction, and vector $\vec{C}$ has a magnitude of 15.0 m and is angled 20.0° counterclockwise from the $-x$ direction. What are (a) the magnitude and (b) the angle (relative to $+x$) of $\vec{B}$?

23P. Prove that two vectors must have equal magnitudes if their sum is perpendicular to their difference. ssm

24P. Find the sum of the following four vectors in (a) unit-vector notation, and as (b) a magnitude and (c) an angle relative to $+x$.

$\vec{P}$: 10.0 m, at 25.0° counterclockwise from $+x$

$\vec{Q}$: 12.0 m, at 10.0° counterclockwise from $+y$

$\vec{R}$: 8.00 m, at 20.0° clockwise from $-y$

$\vec{S}$: 9.00 m, at 40.0° counterclockwise from $-y$

25P. Two vectors of magnitudes a and b make an angle θ with each other when placed tail to tail. Prove, by taking components along two perpendicular axes, that

$$r = \sqrt{a^2 + b^2 + 2ab \cos \theta}$$

gives the magnitude of the sum $\vec{r}$ of the two vectors. ssm

26P. What is the sum of the following four vectors in (a) unit-vector notation, and as (b) a magnitude and (c) an angle? Positive angles are counterclockwise from the positive direction of the x axis; negative angles are clockwise.

$\vec{A} = (2.00 \text{ m})\hat{i} + (3.00 \text{ m})\hat{j}$ $\vec{B}$: 4.00 m, at $+65.0°$

$\vec{C} = (-4.00 \text{ m})\hat{i} - (6.00 \text{ m})\hat{j}$ $\vec{D}$: 5.00 m, at $-235°$

27P. (a) Using unit vectors, write expressions for the four body diagonals (the straight lines from one corner to another through the center) of a cube in terms of its edges, which have length a. (b) Determine the angles that the body diagonals make with the adjacent edges. (c) Determine the length of the body diagonals in terms of a. ssm

SEC. 3-6 Vectors and the Laws of Physics

28E. $\vec{A}$ has the magnitude 12.0 m and is angled 60.0° counterclockwise from the positive direction of the x axis of an xy coordinate system. Also, $\vec{B} = (12.0 \text{ m})\hat{i} + (8.00 \text{ m})\hat{j}$ on that same coordinate system. We now rotate the system counterclockwise about the origin by 20.0° to form an $x'y'$ system. On this new system, what are (a) $\vec{A}$ and (b) $\vec{B}$, both in unit-vector notation?

SEC. 3-7 Multiplying Vectors

29E. A vector $\vec{a}$ of magnitude 10 units and another vector $\vec{b}$ of magnitude 6.0 units differ in directions by 60°. Find (a) the scalar product of the two vectors and (b) the magnitude of the vector product $\vec{a} \times \vec{b}$. ssm

30E. Derive Eq. 3-23 for a scalar product in unit-vector notation.

31P. Use the definition of scalar product, $\vec{a} \cdot \vec{b} = ab \cos \theta$, and the fact that $\vec{a} \cdot \vec{b} = a_x b_x + a_y b_y + a_z b_z$ (see Exercise 30) to calculate the angle between the two vectors given by $\vec{a} = 3.0\hat{i} + 3.0\hat{j} + 3.0\hat{k}$ and $\vec{b} = 2.0\hat{i} + 1.0\hat{j} + 3.0\hat{k}$. ssm ilw www

32P. Derive Eq. 3-30 for a vector product in unit-vector notation.

33P. Show that the area of the triangle contained between $\vec{a}$ and $\vec{b}$ and the red line in Fig. 3-30 is $\frac{1}{2}|\vec{a} \times \vec{b}|$. ssm

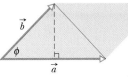

Fig. 3-30 Problem 33.

34P. In the product $\vec{F} = q\vec{v} \times \vec{B}$, take $q = 2$,

$$\vec{v} = 2.0\hat{i} + 4.0\hat{j} + 6.0\hat{k} \quad \text{and} \quad \vec{F} = 4.0\hat{i} - 20\hat{j} + 12\hat{k}.$$

What then is $\vec{B}$ in unit-vector notation if $B_x = B_y$?

35P. (a) Show that $\vec{a} \cdot (\vec{b} \times \vec{a})$ is zero for all vectors $\vec{a}$ and $\vec{b}$. (b) What is the magnitude of $\vec{a} \times (\vec{b} \times \vec{a})$ if there is an angle ϕ between the directions of $\vec{a}$ and $\vec{b}$? ssm

36P. For the following three vectors, what is $3\vec{C} \cdot (2\vec{A} \times \vec{B})$?

$$\vec{A} = 2.00\hat{i} + 3.00\hat{j} - 4.00\hat{k}$$

$$\vec{B} = -3.00\hat{i} + 4.00\hat{j} + 2.00\hat{k} \quad \vec{C} = 7.00\hat{i} - 8.00\hat{j}$$

37P. The three vectors in Fig. 3-31 have magnitudes $a = 3.00$ m, $b = 4.00$ m, and $c = 10.0$ m. What are (a) the x component and (b) the y component of $\vec{a}$; (c) the x component and (d) the y component of $\vec{b}$; and (e) the x component and (f) the y component of $\vec{c}$? If $\vec{c} = p\vec{a} + q\vec{b}$, what are the values of (g) p and (h) q? ilw

38P. Two vectors $\vec{a}$ and $\vec{b}$ have the components, in meters, $a_x = 3.2$, $a_y = 1.6$, $b_x = 0.50$, $b_y = 4.5$. (a) Find the angle between the directions of $\vec{a}$ and $\vec{b}$. There are two vectors in the xy plane that are perpendicular to $\vec{a}$ and have a magnitude of 5.0 m. One, vector $\vec{c}$, has a positive x component and the other, vector $\vec{d}$, a negative x component. What are (b) the x component and (c) the y component of $\vec{c}$, and (d) the x component and (e) the y component of vector $\vec{d}$?

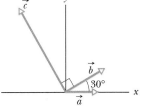

Fig. 3-31 Problem 37.

NEW PROBLEMS

N1. An explorer is caught in a whiteout (in which the snowfall is so thick that the ground cannot be distinguished from the sky) while returning to base camp. He was supposed to travel due north for 5.6 km, but when the snow clears, he discovers that he actually traveled 7.8 km at 50° north of due east. (a) How far and (b) in what direction must he now travel to reach base camp?

N2. A protester carries his sign of protest 40 m along a straight path, then 20 m along a perpendicular path to his left, and then 25 m up a water tower. (a) Choose and describe a coordinate system for this motion. In terms of that system and in unit-vector notation, what is the displacement of the sign from start to end? (b) The sign then falls to the foot of the tower. What is the magnitude of the displacement of the sign from start to this new end?

N3. In a game of lawn chess, where pieces are moved between the centers of squares that are each 1.00 m on edge, a knight is moved in the following way: (1) two squares forward, one square rightward; (2) two squares leftward, one square forward; (3) two squares forward, one square leftward. What are (a) the magnitude and (b) the angle (relative to "forward") of the knight's overall displacement for the series of three moves?

N4. Oasis B is 25 km due east of oasis A. Starting from oasis A, a camel walks 24 km in a direction 15° south of east and then walks 8.0 km due north. How far is the camel then from oasis B?

N5. If $\vec{B}$ is added to $\vec{A}$, the result is $6.0\hat{i} + 1.0\hat{j}$. If $\vec{B}$ is subtracted from $\vec{A}$, the result is $-4.0\hat{i} + 7.0\hat{j}$. What is the magnitude of $\vec{A}$?

N6. If $\vec{B}$ is added to $\vec{C} = 3.0\hat{i} + 4.0\hat{j}$, the result is a vector in the positive direction of the y axis, with a magnitude equal to that of $\vec{C}$. What is the magnitude of $\vec{B}$?

N7. For the vectors in Fig. 3N-1, calculate (a) $\vec{a} \cdot \vec{b}$, (b) $\vec{a} \cdot \vec{c}$, and (c) $\vec{b} \cdot \vec{c}$.

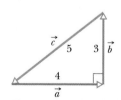

Fig. 3N-1 Problems N7 and N8.

N8. For the vectors in Fig. 3N-1, calculate (a) $\vec{a} \times \vec{b}$, (b) $\vec{a} \times \vec{c}$, and (c) $\vec{b} \times \vec{c}$.

N9. Two vectors, $\vec{r}$ and $\vec{s}$, lie in the xy plane. Their magnitudes are 4.50 and 7.30 units, respectively, and their directions are 320° and 85.0°, respectively, as measured counterclockwise from the positive x axis. What are the values of (a) $\vec{r} \cdot \vec{s}$ and (b) $\vec{r} \times \vec{s}$?

N10. Vector $\vec{A}$ has a magnitude of 6.00 units, vector $\vec{B}$ has a magnitude of 7.00 units, and the dot product $\vec{A} \cdot \vec{B}$ has a value of 14.0. What is the angle between the directions of $\vec{A}$ and $\vec{B}$?

N11. In a meeting of mimes, mime 1 goes through displacement $\vec{d}_1 = (4.0 \text{ m})\hat{i} + (5.0 \text{ m})\hat{j}$ and mime 2 goes through a displacement $\vec{d}_2 = (-3.0 \text{ m})\hat{i} + (4.0 \text{ m})\hat{j}$. What are (a) $\vec{d}_1 \times \vec{d}_2$, (b) $\vec{d}_1 \cdot \vec{d}_2$, (c) $(\vec{d}_1 + \vec{d}_2) \cdot \vec{d}_2$, and (d) the component of $\vec{d}_1$ along the direction of $\vec{d}_2$. (*Hint:* For (d), consider Eq. 3-20 and Fig. 3-19.)

N12. Displacement $\vec{d}_1$ is in the yz plane 63.0° from the positive direction of the y axis, has a positive z component, and has a magnitude of 4.50 m. Displacement $\vec{d}_2$ is in the xz plane 30° from the positive direction of the x axis, has a positive z component, and has magnitude 1.40 m. What are (a) $\vec{d}_1 \cdot \vec{d}_2$, (b) $\vec{d}_1 \times \vec{d}_2$, and (c) the angle between $\vec{d}_1$ and $\vec{d}_2$?

N13. Let $\hat{i}$ be directed to the east, $\hat{j}$ be directed to the north, and $\hat{k}$ be directed upward. What are the values of products (a) $\hat{i} \cdot \hat{k}$, (b) $(-\hat{k}) \cdot (-\hat{j})$, and (c) $\hat{j} \cdot (-\hat{j})$? What are the directions (such as east or down) of products (d) $\hat{k} \times \hat{j}$, (e) $(-\hat{i}) \times (-\hat{j})$, and (f) $(-\hat{k}) \times (-\hat{j})$?

N14. A fire ant, searching for hot sauce in a picnic area, goes through three displacements along level ground: $\vec{d}_1$ for 0.40 m southwest (that is, at 45° from directly south and from directly west), $\vec{d}_2$ for 0.50 m due east (that is, directly east), $\vec{d}_3$ for 0.60 m at 60° north of east (that is 60.0° toward the north from due east). Let the positive x direction be east and the positive y direction be north. What are (a) the x component and (b) the y component of $\vec{d}_1$? What are (c) the x component and (d) the y component of $\vec{d}_2$? What are (e) the x component and (f) the y component of $\vec{d}_3$?

What are (g) the x component, (h) the y component, (i) the magnitude, and (j) the direction of the ant's net displacement? If the ant is to return directly to the starting point, (k) how far and (l) in what direction should it move?

N15. Here are three displacements, each in meters: $\vec{d}_1 = 4.0\hat{i} + 5.0\hat{j} - 6.0\hat{k}$, $\vec{d}_2 = -1.0\hat{i} + 2.0\hat{j} + 3.0\hat{k}$, and $\vec{d}_3 = 4.0\hat{i} + 3.0\hat{j} + 2.0\hat{k}$. (a) What is $\vec{r} = \vec{d}_1 - \vec{d}_2 + \vec{d}_3$? (b) What is the angle between $\vec{r}$ and the positive z axis? (c) What is the component of $\vec{d}_1$ along the direction of $\vec{d}_2$? (d) What is the component of $\vec{d}_1$ that is perpendicular to the direction of $\vec{d}_2$ and in the plane of $\vec{d}_1$ and $\vec{d}_2$? (*Hint:* For (c), consider Eq. 3-20 and Fig. 3-19; for (d), consider Eq. 3-27.)

N16. Vector $\vec{d}_1$ is in the negative direction of a y axis and vector $\vec{d}_2$ is in the positive direction of an x axis. What are the directions of (a) $\vec{d}_2/4$ and (b) $\vec{d}_1/(-4)$? What are the magnitudes of products (c) $\vec{d}_1 \cdot \vec{d}_2$ and (d) $\vec{d}_1 \cdot (\vec{d}_2/4)$? What is the direction of the vector resulting from (e) $\vec{d}_1 \times \vec{d}_2$ and (f) $\vec{d}_2 \times \vec{d}_1$? (g) What are the magnitudes of the vector products in (e) and (f)? What are (h) the magnitude and (i) the direction of $\vec{d}_1 \times (\vec{d}_2/4)$?

N17. A vector $\vec{d}$ has a magnitude 3.0 m and is directed south. What are (a) the magnitude and (b) the direction of the vector $5.0\vec{d}$? What are (c) the magnitude and (d) the direction of the vector $-2.0\vec{d}$?

N18. If $\vec{d}_1 + \vec{d}_2 = 5\vec{d}_3$, $\vec{d}_1 - \vec{d}_2 = 3\vec{d}_3$, and $\vec{d}_3 = 2\hat{i} + 4\hat{j}$, then what are (a) $\vec{d}_1$ and (b) $\vec{d}_2$?

N19. A sailboat sets out from the U.S. side of Lake Erie for a point on the Canadian side, 90.0 km due north. The sailor, however, ends up 50.0 km due east of the starting point. (a) How far and (b) in what direction must the sailor now sail to reach the original destination?

N20. Two beetles run across flat sand, starting at the same point. Beetle 1 runs 0.50 m due east, then 0.80 m at 30° north of due east.

Beetle 2 also makes two runs; the first is 1.6 m at 40° east of due north. What must be (a) the magnitude and (b) the direction of its second run if it is to end up at the new location of beetle 1?

N21. If $\vec{d}_1 = 3\hat{i} - 2\hat{j} + 4\hat{k}$ and $\vec{d}_2 = -5\hat{i} + 2\hat{j} - \hat{k}$, then what is $(\vec{d}_1 + \vec{d}_2) \cdot (\vec{d}_1 \times 4\vec{d}_2)$?

N22. Vectors $\vec{A}$ and $\vec{B}$ lie in an xy plane. $\vec{A}$ has magnitude 8.00 and angle 130°; $\vec{B}$ has components $B_x = -7.72$ and $B_y = -9.20$. (a) What is $5\vec{A} \cdot \vec{B}$? What is $4\vec{A} \times 3\vec{B}$ in (b) unit-vector notation and (c) magnitude-angle notation with spherical coordinates (see Fig. 3N-2)? (d) What is the angle between the directions of $\vec{A}$ and $4\vec{A} \times 3\vec{B}$? (*Hint*: Think a bit before you resort to a calculation.) What is $\vec{A} + 3.00\hat{k}$ in (e) unit-vector notation and (f) magnitude-angle notation with spherical coordinates?

Fig. 3N-2 Problem N22.

N23. Vectors $\vec{A}$ and $\vec{B}$ lie in an xy plane. $\vec{A}$ has magnitude 8.00 and angle 130°; $\vec{B}$ has components $B_x = -7.72$ and $B_y = -9.20$. What

are the angles between the negative direction of the y axis and (a) the direction of $\vec{A}$, (b) the direction of the product $\vec{A} \times \vec{B}$, and (c) the direction of $\vec{A} \times (\vec{B} + 3\hat{k})$?

N24. Vector $\vec{A}$, which is directed along an x axis, is to be added to vector $\vec{B}$, which has a magnitude of 7.0 m. The sum is a third vector that is directed along the y axis, with a magnitude that is 3.0 times that of $\vec{A}$. What is that magnitude of $\vec{A}$?

N25. You are to make four straight-line moves over a flat desert floor, starting at the origin of an xy coordinate system and ending at the xy coordinates (-140 m, 30 m). The x component and y component of your moves are the following, respectively, in meters: (20 and 60), then (b_x and -70), then (-20 and c_y), then (-60 and -70). What are (a) component b_x and (b) component c_y? What are (c) the magnitude and (d) the angle (relative to the positive direction of the x axis) of the overall displacement?

N26. Here are three vectors in meters:

$$\vec{d}_1 = -3.0\hat{i} + 3.0\hat{j} + 2.0\hat{k}$$
$$\vec{d}_2 = -2.0\hat{i} - 4.0\hat{j} + 2.0\hat{k}$$
$$\vec{d}_3 = 2.0\hat{i} + 3.0\hat{j} + 1.0\hat{k}.$$

What results from (a) $\vec{d}_1 \cdot (\vec{d}_2 + \vec{d}_3)$, (b) $\vec{d}_1 \cdot (\vec{d}_2 \times \vec{d}_3)$, and (c) $\vec{d}_1 \times (\vec{d}_2 + \vec{d}_3)$?

4 Motion in Two and Three Dimensions

In 1922, one of the Zacchinis, a famous family of circus performers, was the first human cannonball to be shot across an arena and into a net. To increase the excitement, the family gradually increased the height and distance of the flight until, in 1939 or 1940, Emanuel Zacchini soared over three Ferris wheels and through a horizontal distance of 69 m.

How could he know where to place the net, and how could he be certain he would clear the Ferris wheels?

The answer is in this chapter.

4-1 Moving in Two or Three Dimensions

This chapter extends the material of the preceding two chapters to two and three dimensions. Many of the ideas of Chapter 2, such as position, velocity, and acceleration, are used here, but they are now a little more complex because of the extra dimensions. To keep the notation manageable, we use the vector algebra of Chapter 3. As you read this chapter, you might want to thumb back to those previous chapters to refresh your memory.

4-2 Position and Displacement

One general way of locating a particle (or particle-like object) is with a **position vector** $\vec{r}$, which is a vector that extends from a reference point (usually the origin of a coordinate system) to the particle. In the unit-vector notation of Section 3-4, $\vec{r}$ can be written

$$\vec{r} = x\hat{i} + y\hat{j} + z\hat{k}, \tag{4-1}$$

where $x\hat{i}$, $y\hat{j}$, and $z\hat{k}$ are the vector components of $\vec{r}$, and the coefficients x, y, and z are its scalar components.

The coefficients x, y, and z give the particle's location along the coordinate axes and relative to the origin; that is, the particle has the rectangular coordinates (x, y, z). For instance, Fig. 4-1 shows a particle with position vector

$$\vec{r} = (-3 \text{ m})\hat{i} + (2 \text{ m})\hat{j} + (5 \text{ m})\hat{k}$$

and rectangular coordinates $(-3 \text{ m}, 2 \text{ m}, 5 \text{ m})$. Along the x axis the particle is 3 m from the origin, in the $-\hat{i}$ direction. Along the y axis it is 2 m from the origin, in the $+\hat{j}$ direction. Along the z axis it is 5 m from the origin, in the $+\hat{k}$ direction.

As a particle moves, its position vector changes in such a way that the vector always extends to the particle from the reference point (the origin). If the position vector changes, say, from $\vec{r}_1$ to $\vec{r}_2$ during a certain time interval, then the particle's **displacement** $\Delta\vec{r}$ during that time interval is

$$\Delta\vec{r} = \vec{r}_2 - \vec{r}_1. \tag{4-2}$$

Using the unit-vector notation of Eq. 4-1, we can rewrite this displacement as

$$\Delta\vec{r} = (x_2\hat{i} + y_2\hat{j} + z_2\hat{k}) - (x_1\hat{i} + y_1\hat{j} + z_1\hat{k})$$

or as $\qquad \Delta\vec{r} = (x_2 - x_1)\hat{i} + (y_2 - y_1)\hat{j} + (z_2 - z_1)\hat{k}, \tag{4-3}$

where coordinates (x_1, y_1, z_1) correspond to position vector $\vec{r}_1$ and coordinates (x_2, y_2, z_2) correspond to position vector $\vec{r}_2$. We can also rewrite the displacement by substituting Δx for $(x_2 - x_1)$, Δy for $(y_2 - y_1)$, and Δz for $(z_2 - z_1)$:

$$\Delta\vec{r} = \Delta x\hat{i} + \Delta y\hat{j} + \Delta z\hat{k}. \tag{4-4}$$

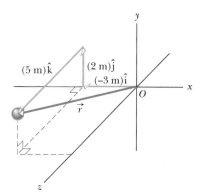

Fig. 4-1 The position vector $\vec{r}$ for a particle is the vector sum of its vector components.

Sample Problem 4-1

In Fig. 4-2, the position vector for a particle is initially

$$\vec{r}_1 = (-3.0 \text{ m})\hat{i} + (2.0 \text{ m})\hat{j} + (5.0 \text{ m})\hat{k}$$

and then later is

$$\vec{r}_2 = (9.0 \text{ m})\hat{i} + (2.0 \text{ m})\hat{j} + (8.0 \text{ m})\hat{k}.$$

What is the particle's displacement $\Delta\vec{r}$ from $\vec{r}_1$ to $\vec{r}_2$?

SOLUTION: The Key Idea is that the displacement $\Delta\vec{r}$ is obtained by subtracting the initial position vector $\vec{r}_1$ from the later position vector $\vec{r}_2$. That is most easily done by components:

$$\Delta\vec{r} = \vec{r}_2 - \vec{r}_1$$
$$= [9.0 - (-3.0)]\hat{i} + [2.0 - 2.0]\hat{j} + [8.0 - 5.0]\hat{k}$$
$$= (12 \text{ m})\hat{i} + (3.0 \text{ m})\hat{k}. \qquad \text{(Answer)}$$

This displacement vector is parallel to the xz plane, because it lacks any y component, a fact that is easier to see in the numerical result than in Fig. 4-2.

✔CHECKPOINT 1:   (a) If a wily bat flies from xyz coordinates $(-2$ m, 4 m, -3 m) to coordinates $(6$ m, -2 m, -3 m), what is its displacement $\Delta \vec{r}$ in unit-vector notation? (b) Is $\Delta \vec{r}$ parallel to one of the three coordinate planes? If so, which plane?

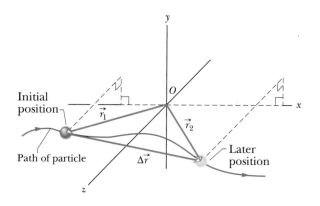

Fig. 4-2 Sample Problem 4-1. The displacement $\Delta \vec{r} = \vec{r}_2 - \vec{r}_1$ extends from the head of the initial position vector $\vec{r}_1$ to the head of the later position vector $\vec{r}_2$.

Sample Problem 4-2

A rabbit runs across a parking lot on which a set of coordinate axes has, strangely enough, been drawn. The coordinates of the rabbit's position as functions of time t are given by

$$x = -0.31t^2 + 7.2t + 28 \qquad (4\text{-}5)$$

and

$$y = 0.22t^2 - 9.1t + 30, \qquad (4\text{-}6)$$

with t in seconds and x and y in meters.

(a) At $t = 15$ s, what is the rabbit's position vector $\vec{r}$ in unit-vector notation and as a magnitude and an angle?

SOLUTION: The Key Idea here is that the x and y coordinates of the rabbit's position, as given by Eqs. 4-5 and 4-6, are the scalar components of the rabbit's position vector $\vec{r}$. Thus, we can write

$$\vec{r}(t) = x(t)\hat{\mathrm{i}} + y(t)\hat{\mathrm{j}}. \qquad (4\text{-}7)$$

(We write $\vec{r}(t)$ rather than $\vec{r}$ because the components are functions of t, and thus $\vec{r}$ is also.)

At $t = 15$ s, the scalar components are

$$x = (-0.31)(15)^2 + (7.2)(15) + 28 = 66 \text{ m}$$

and

$$y = (0.22)(15)^2 - (9.1)(15) + 30 = -57 \text{ m}.$$

Thus, at $t = 15$ s,

$$\vec{r} = (66 \text{ m})\hat{\mathrm{i}} - (57 \text{ m})\hat{\mathrm{j}}, \qquad \text{(Answer)}$$

which is drawn in Fig. 4-3a.

To get the magnitude and angle of $\vec{r}$, we can use a vector-capable calculator, or we can be guided by Eq. 3-6 to write

$$r = \sqrt{x^2 + y^2} = \sqrt{(66 \text{ m})^2 + (-57 \text{ m})^2}$$

$$= 87 \text{ m}, \qquad \text{(Answer)}$$

and

$$\theta = \tan^{-1} \frac{y}{x} = \tan^{-1}\left(\frac{-57 \text{ m}}{66 \text{ m}}\right) = -41°. \qquad \text{(Answer)}$$

(Although $\theta = 139°$ has the same tangent as $-41°$, study of the signs of the components of $\vec{r}$ rules out 139°.)

(b) Graph the rabbit's path for $t = 0$ to $t = 25$ s.

Fig. 4-3 Sample Problem 4-2. (a) A rabbit's position vector $\vec{r}$ at time $t = 15$ s. The scalar components of $\vec{r}$ are shown along the axes. (b) The rabbit's path and its position at five values of t.

SOLUTION: We can repeat part (a) for several values of t and then plot the results. Figure 4-3b shows the plots for five values of t and the path connecting them. We can also use a graphing calculator to make a *parametric graph*; that is, we would have the calculator plot y versus x, where these coordinates are given by Eqs. 4-5 and 4-6 as functions of time t.

4-3 Average Velocity and Instantaneous Velocity

If a particle moves through a displacement $\Delta \vec{r}$ in a time interval Δt, then its **average velocity** $\vec{v}_{avg}$ is

$$\text{average velocity} = \frac{\text{displacement}}{\text{time interval}},$$

or

$$\vec{v}_{avg} = \frac{\Delta \vec{r}}{\Delta t}. \qquad (4\text{-}8)$$

This tells us the direction of $\vec{v}_{avg}$ (the vector on the left side of Eq. 4-8) must be the same as that of the displacement $\Delta \vec{r}$ (the vector on the right side). Using Eq. 4-4, we can write Eq. 4-8 in vector components as

$$\vec{v}_{avg} = \frac{\Delta x \hat{i} + \Delta y \hat{j} + \Delta z \hat{k}}{\Delta t} = \frac{\Delta x}{\Delta t} \hat{i} + \frac{\Delta y}{\Delta t} \hat{j} + \frac{\Delta z}{\Delta t} \hat{k}. \qquad (4\text{-}9)$$

For example, if the particle in Sample Problem 4-1 moves from its initial position to its later position in 2.0 s, then its average velocity during that move is

$$\vec{v}_{avg} = \frac{\Delta \vec{r}}{\Delta t} = \frac{(12 \text{ m})\hat{i} + (3.0 \text{ m})\hat{k}}{2.0 \text{ s}} = (6.0 \text{ m/s})\hat{i} + (1.5 \text{ m/s})\hat{k}.$$

When we speak of the **velocity** of a particle, we usually mean the particle's **instantaneous velocity** $\vec{v}$ at some instant. This $\vec{v}$ is the value that $\vec{v}_{avg}$ approaches in the limit as we shrink the time interval Δt to 0 about that instant. Using the language of calculus, we may write $\vec{v}$ as the derivative

$$\vec{v} = \frac{d\vec{r}}{dt}. \qquad (4\text{-}10)$$

Figure 4-4 shows the path of a particle that is restricted to the xy plane. As the particle travels to the right along the curve, its position vector sweeps to the right. During time interval Δt, the position vector changes from $\vec{r}_1$ to $\vec{r}_2$, and the particle's displacement is $\Delta \vec{r}$.

To find the instantaneous velocity of the particle at, say, instant t_1 (when the particle is at position 1), we shrink interval Δt to 0 about t_1. Three things happen as we do so: (1) Position vector $\vec{r}_2$ in Fig. 4-4 moves toward $\vec{r}_1$ so that $\Delta \vec{r}$ shrinks toward zero. (2) The direction of $\Delta \vec{r}/\Delta t$ (thus of $\vec{v}_{avg}$) approaches the direction of the tangent line to the particle's path at position 1. (3) The average velocity $\vec{v}_{avg}$ approaches the instantaneous velocity $\vec{v}$ at t_1.

In the limit as $\Delta t \to 0$, we have $\vec{v}_{avg} \to \vec{v}$ and, most important here, $\vec{v}_{avg}$ takes on the direction of the tangent line. Thus, $\vec{v}$ has that direction as well:

▶ The direction of the instantaneous velocity $\vec{v}$ of a particle is always tangent to the particle's path at the particle's position.

The result is the same in three dimensions: $\vec{v}$ is always tangent to the particle's path. To write Eq. 4-10 in unit-vector form, we substitute for $\vec{r}$ from Eq. 4-1:

$$\vec{v} = \frac{d}{dt}(x\hat{i} + y\hat{j} + z\hat{k}) = \frac{dx}{dt}\hat{i} + \frac{dy}{dt}\hat{j} + \frac{dz}{dt}\hat{k}.$$

This equation can be simplified somewhat by writing it as

$$\vec{v} = v_x\hat{i} + v_y\hat{j} + v_z\hat{k}, \qquad (4\text{-}11)$$

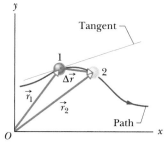

Fig. 4-4 The displacement $\Delta \vec{r}$ of a particle during a time interval Δt, from position 1 with position vector $\vec{r}_1$ at time t_1 to position 2 with position vector $\vec{r}_2$ at time t_2. The tangent to the particle's path at position 1 is shown.

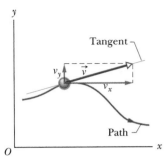

Fig. 4-5 The velocity $\vec{v}$ of a particle, along with the scalar components of $\vec{v}$.

where the scalar components of $\vec{v}$ are

$$v_x = \frac{dx}{dt}, \quad v_y = \frac{dy}{dt}, \quad \text{and} \quad v_z = \frac{dz}{dt}. \tag{4-12}$$

For example, dx/dt is the scalar component of $\vec{v}$ along the x axis. Thus, we can find the scalar components of $\vec{v}$ by differentiating the scalar components of $\vec{r}$.

Figure 4-5 shows a velocity vector $\vec{v}$ and its scalar x and y components. Note that $\vec{v}$ is tangent to the particle's path at the particle's position. *Caution:* When a position vector is drawn as in Figs. 4-1 through 4-4, it is an arrow that extends from one point (a "here") to another point (a "there"). However, when a velocity vector is drawn as in Fig. 4-5, it does *not* extend from one point to another. Rather, it shows the instantaneous direction of travel of a particle located at the tail, and its length (representing the velocity magnitude) can be drawn to any scale.

✔**CHECKPOINT 2:** The figure shows a circular path taken by a particle. If the instantaneous velocity of the particle is $\vec{v} = (2 \text{ m/s})\hat{i} - (2 \text{ m/s})\hat{j}$, through which quadrant is the particle moving when it is traveling (a) clockwise and (b) counterclockwise around the circle? For both cases, draw $\vec{v}$ on the figure.

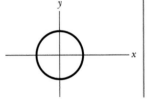

Sample Problem 4-3

For the rabbit in Sample Problem 4-2, find the velocity $\vec{v}$ at time $t = 15$ s, in unit-vector notation and as a magnitude and an angle.

SOLUTION: There are two **Key Ideas** here: (1) We can find the rabbit's velocity $\vec{v}$ by first finding the velocity components. (2) We can find those components by taking derivatives of the components of the rabbit's position vector. Applying the first of Eqs. 4-12 to Eq. 4-5,

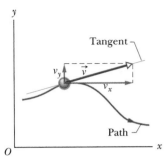

Fig. 4-6 Sample Problem 4-3. The rabbit's velocity $\vec{v}$ at $t = 15$ s. The velocity vector is tangent to the path at the rabbit's position at that instant. The scalar components of $\vec{v}$ are shown.

we find the x component of $\vec{v}$ to be

$$v_x = \frac{dx}{dt} = \frac{d}{dt}(-0.31t^2 + 7.2t + 28)$$
$$= -0.62t + 7.2. \tag{4-13}$$

At $t = 15$ s, this gives $v_x = -2.1$ m/s. Similarly, applying the second of Eqs. 4-12 to Eq. 4-6, we find that the y component is

$$v_y = \frac{dy}{dt} = \frac{d}{dt}(0.22t^2 - 9.1t + 30)$$
$$= 0.44t - 9.1. \tag{4-14}$$

At $t = 15$ s, this gives $v_y = -2.5$ m/s. Equation 4-11 then yields

$$\vec{v} = -(2.1 \text{ m/s})\hat{i} - (2.5 \text{ m/s})\hat{j}, \tag{Answer}$$

which is shown in Fig. 4-6, tangent to the rabbit's path and in the direction the rabbit is running at $t = 15$ s.

To get the magnitude and angle of $\vec{v}$, either we use a vector-capable calculator or we follow Eq. 3-6 to write

$$v = \sqrt{v_x^2 + v_y^2} = \sqrt{(-2.1 \text{ m/s})^2 + (-2.5 \text{ m/s})^2}$$
$$= 3.3 \text{ m/s} \tag{Answer}$$

and

$$\theta = \tan^{-1}\frac{v_y}{v_x} = \tan^{-1}\left(\frac{-2.5 \text{ m/s}}{-2.1 \text{ m/s}}\right)$$
$$= \tan^{-1} 1.19 = -130°. \tag{Answer}$$

(Although $50°$ has the same tangent as $-130°$, inspection of the signs of the velocity components indicates that the desired angle is in the third quadrant, given by $50° - 180° = -130°$.)

4-4 Average Acceleration and Instantaneous Acceleration

When a particle's velocity changes from $\vec{v}_1$ to $\vec{v}_2$ in a time interval Δt, its **average acceleration** $\vec{a}_{\text{avg}}$ during Δt is

$$\frac{\text{average}}{\text{acceleration}} = \frac{\text{change in velocity}}{\text{time interval}},$$

or

$$\vec{a}_{\text{avg}} = \frac{\vec{v}_2 - \vec{v}_1}{\Delta t} = \frac{\Delta \vec{v}}{\Delta t}. \qquad (4\text{-}15)$$

If we shrink Δt to zero about some instant, then in the limit $\vec{a}_{\text{avg}}$ approaches the **instantaneous acceleration** (or **acceleration**) $\vec{a}$ at that instant; that is,

$$\vec{a} = \frac{d\vec{v}}{dt}. \qquad (4\text{-}16)$$

If the velocity changes in *either* magnitude *or* direction (or both), the particle must have an acceleration.

We can write Eq. 4-16 in unit-vector form by substituting for $\vec{v}$ from Eq. 4-11 to obtain

$$\vec{a} = \frac{d}{dt}(v_x \hat{\text{i}} + v_y \hat{\text{j}} + v_z \hat{\text{k}})$$

$$= \frac{dv_x}{dt} \hat{\text{i}} + \frac{dv_y}{dt} \hat{\text{j}} + \frac{dv_z}{dt} \hat{\text{k}}.$$

We can rewrite this as

$$\vec{a} = a_x \hat{\text{i}} + a_y \hat{\text{j}} + a_z \hat{\text{k}}, \qquad (4\text{-}17)$$

where the scalar components of $\vec{a}$ are

$$a_x = \frac{dv_x}{dt}, \quad a_y = \frac{dv_y}{dt}, \quad \text{and} \quad a_z = \frac{dv_z}{dt}. \qquad (4\text{-}18)$$

Thus, we can find the scalar components of $\vec{a}$ by differentiating the scalar components of $\vec{v}$.

Figure 4-7 shows an acceleration vector $\vec{a}$ and its scalar components for a particle moving in two dimensions. *Caution:* When an acceleration vector is drawn as in Fig. 4-7, it does *not* extend from one position to another. Rather, it shows the direction of acceleration for a particle located at its tail, and its length (representing the acceleration magnitude) can be drawn to any scale.

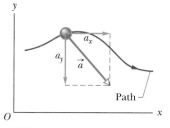

Fig. 4-7 The acceleration $\vec{a}$ of a particle and the scalar components of $\vec{a}$.

Sample Problem 4-4

For the rabbit in Sample Problems 4-2 and 4-3, find the acceleration $\vec{a}$ at time $t = 15$ s, in unit-vector notation and as a magnitude and an angle.

SOLUTION: There are two Key Ideas here: (1) We can find the rabbit's acceleration $\vec{a}$ by first finding the acceleration components. (2) We can find those components by taking derivatives of the rabbit's velocity components. Applying the first of Eqs. 4-18 to Eq. 4-13, we find the x component of $\vec{a}$ to be

$$a_x = \frac{dv_x}{dt} = \frac{d}{dt}(-0.62t + 7.2) = -0.62 \text{ m/s}^2.$$

Similarly, applying the second of Eqs. 4-18 to Eq. 4-14 yields the y component as

$$a_y = \frac{dv_y}{dt} = \frac{d}{dt}(0.44t - 9.1) = 0.44 \text{ m/s}^2.$$

We see that the acceleration does not vary with time (it is a constant) because the time variable t does not appear in the expression for either acceleration component. Equation 4-17 then yields

$$\vec{a} = (-0.62 \text{ m/s}^2)\hat{\text{i}} + (0.44 \text{ m/s}^2)\hat{\text{j}}, \qquad (\text{Answer})$$

which is shown superimposed on the rabbit's path in Fig. 4-8.

To get the magnitude and angle of $\vec{a}$, either we use a vector-capable calculator or we follow Eq. 3-6. For the magnitude we have

$$a = \sqrt{a_x^2 + a_y^2} = \sqrt{(-0.62 \text{ m/s}^2)^2 + (0.44 \text{ m/s}^2)^2}$$
$$= 0.76 \text{ m/s}^2.$$ (Answer)

For the angle we have

$$\theta = \tan^{-1} \frac{a_y}{a_x} = \tan^{-1} \left(\frac{0.44 \text{ m/s}^2}{-0.62 \text{ m/s}^2} \right) = -35°.$$

However, this last result, which is displayed on a calculator, indicates that $\vec{a}$ is directed to the right and downward in Fig. 4-8. Yet, we know from the components above that $\vec{a}$ must be directed to the left and upward. To find the other angle that has the same tangent as $-35°$, but which is not displayed on a calculator, we add $180°$:

$$-35° + 180° = 145°.$$ (Answer)

This *is* consistent with the components of $\vec{a}$. Note that $\vec{a}$ has the same magnitude and direction throughout the rabbit's run because, as we noted previously, the acceleration is constant.

✔**CHECKPOINT 3:**   Here are four descriptions of the position (in meters) of a hockey puck as it moves in the xy plane:

(1) $x = -3t^2 + 4t - 2$ and $y = 6t^2 - 4t$
(2) $x = -3t^3 - 4t$ and $y = -5t^2 + 6$
(3) $\vec{r} = 2t^2\hat{i} - (4t + 3)\hat{j}$
(4) $\vec{r} = (4t^3 - 2t)\hat{i} + 3\hat{j}$

For each description, determine whether the x and y components of the puck's acceleration are constant, and whether the acceleration $\vec{a}$ is constant.

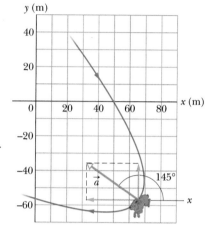

Fig. 4-8 Sample Problem 4-4. The acceleration $\vec{a}$ of the rabbit at $t = 15$ s. The rabbit happens to have this same acceleration at all points of its path.

Sample Problem 4-5

A particle with velocity $\vec{v}_0 = -2.0\hat{i} + 4.0\hat{j}$ (in meters per second) at $t = 0$ undergoes a constant acceleration $\vec{a}$ of magnitude $a = 3.0 \text{ m/s}^2$ at an angle $\theta = 130°$ from the positive direction of the x axis. What is the particle's velocity $\vec{v}$ at $t = 5.0$ s, in unit-vector notation and as a magnitude and an angle?

SOLUTION: We first note that this is two-dimensional motion, in the xy plane. Then there are two Key Ideas here. One is that, because the acceleration is constant, Eq. 2-11 ($v = v_0 + at$) applies. The second is that, because Eq. 2-11 applies only to straight-line motion, we must apply it separately for motion parallel to the x axis and motion parallel to the y axis; that is, we must find the velocity components v_x and v_y from the equations

$$v_x = v_{0x} + a_x t \quad \text{and} \quad v_y = v_{0y} + a_y t.$$

In these equations, v_{0x} ($= -2.0$ m/s) and v_{0y} ($= 4.0$ m/s) are the x and y components of $\vec{v}_0$, and a_x and a_y are the x and y components of $\vec{a}$. To find a_x and a_y, we resolve $\vec{a}$ either with a vector-capable calculator or with Eqs. 3-5:

$$a_x = a \cos \theta = (3.0 \text{ m/s}^2)(\cos 130°) = -1.93 \text{ m/s}^2,$$
$$a_y = a \sin \theta = (3.0 \text{ m/s}^2)(\sin 130°) = +2.30 \text{ m/s}^2.$$

When these values are inserted into the equations for v_x and v_y, we find that, at time $t = 5.0$ s,

$$v_x = -2.0 \text{ m/s} + (-1.93 \text{ m/s}^2)(5.0 \text{ s}) = -11.65 \text{ m/s},$$
$$v_y = 4.0 \text{ m/s} + (2.30 \text{ m/s}^2)(5.0 \text{ s}) = 15.50 \text{ m/s}.$$

Thus, at $t = 5.0$ s, we have, after rounding,

$$\vec{v} = (-12 \text{ m/s})\hat{i} + (16 \text{ m/s})\hat{j}.$$ (Answer)

Either using a vector-capable calculator or following Eq. 3-6, we find that the magnitude and angle of $\vec{v}$ are

$$v = \sqrt{v_x^2 + v_y^2} = 19.4 \approx 19 \text{ m/s}$$ (Answer)

and

$$\theta = \tan^{-1} \frac{v_y}{v_x} = 127° \approx 130°.$$ (Answer)

Check the last line with your calculator. Does $127°$ appear on the display, or does $-53°$ appear? Now sketch the vector $\vec{v}$ with its components to see which angle is reasonable.

✔**CHECKPOINT 4:**   If the position of a hobo's marble is given by $\vec{r} = (4t^3 - 2t)\hat{i} + 3\hat{j}$, with $\vec{r}$ in meters and t in seconds, what must be the units of the coefficients 4, -2, and 3?

4-5 Projectile Motion

We next consider a special case of two-dimensional motion: A particle moves in a vertical plane with some initial velocity $\vec{v}_0$ but its acceleration is always the free-fall acceleration $\vec{g}$, which is downward. Such a particle is called a **projectile** (meaning that it is projected or launched) and its motion is called **projectile motion.** A projectile might be a golf ball (Fig. 4-9) or a baseball in flight, but it is not an airplane or a duck in flight. Our goal here is to analyze projectile motion using the tools for two-dimensional motion in Sections 4-2 through 4-4 and making the assumption that air has no effect on the projectile.

Figure 4-10, which is analyzed in the next section, shows the path followed by a projectile when the air has no effect. The projectile is launched with an initial velocity $\vec{v}_0$ that can be written as

$$\vec{v}_0 = v_{0x}\hat{i} + v_{0y}\hat{j}. \tag{4-19}$$

The components v_{0x} and v_{0y} can then be found if we know the angle θ_0 between $\vec{v}_0$ and the positive x direction:

$$v_{0x} = v_0 \cos\theta_0 \quad \text{and} \quad v_{0y} = v_0 \sin\theta_0. \tag{4-20}$$

During its two-dimensional motion, the projectile's position vector $\vec{r}$ and velocity vector $\vec{v}$ change continuously, but its acceleration vector $\vec{a}$ is constant and *always* directed vertically downward. The projectile has *no* horizontal acceleration.

Projectile motion, like that in Figs. 4-9 and 4-10, looks complicated, but we have the following simplifying feature (known from experiment):

> In projectile motion, the horizontal motion and the vertical motion are independent of each other; that is, neither motion affects the other.

This feature allows us to break up a problem involving two-dimensional motion into two separate and easier one-dimensional problems, one for the horizontal motion (with *zero acceleration*) and one for the vertical motion (with *constant downward acceleration*). Here are two experiments that show that the horizontal motion and the vertical motion are independent.

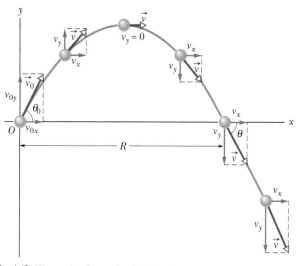

Fig. 4-10 The path of a projectile that is launched at $x_0 = 0$ and $y_0 = 0$, with an initial velocity $\vec{v}_0$. The initial velocity and the velocities at various points along its path are shown, along with their components. Note that the horizontal velocity component remains constant but the vertical velocity component changes continuously. The *range R* is the horizontal distance the projectile has traveled *when it returns to its launch height.*

Fig. 4-9 A stroboscopic photograph of an orange golf ball bouncing off a hard surface. Between impacts, the ball has projectile motion.

Fig. 4-11 One ball is released from rest at the same instant that another ball is shot horizontally to the right. Their vertical motions are identical.

Two Golf Balls

Figure 4-11 is a stroboscopic photograph of two golf balls, one simply released and the other shot horizontally by a spring. The golf balls have the same vertical motion, both falling through the same vertical distance in the same interval of time. *The fact that one ball is moving horizontally while it is falling has no effect on its vertical motion;* that is, the horizontal and vertical motions are independent.

A Great Student Rouser

Figure 4-12 shows a demonstration that has enlivened many a physics lecture. It involves a blow gun G, using a ball as a projectile. The target is a can suspended from a magnet M, and the tube of the blow gun is aimed directly at the can. The experiment is arranged so that the magnet releases the can just as the ball leaves the blow gun.

If g (the magnitude of the free-fall acceleration) were zero, the ball would follow the straight line shown in Fig. 4-12 and the can would float in place after the magnet released it. The ball would certainly hit the can.

However, g is *not* zero. The ball *still* hits the can! As Fig. 4-12 shows, during the time of flight of the ball, both ball and can fall the same distance h from their zero-g locations. The harder the demonstrator blows, the greater the ball's initial speed, the shorter the time of flight, and the smaller the value of h.

4-6 Projectile Motion Analyzed

Now we are ready to analyze projectile motion, horizontally and vertically.

The Horizontal Motion

Because there is *no acceleration* in the horizontal direction, the horizontal component v_x of the projectile's velocity remains unchanged from its initial value v_{0x} throughout the motion, as demonstrated in Fig. 4-13. At any time t, the projectile's horizontal displacement $x - x_0$ from an initial position x_0 is given by Eq. 2-15 with $a = 0$, which we write as

$$x - x_0 = v_{0x}t.$$

Because $v_{0x} = v_0 \cos \theta_0$, this becomes

$$x - x_0 = (v_0 \cos \theta_0)t. \tag{4-21}$$

The Vertical Motion

The vertical motion is the motion we discussed in Section 2-8 for a particle in free fall. Most important is that the acceleration is constant. Thus, the equations of Table 2-1 apply, provided we substitute $-g$ for a and switch to y notation. Then, for example, Eq. 2-15 becomes

$$y - y_0 = v_{0y}t - \tfrac{1}{2}gt^2$$
$$= (v_0 \sin \theta_0)t - \tfrac{1}{2}gt^2, \tag{4-22}$$

where the initial vertical velocity component v_{0y} is replaced with the equivalent $v_0 \sin \theta_0$. Similarly, Eqs. 2-11 and 2-16 become

$$v_y = v_0 \sin \theta_0 - gt \tag{4-23}$$

and

$$v_y^2 = (v_0 \sin \theta_0)^2 - 2g(y - y_0). \tag{4-24}$$

As is illustrated in Fig. 4-10 and Eq. 4-23, the vertical velocity component behaves just as for a ball thrown vertically upward. It is directed upward initially

Fig. 4-12 The projectile ball always hits the falling can. Each falls a distance h from where it would be were there no free-fall acceleration.

Fig. 4-13 The vertical component of this skateboarder's velocity is changing, but not the horizontal component, which matches the skateboard's velocity. As a result, the skateboard stays underneath him, allowing him to land on it.

and its magnitude steadily decreases to zero, *which marks the maximum height of the path*. The vertical velocity component then reverses direction, and its magnitude becomes larger with time.

The Equation of the Path

We can find the equation of the projectile's path (its **trajectory**) by eliminating t between Eqs. 4-21 and 4-22. Solving Eq. 4-21 for t and substituting into Eq. 4-22, we obtain, after a little rearrangement,

$$y = (\tan \theta_0)x - \frac{gx^2}{2(v_0 \cos \theta_0)^2} \quad \text{(trajectory)}. \tag{4-25}$$

This is the equation of the path shown in Fig. 4-10. In deriving it, for simplicity we let $x_0 = 0$ and $y_0 = 0$ in Eqs. 4-21 and 4-22, respectively. Because g, θ_0, and v_0 are constants, Eq. 4-25 is of the form $y = ax + bx^2$, in which a and b are constants. This is the equation of a parabola, so the path is *parabolic*.

The Horizontal Range

The *horizontal range R* of the projectile, as Fig. 4-10 shows, is the *horizontal* distance the projectile has traveled when it returns to its initial (launch) height. To find range R, let us put $x - x_0 = R$ in Eq. 4-21 and $y - y_0 = 0$ in Eq. 4-22, obtaining

$$R = (v_0 \cos \theta_0)t$$

and

$$0 = (v_0 \sin \theta_0)t - \tfrac{1}{2}gt^2.$$

Eliminating t between these two equations yields

$$R = \frac{2v_0^2}{g} \sin \theta_0 \cos \theta_0.$$

Using the identity $\sin 2\theta_0 = 2 \sin \theta_0 \cos \theta_0$ (see Appendix E), we obtain

$$R = \frac{v_0^2}{g} \sin 2\theta_0. \tag{4-26}$$

Caution: This equation does *not* give the horizontal distance traveled by a projectile when the final height is not the launch height.

Note that R in Eq. 4-26 has its maximum value when $\sin 2\theta_0 = 1$, which corresponds to $2\theta_0 = 90°$ or $\theta_0 = 45°$.

> The horizontal range R is maximum for a launch angle of 45°.

The Effects of the Air

We have assumed that the air through which the projectile moves has no effect on its motion. However, in many situations, the disagreement between our calculations and the actual motion of the projectile can be large because the air resists (or opposes) the motion. Figure 4-14, for example, shows two paths for a fly ball that leaves the

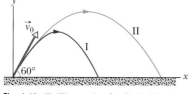

Fig. 4-14 (I) The path of a fly ball, calculated by taking air resistance into account. (II) The path the ball would follow in a vacuum, calculated by the methods of this chapter. See Table 4-1 for corresponding data. (Adapted from "The Trajectory of a Fly Ball," by Peter J. Brancazio, *The Physics Teacher*, January 1985.)

TABLE 4-1 **Two Fly Balls**[a]

	Path I (Air)	Path II (Vacuum)
Range	98.5 m	177 m
Maximum height	53.0 m	76.8 m
Time of flight	6.6 s	7.9 s

[a]See Fig. 4-14. The launch angle is 60° and the launch speed is 44.7 m/s.

bat at an angle of $60°$ with the horizontal and an initial speed of 44.7 m/s. Path I (the baseball player's fly ball) is a calculated path that approximates normal conditions of play, in air. Path II (the physics professor's fly ball) is the path that the ball would follow in a vacuum.

✓**CHECKPOINT 5:**   A fly ball is hit to the outfield. During its flight (ignore the effects of the air), what happens to its (a) horizontal and (b) vertical components of velocity? What are the (c) horizontal and (d) vertical components of its acceleration during its ascent and its descent, and at the topmost point of its flight?

Sample Problem 4-6

In Fig. 4-15, a rescue plane flies at 198 km/h ($= 55.0$ m/s) and a constant elevation of 500 m toward a point directly over a boating accident victim struggling in the water. The pilot wants to release a rescue capsule so that it hits the water very close to the victim. (a) What should be the angle ϕ of the pilot's line of sight to the victim when the release is made?

SOLUTION: The Key Idea here is that, once released, the capsule is a projectile, so its horizontal and vertical motions are independent and can be considered separately (we need not consider the actual curved path of the capsule). Figure 4-15 includes a coordinate system with its origin at the point of release, and we see there that ϕ is given by

$$\phi = \tan^{-1}\frac{x}{h}, \qquad (4\text{-}27)$$

where x is the horizontal coordinate of the victim at release (and of the capsule when it hits the water), and h is the height of the plane. That height is 500 m, so we need only x in order to find ϕ. We should be able to find x with Eq. 4-21:

$$x - x_0 = (v_0 \cos \theta_0)t. \qquad (4\text{-}28)$$

Here we know $x_0 = 0$ because the origin is placed at the point of release. Because the capsule is *released* and not shot from the plane, its initial velocity $\vec{v}_0$ is equal to the plane's velocity. Thus, we know also that the initial velocity has magnitude $v_0 = 55.0$ m/s and angle $\theta_0 = 0°$ (measured relative to the positive direction of the x axis). However, we do not know the time t the capsule takes to move from the plane to the victim.

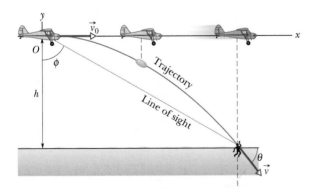

Fig. 4-15 Sample Problem 4-6. A plane drops a rescue capsule while moving at constant velocity in level flight. While the capsule is falling, its horizontal velocity component remains equal to the velocity of the plane.

To find t, we next consider the vertical motion and specifically Eq. 4-22:

$$y - y_0 = (v_0 \sin \theta_0)t - \tfrac{1}{2}gt^2. \qquad (4\text{-}29)$$

Here the vertical displacement $y - y_0$ of the capsule is -500 m (the negative value indicates that the capsule moves *downward*). Putting this and other known values into Eq. 4-29 gives us

$$-500 \text{ m} = (55.0 \text{ m/s})(\sin 0°)t - \tfrac{1}{2}(9.8 \text{ m/s}^2)t^2.$$

Solving for t, we find $t = 10.1$ s. Using that value in Eq. 4-28 yields

$$x - 0 = (55.0 \text{ m/s})(\cos 0°)(10.1 \text{ s}),$$

or

$$x = 555.5 \text{ m}.$$

Then Eq. 4-27 gives us

$$\phi = \tan^{-1}\frac{555.5 \text{ m}}{500 \text{ m}} = 48°. \qquad \text{(Answer)}$$

(b) As the capsule reaches the water, what is its velocity $\vec{v}$ in unit-vector notation and as a magnitude and an angle?

SOLUTION: Again, we need the Key Idea that the horizontal and vertical motions of the capsule are independent during the capsule's flight. In particular, the horizontal and vertical components of the capsule's velocity are independent of each other.

A second Key Idea is that the horizontal component of velocity v_x does not change from its initial value $v_{0x} = v_0 \cos \theta_0$ because there is no horizontal acceleration. Thus, when the capsule reaches the water,

$$v_x = v_0 \cos \theta_0 = (55.0 \text{ m/s})(\cos 0°) = 55.0 \text{ m/s}.$$

A third Key Idea is that the vertical component of velocity v_y changes from its initial value $v_{0y} = v_0 \sin \theta_0$ because there is a vertical acceleration. Using Eq. 4-23 and the capsule's time of fall $t = 10.1$ s, we find that when the capsule reaches the water,

$$\begin{aligned} v_y &= v_0 \sin \theta_0 - gt \\ &= (55.0 \text{ m/s})(\sin 0°) - (9.8 \text{ m/s}^2)(10.1 \text{ s}) \\ &= -99.0 \text{ m/s}. \end{aligned}$$

Thus, when the capsule reaches the water it has the velocity

$$\vec{v} = (55.0 \text{ m/s})\hat{i} - (99.0 \text{ m/s})\hat{j}. \qquad \text{(Answer)}$$

Using either Eq. 3-6 as a guide or a vector-capable calculator, we find that the magnitude and the angle of $\vec{v}$ are

$$v = 113 \text{ m/s} \quad \text{and} \quad \theta = -61°. \qquad \text{(Answer)}$$

Sample Problem 4-7

Figure 4-16 shows a pirate ship 560 m from a fort defending the harbor entrance of an island. A defense cannon, located at sea level, fires balls at initial speed $v_0 = 82$ m/s.

(a) At what angle θ_0 from the horizontal must a ball be fired to hit the ship?

SOLUTION: The Key Idea here is obvious: A fired cannonball is a projectile and the projectile equations apply. We want an equation that relates the launch angle θ_0 to the horizontal displacement of the ball from the cannon to the ship.

A second Key Idea is that, because the cannon and the ship are at the same height, the horizontal displacement is the range. We can then relate the launch angle θ_0 to the range R with Eq. 4-26,

$$R = \frac{v_0^2}{g} \sin 2\theta_0, \tag{4-30}$$

which gives us

$$2\theta_0 = \sin^{-1} \frac{gR}{v_0^2} = \sin^{-1} \frac{(9.8 \text{ m/s}^2)(560 \text{ m})}{(82 \text{ m/s})^2}$$
$$= \sin^{-1} 0.816. \tag{4-31}$$

The inverse function $\sin^{-1}$ always has two possible solutions. One solution (here, 54.7°) is displayed by a calculator; we subtract it from 180° to get the other solution (here, 125.3°). Thus, Eq. 4-31 gives us

$$\theta_0 = \tfrac{1}{2}(54.7°) \approx 27° \qquad \text{(Answer)}$$

and

$$\theta_0 = \tfrac{1}{2}(125.3°) \approx 63°. \qquad \text{(Answer)}$$

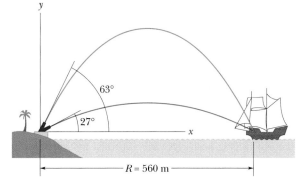

Fig. 4-16 Sample Problem 4-7. At this range, the cannon can hit the pirate ship at two elevation angles of the cannon.

The commandant of the fort can elevate the cannon to either of these two angles and (if only there were no intervening air!) hit the pirate ship.

(b) How far should the pirate ship be from the cannon if it is to be beyond the maximum range of the cannonballs?

SOLUTION: We have seen that maximum range corresponds to an elevation angle θ_0 of 45°. Thus, from Eq. 4-30 with $\theta_0 = 45°$,

$$R = \frac{v_0^2}{g} \sin 2\theta_0 = \frac{(82 \text{ m/s})^2}{9.8 \text{ m/s}^2} \sin(2 \times 45°)$$
$$= 686 \text{ m} \approx 690 \text{ m}. \qquad \text{(Answer)}$$

As the pirate ship sails away, the two elevation angles at which the ship can be hit draw together, eventually merging at $\theta_0 = 45°$ when the ship is 690 m away. Beyond that distance the ship is safe.

Sample Problem 4-8

Figure 4-17 illustrates the flight of Emanuel Zacchini over three Ferris wheels, located as shown and each 18 m high. Zacchini is launched with speed $v_0 = 26.5$ m/s, at an angle $\theta_0 = 53°$ up from the horizontal and with an initial height of 3.0 m above the ground. The net in which he is to land is at the same height.

(a) Does he clear the first Ferris wheel?

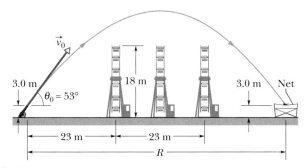

Fig. 4-17 Sample Problem 4-8. The flight of a human cannonball over three Ferris wheels and into a net.

SOLUTION: A Key Idea here is that Zacchini is a human projectile, so we can use the projectile equations. To do so, we place the origin of an xy coordinate system at the cannon muzzle. Then $x_0 = 0$ and $y_0 = 0$ and we want his height y when $x = 23$ m, but we do not know the time t when he reaches that height. To relate y to x without t, we use Eq. 4-25:

$$y = (\tan \theta_0)x - \frac{gx^2}{2(v_0 \cos \theta_0)^2}$$
$$= (\tan 53°)(23 \text{ m}) - \frac{(9.8 \text{ m/s}^2)(23 \text{ m})^2}{2(26.5 \text{ m/s})^2(\cos 53°)^2}$$
$$= 20.3 \text{ m}.$$

Since he begins 3.0 m off the ground, he clears the Ferris wheel by about 5.3 m.

(b) If he reaches his maximum height when he is over the middle Ferris wheel, what is his clearance above it?

SOLUTION: A Key Idea here is that the vertical component v_y of his velocity is zero when he reaches his maximum height. Since Eq. 4-24 relates v_y and his height y, we write

$$v_y^2 = (v_0 \sin \theta_0)^2 - 2gy = 0.$$

Solving for y gives us

$$y = \frac{(v_0 \sin \theta_0)^2}{2g} = \frac{(26.5 \text{ m/s})^2 (\sin 53°)^2}{(2)(9.8 \text{ m/s}^2)} = 22.9 \text{ m},$$

which means that he clears the middle Ferris wheel by 7.9 m.

(c) How far from the cannon should the center of the net be positioned?

SOLUTION: The additional Key Idea here is that, because Zacchini's initial and landing heights are the same, the horizontal distance from cannon muzzle to net is his horizontal range for the flight. From Eq. 4-26, we find

$$R = \frac{v_0^2}{g} \sin 2\theta_0 = \frac{(26.5 \text{ m/s})^2}{9.8 \text{ m/s}^2} \sin 2(53°)$$

$$= 69 \text{ m}. \qquad \text{(Answer)}$$

We can now answer the questions that opened this chapter: How could Zacchini know where to place the net, and how could he be certain he would clear the Ferris wheels? He (or someone) did the calculations as we have here. Although he could not take into account the complicated effects of the air on his flight, Zacchini knew that the air would slow him and thus decrease his range from the calculated value, so he used a wide net and biased it toward the cannon. He was then relatively safe whether the effects of the air in a particular flight happened to slow him considerably or very little. Still, the variability of this factor of air effects must have played on his imagination before each flight.

Zacchini still faced a subtle danger: Even for shorter flights, his propulsion through the cannon was so severe that he underwent a momentary blackout. If he landed during the blackout, he could break his neck. To avoid this, he had trained himself to awake quickly. Indeed, not waking up in time presents the only real danger to a human cannonball in the short flights today.

PROBLEM-SOLVING TACTICS

Tactic 1: *Numbers Versus Algebra*
One way to avoid rounding errors and other numerical errors is to solve problems algebraically, substituting numbers only in the final step. That is easy to do in Sample Problems 4-6 to 4-8, and that is the way experienced problem solvers operate. In these early chapters, however, we prefer to solve most problems in parts, to give you a firmer numerical grasp of what is going on. Later we shall stick to the algebra longer.

4-7 Uniform Circular Motion

A particle is in **uniform circular motion** if it travels around a circle or a circular arc at constant (*uniform*) speed. Although the speed does not vary, *the particle is accelerating.* That fact may be surprising because we often think of acceleration (a change in velocity) as an increase or decrease in speed. However, actually velocity is a vector, not a scalar. Thus, even if a velocity changes only in direction, there is still an acceleration, and that is what happens in uniform circular motion.

Figure 4-18 shows the relation between the velocity and acceleration vectors at various stages during uniform circular motion. Both vectors have constant magnitude as the motion progresses, but their directions change continuously. The velocity is always directed tangent to the circle in the direction of motion. The acceleration is always directed *radially inward.* Because of this, the acceleration associated with uniform circular motion is called a **centripetal** (meaning "center seeking") **acceleration.** As we prove next, the magnitude of this acceleration $\vec{a}$ is

$$a = \frac{v^2}{r} \qquad \text{(centripetal acceleration)}, \qquad (4\text{-}32)$$

where r is the radius of the circle and v is the speed of the particle.

In addition, during this acceleration at constant speed, the particle travels the circumference of the circle (a distance of $2\pi r$) in time

$$T = \frac{2\pi r}{v} \qquad \text{(period)}. \qquad (4\text{-}33)$$

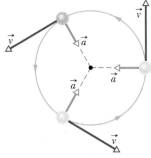

Fig. 4-18 Velocity and acceleration vectors for a particle in counterclockwise uniform circular motion. Both have constant magnitude but vary continuously in direction.

T is called the *period of revolution,* or simply the *period,* of the motion. It is, in general, the time for a particle to go around a closed path exactly once.

Proof of Eq. 4-32

To find the magnitude and direction of the acceleration for uniform circular motion, we consider Fig. 4-19. In Fig. 4-19a, particle p moves at constant speed v around a circle of radius r. At the instant shown, p has coordinates x_p and y_p.

Recall from Section 4-3 that the velocity $\vec{v}$ of a moving particle is always tangent to the particle's path at the particle's position. In Fig. 4-19a, that means $\vec{v}$ is perpendicular to a radius r drawn to the particle's position. Then the angle θ that $\vec{v}$ makes with a vertical at p equals the angle θ that radius r makes with the x axis.

The scalar components of $\vec{v}$ are shown in Fig. 4-19b. With them, we can write the velocity $\vec{v}$ as

$$\vec{v} = v_x\hat{i} + v_y\hat{j} = (-v\sin\theta)\hat{i} + (v\cos\theta)\hat{j}. \tag{4-34}$$

Now, using the right triangle in Fig. 4-19a, we can replace $\sin\theta$ with y_p/r and $\cos\theta$ with x_p/r to write

$$\vec{v} = \left(-\frac{vy_p}{r}\right)\hat{i} + \left(\frac{vx_p}{r}\right)\hat{j}. \tag{4-35}$$

To find the acceleration $\vec{a}$ of particle p, we must take the time derivative of this equation. Noting that speed v and radius r do not change with time, we obtain

$$\vec{a} = \frac{d\vec{v}}{dt} = \left(-\frac{v}{r}\frac{dy_p}{dt}\right)\hat{i} + \left(\frac{v}{r}\frac{dx_p}{dt}\right)\hat{j}. \tag{4-36}$$

Now note that the rate dy_p/dt at which y_p changes is equal to the velocity component v_y. Similarly, $dx_p/dt = v_x$, and, again from Fig. 4-19b, we see that $v_x = -v\sin\theta$ and $v_y = v\cos\theta$. Making these substitutions in Eq. 4-36, we find

$$\vec{a} = \left(-\frac{v^2}{r}\cos\theta\right)\hat{i} + \left(-\frac{v^2}{r}\sin\theta\right)\hat{j}. \tag{4-37}$$

This vector and its components are shown in Fig. 4-19c. Following Eq. 3-6, we find that the magnitude of $\vec{a}$ is

$$a = \sqrt{a_x^2 + a_y^2} = \frac{v^2}{r}\sqrt{(\cos\theta)^2 + (\sin\theta)^2} = \frac{v^2}{r},$$

as we wanted to prove. To orient $\vec{a}$, we can find the angle ϕ shown in Fig. 4-19c:

$$\tan\phi = \frac{a_y}{a_x} = \frac{-(v^2/r)\sin\theta}{-(v^2/r)\cos\theta} = \tan\theta.$$

Thus, $\phi = \theta$, which means that $\vec{a}$ is directed along the radius r of Fig. 4-19a, toward the circle's center, as we wanted to prove.

(a)

(b)

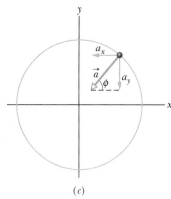

(c)

Fig. 4-19 Particle p moves in counterclockwise uniform circular motion. (a) Its position and velocity $\vec{v}$ at a certain instant. (b) Velocity $\vec{v}$ and its components. (c) The particle's acceleration $\vec{a}$ and its components.

✓**CHECKPOINT 6:** An object moves at constant speed along a circular path in a horizontal xy plane, with the center at the origin. When the object is at $x = -2$ m, its velocity is $-(4$ m/s$)\hat{j}$. Give the object's (a) velocity and (b) acceleration when it is at $y = 2$ m.

Sample Problem 4-9

"Top gun" pilots have long worried about taking a turn too tightly. As a pilot's body undergoes centripetal acceleration, with the head toward the center of curvature, the blood pressure in the brain decreases, leading to loss of brain function.

There are several warning signs to signal a pilot to ease up: when the centripetal acceleration is $2g$ or $3g$, the pilot feels heavy. At about $4g$, the pilot's vision switches to black and white and narrows to "tunnel vision." If that acceleration is sustained or increased, vision ceases and, soon after, the pilot is unconscious—a condition known as *g*-LOC for "*g*-induced loss of consciousness."

What is the centripetal acceleration, in *g* units, of a pilot flying an F-22 at speed $v = 2500$ km/h (694 m/s) through a circular arc with radius of curvature $r = 5.80$ km?

SOLUTION: The Key Idea here is that although the pilot's speed is constant, the circular path requires a (centripetal) acceleration, with

magnitude given by Eq. 4-32:

$$a = \frac{v^2}{r} = \frac{(694 \text{ m/s})^2}{5800 \text{ m}} = 83.0 \text{ m/s}^2 = 8.5g. \quad \text{(Answer)}$$

If an unwary pilot caught in a dogfight puts the aircraft into such a tight turn, the pilot goes into *g*-LOC almost immediately, with no warning signs to signal the danger.

4-8 Relative Motion in One Dimension

Fig. 4-20 Alex (frame A) and Barbara (frame B) watch car P, as both B and P move at different velocities along the common x axes of the two frames. At the instant shown, x_{BA} is the coordinate of B in the A frame. Also, P is at coordinate x_{PB} in the B frame and coordinate $x_{PA} = x_{PB} + x_{BA}$ in the A frame.

Suppose you see a duck flying north at, say, 30 km/h. To another duck flying alongside, the first duck seems to be stationary. In other words, the velocity of a particle depends on the **reference frame** of whoever is observing or measuring the velocity. For our purposes, a reference frame is the physical object to which we attach our coordinate system. In everyday life, that object is the ground. For example, the speed listed on a speeding ticket is always measured relative to the ground. The speed relative to the police officer would be different if the officer were moving while making the speed measurement.

Suppose that Alex (at the origin of frame A) is parked by the side of a highway, watching car P (the "particle") speed past. Barbara (at the origin of frame B) is driving along the highway at constant speed and is also watching car P. Suppose that, as in Fig. 4-20, they both measure the position of the car at a given moment. From the figure we see that

$$x_{PA} = x_{PB} + x_{BA}. \tag{4-38}$$

The equation is read: "The coordinate x_{PA} of P as measured by A *is equal to* the coordinate x_{PB} of P as measured by B *plus* the coordinate x_{BA} of B as measured by A." Note how this reading is supported by the sequence of the subscripts.

Taking the time derivative of Eq. 4-38, we obtain

$$\frac{d}{dt}(x_{PA}) = \frac{d}{dt}(x_{PB}) + \frac{d}{dt}(x_{BA}),$$

or (because $v = dx/dt$)

$$v_{PA} = v_{PB} + v_{BA}. \tag{4-39}$$

This equation is read: "The velocity v_{PA} of P as measured by A *is equal to* the velocity v_{PB} of P as measured by B *plus* the velocity v_{BA} of B as measured by A." The term v_{BA} is the velocity of frame B relative to frame A. (Because the motions are along a single axis, we can use components along that axis in Eq. 4-39 and omit overhead vector arrows.)

Here we consider only frames that move at constant velocity relative to each other. In our example, this means that Barbara (frame B) will drive always at constant velocity v_{BA} relative to Alex (frame A). Car P (the moving particle), however, may speed up, slow down, come to rest, or reverse direction (that is, it can accelerate).

To relate an acceleration of P as measured by Barbara and by Alex, we take the time derivative of Eq. 4-39:

$$\frac{d}{dt}(v_{PA}) = \frac{d}{dt}(v_{PB}) + \frac{d}{dt}(v_{BA}).$$

Because v_{BA} is constant, the last term is zero and we have

$$a_{PA} = a_{PB}. \tag{4-40}$$

In other words,

> Observers on different frames of reference (that move at constant velocity relative to each other) will measure the same acceleration for a moving particle.

✔**CHECKPOINT 7:**   The table here gives velocities (km/h) for Barbara and car P of Fig. 4-20 for three situations. For each, what is the missing value and how is the distance between Barbara and car P changing?

Situation	v_{BA}	v_{PA}	v_{PB}
(a)	+50	+50	
(b)	+30		+40
(c)		+60	−20

Sample Problem 4-10

For the situation of Fig. 4-20 and this section, Barbara's velocity relative to Alex is a constant v_{BA} = 52 km/h and car P is moving in the negative direction of the x axis.

(a) If Alex measures a constant velocity v_{PA} = −78 km/h for car P, what velocity v_{PB} will Barbara measure?

SOLUTION: The Key Idea here is that we can attach a frame of reference A to Alex and another frame of reference B to Barbara. Further, because these two frames move at constant velocity relative to each other, along a single axis, we can use Eq. 4-39 to relate v_{PB} and v_{PA}. We find

$$v_{PA} = v_{PB} + v_{BA}$$

or $-78 \text{ km/h} = v_{PB} + 52 \text{ km/h}.$

Thus, $v_{PB} = -130 \text{ km/h}.$ (Answer)

If car P were connected to Barbara's car by a cord wound on a spool, the cord would be unwinding at a speed of 130 km/h as the two cars separated.

(b) If car P brakes to a stop relative to Alex (and thus the ground) in time t = 10 s at constant acceleration, what is its acceleration a_{PA} relative to Alex?

SOLUTION: The Key Idea here is that, to calculate the acceleration of car P *relative to Alex*, we must use the car's velocities *relative to Alex*. Because the acceleration is constant, we can use Eq. 2-11 ($v = v_0 + at$) to relate the acceleration to the initial and final velocities of P. The initial velocity of P relative to Alex is v_{PA} = −78 km/h and the final velocity is 0. Thus, Eq. 2-11 gives us

$$a_{PA} = \frac{v - v_0}{t} = \frac{0 - (-78 \text{ km/h})}{10 \text{ s}} \frac{1 \text{ m/s}}{3.6 \text{ km/h}}$$
$$= 2.2 \text{ m/s}^2. \tag{Answer}$$

(c) What is the acceleration a_{PB} of car P relative to Barbara during the braking?

SOLUTION: The Key Idea is that now, to calculate the acceleration of car P *relative to Barbara,* we must use the car's velocities *relative to Barbara.* We know the initial velocity of P relative to Barbara from part (a) (v_{PB} = −130 km/h). The final velocity of P relative to Barbara is −52 km/h (this is the velocity of the stopped car relative to the moving Barbara). Thus, again from Eq. 2-11,

$$a_{PB} = \frac{v - v_0}{t} = \frac{-52 \text{ km/h} - (-130 \text{ km/h})}{10 \text{ s}} \frac{1 \text{ m/s}}{3.6 \text{ km/h}}$$
$$= 2.2 \text{ m/s}^2. \tag{Answer}$$

We should have foreseen this result: Because Alex and Barbara have a constant relative velocity, they must measure the same acceleration for the car.

4-9 Relative Motion in Two Dimensions

Now we turn from relative motion in one dimension to relative motion in two (and, by extension, in three) dimensions. In Fig. 4-21, our two observers are again watching a moving particle P from the origins of reference frames A and B, while B moves at a constant velocity $\vec{v}_{BA}$ relative to A. (The corresponding axes of these two frames remain parallel.)

Figure 4-21 shows a certain instant during the motion. At that instant, the position vector of B relative to A is $\vec{r}_{BA}$. Also, the position vectors of particle P are

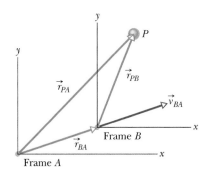

Fig. 4-21 Frame B has the constant two-dimensional velocity $\vec{v}_{BA}$ relative to frame A. The position vector of B relative to A is $\vec{r}_{BA}$. The position vectors of particle P are $\vec{r}_{PA}$ relative to A and $\vec{r}_{PB}$ relative to B.

$\vec{r}_{PA}$ relative to A and $\vec{r}_{PB}$ relative to B. From the arrangement of heads and tails of those three position vectors, we can relate the vectors with

$$\vec{r}_{PA} = \vec{r}_{PB} + \vec{r}_{BA}. \qquad (4\text{-}41)$$

By taking the time derivative of this equation, we can relate the velocities $\vec{v}_{PA}$ and $\vec{v}_{PB}$ of particle P relative to our observers. We get

$$\vec{v}_{PA} = \vec{v}_{PB} + \vec{v}_{BA}. \qquad (4\text{-}42)$$

By taking the time derivative of this relation, we can relate the accelerations $\vec{a}_{PA}$ and $\vec{a}_{PB}$ of the particle P relative to our observers. However, note that because $\vec{v}_{BA}$ is constant, its time derivative is zero. Thus, we get

$$\vec{a}_{PA} = \vec{a}_{PB}. \qquad (4\text{-}43)$$

As for one-dimensional motion, we have the following rule: Observers on different frames of reference that move at constant velocity relative to each other will measure the *same* acceleration for a moving particle.

Sample Problem 4-11

In Fig. 4-22a, a plane moves due east (directly toward the east) while the pilot points the plane somewhat south of east, toward a steady wind that blows to the northeast. The plane has velocity $\vec{v}_{PW}$ relative to the wind, with an airspeed (speed relative to the wind) of 215 km/h, directed at angle θ south of east. The wind has velocity $\vec{v}_{WG}$ relative to the ground, with a speed of 65.0 km/h, directed 20.0° east of north. What is the magnitude of the velocity $\vec{v}_{PG}$ of the plane relative to the ground, and what is θ?

SOLUTION: The **Key Idea** is that the situation is like the one in Fig. 4-21. Here the moving particle P is the plane, frame A is attached to the ground (call it G), and frame B is "attached" to the wind (call it W). We need to construct a vector diagram like that in Fig. 4-21 but this time using the three velocity vectors.

First construct a sentence that relates the three vectors:

$$\begin{array}{ccc} \text{velocity of plane} & \text{velocity of plane} & \text{velocity of wind} \\ \text{relative to ground} = & \text{relative to wind} + & \text{relative to ground.} \\ (PG) & (PW) & (WG) \end{array}$$

This relation can be drawn as in Fig. 4-22b and written in vector notation as

$$\vec{v}_{PG} = \vec{v}_{PW} + \vec{v}_{WG}. \qquad (4\text{-}44)$$

We want the magnitude of the first vector and the direction of the second vector. With unknowns in two vectors, we cannot solve Eq. 4-44 directly on a vector-capable calculator. Instead, we need to resolve the vectors into components on the coordinate system of Fig. 4-22b, and then solve Eq. 4-44 axis by axis (see Section 3-5). For the y components, we find

$$v_{PG,y} = v_{PW,y} + v_{WG,y}$$

or $0 = -(215 \text{ km/h}) \sin \theta + (65.0 \text{ km/h})(\cos 20.0°).$

Solving for θ gives us

$$\theta = \sin^{-1} \frac{(65.0 \text{ km/h})(\cos 20.0°)}{215 \text{ km/h}} = 16.5°. \qquad \text{(Answer)}$$

(a)

(b)

Fig. 4-22 Sample Problem 4-11. To travel due east, the plane must head somewhat into the wind.

Similarly, for the x components we find

$$v_{PG,x} = v_{PW,x} + v_{WG,x}.$$

Here, because $\vec{v}_{PG}$ is parallel to the x axis, the component $v_{PG,x}$ is equal to the magnitude v_{PG}. Substituting this and $\theta = 16.5°$, we find

$$v_{PG} = (215 \text{ km/h})(\cos 16.5°) + (65.0 \text{ km/h})(\sin 20.0°)$$
$$= 228 \text{ km/h}. \qquad \text{(Answer)}$$

✔**CHECKPOINT 8:** In this sample problem, suppose the pilot turns the plane to point it due east but without changing the air speed. Do the following magnitudes increase, decrease, or remain the same: (a) $v_{PG,y}$, (b) $v_{PG,x}$, and (c) v_{PG}? (You can answer without computation.)

REVIEW & SUMMARY

Position Vector   The location of a particle relative to the origin of a coordinate system is given by a *position vector* $\vec{r}$, which in unit-vector notation is

$$\vec{r} = x\hat{i} + y\hat{j} + z\hat{k}. \tag{4-1}$$

Here $x\hat{i}$, $y\hat{j}$, and $z\hat{k}$ are the *vector components* of position vector $\vec{r}$, and x, y, and z are its *scalar components* (as well as the coordinates of the particle). A position vector is described by a magnitude and one or two angles for orientation, or by its vector or scalar components.

Displacement   If a particle moves so that its position vector changes from $\vec{r}_1$ to $\vec{r}_2$, then the particle's *displacement* $\Delta\vec{r}$ is

$$\Delta\vec{r} = \vec{r}_2 - \vec{r}_1. \tag{4-2}$$

The displacement can also be written as

$$\Delta\vec{r} = (x_2 - x_1)\hat{i} + (y_2 - y_1)\hat{j} + (z_2 - z_1)\hat{k} \tag{4-3}$$

$$= \Delta x\hat{i} + \Delta y\hat{j} + \Delta z\hat{k}, \tag{4-4}$$

where coordinates (x_1, y_1, z_1) correspond to position vector $\vec{r}_1$ and coordinates (x_2, y_2, z_2) correspond to position vector $\vec{r}_2$.

Average Velocity and (Instantaneous) Velocity   If a particle undergoes a displacement $\Delta\vec{r}$ in time Δt, its *average velocity* $\vec{v}_{avg}$ for that time interval is

$$\vec{v}_{avg} = \frac{\Delta\vec{r}}{\Delta t}. \tag{4-8}$$

As Δt in Eq. 4-8 is shrunk to 0, $\vec{v}_{avg}$ reaches a limit called the *velocity* or *instantaneous velocity* $\vec{v}$:

$$\vec{v} = \frac{d\vec{r}}{dt}, \tag{4-10}$$

which can be rewritten in unit-vector notation as

$$\vec{v} = v_x\hat{i} + v_y\hat{j} + v_z\hat{k}, \tag{4-11}$$

where $v_x = dx/dt$, $v_y = dy/dt$, and $v_z = dz/dt$. The instantaneous velocity $\vec{v}$ of a particle is always directed along the tangent to the particle's path at the particle's position.

Average Acceleration and (Instantaneous) Acceleration   If a particle's velocity changes from $\vec{v}_1$ to $\vec{v}_2$ in time interval Δt, its *average acceleration* during Δt is

$$\vec{a}_{avg} = \frac{\vec{v}_2 - \vec{v}_1}{\Delta t} = \frac{\Delta\vec{v}}{\Delta t}. \tag{4-15}$$

As Δt in Eq. 4-15 is shrunk to 0, $\vec{a}_{avg}$ reaches a limiting value called the *acceleration* or *instantaneous acceleration* $\vec{a}$:

$$\vec{a} = \frac{d\vec{v}}{dt}. \tag{4-16}$$

In unit-vector notation,

$$\vec{a} = a_x\hat{i} + a_y\hat{j} + a_z\hat{k}, \tag{4-17}$$

where $a_x = dv_x/dt$, $a_y = dv_y/dt$, and $a_z = dv_z/dt$.

Projectile Motion   *Projectile motion* is the motion of a particle that is launched with an initial velocity $\vec{v}_0$. During its flight, the particle's horizontal acceleration is zero and its vertical acceleration is the free-fall acceleration $-g$. (Upward is taken to be a positive direction.) If $\vec{v}_0$ is expressed as a magnitude (the speed v_0) and an angle θ_0, the particle's equations of motion along the horizontal x axis and vertical y axis are

$$x - x_0 = (v_0 \cos \theta_0)t, \tag{4-21}$$

$$y - y_0 = (v_0 \sin \theta_0)t - \tfrac{1}{2}gt^2, \tag{4-22}$$

$$v_y = v_0 \sin \theta_0 - gt, \tag{4-23}$$

$$v_y^2 = (v_0 \sin \theta_0)^2 - 2g(y - y_0). \tag{4-24}$$

The **trajectory** (path) of a particle in projectile motion is parabolic and is given by

$$y = (\tan \theta_0)x - \frac{gx^2}{2(v_0 \cos \theta_0)^2}, \tag{4-25}$$

where the origin has been chosen so that x_0 and y_0 of Eqs. 4-21 to 4-24 are zero. The particle's **horizontal range** R, which is the horizontal distance from the launch point to the point at which the particle returns to the launch height, is

$$R = \frac{v_0^2}{g} \sin 2\theta_0. \tag{4-26}$$

Uniform Circular Motion   If a particle travels along a circle or circular arc with radius r at constant speed v, it is in *uniform circular motion* and has an acceleration $\vec{a}$ of magnitude

$$a = \frac{v^2}{r}. \tag{4-32}$$

The direction of $\vec{a}$ is toward the center of the circle or circular arc, and $\vec{a}$ is said to be *centripetal*. The time for the particle to complete a circle is

$$T = \frac{2\pi r}{v}. \tag{4-33}$$

T is called the *period of revolution*, or simply the *period*, of the motion.

Relative Motion   When two frames of reference A and B are moving relative to each other at constant velocity, the velocity of a particle P as measured by an observer in frame A usually differs from that measured from frame B. The two measured velocities are related by

$$\vec{v}_{PA} = \vec{v}_{PB} + \vec{v}_{BA}, \tag{4-42}$$

in which $\vec{v}_{BA}$ is the velocity of B with respect to A. Both observers measure the same acceleration for the particle; that is,

$$\vec{a}_{PA} = \vec{a}_{PB}. \tag{4-43}$$

QUESTIONS

1. Figure 4-23 shows the initial position i and the final position f of a particle. What are the particle's (a) initial position vector $\vec{r}_i$ and (b) final position vector $\vec{r}_f$, both in unit-vector notation? (c) What is the x component of the particle's displacement $\Delta\vec{r}$?

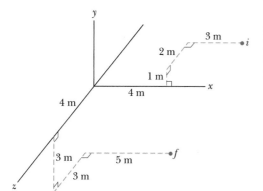

Fig. 4-23
Question 1.

2. Here are four descriptions for the velocity of a hockey puck in the xy plane, all in meters per second:
(1) $v_x = -3t^2 + 4t - 2$ and $v_y = 6t - 4$
(2) $v_x = -3$ and $v_y = -5t^2 + 6$
(3) $\vec{v} = 2t^2\hat{i} - (4t + 3)\hat{j}$
(4) $\vec{v} = -2t\hat{i} + 3\hat{j}$
(a) For each description, are the x and y components of the acceleration constant, and is the acceleration vector $\vec{a}$ constant? (b) In description (4), if $\vec{v}$ is in meters per second and t is in seconds, what must be the units of the coefficients -2 and 3?

3. Figure 4-24 shows three situations in which identical projectiles are launched from the ground (at the same level) at identical initial speeds and angles. The projectiles do not land on the same terrain, however. Rank the situations according to the final speeds of the projectiles just before they land, greatest first.

(a)   (b)   (c)

Fig. 4-24 Question 3.

4. At a certain instant, a fly ball has velocity $\vec{v} = 25\hat{i} - 4.9\hat{j}$ (the x axis is horizontal, the y axis is upward, and $\vec{v}$ is in meters per second). Has the ball passed the highest point of its trajectory?

5. You are to launch a rocket, from just above the ground, with one of the following initial velocity vectors: (1) $\vec{v}_0 = 20\hat{i} + 70\hat{j}$, (2) $\vec{v}_0 = -20\hat{i} + 70\hat{j}$, (3) $\vec{v}_0 = 20\hat{i} - 70\hat{j}$, (4) $\vec{v}_0 = -20\hat{i} - 70\hat{j}$. In your coordinate system, x runs along level ground and y increases upward. (a) Rank the vectors according to the launch speed of the projectile, greatest first. (b) Rank the vectors according to the time of flight of the projectile, greatest first.

6. A mud ball is launched 2 m above the ground with initial velocity $\vec{v}_0 = (2\hat{i} + 4\hat{j})$ m/s. What is its velocity just before it lands on a surface that is 2 m above the ground?

7. In Fig. 4-25, a cream tangerine is thrown up past windows 1, 2, and 3, which are identical in size and regularly spaced vertically. Rank those three windows according to (a) the time the cream tangerine takes to pass them and (b) the average speed of the cream tangerine during the passage, greatest first.

The cream tangerine then moves down past windows 4, 5, and 6, which are identical in size and irregularly spaced horizontally. Rank those three windows according to (c) the time the cream tangerine takes to pass them and (d) the average speed of the cream tangerine during the passage, greatest first.

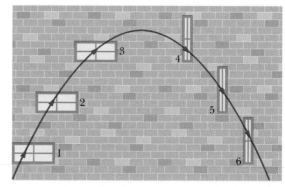

Fig. 4-25 Question 7.

8. An airplane flying horizontally at a constant speed of 350 km/h over level ground releases a bundle of food supplies. Ignore the effect of the air on the bundle. What are the bundle's initial (a) vertical and (b) horizontal components of velocity? (c) What is its horizontal component of velocity just before hitting the ground? (d) If the airplane's speed were, instead, 450 km/h, would the time of fall be larger, smaller, or the same?

9. You throw a ball with a launch velocity of $\vec{v}_i = (3$ m/s$)\hat{i} + (4$ m/s$)\hat{j}$ toward a wall, where it hits at height h_1 in time t_1 after the launch (Fig. 4-26). Suppose that the launch velocity were, instead, $\vec{v}_i = (5$ m/s$)\hat{i} + (4$ m/s$)\hat{j}$. (a) Would the time taken by the ball to reach the wall be greater than, less than, or equal to t_1, or is the question unanswerable without more information? (b) Would the height at which the ball hits be greater than, less than, or equal to h_1, or is the question unanswerable?

Suppose, instead, that the launch velocity were $\vec{v}_i = (3$ m/s$)\hat{i} + (5$ m/s$)\hat{j}$. (c) Would the time taken by the ball to reach the wall be greater than, less than, or equal to t_1, or is the question unanswerable? (d) Would the height at which the ball hits be greater than, less than, or equal to h_1, or is the question unanswerable?

Fig. 4-26 Question 9.

10. Figure 4-27 shows three paths for a football kicked from ground level. Ignoring the effects of air

Fig. 4-27 Question 10.

on the flight, rank the paths according to (a) time of flight, (b) initial vertical velocity component, (c) initial horizontal velocity component, and (d) initial speed, greatest first.

11. Figure 4-28 shows the velocity and acceleration of a particle at a particular instant in three situations. In which situation is (a) the speed of the particle increasing, (b) the speed decreasing, (c) the speed not changing, (d) $\vec{v} \cdot \vec{a}$ positive, (e) $\vec{v} \cdot \vec{a}$ negative, and (f) $\vec{v} \cdot \vec{a} = 0$?

Fig. 4-28 Question 11.

12. Figure 4-29 shows four tracks (either half- or quarter-circles) that can be taken by a train, which moves at a constant speed. Rank the tracks according to the magnitude of a train's acceleration on the curved portion, greatest first.

13. (a) Is it possible to be accelerating while traveling at constant speed? Is it possible to round a curve with (b) zero acceleration and (c) a constant magnitude of acceleration?

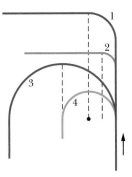

Fig. 4-29 Question 12.

EXERCISES & PROBLEMS

ssm   Solution is in the Student Solutions Manual.
www   Solution is available on the World Wide Web at:
            http://www.wiley.com/college/hrw
ilw   Solution is available on the Interactive LearningWare.

SEC. 4-2   Position and Displacement

1E. A watermelon seed has the following coordinates: $x = -5.0$ m, $y = 8.0$ m, and $z = 0$ m. Find its position vector (a) in unit-vector notation and as (b) a magnitude and (c) an angle relative to the positive direction of the x axis. (d) Sketch the vector on a right-handed coordinate system. If the seed is moved to the xyz coordinates (3.00 m, 0 m, 0 m), what is its displacement (e) in unit-vector notation and as (f) a magnitude and (g) an angle relative to the positive direction of the x axis?

2E. The position vector for an electron is $\vec{r} = (5.0 \text{ m})\hat{i} - (3.0 \text{ m})\hat{j} + (2.0 \text{ m})\hat{k}$. (a) Find the magnitude of $\vec{r}$. (b) Sketch the vector on a right-handed coordinate system.

3E. The position vector for a proton is initially $\vec{r} = 5.0\hat{i} - 6.0\hat{j} + 2.0\hat{k}$ and then later is $\vec{r} = -2.0\hat{i} + 6.0\hat{j} + 2.0\hat{k}$, all in meters. (a) What is the proton's displacement vector, and (b) to what plane is that vector parallel?

4P. A radar station detects an airplane approaching directly from the east. At first observation, the range to the plane is 360 m at 40° above the horizon. The airplane is tracked for another 123° in the vertical east–west plane, the range at final contact being 790 m. See Fig. 4-30. Find the displacement of the airplane during the period of observation.

Fig. 4-30 Problem 4.

SEC. 4-3   Average Velocity and Instantaneous Velocity

5E. A train moving at a constant speed of 60.0 km/h moves east for 40.0 min, then in a direction 50.0° east of north for 20.0 min, and finally west for 50.0 min. What is the average velocity of the train during this trip? ssm

6E. An ion's position vector is initially $\vec{r} = 5.0\hat{i} - 6.0\hat{j} + 2.0\hat{k}$, and 10 s later it is $\vec{r} = -2.0\hat{i} + 8.0\hat{j} - 2.0\hat{k}$, all in meters. What is its average velocity during the 10 s?

7P. The position of an electron is given by $\vec{r} = 3.00t\hat{i} - 4.00t^2\hat{j} + 2.00\hat{k}$, with t in seconds and $\vec{r}$ in meters. (a) What is the electron's velocity $\vec{v}(t)$? At $t = 2.00$ s, what is $\vec{v}$ (b) in unit-vector notation and as (c) a magnitude and (d) an angle relative to the positive direction of the x axis?

8P. Oasis A is 90 km west of oasis B. A camel leaves oasis A and during a 50 h period walks 75 km in a direction 37° north of east. The camel then walks toward the south a distance of 65 km in a 35 h period, after which it rests for 5.0 h. (a) What is the camel's displacement with respect to oasis A after resting? (b) What is the camel's average velocity from the time it leaves oasis A until it finishes resting? (c) What is the camel's average speed from the time it leaves oasis A until it finishes resting? (d) If the camel is able to go without water for five days (120 h), what must its average velocity be after resting if it is to reach oasis B just in time?

SEC. 4-4   Average Acceleration and Instantaneous Acceleration

9E. A particle moves so that its position (in meters) as a function of time (in seconds) is $\vec{r} = \hat{i} + 4t^2\hat{j} + t\hat{k}$. Write expressions for (a) its velocity and (b) its acceleration as functions of time. ssm

10E. A proton initially has $\vec{v} = 4.0\hat{i} - 2.0\hat{j} + 3.0\hat{k}$ and then 4.0 s later has $\vec{v} = -2.0\hat{i} - 2.0\hat{j} + 5.0\hat{k}$ (in meters per second). For that 4.0 s, what is the proton's average acceleration $\vec{a}_{\text{avg}}$ (a) in unit-vector notation and (b) as a magnitude and a direction?

11E. The position $\vec{r}$ of a particle moving in an xy plane is given by $\vec{r} = (2.00t^3 - 5.00t)\hat{i} + (6.00 - 7.00t^4)\hat{j}$, with $\vec{r}$ in meters and t in seconds. Calculate (a) $\vec{r}$, (b) $\vec{v}$, and (c) $\vec{a}$ for $t = 2.00$ s. (d) What is the orientation of a line that is tangent to the particle's path at $t = 2.00$ s?

12E. An iceboat sails across the surface of a frozen lake with constant acceleration produced by the wind. At a certain instant the boat's velocity is $(6.30\hat{i} - 8.42\hat{j})$ m/s. Three seconds later, because of a wind shift, the boat is instantaneously at rest. What is its average acceleration for this 3 s interval?

13P. A particle leaves the origin with an initial velocity $\vec{v} = (3.00\hat{i})$ m/s and a constant acceleration $\vec{a} = (-1.00\hat{i} - 0.500\hat{j})$ m/s². When the particle reaches its maximum x coordinate, what are (a) its velocity and (b) its position vector? ssm ilw www

14P. The velocity $\vec{v}$ of a particle moving in the xy plane is given by $\vec{v} = (6.0t - 4.0t^2)\hat{i} + 8.0\hat{j}$, with $\vec{v}$ in meters per second and t (> 0) in seconds. (a) What is the acceleration when $t = 3.0$ s? (b) When (if ever) is the acceleration zero? (c) When (if ever) is the velocity zero? (d) When (if ever) does the speed equal 10 m/s?

15P. A particle starts from the origin at $t = 0$ with a velocity of $8.0\hat{j}$ m/s and moves in the xy plane with a constant acceleration of $(4.0\hat{i} + 2.0\hat{j})$ m/s². At the instant the particle's x coordinate is 29 m, what are (a) its y coordinate and (b) its speed?

16P. Particle A moves along the line $y = 30$ m with a constant velocity $\vec{v}$ of magnitude 3.0 m/s and directed parallel to the positive x axis (Fig. 4-31). Particle B starts at the origin with zero speed and constant acceleration $\vec{a}$ (of magnitude 0.40 m/s²) at the same instant that particle A passes the y axis. What angle θ between $\vec{a}$ and the positive y axis would result in a collision between these two particles? (If your computation involves an equation with a term such as t^4, substitute $u = t^2$ and then consider solving the resulting quadratic equation to get u.)

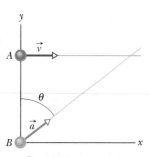

Fig. 4-31 Problem 16.

SEC. 4-6 Projectile Motion Analyzed

In some of these problems, exclusion of the effects of the air is unwarranted but helps simplify the calculations.

17E. A rifle is aimed horizontally at a target 30 m away. The bullet hits the target 1.9 cm below the aiming point. What are (a) the bullet's time of flight and (b) its speed as it emerges from the rifle? ssm

18E. A small ball rolls horizontally off the edge of a tabletop that is 1.20 m high. It strikes the floor at a point 1.52 m horizontally away from the edge of the table. (a) How long is the ball in the air? (b) What is its speed at the instant it leaves the table?

19E. A baseball leaves a pitcher's hand horizontally at a speed of 161 km/h. The distance to the batter is 18.3 m. (Ignore the effect of air resistance.) (a) How long does the ball take to travel the first half of that distance? (b) The second half? (c) How far does the ball fall freely during the first half? (d) During the second half? (e) Why aren't the quantities in (c) and (d) equal?

20E. A dart is thrown horizontally with an initial speed of 10 m/s toward point P, the bull's-eye on a dart board. It hits at point Q on the rim, vertically below P, 0.19 s later. (a) What is the distance PQ? (b) How far away from the dart board is the dart released?

21E. An electron, with an initial horizontal velocity of magnitude 1.00×10^9 cm/s, travels into the region between two horizontal metal plates that are electrically charged. In that region, it travels a horizontal distance of 2.00 cm and has a constant downward acceleration of magnitude 1.00×10^{17} cm/s² due to the charged plates. Find (a) the time required by the electron to travel the 2.00 cm and (b) the vertical distance it travels during that time. Also find the magnitudes of the (c) horizontal and (d) vertical velocity components of the electron as it emerges. ssm

22E. In the 1991 World Track and Field Championships in Tokyo, Mike Powell (Fig. 4-32) jumped 8.95 m, breaking the 23-year long-jump record set by Bob Beamon by a full 5 cm. Assume that Powell's speed on takeoff was 9.5 m/s (about equal to that of a sprinter) and that $g = 9.80$ m/s² in Tokyo. How much less was Powell's horizontal range than the maximum possible horizontal range (neglecting the effects of air) for a particle launched at the same speed of 9.5 m/s?

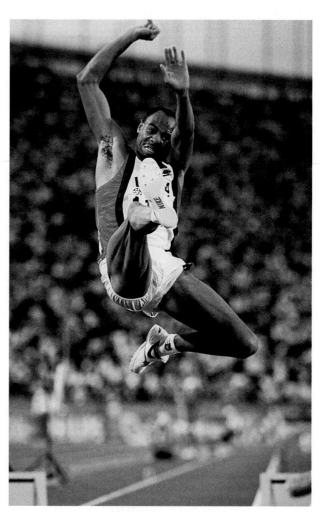

Fig. 4-32 Exercise 22.

23E. A stone is catapulted at time $t = 0$, with an initial velocity of magnitude 20.0 m/s and at an angle of 40.0° above the horizontal. What are the magnitudes of the (a) horizontal and (b) vertical components of its displacement from the catapult site at $t = 1.10$ s? Repeat for the (c) horizontal and (d) vertical components at $t = 1.80$ s, and for the (e) horizontal and (f) vertical components at $t = 5.00$ s. ssm

24P. A golf ball is struck at ground level. The speed of the golf ball as a function of the time is shown in Fig. 4-33, where $t = 0$ at the instant the ball is struck. (a) How far does the golf ball travel horizontally before returning to ground level? (b) What is the maximum height above ground level attained by the ball?

Fig. 4-33 Problem 24.

25P. A rifle that shoots bullets at 460 m/s is to be aimed at a target 45.7 m away and level with the rifle. How high above the target must the rifle barrel be pointed so that the bullet hits the target? ssm

26P. The pitcher in a slow-pitch softball game releases the ball at a point 3.0 ft above ground level. A stroboscopic plot of the position of the ball is shown in Fig. 4-34, where the readings are 0.25 s apart and the ball is released at $t = 0$. (a) What is the initial speed of the ball? (b) What is the speed of the ball at the instant it reaches its maximum height above ground level? (c) What is that maximum height?

Fig. 4-34 Problem 26.

27P. Show that the maximum height reached by a projectile is $y_{max} = (v_0 \sin \theta_0)^2/2g$. ssm www

28P. You throw a ball toward a wall with a speed of 25.0 m/s and at an angle of 40.0° above the horizontal (Fig. 4-35). The wall is 22.0 m from the release point of the ball. (a) How far above the release point does the ball hit the wall? (b) What are the horizontal and vertical components of its velocity as it hits the wall? (c) When it hits, has it passed the highest point on its trajectory?

Fig. 4-35 Problem 28.

29P. A ball is shot from the ground into the air. At a height of 9.1 m, its velocity is observed to be $\vec{v} = 7.6\hat{i} + 6.1\hat{j}$ in meters per second ($\hat{i}$ horizontal, $\hat{j}$ upward). (a) To what maximum height does the ball rise? (b) What total horizontal distance does the ball travel? What are (c) the magnitude and (d) the direction of the ball's velocity just before it hits the ground? ilw

30P. Two seconds after being projected from ground level, a projectile is displaced 40 m horizontally and 53 m vertically above its point of projection. What are the (a) horizontal and (b) vertical components of the initial velocity of the projectile? (c) At the instant the projectile achieves its maximum height above ground level, how far is it displaced horizontally from its point of projection?

31P. A football player punts the football so that it will have a "hang time" (time of flight) of 4.5 s and land 46 m away. If the ball leaves the player's foot 150 cm above the ground, what must be (a) the magnitude and (b) the direction of the ball's initial velocity? ssm

32P. A lowly high diver pushes off horizontally with a speed of 2.00 m/s from the edge of a platform that is 10.0 m above the surface of the water. (a) At what horizontal distance from the edge of the platform is the diver 0.800 s after pushing off? (b) At what vertical distance above the surface of the water is the diver just then? (c) At what horizontal distance from the edge of the platform does the diver strike the water?

33P. A certain airplane has a speed of 290.0 km/h and is diving at an angle of 30.0° below the horizontal when the pilot releases a radar decoy (Fig. 4-36). The horizontal distance between the release point and the point where the decoy strikes the ground is 700 m. (a) How long is the decoy in the air? (b) How high was the released point? ilw

Fig. 4-36 Problem 33.

34P. The launching speed of a certain projectile is five times the speed it has at its maximum height. Calculate the elevation angle θ_0 at launching.

35P. A ball rolls horizontally off the top of a stairway with a speed of 1.52 m/s. The steps are 20.3 cm high and 20.3 cm wide. Which step does the ball hit first? ssm

36P. A soccer ball is kicked from the ground with an initial speed of 19.5 m/s at an upward angle of 45°. A player 55 m away in the direction of the kick starts running to meet the ball at that instant. What must be his average speed if he is to meet the ball just before it hits the ground? Neglect air resistance.

37P. An airplane, diving at an angle of 53.0° with the vertical, releases a projectile at an altitude of 730 m. The projectile hits the ground 5.00 s after being released. (a) What is the speed of the aircraft? (b) How far did the projectile travel horizontally during its flight? What were the (c) horizontal and (d) vertical components of its velocity just before striking the ground? ssm

38P. For women's volleyball the top of the net is 2.24 m above the floor and the court measures 9.0 m by 9.0 m on each side of the net. Using a jump serve, a player strikes the ball at a point that is 3.0 m above the floor and a horizontal distance of 8.0 m from the net. If the initial velocity of the ball is horizontal, (a) what minimum magnitude must it have if the ball is to clear the net and (b) what maximum magnitude can it have if the ball is to strike the floor inside the back line on the other side of the net?

39P. A batter hits a pitched ball when the center of the ball is 1.22 m above the ground. The ball leaves the bat at an angle of 45° with the ground. With that launch, the ball should have a horizontal range (returning to the *launch* level) of 107 m. (a) Does the ball clear a 7.32-m-high fence that is 97.5 m horizontally from the launch point? (b) Either way, find the distance between the top of the fence and the center of the ball when the ball reaches the fence. ssm www

40P. During a tennis match, a player serves the ball at 23.6 m/s, with the center of the ball leaving the racquet horizontally 2.37 m above the court surface. The net is 12 m away and 0.90 m high. When the ball reaches the net, (a) does the ball clear it and (b) what is the distance between the center of the ball and the top of the net? Suppose that, instead, the ball is served as before but now it leaves the racquet at 5.00° below the horizontal. When the ball reaches the net, (c) does the ball clear it and (d) what now is the distance between the center of the ball and the top of the net?

41P. A football kicker can give the ball an initial speed of 25 m/s. Within what two elevation angles must he kick the ball to score a field goal from a point 50 m in front of goalposts whose horizontal bar is 3.44 m above the ground? (If you want to work this out algebraically, use $\sin^2 \theta + \cos^2 \theta = 1$ to get a relation between $\tan^2 \theta$ and $1/\cos^2 \theta$, substitute, and then solve the resulting quadratic equation.) ssm

SEC. 4-7 Uniform Circular Motion

42E. What is the magnitude of the acceleration of a sprinter running at 10 m/s when rounding a turn with a radius of 25 m?

43E. An Earth satellite moves in a circular orbit 640 km above Earth's surface with a period of 98.0 min. What are (a) the speed and (b) the magnitude of the centripetal acceleration of the satellite? ssm

44E. A rotating fan completes 1200 revolutions every minute. Consider the tip of a blade, at a radius of 0.15 m. (a) Through what distance does the tip move in one revolution? What are (b) the tip's speed and (c) the magnitude of its acceleration? (d) What is the period of the motion?

45E. An astronaut is rotated in a horizontal centrifuge at a radius of 5.0 m. (a) What is the astronaut's speed if the centripetal acceleration has a magnitude of 7.0g? (b) How many revolutions per minute are required to produce this acceleration? (c) What is the period of the motion? ssm

46P. A carnival merry-go-round rotates about a vertical axis at a constant rate. A passenger standing on the edge of the merry-go-round has a constant speed of 3.66 m/s. For each of the following instantaneous situations, state how far the passenger is from the center of the merry-go-round, and in which direction. (a) The passenger has an acceleration of 1.83 m/s², east. (b) The passenger has an acceleration of 1.83 m/s², south.

47P. (a) What is the magnitude of the centripetal acceleration of an object on Earth's equator owing to the rotation of Earth? (b) What would the period of rotation of Earth have to be for objects on the equator to have a centripetal acceleration with a magnitude of 9.8 m/s²? ssm www

48P. The fast French train known as the TGV (Train à Grande Vitesse) has a scheduled average speed of 216 km/h. (a) If the train goes around a curve at that speed and the magnitude of the acceleration experienced by the passengers is to be limited to 0.050g, what is the smallest radius of curvature for the track that can be tolerated? (b) At what speed must the train go around a curve with a 1.00 km radius to be at the acceleration limit?

49P. A carnival Ferris wheel has a 15 m radius and completes five turns about its horizontal axis every minute. (a) What is the period of the motion? What is the centripetal acceleration of a passenger at (b) the highest point and (c) the lowest point, assuming the passenger is at a 15 m radius? ssm ilw

50P. When a large star becomes a *supernova,* its core may be compressed so tightly that it becomes a *neutron star,* with a radius of about 20 km (about the size of the San Francisco area). If a neutron star rotates once every second, (a) what is the speed of a particle on the star's equator and (b) what is the magnitude of the particle's centripetal acceleration? (c) If the neutron star rotates faster, do the answers to (a) and (b) increase, decrease, or remain the same?

51P. A boy whirls a stone in a horizontal circle of radius 1.5 m and at height 2.0 m above level ground. The string breaks, and the stone flies off horizontally and strikes the ground after traveling a horizontal distance of 10 m. What is the magnitude of the centripetal acceleration of the stone while in circular motion? ssm

52P. A particle *P* travels with constant speed on a circle of radius $r = 3.00$ m (Fig. 4-37) and completes one revolution in 20.0 s. The particle passes through *O* at time $t = 0$. State the following vectors in magnitude-angle notation (angle relative to the positive direction of *x*). With respect to *O*, find the particle's position vector at the times *t* of (a) 5.00 s, (b) 7.50 s, and (c) 10.0 s.

  (d) For the 5.00 s interval from the end of the fifth second to the end of the tenth second, find the particle's displacement. (e) For the same interval, find its average velocity. Find its velocity at (f) the beginning and (g) the end of that 5.00 s interval. Next, find the acceleration at (h) the beginning and (i) the end of that interval.

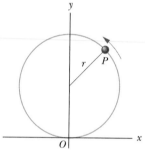

Fig. 4-37 Problem 52.

SEC. 4-8 Relative Motion in One Dimension

53E. A cameraman on a pickup truck is traveling westward at 20 km/h while he videotapes a cheetah that is moving westward 30 km/h faster than the truck. Suddenly, the cheetah stops, turns, and then runs at 45 km/h eastward, as measured by a suddenly nervous crew member who stands alongside the cheetah's path. The change in the animal's velocity takes 2.0 s. What is its acceleration from the perspective of (a) the cameraman and (b) the nervous crew member?

54E. A boat is traveling upstream at 14 km/h with respect to the water of a river. The water is flowing at 9 km/h with respect to the ground. (a) What is the velocity of the boat with respect to the ground? (b) A child on the boat walks from front to rear at 6 km/h with respect to the boat. What is the child's velocity with respect to the ground?

55P. A person walks up a stalled 15-m-long escalator in 90 s. When standing on the same escalator, now moving, the person is carried up in 60 s. How much time would it take that person to walk up the moving escalator? Does the answer depend on the length of the escalator? ssm

SEC. 4-9   Relative Motion in Two Dimensions

56E. In rugby a player can legally pass the ball to a teammate as long as the pass is not "forward" (it must not have a velocity component parallel to the length of the field and directed toward the other team's goal). Suppose a player runs parallel to the field's length and toward the other team's goal with a speed of 4.0 m/s while he passes the ball with a speed of 6.0 m/s relative to himself. What is the smallest angle from the forward direction that keeps the pass legal?

57E. Snow is falling vertically at a constant speed of 8.0 m/s. At what angle from the vertical do the snowflakes appear to be falling as viewed by the driver of a car traveling on a straight, level road with a speed of 50 km/h? ssm

58E. Two highways intersect as shown in Fig. 4-38. At the instant shown, a police car P is 800 m from the intersection and moving at 80 km/h. Motorist M is 600 m from the intersection and moving at 60 km/h. (a) In unit-vector notation, what is the velocity of the motorist with respect to the police car? (b) For the instant shown in Fig. 4-38, how does the direction of the velocity found in (a) compare to the line of sight between the two cars? (c) If the cars maintain their velocities, do the answers to (a) and (b) change as the cars move nearer the intersection?

Fig. 4-38 Exercise 58.

59P. A train travels due south at 30 m/s (relative to the ground) in a rain that is blown toward the south by the wind. The path of each raindrop makes an angle of 70° with the vertical, as measured by an observer stationary on the ground. An observer on the train, however, sees the drops fall perfectly vertically. Determine the speed of the raindrops relative to the ground. ssm

60P. Ship A is located 4.0 km north and 2.5 km east of ship B. Ship A has a velocity of 22 km/h toward the south and ship B has a velocity of 40 km/h in a direction 37° north of east. (a) What is the velocity of A relative to B? (Express your answer in terms of the unit vectors $\hat{i}$ and $\hat{j}$, where $\hat{i}$ is toward the east.) (b) Write an expression (in terms of $\hat{i}$ and $\hat{j}$) for the position of A relative to B as a function of t, where $t = 0$ when the ships are in the positions described above. (c) At what time is the separation between the ships least? (d) What is that least separation?

61P. Two ships, A and B, leave port at the same time. Ship A travels northwest at 24 knots and ship B travels at 28 knots in a direction

40° west of south. (1 knot = 1 nautical mile per hour; see Appendix D.) (a) What are the magnitude and direction of the velocity of ship A relative to B? (b) After what time will the ships be 160 nautical miles apart? (c) What will be the bearing of B (the direction of the position of B) relative to A at that time? ssm ilw

62P. A wooden boxcar is moving along a straight railroad track at speed v_1. A sniper fires a bullet (initial speed v_2) at it from a high-powered rifle. The bullet passes through both lengthwise walls of the car, its entrance and exit holes being exactly opposite each other as viewed from within the car. From what direction, relative to the track, is the bullet fired? Assume that the bullet is not deflected upon entering the car, but that its speed decreases by 20%. Take $v_1 = 85$ km/h and $v_2 = 650$ m/s. (Why don't you need to know the width of the boxcar?)

63P. A 200-m-wide river has a uniform flow speed of 1.1 m/s through a jungle and toward the east. An explorer wishes to leave a small clearing on the south bank and cross the river in a power-boat that moves at a constant speed of 4.0 m/s with respect to the water. There is a clearing on the north bank 82 m upstream from a point directly opposite the clearing on the south bank. (a) In what direction must the boat be pointed in order to travel in a straight line and land in the clearing on the north bank? (b) How long will the boat take to cross the river and land in the clearing?

Additional Problem

64. *Curtain of death.* A large metallic asteroid strikes Earth and quickly digs a crater into the rocky material below ground level by launching rocks upward and outward. The following table gives five pairs of launch speeds and angles (from the horizontal) for such rocks, based on a model of crater formation. (Other rocks, with intermediate speeds and angles, are also launched.) Suppose that you are at $x = 20$ km when the asteroid strikes the ground at time $t = 0$ and position $x = 0$ (Fig. 4-39). (a) At $t = 20$ s, what are the x and y coordinates of the rocks headed in your direction from launches A through E? (b) Plot these coordinates and then sketch a curve through the points to include rocks with intermediate launch speeds and angles. The curve should give you an idea of what you would see as you look up into the approaching rocks and what dinosaurs must have seen during asteroid strikes long ago.

Launch	Speed (m/s)	Angle (degrees)
A	520	14.0
B	630	16.0
C	750	18.0
D	870	20.0
E	1000	22.0

Fig. 4-39 Problem 64.

NEW PROBLEMS

N1. You are kidnapped by armed political-science majors (who are upset because you told them that political science is not a real science). Although blindfolded, you can tell the speed of their car (by the whine of the engine), the time of travel (by mentally counting off seconds), and the direction of travel (by turns along the rectangular street system). From these clues, you know that you are taken along the following course: 50 km/h for 2.0 min, turn 90° to the right, 20 km/h for 4.0 min, turn 90° to the right, 20 km/h for 60 s, turn 90° to the left, 50 km/h for 60 s, turn 90° to the right, 20 km/h for 2.0 min, turn 90° to the left, 50 km/h for 30 s. At that point, (a) how far are you from your starting point and (b) in what direction relative to your initial direction of travel are you?

N2. In Fig. 4N-1a, a sled moves in the negative x direction at speed v_s while a ball of ice is shot from the sled with a velocity $\vec{v}_0 = v_{0x}\hat{i} + v_{0y}\hat{j}$ relative to the sled. When the ball lands, its horizontal displacement Δx_{bg} relative to the ground (from its launch position to its landing position) is measured. Figure 4N-1b gives Δx_{bg} as a function of v_s. Assume it lands at approximately its launch height. What are the values of (a) v_{0x} and (b) v_{0y}?

The ball's displacement Δx_{bs} relative to the sled can also be measured. Assume that the sled's velocity is not changed when the ball is shot. What is Δx_{bs} when v_s is (c) 5.0 m/s and (d) 15 m/s?

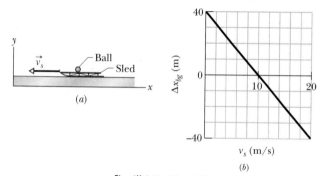

(a)

(b)

Fig. 4N-1 Problem N2.

N3. A third baseman wishes to throw to first base, 127 ft distant. His best throwing speed is 85 mi/h. (a) If he throws the ball horizontally 3.0 ft above the ground, how far from first base will it hit the ground? (b) From the same initial height, at what upward angle must the third baseman throw the ball if the first baseman is to catch it 3.0 ft above the ground? (c) What will be the time of flight in that case?

N4. A cat rides a merry-go-round while turning with uniform circular motion. At time $t_1 = 2.00$ s, the cat's velocity is

$$\vec{v}_1 = (3.00 \text{ m/s})\hat{i} + (4.00 \text{ m/s})\hat{j},$$

measured on a horizontal xy coordinate system. At time $t_2 = 5.00$ s, its velocity is

$$\vec{v}_2 = (-3.00 \text{ m/s})\hat{i} + (-4.00 \text{ m/s})\hat{j}.$$

What are (a) the magnitude of the cat's centripetal acceleration and (b) the cat's average acceleration during the time interval $t_2 - t_1$?

N5. In a detective story, a body is found 4.6 m from the base of a building and 24 m below an open window. (a) Assuming the victim left that window horizontally, what was the victim's speed just then? (b) Would you guess the death to be accidental? Explain your answer.

N6. In the overhead view of Fig. 4N-2, Jeeps P and B race along straight lines, across flat terrain, and past stationary border guard A. Relative to the guard, Jeep B travels at a constant speed of 20.0 m/s, at the angle $\theta_2 = 30.0°$. Relative to the guard, Jeep P has accelerated from rest at a constant rate of 0.40 m/s^2 at the angle $\theta_1 = 60.0°$. At a certain time during the acceleration, it has a speed of 40 m/s. At that time, and in magnitude-angle notation, what are (a) the velocity of Jeep P relative to Jeep B and (b) the acceleration of Jeep P relative to Jeep B?

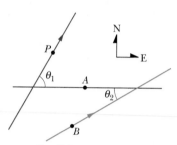

Fig. 4N-2 Problem N6.

N7. In Galileo's *Two New Sciences*, the author states that "for elevations [angles of projection] which exceed or fall short of 45° by equal amounts, the ranges are equal. . . ." Prove this statement. (See Fig. 4N-3.)

Fig. 4N-3 Problem N7.

N8. A particle moves horizontally in uniform circular motion, over a horizontal xy plane. At one instant, it moves through the point at coordinates (4.00 m, 4.00 m) with a velocity of $-5.00\hat{i}$ m/s and an acceleration of $+12.5\hat{j}$ m/s^2. What are the coordinates of the center of the circular path?

N9. A hang glider is 7.5 m above ground level with a velocity of 8.0 m/s at an angle of 30° below the horizontal and a constant acceleration of 1.0 m/s^2, up. (a) Assume $t = 0$ at the instant just described and write an equation for the elevation y of the hang glider as a function of t, with $y = 0$ at ground level. (b) Use the equation to determine the value of t when $y = 0$. (c) Explain why there are two solutions to part (b). Which one represents the time it takes the hang glider to reach ground level? (d) How far does the hang glider travel horizontally during the interval between $t = 0$ and the time it reaches the ground? (e) For the same initial position and velocity, what constant acceleration will cause the hang glider to reach ground level with zero velocity? Express your answer in terms of the unit vectors $\hat{i}$ (horizontal, in the direction of travel) and $\hat{j}$ (upward).

N10. A ball is to be shot from level ground toward a wall at distance x (Fig. 4N-4a). Figure 4N-4b shows the y component v_y of the ball's velocity just as it would reach the wall, as a function of that distance x. What is the launch angle θ_0 of the ball?

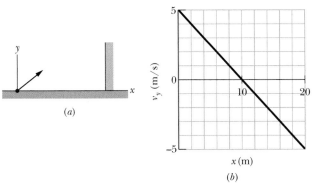

Fig. 4N-4 Problem N10.

N11. A sprinter runs at 9.2 m/s around a circular track with a centripetal acceleration of magnitude 3.8 m/s². (a) What is the track radius? (b) What is the period of the motion?

N12. The position vector $\vec{r} = 5.00t\hat{i} + (et + ft^2)\hat{j}$ locates a particle as a function of time t. Vector $\vec{r}$ is in meters, t is in seconds, and factors e and f are constants. Figure 4N-5 gives the angle θ of the particle's direction of travel as a function of time (θ is measured from the positive direction of the x axis). What are (a) factor e and (b) factor f, including units?

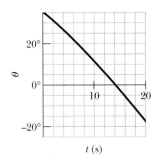

Fig. 4N-5 Problem N12.

N13. After flying for 15 min in a wind blowing 42 km/h at an angle of 20° south of east, an airplane pilot is over a town that is 55 km due north of the starting point. What is the speed of the airplane relative to the air?

N14. A ball is to be shot from level ground with a certain speed. Figure 4N-6 shows the range R it will have versus the launch angle θ_0 at which it can be launched. The choice of θ_0 determines the flight time; let t_{max} represent the maximum flight time. What is the least speed the ball will have during its flight if θ_0 is chosen such that the flight time is $0.500t_{max}$?

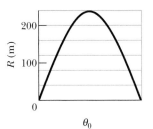

Fig. 4N-6 Problem N14.

N15. At one instant a butterfly has the position vector $\vec{D}_i = (2.00 \text{ m})\hat{i} + (3.00 \text{ m})\hat{j} + (1.00 \text{ m})\hat{k}$ relative to the base of a birdbath. Then 40.0 s later, the butterfly has the position vector $\vec{D}_f = (3.00 \text{ m})\hat{i} + (1.00 \text{ m})\hat{j} + (2.00 \text{ m})\hat{k}$. Find (if possible) (a) the displacement in unit-vector notation, (b) the magnitude of the displacement, (c) the average velocity in unit-vector notation, and (d) the average speed of the butterfly during the 40.0 s interval.

N16. Figure 4N-7 gives the path of a squirrel moving about on level ground, from point A (at time $t = 0$), to points B (at $t = 5.00$ min), C (at $t = 10.0$ min), and finally D (at $t = 15.0$ min). Consider the average velocities of the squirrel from point A to each of the other three points. (a) Of those three average velocities, which has the least magnitude, and what is the average velocity in magnitude-angle notation? (b) Which has the greatest magnitude, and what is the average velocity in magnitude-angle notation?

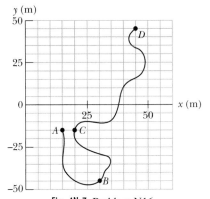

Fig. 4N-7 Problem N16.

N17. A light plane attains an airspeed of 500 km/h. The pilot sets out for a destination 800 km to the north but discovers that the plane must be headed 20.0° east of north to fly there directly. The plane arrives in 2.00 h. What was the wind velocity?

N18. At time $t_1 = 2.00$ s, the acceleration of a particle in counterclockwise circular motion is $(6.00 \text{ m/s}^2)\hat{i} + (4.00 \text{ m/s}^2)\hat{j}$. It moves at constant speed. At time $t_2 = 5.00$ s, its acceleration is $(4.00 \text{ m/s}^2)\hat{i} + (-6.00 \text{ m/s}^2)\hat{j}$. What is the radius of the path taken by the particle?

N19. A baseball is hit at ground level. The ball reaches its maximum height above ground level 3.0 s after being hit. Then 2.5 s after reaching its maximum height, the ball barely clears a fence that is 97.5 m from where it was hit. Assume the ground is level. (a) What maximum height above ground level is reached by the ball? (b) How high is the fence? (c) How far beyond the fence does the ball strike the ground?

N20. A centripetal-acceleration addict rides in uniform circular motion with period $T = 2.0$ s and radius $r = 3.00$ m. At one instant his acceleration is $\vec{a} = (6.00 \text{ m/s}^2)\hat{i} + (-4.00 \text{ m/s}^2)\hat{j}$. At that instant, what are the values of (a) $\vec{v} \cdot \vec{a}$ and (b) $\vec{r} \times \vec{a}$?

N21. The position $\vec{r}$ of a particle moving in the xy plane is given by $\vec{r} = 2t\hat{i} + 2 \sin[(\pi/4 \text{ rad/s})t]\hat{j}$, where $\vec{r}$ is in meters and t is in seconds. (a) Calculate the x and y components of the particle's position at $t = 0$, 1.0, 2.0, 3.0, and 4.0 s and sketch the particle's path in the xy plane for the interval $0 \le t \le 4.0$ s. (b) Calculate the components of the particle's velocity at $t = 1.0$, 2.0, and 3.0 s. Show that the velocity is tangent to the path of the particle and in the direction the particle is moving at each time by drawing the velocity vectors on the plot of the particle's path in part (a). (c) Calculate the components of the particle's acceleration at $t = 1.0$, 2.0, and 3.0 s.

N22. A suspicious-looking man runs as fast as he can along a moving sidewalk from one end to the other, taking 2.50 s. Then security agents appear and the man runs as fast as he can back along the sidewalk to his starting point, taking 10.0 s. What is the ratio of the man's running speed to the sidewalk's speed?

N23. A particle moves along a circular path over a horizontal xy coordinate system, at constant speed. At time $t_1 = 4.00$ s, it is at point (5.00 m, 6.00 m) with velocity $(3.00 \text{ m/s})\hat{j}$ and acceleration in the positive x direction. At time $t_2 = 10.0$ s, it has velocity $(-3.00 \text{ m/s})\hat{i}$ and acceleration in the positive y direction. What are the coordinates of the center of the circular path?

N24. In Fig. 4N-8, a lump of wet putty moves in uniform circular motion as it rides at a radius of 20.0 cm on the rim of a wheel rotating counterclockwise with a period of 5.00 ms. The lump then happens to fly off the rim at the 5:00 position (as if on a clock face). It leaves the rim at a height of $h = 1.20$ m from the floor and at a distance $d = 2.50$ m from a wall. At what height on the wall does the lump hit?

Fig. 4N-8 Problem N24.

N25. Long flights at mid latitudes in the northern hemisphere encounter the jet stream, a flow of rapidly moving air that is generally eastward. The speed of an aircraft relative to Earth's surface can be affected by that flow. If the pilot maintains a certain speed relative to the air (the *airspeed* of the aircraft), then the speed relative to the surface (*groundspeed*) is considerably more when the pilot

flies in the direction of the jet stream than in the opposite direction. Suppose an outgoing flight and its corresponding return flight are scheduled between two cities separated by 4000 km, with the outgoing flight in the direction of the jet stream and the return flight in the opposite direction. The airline computer advises the pilots to fly at an airspeed of 1000 km/h, for which the difference in flight times for the outgoing and return flights should be 70.0 min. What speed for the jet stream is the computer using?

N26. The acceleration of a particle on a horizontal xy plane is given by $\vec{a} = 3t\hat{i} + 4t\hat{j}$, where $\vec{a}$ is in meters per second-squared and t is in seconds. At $t = 0$, the particle has the position vector $\vec{r} = (20.0 \text{ m})\hat{i} + (40.0 \text{ m})\hat{j}$ and the velocity vector $\vec{v} = (5.00 \text{ m/s})\hat{i} + (2.00 \text{ m/s})\hat{j}$. (a) What is the position vector of the particle at $t = 4.00$ s, and (b) what is its direction just then?

N27. Figure 4N-9 shows the path taken by my drunk skunk over level ground, from initial point i to final point f. The angles are $\theta_1 = 30.0°$, $\theta_2 = 50.0°$, and $\theta_3 = 80.0°$, and the distances are $d_1 = 5.00$ m, $d_2 = 8.00$ m, and $d_3 = 12.0$ m. In magnitude-angle notation, what is the skunk's displacement from i to f?

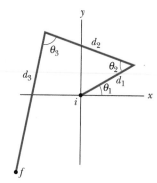

Fig. 4N-9 Problem N27.

N28. A particle is in uniform circular motion about the origin of an xy coordinate system, moving clockwise with a period of 7.00 s. At one instant, its position vector (from the origin) is $\vec{r} = (2.00 \text{ m})\hat{i} - (3.00 \text{ m})\hat{j}$. At that instant, what is its velocity in unit-vector notation?

N29. You are to ride a jet-cycle over a lake, starting from rest at point i: First, moving at 30° north of due east:

1. increase your speed at 0.400 m/s² for 6.00 s
2. with whatever speed you then have, move for 8.00 s
3. then slow at 0.400 m/s² for 6.00 s.

Immediately next, moving due west:

4. increase your speed at 0.400 m/s² for 5.00 s
5. with whatever speed you then have, move for 10.0 s
6. then slow at 0.400 m/s² until you stop.

In magnitude-angle notation, what then is your average velocity for the trip from point i?

N30. In Fig. 4N-10, a ball is shot directly upward from the ground with an initial speed of $v_0 = 7.00$ m/s. Simultaneously, a construction elevator cab begins to move upward from the ground with a constant speed $v_c = 3.00$ m/s. What maximum height does the ball reach relative to (a) the ground and (b) the floor of the elevator cab? What is the rate at which the speed of the ball changes relative to (c) the ground and (d) the cab floor?

Fig. 4N-10 Problem N30.

N31. A golfer tees off from the top of a rise, giving the golf ball an initial velocity of 43 m/s at an angle of 30° above the horizontal. The ball strikes the fairway a horizontal distance of 180 m from the tee. Assume the fairway is level. (a) How high is the rise above the fairway? (b) What is the speed of the ball as it strikes the fairway?

N32. In Fig. 4N-11, a ball is launched with a velocity of 10 m/s, at an angle of 50° to the horizontal. The launch point is at the base of a ramp of horizontal length $d_1 = 6.00$ m and height $d_2 = 3.60$ m. A plateau is located at the top of the ramp. (a) Does the ball land on the ramp or the plateau? When it lands, what are (b) the magnitude and (c) the angle of its displacement from the launch point?

Fig. 4N-11 Problem N32.

N33. A purse at radius 2.00 m and a wallet at radius 3.00 m travel in uniform circular motion on the floor of a merry-go-round as the ride turns. They are on the same radial line. At one instant, the acceleration of the purse is $(2.00 \text{ m/s}^2)\hat{i} + (4.00 \text{ m/s}^2)\hat{j}$. At that time and in unit-vector notation, what is the acceleration of the wallet?

N34. *A graphing surprise*: At time $t = 0$, a burrito is launched from level ground, with an initial speed of 16.0 m/s, at an initial angle θ_0. Imagine a position vector $\vec{r}$ continuously directed from the launching point to the burrito during the flight. Graph the magnitude r of the position vector for launch angles (a) $\theta_0 = 40°$ and (b) $\theta_0 = 80°$. For $\theta_0 = 40°$, (c) when does r reach its maximum value, (d) what is that value, and (e) where is the burrito? For $\theta_0 = 80°$, (f) when does r reach its maximum value, (g) what is that value, and (h) where is the burrito?

N35. Figure 4N-12 shows the straight path of a particle across an xy coordinate system as it is accelerated from rest during time interval Δt_1. The acceleration is constant. The xy coordinates for point A are (4.00 m, 6.00 m); those for point B are (12.0 m, 18.0 m). (a) What is the ratio a_y/a_x of the acceleration components? (b) What are the coordinates of the particle if the motion is continued for another interval of Δt_1?

Fig. 4N-12 Problem N35.

N36. A cart is propelled over an xy plane with acceleration components $a_x = 4.0 \text{ m/s}^2$ and $a_y = -2.0 \text{ m/s}^2$. Its initial velocity has components $v_{0x} = 8$ m/s and $v_{0y} = 12$ m/s. In unit-vector notation, what is the velocity of the cart when it reaches its greatest y coordinate?

N37. A moderate wind accelerates a smooth pebble over a horizontal xy plane with a constant acceleration

$$\vec{a} = (5.00 \text{ m/s}^2)\hat{i} + (7.00 \text{ m/s}^2)\hat{j}.$$

At time $t = 0$, its velocity is $(4.00 \text{ m/s})\hat{i}$. In magnitude-angle notation, what is its velocity when it has been displaced by 12.0 m parallel to the x axis?

5 Force and Motion—I

On April 4, 1974, John Massis of Belgium managed to move two passenger cars belonging to New York's Long Island Railroad. He did so by clamping his teeth down on a bit that was attached to the cars with a rope and then leaning backward while pressing his feet against the railway ties. The cars together weighed about 80 tons.

Did Massis have to pull with superhuman force to accelerate them?

The answer is in this chapter.

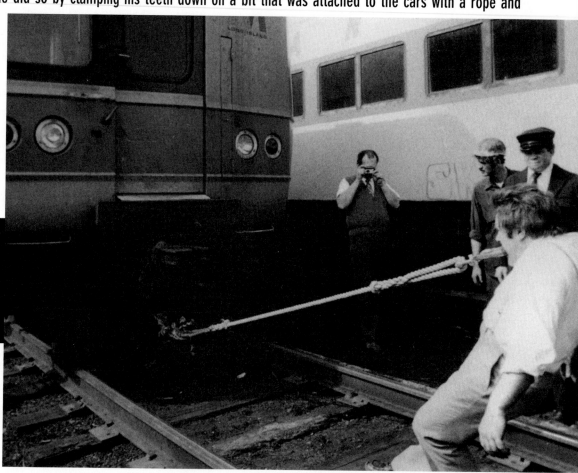

5-1 What Causes an Acceleration?

If you see the velocity of a particle-like body change in either magnitude or direction, you know that something must have *caused* that change (that acceleration). Indeed, out of common experience, you know that the change in velocity must be due to an interaction between the body and something in its surroundings. For example, if you see a hockey puck that is sliding across an ice rink suddenly stop or suddenly change direction, you will suspect that the puck encountered a slight ridge on the ice surface.

An interaction that can cause an acceleration of a body is called a **force**, which is, loosely speaking, a push or pull on the body—the force is said to *act* on the body. For example, the bump on the hockey puck by the ridge is a push on the puck, causing an acceleration. The relationship between a force and the acceleration it causes was first understood by Isaac Newton (1642–1727) and is the subject of this chapter. The study of that relationship, as Newton presented it, is called *Newtonian mechanics.* We shall focus on its three primary laws of motion.

Newtonian mechanics does not apply to all situations. If the speeds of the interacting bodies are very large—an appreciable fraction of the speed of light—we must replace Newtonian mechanics with Einstein's special theory of relativity, which holds at any speed, including those near the speed of light. If the interacting bodies are on the scale of atomic structure (for example, they might be electrons within an atom), we must replace Newtonian mechanics with quantum mechanics. Physicists now view Newtonian mechanics as a special case of these two more comprehensive theories. Still, it is a very important special case because it applies to the motion of objects ranging in size from the very small (almost on the scale of atomic structure) to astronomical (objects such as galaxies and clusters of galaxies).

5-2 Newton's First Law

Before Newton formulated his mechanics, it was thought that some influence, a "force," was needed to keep a body moving at constant velocity. Similarly, a body was thought to be in its "natural state" when it was at rest. For it to move with constant velocity, it seemingly had to be propelled in some way, by a push or a pull. Otherwise, it would "naturally" stop moving.

These ideas were reasonable. If you send a puck sliding across a wooden floor, it does indeed slow and then stop. If you want to make it move across the floor with constant velocity, you have to continuously pull or push it.

Send a puck sliding over the ice of a skating rink, however, and it goes a lot farther. You can imagine longer and more slippery surfaces, over which the puck would slide farther and farther. In the limit you can think of a long, extremely slippery surface (said to be a **frictionless surface**), over which the puck would hardly slow. (We can in fact come close to this situation in the laboratory, by sending a puck sliding over a horizontal air table, across which it moves on a film of air.)

From these observations, we can conclude that a body will keep moving with constant velocity if no force acts on it. That leads us to the first of Newton's three laws of motion:

> **Newton's First Law:** If no force acts on a body, then the body's velocity cannot change; that is, the body cannot accelerate.

In other words, if the body is at rest, it stays at rest. If it is moving, it will continue to move with the same velocity (same magnitude *and* same direction).

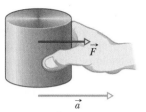

Fig. 5-1 A force $\vec{F}$ on the standard kilogram gives that body an acceleration $\vec{a}$.

5-3 Force

We now wish to define the unit of force. We know that a force can cause the acceleration of a body. Thus, we shall define the unit of force in terms of the acceleration that a force gives to a standard reference body. As the standard body, we shall use (or rather imagine that we use) the standard kilogram of Fig. 1-4. This body has been assigned, exactly and by definition, a mass of 1 kg.

We put the standard body on a horizontal frictionless table and pull the body to the right (Fig. 5-1) so that by trial and error, it eventually experiences a measured acceleration of 1 m/s². We then declare, as a matter of definition, that the force we are exerting on the standard body has a magnitude of 1 newton (abbreviated N).

We can exert a 2 N force on our standard body by pulling it so that its measured acceleration is 2 m/s², and so on. Thus in general, if our standard body of 1 kg mass has an acceleration of magnitude a, we know that force F must be acting on it and that the magnitude of the force (in newtons) is equal to the magnitude of the acceleration (in meters per second per second).

Thus, a force is measured by the acceleration it produces. However, acceleration is a vector quantity, with both magnitude and direction. Is force also a vector quantity? We can easily assign a direction to a force (just assign the direction of the acceleration), but that is not sufficient. We must prove by experiment that forces are vector quantities. Actually, that has been done: forces are indeed vector quantities; they have magnitudes and directions and they combine according to the vector rules of Chapter 3.

This means that when two or more forces act on a body, we can find their **net force** or **resultant force** by adding the individual forces vectorially. A single force with the magnitude and direction of the net force has the same effect on the body as all the individual forces together. This fact is called the **principle of superposition for forces.** The world would be quite strange if, for example, you and a friend were to pull on the standard body in the same direction, each with a force of 1 N, and yet somehow the net pull was 14 N.

In this book, forces are most often represented with a vector symbol such as $\vec{F}$, and a net force is represented with the vector symbol $\vec{F}_{net}$. As with other vectors, a force or a net force can have components along coordinate axes. When forces act only along a single axis, they are single-component forces. Then we can drop the overhead arrows on the force symbols and just use signs to indicate the directions of the forces along that axis.

Instead of what was previously given, the proper statement of Newton's First Law is in terms of a *net* force:

> **Newton's First Law:** If no *net* force acts on a body ($\vec{F}_{net} = 0$), then the body's velocity cannot change; that is, the body cannot accelerate.

There may be multiple forces acting on a body, but if their net (or resultant) force is zero, then the body cannot accelerate.

Inertial Reference Frames

Newton's first law is not true in all reference frames, but we can always find reference frames in which it (and the rest of Newtonian mechanics) is true. Such frames are called **inertial reference frames,** or simply **inertial frames.**

> An inertial reference frame is one in which Newton's laws hold.

For example, we can assume that the ground is an inertial frame provided that we can neglect Earth's actual astronomical motions (such as its rotation).

Fig. 5-2 (*a*) The path of a puck that is sent sliding due south along a long strip of frictionless ice, as seen by an observer on the ground. (*b*) The ground beneath the southward sliding puck rotates to the east as Earth rotates.

That assumption works well if we, say, send a puck sliding along a short strip of frictionless ice—an observer on the ground would find that the puck's motion obeys the laws of Newtonian mechanics. However, suppose we made the strip very long, extending, say, from north to south. Then an observer on the ground would find that the puck accelerates slightly toward the west as it moves south (Fig. 5-2*a*); yet, that observer would not be able to find a force that causes the westward acceleration. In this case, the ground is a **noninertial frame** because, for the puck's long path, Earth's rotation cannot be neglected. The surprising westward acceleration of the sliding puck relative to the ground is actually due to the eastward rotation of the ground beneath the puck (Fig. 5-2*b*).

In this book we usually assume that the ground is an inertial frame and that measurements of forces and accelerations are made from it. If measurements are made in, say, an elevator that is accelerating relative to the ground, then the measurements are being made in a noninertial frame and the results can be surprising. We see an example of this in Sample Problem 5-8.

✔**CHECKPOINT 1:** Which of the figure's six arrangements correctly show the vector addition of forces $\vec{F}_1$ and $\vec{F}_2$ to yield the third vector, which is meant to represent their net force $\vec{F}_{net}$?

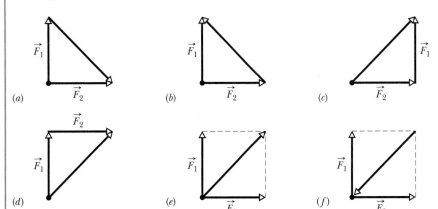

5-4 Mass

Everyday experience tells us that a given force produces different magnitudes of acceleration for different bodies. Put a baseball and a bowling ball on the floor and give both the same sharp kick. Even if you don't actually do this, you know the result: The baseball receives a noticeably larger acceleration than the bowling ball. The two accelerations differ because the mass of the baseball differs from the mass of the bowling ball—but what, exactly, is mass?

We can explain how to measure mass by imagining a series of experiments in an inertial frame. In the first experiment we exert a force on our standard body, whose mass m_0 is defined to be 1.0 kg. Suppose that the standard body accelerates at 1.0 m/s^2. We can then say the force on that body is 1.0 N.

We next apply that same force (we would need some way of being certain it is the same force) to a second body, body X, whose mass is not known. Suppose we find that this body X accelerates at 0.25 m/s^2. We know that a *less massive* baseball receives a *greater acceleration* than a more massive bowling ball when the same force (kick) is applied to both. Let us then make the following conjecture: The ratio of the masses of two bodies is equal to the inverse of the ratio of their accelerations when the same force is applied to both. For body X and the standard body, this tells us that

$$\frac{m_X}{m_0} = \frac{a_0}{a_X}.$$

Solving for m_X yields

$$m_X = m_0 \frac{a_0}{a_X} = (1.0 \text{ kg}) \frac{1.0 \text{ m/s}^2}{0.25 \text{ m/s}^2} = 4.0 \text{ kg}.$$

Our conjecture will be useful, of course, only if it continues to hold when we change the applied force to other values. For example, if we apply an 8.0 N force to the standard body, we obtain an acceleration of 8.0 m/s². When the 8.0 N force is applied to body X, we obtain an acceleration of 2.0 m/s². Our conjecture then gives us

$$m_X = m_0 \frac{a_0}{a_X} = (1.0 \text{ kg}) \frac{8.0 \text{ m/s}^2}{2.0 \text{ m/s}^2} = 4.0 \text{ kg},$$

consistent with our first experiment. Many experiments yielding similar results indicate that our conjecture provides a consistent and reliable means of assigning a mass to any given body.

Our measurement experiments indicate that mass is an *intrinsic* characteristic of a body—that is, a characteristic that automatically comes with the existence of the body. They also indicate that mass is a scalar quantity. However, the nagging question remains: What, exactly, is mass?

Since the word *mass* is used in everyday English, we should have some intuitive understanding of it, maybe something that we can physically sense. Is it a body's size, weight, or density? The answer is no, although those characteristics are sometimes confused with mass. We can say only that *the mass of a body is the characteristic that relates a force on the body to the resulting acceleration*. Mass has no more familiar definition; you can have a physical sensation of mass only when you attempt to accelerate a body, as in the kicking of a baseball or a bowling ball.

5-5 Newton's Second Law

All the definitions, experiments, and observations that we have discussed so far can be summarized in one neat statement:

▶ **Newton's Second Law:** The net force on a body is equal to the product of the body's mass and the acceleration of the body.

In equation form,

$$\vec{F}_{net} = m\vec{a} \qquad \text{(Newton's second law).} \qquad (5\text{-}1)$$

This equation is simple, but we must use it cautiously. First, we must be certain about which body we are applying it to. Then $\vec{F}_{net}$ must be the vector sum of *all* the forces that act on *that* body. Only forces that act on *that* body are to be included in the vector sum, not forces acting on other bodies that might be involved in the given situation. For example, if you are in a rugby scrum, the net force on *you* is the vector sum of all the pushes and pulls on *your* body. It does not include any push or pull on another player from you.

Like other vector equations, Eq. 5-1 is equivalent to three component equations, one written for each axis of an *xyz* coordinate system:

$$F_{net,x} = ma_x, \quad F_{net,y} = ma_y, \quad \text{and} \quad F_{net,z} = ma_z. \qquad (5\text{-}2)$$

TABLE 5-1   Units in Newton's Second Law (Eqs. 5-1 and 5-2)

System	Force	Mass	Acceleration
SI	newton (N)	kilogram (kg)	m/s²
CGS[a]	dyne	gram (g)	cm/s²
British[b]	pound (lb)	slug	ft/s²

[a]1 dyne = 1 g·cm/s².

[b]1 lb = 1 slug·ft/s².

Each of these equations relates the net force component along an axis to the acceleration along that same axis. For example, the first equation tells us that the sum of all the force components along the x axis causes the x component a_x of the body's acceleration, but causes no acceleration in the y and z directions. Turned around, the acceleration component a_x is caused only by the sum of the force components along the x axis. In general,

> The acceleration component along a given axis is caused only by the sum of the force components along that *same* axis, and not by force components along any other axis.

Equation 5-1 tells us that if the net force on a body is zero, the body's acceleration $\vec{a} = 0$. If the body is at rest, it stays at rest; if it is moving, it continues to move at constant velocity. In such cases, any forces on the body *balance* one another, and both the forces and the body can be said to be in *equilibrium*. Commonly, the forces are also said to *cancel* one another, but the term "cancel" is tricky. It does *not* mean that the forces cease to exist (canceling forces is not like canceling dinner reservations). The forces still act on the body.

For SI units, Eq. 5-1 tells us that

$$1 \text{ N} = (1 \text{ kg})(1 \text{ m/s}^2) = 1 \text{ kg} \cdot \text{m/s}^2. \tag{5-3}$$

Some force units in other systems of units are given in Table 5-1 and Appendix D.

To solve problems with Newton's second law, we often draw a **free-body diagram** in which the only body shown is the one for which we are summing forces. A sketch of the body itself is preferred by some teachers but, to save space in these chapters, we shall usually represent the body with a dot. Also, each force on the body is drawn as a vector arrow with its tail on the body. A coordinate system is usually included, and the acceleration of the body is sometimes shown with a vector arrow (labeled as an acceleration).

A collection of two or more bodies is called a **system,** and any force on the bodies inside the system from bodies outside the system is called an **external force.** If the bodies are rigidly connected, then we can treat the system as one composite body, and the net force $\vec{F}_{net}$ on it is the vector sum of all external forces. (We do not include **internal forces**—that is, forces between two bodies inside the system.) For example, a connected railroad engine and car form a system. If, say, a tow line pulls on the front of the engine, then the force due to the tow line acts on the whole engine–car system. Just as for a single body, we can relate the net external force on a system to its acceleration with Newton's second law, $\vec{F}_{net} = m\vec{a}$, where m is the total mass of the system.

CHECKPOINT 2:   The figure here shows two horizontal forces acting on a block that is on a frictionless floor. Assume that a third horizontal force $\vec{F}_3$ also acts on the block. What are the magnitude and direction of $\vec{F}_3$ when the block is (a) stationary and (b) moving to the left with a constant speed of 5 m/s?

3 N ← □ → 5 N

Sample Problem 5-1

In Figs. 5-3a to c, one or two forces act on a puck that moves over frictionless ice and along an x axis, in one-dimensional motion. The puck's mass is $m = 0.20$ kg. Forces $\vec{F}_1$ and $\vec{F}_2$ are directed along the axis and have magnitudes $F_1 = 4.0$ N and $F_2 = 2.0$ N. Force $\vec{F}_3$ is directed at angle $\theta = 30°$ and has magnitude $F_3 = 1.0$ N. In each situation, what is the acceleration of the puck?

SOLUTION: The Key Idea in each situation is that we can relate the acceleration $\vec{a}$ to the net force $\vec{F}_{net}$ acting on the puck with Newton's second law, $\vec{F}_{net} = m\vec{a}$. However, because the motion is along only the x axis, we can simplify each situation by writing the second law for x components only:

$$F_{net,x} = ma_x. \qquad (5\text{-}4)$$

The free-body diagrams for the three situations are given in Figs. 5-3d to f, where the puck is represented by a dot.

For Fig. 5-3d, where only one horizontal force acts, Eq. 5-4 gives us

$$F_1 = ma_x,$$

which, with given data, yields

$$a_x = \frac{F_1}{m} = \frac{4.0 \text{ N}}{0.20 \text{ kg}} = 20 \text{ m/s}^2. \qquad \text{(Answer)}$$

The positive answer indicates that the acceleration is in the positive direction of the x axis.

In Fig. 5-3e, two horizontal forces act on the puck, $\vec{F}_1$ in the positive direction of x and $\vec{F}_2$ in the negative direction. Now Eq. 5-4 gives us

$$F_1 - F_2 = ma_x,$$

which, with given data, yields

$$a_x = \frac{F_1 - F_2}{m} = \frac{4.0 \text{ N} - 2.0 \text{ N}}{0.20 \text{ kg}} = 10 \text{ m/s}^2. \qquad \text{(Answer)}$$

Thus, the net force accelerates the puck in the positive direction of the x axis.

In Fig. 5-3f, force $\vec{F}_3$ is not directed along the direction of the puck's acceleration; only the x component $F_{3,x}$ is. (Force $\vec{F}_3$ is two-dimensional but the motion is only one-dimensional.) Thus, we

(a)   (b)   (c)

(d)   (e)   (f)

Fig. 5-3 Sample Problem 5-1. (a)–(c) In three situations, forces act on a puck that moves on frictionless ice along an x axis. (d)–(f) Free-body diagrams for the three situations.

write Eq. 5-4 as

$$F_{3,x} - F_2 = ma_x.$$

From the figure, we see that $F_{3,x} = F_3 \cos\theta$. Solving for the acceleration and substituting for $F_{3,x}$ yield

$$a_x = \frac{F_{3,x} - F_2}{m} = \frac{F_3 \cos\theta - F_2}{m}$$

$$= \frac{(1.0 \text{ N})(\cos 30°) - 2.0 \text{ N}}{0.20 \text{ kg}} = -5.7 \text{ m/s}^2. \qquad \text{(Answer)}$$

Thus, the net force accelerates the puck in the negative direction of the x axis.

✔CHECKPOINT 3:   The figure shows overhead views of four situations in which two forces accelerate the same block across a frictionless floor. Rank the situations according to the magnitudes of (a) the net force on the block and (b) the acceleration of the block, greatest first.

Sample Problem 5-2

In the overhead view of Fig. 5-4a, a 2.0 kg cookie tin is accelerated at 3.0 m/s² in the direction shown by $\vec{a}$, over a frictionless horizontal surface. The acceleration is caused by three horizontal forces, only two of which are shown: $\vec{F}_1$ of magnitude 10 N and $\vec{F}_2$ of magnitude 20 N. What is the third force $\vec{F}_3$ in unit-vector notation and as a magnitude and an angle?

SOLUTION: One Key Idea here is that the net force $\vec{F}_{net}$ on the tin is the sum of the three forces and is related to the acceleration $\vec{a}$ of the tin via Newton's second law ($\vec{F}_{net} = m\vec{a}$). Thus,

$$\vec{F}_1 + \vec{F}_2 + \vec{F}_3 = m\vec{a},$$

which gives us

$$\vec{F}_3 = m\vec{a} - \vec{F}_1 - \vec{F}_2. \qquad (5\text{-}5)$$

A second Key Idea is that this is a two-dimensional problem; so we *cannot* find $\vec{F}_3$ merely by substituting the magnitudes for the vector quantities on the right side of Eq. 5-5. Instead, we must vectorially add $m\vec{a}$, $-\vec{F}_1$ (the reverse of $\vec{F}_1$), and $-\vec{F}_2$ (the reverse of $\vec{F}_2$), as is shown in Fig. 5-4b. This addition can be done directly on a vector-capable calculator because we know both magnitude and angle for each of these three vectors. However, here we shall evaluate the right side of Eq. 5-5 in terms of components, first along the x axis and then along the y axis.

Along the x axis we have

$$F_{3,x} = ma_x - F_{1,x} - F_{2,x}$$

$$= m(a \cos 50°) - F_1 \cos(-150°) - F_2 \cos 90°.$$

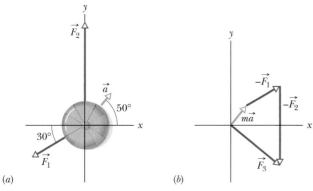

(a)   (b)

Fig. 5-4 Sample Problem 5-2. (a) An overhead view of two of three horizontal forces that act on a cookie tin, resulting in acceleration $\vec{a}$. $\vec{F_3}$ is not shown. (b) An arrangement of vectors $m\vec{a}$, $-\vec{F_1}$, and $-\vec{F_2}$ to find force $\vec{F_3}$.

Then, substituting known data, we find

$$F_{3,x} = (2.0 \text{ kg})(3.0 \text{ m/s}^2) \cos 50° - (10 \text{ N}) \cos(-150°)$$
$$- (20 \text{ N}) \cos 90°$$
$$= 12.5 \text{ N}.$$

Similarly, along the y axis we find

$$F_{3,y} = ma_y - F_{1,y} - F_{2,y}$$
$$= m(a \sin 50°) - F_1 \sin(-150°) - F_2 \sin 90°$$
$$= (2.0 \text{ kg})(3.0 \text{ m/s}^2) \sin 50° - (10 \text{ N}) \sin(-150°)$$
$$- (20 \text{ N}) \sin 90°$$
$$= -10.4 \text{ N}.$$

Thus, in unit-vector notation, we have

$$\vec{F_3} = F_{3,x}\hat{i} + F_{3,y}\hat{j} = (12.5 \text{ N})\hat{i} - (10.4 \text{ N})\hat{j}$$
$$\approx (13 \text{ N})\hat{i} - (10 \text{ N})\hat{j}. \qquad \text{(Answer)}$$

We can now use a vector-capable calculator to get the magnitude and the angle of $\vec{F_3}$. We can also use Eq. 3-6 to obtain the magnitude and the angle (from the positive direction of the x axis) as

$$F_3 = \sqrt{F_{3,x}^2 + F_{3,y}^2} = 16 \text{ N}$$

and

$$\theta = \tan^{-1} \frac{F_{3,y}}{F_{3,x}} = -40°. \qquad \text{(Answer)}$$

Sample Problem 5-3

In a two-dimensional tug-of-war, Alex, Betty, and Charles pull horizontally on an automobile tire at the angles shown in the overhead view of Fig. 5-5a. The tire remains stationary in spite of the three pulls. Alex pulls with force $\vec{F_A}$ of magnitude 220 N, and Charles pulls with force $\vec{F_C}$ of magnitude 170 N. The direction of $\vec{F_C}$ is not given. What is the magnitude of Betty's force $\vec{F_B}$?

SOLUTION: Because the three forces pulling on the tire do not accelerate the tire, the tire's acceleration is $\vec{a} = 0$ (that is, the forces are in equilibrium). The Key Idea here is that we can relate that acceleration to the net force $\vec{F}_{\text{net}}$ on the tire with Newton's second law ($\vec{F}_{\text{net}} = m\vec{a}$), which we can write as

$$\vec{F_A} + \vec{F_B} + \vec{F_C} = m(0) = 0,$$

or

$$\vec{F_B} = -\vec{F_A} - \vec{F_C}. \qquad (5\text{-}6)$$

The free-body diagram for the tire is shown in Fig. 5-5b, where we have conveniently centered a coordinate system on the tire and assigned ϕ to the angle of $\vec{F_C}$.

We want to solve for the magnitude of $\vec{F_B}$. Although we know both magnitude and direction for $\vec{F_A}$, we know only the magnitude of $\vec{F_C}$ and not its direction. Thus, with unknowns on both sides of Eq. 5-6, we cannot directly solve it on a vector-capable calculator. Instead we must rewrite Eq. 5-6 in terms of components for either the x or the y axis. Since $\vec{F_B}$ is directed along the y axis, we choose that axis and write

$$F_{By} = -F_{Ay} - F_{Cy}.$$

Evaluating these components with their angles and using the angle $133°\ (= 180° - 47.0°)$ for $\vec{F_A}$, we obtain

$$F_B \sin(-90°) = -F_A \sin 133° - F_C \sin \phi,$$

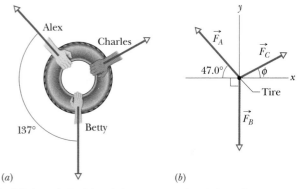

(a)   (b)

Fig. 5-5 Sample Problem 5-3. (a) An overhead view of three people pulling on a tire. (b) A free-body diagram for the tire.

which, with the given data for the magnitudes, yields

$$-F_B = -(220 \text{ N})(\sin 133°) - (170 \text{ N}) \sin \phi. \qquad (5\text{-}7)$$

However, we do not know ϕ.

We can find it by rewriting Eq. 5-6 for the x axis as

$$F_{Bx} = -F_{Ax} - F_{Cx}$$

and then as

$$F_B \cos(-90°) = -F_A \cos 133° - F_C \cos \phi,$$

which gives us

$$0 = -(220 \text{ N})(\cos 133°) - (170 \text{ N}) \cos \phi$$

and

$$\phi = \cos^{-1} -\frac{(220 \text{ N})(\cos 133°)}{170 \text{ N}} = 28.04°.$$

Inserting this into Eq. 5-7, we find

$$F_B = 241 \text{ N}. \qquad \text{(Answer)}$$

PROBLEM-SOLVING TACTICS

Tactic 1: *Dimensions and Vectors*
Many students fail to grasp the second Key Idea in Sample Problem 5-2, and that failure haunts them through the rest of this book. When you are dealing with forces, you cannot just add or subtract their magnitudes to find their net force unless they happen to be directed *along the same axis*. If they are not, you must use vector addition, either by means of a vector-capable calculator or by finding components along axes, as is done in Sample Problem 5-2.

Tactic 2: *Reading Force Problems*
Read the problem statement several times until you have a good mental picture of what the situation is, what data are given, and what is requested. If you know what the problem is about but don't know what to do next, put the problem aside and reread the text. If you are hazy about Newton's second law, reread that section. Study the sample problems. And remember, solving physics problems (like repairing cars and designing computer chips) takes training—you were not born with the ability.

Tactic 3: *Draw Two Types of Figures*
You may need two figures. One is a rough sketch of the actual situation. When you draw the forces, place the tail of each force vector either on the boundary of or within the body on which that force acts. The other figure is a free-body diagram: the forces on a *single* body are drawn, with the body represented by a dot or a sketch. Place the tail of each force vector on the dot or sketch.

Tactic 4: *What Is Your System?*
If you are using Newton's second law, you must know what body or system you are applying it to. In Sample Problem 5-1 it is the puck (not the ice). In Sample Problem 5-2, it is the cookie tin. In Sample Problem 5-3, it is the tire (not the people).

Tactic 5: *Choose Your Axes Wisely*
In Sample Problem 5-3, we saved a lot of work by choosing one of our coordinate axes to coincide with one of the forces (the *y* axis with $\vec{F}_B$).

5-6   Some Particular Forces

The Gravitational Force

A **gravitational force** $\vec{F}_g$ on a body is a pull that is directed toward a second body. In these early chapters, we do not discuss the nature of this force and we usually consider situations in which the second body is Earth. Thus, when we speak of *the* gravitational force $\vec{F}_g$ on a body, we usually mean a force that pulls on it directly toward the center of Earth—that is, directly down toward the ground. We shall assume that the ground is an inertial frame.

Suppose that the body, of mass *m*, is in free fall with the free-fall acceleration of magnitude *g*. Then, if we neglect the effects of the air, the only force acting on the body is the gravitational force $\vec{F}_g$. We can relate this downward force and downward acceleration with Newton's second law ($\vec{F} = m\vec{a}$). We place a vertical *y* axis along the body's path, with the positive direction upward. For this axis, Newton's second law can be written in the form $F_{\text{net},y} = ma_y$, which, in our situation, becomes

$$-F_g = m(-g)$$

or
$$F_g = mg. \tag{5-8}$$

In other words, the magnitude of the gravitational force is equal to the product *mg*.

This same gravitational force, with the same magnitude, still acts on the body even when the body is not in free fall but, say, at rest on a pool table or moving across the table. (For the gravitational force to disappear, Earth would have to disappear.)

We can write Newton's second law for the gravitational force in these vector forms:

$$\vec{F}_g = -F_g\hat{j} = -mg\hat{j} = m\vec{g}, \tag{5-9}$$

where $\hat{j}$ is the unit vector that points upward along a *y* axis, directly away from the ground, and $\vec{g}$ is the free-fall acceleration (written as a vector), directed downward.

Weight

The **weight** *W* of a body is the magnitude of the net force required to prevent the body from falling freely, as measured by someone on the ground. For example, to

$$\vec{F}_{gL} = m_L\vec{g} \qquad\qquad \vec{F}_{gR} = m_R\vec{g}$$

Fig. 5-6 An equal-arm balance. When the device is in balance, the gravitational force $\vec{F}_{gL}$ on the body being weighed (on the left pan) and the total gravitational force $\vec{F}_{gR}$ on the reference bodies (on the right pan) are equal. Thus, the mass of the body being weighed is equal to the total mass of the reference bodies.

keep a ball at rest in your hand while you stand on the ground, you must provide an upward force to balance the gravitational force on the ball from Earth. Suppose the magnitude of the gravitational force is 2.0 N. Then the magnitude of your upward force must be 2.0 N, and thus, the weight W of the ball is 2.0 N. We also say that the ball *weighs* 2.0 N, and speak about the ball *weighing* 2.0 N.

A ball with a weight of 3.0 N would require a greater force from you—namely, a 3.0 N force—to keep it at rest. The reason is that the gravitational force you must balance has a greater magnitude—namely, 3.0 N. We say that this second ball is *heavier* than the first ball.

Now let us generalize the situation. Consider a body that has an acceleration $\vec{a}$ of zero relative to the ground, which we again assume to be an inertial frame. Two forces act on the body: a downward gravitational force $\vec{F}_g$ and a balancing upward force of magnitude W. We can write Newton's second law for a vertical y axis, with the positive direction upward, as

$$F_{\text{net},y} = ma_y.$$

In our situation, this becomes

$$W - F_g = m(0) \qquad\qquad (5\text{-}10)$$

or
$$W = F_g \qquad \text{(weight, with ground as inertial frame).} \qquad (5\text{-}11)$$

This equation tells us (assuming the ground is an inertial frame) that

> The weight W of a body is equal to the magnitude F_g of the gravitational force on the body.

Substituting mg for F_g from Eq. 5-8, we find

$$W = mg \qquad \text{(weight),} \qquad (5\text{-}12)$$

which relates a body's weight to its mass.

To *weigh* a body means to measure its weight. One way to do this is to place the body on one of the pans of an equal-arm balance (Fig. 5-6) and then add reference bodies (whose masses are known) on the other pan until we strike a balance (so that the gravitational forces on the two sides match). The masses on the pans then match, and we know the mass m of the body. If we know the value of g for the location of the balance, we can find the weight of the body with Eq. 5-12.

We can also weigh a body with a spring scale (Fig. 5-7). The body stretches a spring, moving a pointer along a scale that has been calibrated and marked in either mass or weight units. (Most bathroom scales in the United States work this way and are marked in the force unit pounds.) If the scale is marked in mass units, it is accurate only where the value of g is the same as where the scale was calibrated.

The weight of a body must be measured when the body is not accelerating vertically relative to the ground. For example, you can measure your weight on a scale in your bathroom or on a fast train. However, if you repeat the measurement with the scale in an accelerating elevator, the reading on the scale differs from your weight because of the acceleration. Such a measurement is called an *apparent weight*.

Caution: The weight of a body is not the mass of the body. Weight is the magnitude of a force and is related to the mass by Eq. 5-12. If you move a body to a point where the value of g is different, the body's mass (an intrinsic property of the body) is not different but the weight is. For example, the weight of a bowling ball with a mass of 7.2 kg is 71 N on Earth but would be only 12 N on the Moon. The mass is the same on Earth and Moon, but the free-fall acceleration on the Moon is only 1.7 m/s².

Scale marked in weight or mass units

$$\vec{F}_g = m\vec{g}$$

Fig. 5-7 A spring scale. The reading is proportional to the *weight* of the object placed on the pan, and the scale gives that weight if marked in weight units. If, instead, it is marked in mass units, the reading is accurate only where the free-fall acceleration g is the same as where the scale was calibrated.

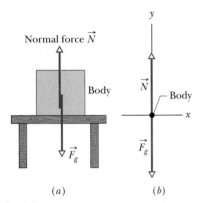

Fig. 5-8 (a) A body resting on a tabletop experiences a normal force $\vec{N}$ perpendicular to the tabletop. (b) The corresponding free-body diagram for the body.

The Normal Force

If you stand on a mattress, Earth pulls you downward, but you are stationary. The reason is that the mattress, because it deforms downward due to you, pushes up on you. Similarly, if you stand on a floor, it deforms (it is compressed, bent, or buckled ever so slightly), and it pushes up on you. Even a seemingly rigid concrete floor does this (if it is not sitting directly on the ground, enough people on the floor could actually break it).

The push on you from the mattress or floor is called a **normal force** and usually symbolized as $\vec{N}$. The name comes from the mathematical term *normal*, meaning perpendicular: The force on you from, say, the floor is perpendicular to the floor.

▶ When a body presses against a surface, the surface (even a seemingly rigid surface) deforms and pushes on the body with a normal force $\vec{N}$ that is perpendicular to the surface.

Figure 5-8a shows an example. A block of mass m lies on a table's horizontal surface and presses down on the table, deforming the table somewhat because of the gravitational force $\vec{F}_g$ on the block. The table pushes up on the block with normal force $\vec{N}$. The free-body diagram for the block is given in Fig. 5-8b. Forces $\vec{F}_g$ and $\vec{N}$ are the only two forces on the block and they are vertical. Thus, for the block we can write Newton's second law for a positive-upward y axis ($F_{net,y} = ma_y$) as

$$N - F_g = ma_y.$$

From Eq. 5-8, we substitute mg for F_g, finding

$$N - mg = ma_y.$$

Then the magnitude of the normal force is

$$N = mg + ma_y = m(g + a_y) \tag{5-13}$$

for any vertical acceleration a_y of the table and block (they might be in an accelerating elevator). If the table and block are not accelerating relative to the ground, then $a_y = 0$ and Eq. 5-13 yields

$$N = mg. \tag{5-14}$$

✔CHECKPOINT 4:   In Fig. 5-8, is the magnitude of the normal force $\vec{N}$ greater than, less than, or equal to mg if the body and table are in an elevator that is moving upward (a) at constant speed and (b) at increasing speed?

Friction

If we slide or attempt to slide a body over a surface, the motion is resisted by a bonding between the body and the surface. (We discuss this bonding more in the next chapter.) The resistance is considered to be a single force $\vec{f}$, called the **frictional force,** or simply **friction.** This force is directed along the surface, opposite the direction of the intended motion (Fig. 5-9). Sometimes, to simplify a situation, friction is assumed to be negligible (the surface is *frictionless*).

Fig. 5-9 A frictional force $\vec{f}$ opposes the attempted slide of a body over a surface.

Tension

When a cord (or a rope, cable, or other such object) is attached to a body and pulled taut, the cord pulls on the body with a force $\vec{T}$ directed away from the body and along the cord (Fig. 5-10a). The force is often called a *tension force* because the

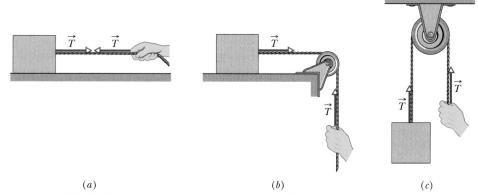

(a)   (b)   (c)

Fig. 5-10 (a) The cord, pulled taut, is under tension. If its mass is negligible, it pulls on the body and the hand with force $\vec{T}$, even if it runs around a massless, frictionless pulley as in (b) and (c).

cord is said to be in a state of *tension* (or to be *under tension*), which means that it is being pulled taut. The *tension in the cord* is the magnitude T of the force on the body. For example, if the force on the body has magnitude $T = 50$ N, then the tension in the cord is 50 N.

A cord is often said to be *massless* (meaning its mass is negligible compared to the body's mass) and unstretchable. The cord then exists only as a connection between two bodies. It pulls on both bodies with the same magnitude T, even if the bodies and the cord are accelerating and even if the cord runs around a *massless, frictionless pulley* (Figs. 5-10b and c). Such a pulley has negligible mass compared to the bodies and has negligible friction on its axle opposing its rotation. If the cord wraps halfway around a pulley, as in Fig. 5-10c, then the net force on the pulley from the cord has the magnitude 2T.

✓CHECKPOINT 5: The body that is suspended by a rope in Fig. 5-10c has a weight of 75 N. Is T equal to, greater than, or less than 75 N when the body is moving upward (a) at constant speed, (b) at increasing speed, and (c) at decreasing speed?

PROBLEM-SOLVING TACTICS

Tactic 6: *Normal Force*
Equation 5-14 for the normal force on a body holds only when $\vec{N}$ is directed upward and the body's vertical acceleration is zero, so we do *not* apply it for other orientations of $\vec{N}$ or when the vertical acceleration is not zero. Instead, we must derive a new expression for $\vec{N}$ from Newton's second law.

We are free to move $\vec{N}$ around in a figure as long as we main-

tain its orientation. For example, in Fig. 5-8a we can slide it downward so that its head is at the boundary of the body and the tabletop. However, $\vec{N}$ is least likely to be misinterpreted if its tail is at that boundary or somewhere within the body (as shown). An even better technique is to draw a free-body diagram as in Fig. 5-8b, with the tail of $\vec{N}$ directly on the dot or sketch representing the body.

Sample Problem 5-4

Let us return to John Massis and the railroad cars, and assume that Massis pulled (with his teeth) on his end of the rope with a constant force that was 2.5 times his body weight, at an angle θ of 30° from the horizontal. His mass m was 80 kg. The weight W of the cars was 700 kN, and he moved them 1.0 m along the rails. Assume that the rolling wheels encountered no retarding force from the rails. What was the speed of the cars at the end of the pull?

SOLUTION: A Key Idea here is that, from Newton's second law, the constant horizontal force on the cars from Massis causes a constant horizontal acceleration of the cars. Because the acceleration is constant and the motion is one-dimensional, we can use the equations of Table 2-1 to find the velocity v at the end of the pulling distance $d = 1.0$ m. We need an equation that contains v. Let us try Eq. 2-16,

$$v^2 = v_0^2 + 2a(x - x_0),\qquad (5\text{-}15)$$

and place an x axis along the direction of motion, as shown in the free-body diagram of Fig. 5-11. We know that the initial velocity v_0 is 0 and the displacement $x - x_0$ is $d = 1.0$ m. However, we do not know the acceleration a along the x axis.

A second Key Idea is that we can relate a to the force on the cars from the rope by using Newton's second law. We can write that law for the x axis in Fig. 5-11 as $F_{net,x} = ma_x$ or, here,

$$F_{net,x} = Ma, \qquad (5\text{-}16)$$

where M is the mass of the cars. The only force on the cars along the x axis is the horizontal component $T \cos \theta$ of the tension force $\vec{T}$ on the cars from the rope pulled by Massis. Thus, Eq. 5-16 becomes

$$T \cos \theta = Ma. \qquad (5\text{-}17)$$

We know that T is 2.5 times the body weight of Massis. From Eq. 5-12, his weight is equal to mg, so we have

$$T = 2.5mg = (2.5)(80 \text{ kg})(9.8 \text{ m/s}^2) = 1960 \text{ N},$$

which is the force a good middle-weight weight lifter can produce—and far from a superhuman force.

We still need M in order to evaluate Eq. 5-17 for a. To find M, we again use Eq. 5-12, but now with the weight W of the cars:

$$M = \frac{W}{g} = \frac{7.0 \times 10^5 \text{ N}}{9.8 \text{ m/s}^2} = 7.143 \times 10^4 \text{ kg}.$$

Fig. 5-11 Sample Problem 5-4. Free-body diagram for the passenger cars pulled by Massis. The vectors are not drawn to scale; the force $\vec{T}$ on the cars from the rope is much smaller than the normal force $\vec{N}$ on the cars from the rails and the gravitational force $\vec{F}_g$ on the cars.

By rearranging Eq. 5-17 and substituting for T, M, and θ, we get

$$a = \frac{T \cos \theta}{M} = \frac{(1960 \text{ N})(\cos 30°)}{7.143 \times 10^4 \text{ kg}} = 0.02376 \text{ m/s}^2.$$

Substituting this value and the other known values into Eq. 5-15 now gives us

$$v^2 = 0 + 2(0.02376 \text{ m/s}^2)(1.0 \text{ m})$$

and

$$v = 0.22 \text{ m/s}. \qquad \text{(Answer)}$$

Massis would have done better if the rope had been attached higher on the car, so that it was horizontal. Can you see why?

5-7 Newton's Third Law

Two bodies are said to *interact* when they push or pull on each other—that is, when a force acts on each body due to the other body. For example, suppose that you position a book B so it leans against a crate C (Fig. 5-12a). Then the book and crate interact: There is a horizontal force $\vec{F}_{BC}$ on the book from the crate (or due to the crate) and a horizontal force $\vec{F}_{CB}$ on the crate from the book (or due to the book). This pair of forces is shown in Fig. 5-12b. Newton's third law states that

> **Newton's Third Law:** When two bodies interact, the forces on the bodies from each other are always equal in magnitude and opposite in direction.

Fig. 5-12 (a) Book B leans against crate C. (b) According to Newton's third law, the force $\vec{F}_{BC}$ on the book from the crate has the same magnitude but the opposite direction of the force $\vec{F}_{CB}$ on the crate from the book.

For the book and crate, we can write this law as the scalar relation

$$F_{BC} = F_{CB} \qquad \text{(equal magnitudes)}$$

or as the vector relation

$$\vec{F}_{BC} = -\vec{F}_{CB} \qquad \text{(equal magnitudes and opposite directions),}$$

where the minus sign means that these two forces are in opposite directions. We can call the forces between two interacting bodies a **third-law force pair.** When any two bodies interact in any situation, a third-law force pair is present. The book and crate in Fig. 5-12a are stationary, but the third law would still hold if they were moving and even if they were accelerating.

As another example, let us find the third-law force pairs involving the cantaloupe in Fig. 5-13a, which lies on a table that stands on Earth. The cantaloupe interacts with the table and with Earth (this time, there are three bodies whose interactions we must sort out).

Cantaloupe C

Table T

Earth E

(a)

$\vec{F}_{CT}$ (normal force from table)

$\vec{F}_{CE}$ (gravitational force)

(b)

Fig. 5-13 (a) A cantaloupe lies on a table that stands on Earth. (b) The forces *on the cantaloupe* are $\vec{F}_{CT}$ and $\vec{F}_{CE}$. (c) The third-law force pair for the cantaloupe–Earth interaction. (d) The third-law force pair for the cantaloupe–table interaction.

Let's first focus only on the cantaloupe (Fig. 5-13b). Force $\vec{F}_{CT}$ is the normal force on the cantaloupe from the table, and force $\vec{F}_{CE}$ is the gravitational force on the cantaloupe due to Earth. Are they a third-law force pair? No, they are forces on a single body, the cantaloupe, and not on two interacting bodies.

To find a third-law pair, we should focus not on the cantaloupe but on the interaction between the cantaloupe and one of the other two bodies. First, in the cantaloupe–Earth interaction (Fig. 5-13c), Earth pulls on the cantaloupe with a gravitational force $\vec{F}_{CE}$ and the cantaloupe pulls on Earth with a gravitational force $\vec{F}_{EC}$. Are these forces a third-law force pair? Yes, they are forces on two interacting bodies, the force on each due to the other. Thus, by Newton's third law,

$$\vec{F}_{CE} = -\vec{F}_{EC} \qquad \text{(cantaloupe–Earth interaction).}$$

Cantaloupe

$\vec{F}_{CE}$

$\vec{F}_{EC}$

Earth

(c)

Next, in the cantaloupe–table interaction, the force on the cantaloupe from the table is $\vec{F}_{CT}$ and the force on the table from the cantaloupe is $\vec{F}_{TC}$ (Fig. 5-13d). These forces are also a third-law force pair, and so

$$\vec{F}_{CT} = -\vec{F}_{TC} \qquad \text{(cantaloupe–table interaction).}$$

✔**CHECKPOINT 6:** Suppose that the cantaloupe and table of Fig. 5-13 are in an elevator cab that begins to accelerate upward. (a) Do the magnitudes of forces $\vec{F}_{TC}$ and $\vec{F}_{CT}$ increase, decrease, or stay the same? (b) Are those two forces still equal in magnitude and opposite in direction? (c) Do the magnitudes of forces $\vec{F}_{CE}$ and $\vec{F}_{EC}$ increase, decrease, or stay the same? (d) Are those two forces still equal in magnitude and opposite in direction?

$\vec{F}_{CT}$

$\vec{F}_{TC}$

(d)

5-8 Applying Newton's Laws

The rest of this chapter consists of sample problems. You should pore over them, learning not just their particular answers but, instead, the procedures they show for attacking a problem. Especially important is knowing how to translate a sketch of a situation into a free-body diagram with appropriate axes, so that Newton's laws can be applied. We begin with a sample problem that is worked out in exhaustive detail, using a question-and-answer format.

Sample Problem 5-5

Figure 5-14 shows a block S (the *sliding block*) with mass M = 3.3 kg. The block is free to move along a horizontal frictionless surface such as an air table. This first block is connected by a cord that wraps over a frictionless pulley to a second block H (the *hanging block*), with mass m = 2.1 kg. The cord and pulley have negligible masses compared to the blocks (they are "massless"). The hanging block H falls as the sliding block S accelerates to the right. Find (a) the acceleration of the sliding block, (b) the acceleration of the hanging block, and (c) the tension in the cord.

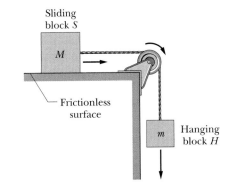

Sliding block S

M

Frictionless surface

m Hanging block H

Fig. 5-14 Sample Problem 5-5. A block S of mass M is connected to a block H of mass m by a cord that wraps over a pulley.

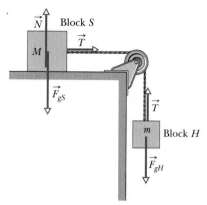

Fig. 5-15 The forces acting on the two blocks of Fig. 5-14.

Q *What is this problem all about?*

You are given two bodies, the sliding block and the hanging block, *and also Earth,* which pulls on both bodies. (Without Earth, nothing would happen here.) A total of five forces act on the blocks, as shown in Fig. 5-15:

1. The cord pulls to the right on sliding block S with a force of magnitude T.

2. The cord pulls upward on hanging block H with a force of the same magnitude T. This upward force keeps the hanging block from falling freely.

3. Earth pulls down on sliding block S with the gravitational force $\vec{F}_{gS}$, which has a magnitude equal to Mg.

4. Earth pulls down on hanging block H with the gravitational force $\vec{F}_{gH}$, which has a magnitude equal to mg.

5. The table pushes up on sliding block S with a normal force $\vec{N}$.

There is another thing that you should note. We assume that the cord does not stretch, so that if block H falls 1 mm in a certain time, block S moves 1 mm to the right in that same time. This means that the blocks move together and their accelerations have the same magnitude a.

Q *How do I classify this problem? Should it suggest a particular law of physics to me?*

Yes. Forces, masses, and accelerations are involved, and they should suggest Newton's second law of motion, $\vec{F}_{net} = m\vec{a}$. That is our starting Key Idea.

Q *If I apply Newton's second law to this problem, to what body should I apply it?*

We focus on two bodies in this problem, the sliding block and the hanging block. Although they are *extended objects* (they are not points), we can still treat each block as a particle because every small part of it (every atom, say) moves in exactly the same way. A second Key Idea is to apply Newton's second law separately to each block.

Q *What about the pulley?*

We cannot represent the pulley as a particle because different parts of it move in different ways. When we discuss rotation, we shall deal with pulleys in detail. Meanwhile, we eliminate the pul-

ley from consideration by assuming its mass is negligible compared with the masses of the two blocks. Then its function is just to change the cord's orientation.

Q *OK. Now how do I apply $\vec{F}_{net} = m\vec{a}$ to the sliding block?*

Represent block S as a particle of mass M and draw *all* the forces that act *on* it, as in Fig. 5-16a. This is the block's *free-body diagram*. There are three forces. Next, draw a set of axes. It makes sense to draw the x axis parallel to the table, in the direction in which the block moves.

Q *Thanks, but you still haven't told me how to apply $\vec{F}_{net} = m\vec{a}$ to the sliding block. All you have done is explain how to draw a free-body diagram.*

You are right, and here's the third Key Idea: The expression $\vec{F}_{net} = M\vec{a}$ is a vector equation, so we can write it as three component equations:

$$F_{net,x} = Ma_x \qquad F_{net,y} = Ma_y \qquad F_{net,z} = Ma_z \quad (5\text{-}18)$$

in which $F_{net,x}$, $F_{net,y}$, and $F_{net,z}$ are the components of the net force along the three axes. Now we apply each component equation to its corresponding direction.

Because block S does not accelerate vertically, $F_{net,y} = Ma_y$ becomes

$$N - F_{gS} = 0 \quad \text{or} \quad N = F_{gS}.$$

Thus in the y direction, the magnitude of the normal force on block S is equal to the magnitude of the gravitational force on that block. No force acts in the z direction, which is perpendicular to the page.

In the x direction, there is only one force component, which is T. Thus, $F_{net,x} = Ma_x$ becomes

$$T = Ma. \quad (5\text{-}19)$$

This equation contains two unknowns, T and a, so we cannot yet solve it. Recall, however, that we have not said anything about the hanging block.

Q *I agree. How do I apply $\vec{F}_{net} = m\vec{a}$ to the hanging block?*

We apply it just as we did for block S: Draw a free-body diagram for block H, as in Fig. 5-16b. Then apply $\vec{F}_{net} = m\vec{a}$ in component form. This time, because the acceleration is along the y axis, we use the second of Eqs. 5-18 ($F_{net,y} = ma_y$) to write

$$T - F_{gH} = ma_y.$$

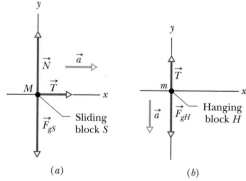

(a)   *(b)*

Fig. 5-16 (a) A free-body diagram for block S of Fig. 5-14. (b) A free-body diagram for the hanging block H of Fig. 5-14.

We can now substitute mg for F_{gH} and $-a$ for a_y (negative because block H accelerates downward, in the negative direction of the y axis). We find

$$T - mg = -ma. \qquad (5\text{-}20)$$

Now note that Eqs. 5-19 and 5-20 are simultaneous equations with the same two unknowns, T and a. Subtracting these equations eliminates T. Then solving for a yields

$$a = \frac{m}{M + m}\, g. \qquad (5\text{-}21)$$

Substituting this result into Eq. 5-19 yields

$$T = \frac{Mm}{M + m}\, g. \qquad (5\text{-}22)$$

Putting in the numbers gives, for these two quantities,

$$a = \frac{m}{M + m}\, g = \frac{2.1 \text{ kg}}{3.3 \text{ kg} + 2.1 \text{ kg}}\,(9.8 \text{ m/s}^2)$$
$$= 3.8 \text{ m/s}^2 \qquad \text{(Answer)}$$

and
$$T = \frac{Mm}{M + m}\, g = \frac{(3.3 \text{ kg})(2.1 \text{ kg})}{3.3 \text{ kg} + 2.1 \text{ kg}}\,(9.8 \text{ m/s}^2)$$
$$= 13 \text{ N}. \qquad \text{(Answer)}$$

Q *The problem is now solved, right?*

That's a fair question, but the problem is not really finished until we have examined the results to see whether they make sense. (If you made these calculations on the job, wouldn't you want to see whether they made sense before you turned them in?)

Look first at Eq. 5-21. Note that it is dimensionally correct and that the acceleration a will always be less than g. This is as it must be, because the hanging block is not in free fall. The cord pulls upward on it.

Look now at Eq. 5-22, which we can rewrite in the form

$$T = \frac{M}{M + m}\, mg. \qquad (5\text{-}23)$$

In this form, it is easier to see that this equation is also dimensionally correct, because both T and mg have dimensions of forces. Equation 5-23 also lets us see that the tension in the cord is always less than mg, and thus always less than the gravitational force on the hanging block. That is a comforting thought because, if T were *greater* than mg, the hanging block would accelerate upward.

We can also check the results by studying special cases, in which we can guess what the answers must be. A simple example is to put $g = 0$, as if the experiment were carried out in interstellar space. We know that in that case, the blocks would not move from rest, there would be no forces on the ends of the cord, and so there would be no tension in the cord. Do the formulas predict this? Yes, they do. If you put $g = 0$ in Eqs. 5-21 and 5-22, you find $a = 0$ and $T = 0$. Two more special cases that you might try are $M = 0$ and $m \rightarrow \infty$.

Sample Problem 5-6

In Fig. 5-17a, a block B of mass $M = 15.0$ kg hangs by a cord from a knot K of mass m_K, which hangs from a ceiling by means of two other cords. The cords have negligible mass, and the magnitude of the gravitational force on the knot is negligible compared to the gravitational force on the block. What are the tensions in the three cords?

SOLUTION: Let's start with the block because it has only one attached cord. The free-body diagram in Fig. 5-17b shows the forces on the block: gravitational force $\vec{F}_g$ (with a magnitude of Mg) and force T_3 from the attached cord. A Key Idea is that we can relate these forces to the acceleration of the block via Newton's second law ($\vec{F}_{net} = m\vec{a}$). Because the forces are both vertical, we choose the vertical component version of the law ($F_{net,y} = ma_y$) and write

$$T_3 - F_g = Ma_y.$$

Substituting Mg for F_g and 0 for the block's acceleration a_y, we find

$$T_3 - Mg = M(0) = 0.$$

This means that the two forces on the block are in equilibrium. Substituting for M ($= 15.0$ kg) and g and solving for T_3 yield

$$T_3 = 147 \text{ N}. \qquad \text{(Answer)}$$

We next consider the knot in the free-body diagram of Fig. 5-17c, where the negligible gravitational force on the knot is not included. The Key Idea here is that we can relate the three other forces acting on the knot to the acceleration of the knot via New-

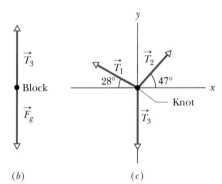

Fig. 5-17 Sample Problem 5-6. (*a*) A block of mass M hangs from three cords by means of a knot. (*b*) A free-body diagram for the block. (*c*) A free-body diagram for the knot.

ton's second law ($\vec{F}_{net} = m\vec{a}$) by writing

$$\vec{T}_1 + \vec{T}_2 + \vec{T}_3 = m_K \vec{a}_K.$$

Substituting 0 for the knot's acceleration $\vec{a}_K$ yields

$$\vec{T}_1 + \vec{T}_2 + \vec{T}_3 = 0, \tag{5-24}$$

which means that the three forces on the knot are in equilibrium. Although we know both magnitude and angle for $\vec{T}_3$, we know only the angles and not the magnitudes for $\vec{T}_1$ and $\vec{T}_2$. With unknowns in two vectors, we cannot solve Eq. 5-24 for $\vec{T}_1$ or $\vec{T}_2$ directly on a vector-capable calculator.

Instead we rewrite Eq. 5-24 in terms of components along the x and y axes. For the x axis, we have

$$T_{1x} + T_{2x} + T_{3x} = 0,$$

which, using the given data, yields

$$-T_1 \cos 28° + T_2 \cos 47° + 0 = 0. \tag{5-25}$$

(For the first term, we have two choices, either the one shown or the equivalent $T_1 \cos 152°$, where 152° is the angle from the positive direction of the x axis.)

Similarly, for the y axis we rewrite Eq. 5-24 as

$$T_{1y} + T_{2y} + T_{3y} = 0,$$

or

$$T_1 \sin 28° + T_2 \sin 47° - T_3 = 0.$$

Substituting our previous result for T_3 then gives us

$$T_1 \sin 28° + T_2 \sin 47° - 147 \text{ N} = 0. \tag{5-26}$$

We cannot solve Eq. 5-25 or Eq. 5-26 separately because each contains two unknowns, but we can solve them simultaneously because they contain the same two unknowns. Doing so (either by substitution, by adding or subtracting the equations appropriately, or by using the equation-solving capability of a calculator), we discover

$$T_1 = 104 \text{ N} \quad \text{and} \quad T_2 = 134 \text{ N}. \tag{Answer}$$

Thus, the tensions in the cords are 104 N in cord 1, 134 N in cord 2, and 147 N in cord 3.

Sample Problem 5-7

In Fig. 5-18a, a cord holds stationary a block of mass $m = 15$ kg, on a frictionless plane that is inclined at angle $\theta = 27°$.

(a) What are the magnitudes of the force $\vec{T}$ on the block from the cord and the normal force $\vec{N}$ on the block from the plane?

SOLUTION: Those two forces and the gravitational force $\vec{F}_g$ on the block are shown in the block's free-body diagram of Fig. 5-18b. Only these three forces act on the block. A Key Idea is that we can relate them to the block's acceleration via Newton's second law ($\vec{F}_{net} = m\vec{a}$), which we write as

$$\vec{T} + \vec{N} + \vec{F}_g = m\vec{a}.$$

Substituting 0 for the block's acceleration $\vec{a}$ yields

$$\vec{T} + \vec{N} + \vec{F}_g = 0, \tag{5-27}$$

which says that the three forces are in equilibrium.

With two unknown vectors in Eq. 5-27, we cannot solve it for either vector directly on a vector-capable calculator. So, we must rewrite it in terms of components. We use a coordinate system with its x axis parallel to the plane, as shown in Fig. 5-18b; then two forces ($\vec{N}$ and $\vec{T}$) line up with the axes, making their components easy to find. To find the components of the gravitational force $\vec{F}_g$, we first note that the angle θ of the plane is also the angle between the y axis and the direction of $\vec{F}_g$ (see Fig. 5-18c). Component F_{gx} is then $-F_g \sin \theta$, which is equal to $-mg \sin \theta$; and component F_{gy} is then $-F_g \cos \theta$, which is equal to $-mg \cos \theta$.

Now, writing the x component version of Eq. 5-27 yields

$$T + 0 - mg \sin \theta = 0,$$

from which

$$
\begin{aligned}
T &= mg \sin \theta \\
&= (15 \text{ kg})(9.8 \text{ m/s}^2)(\sin 27°) \\
&= 67 \text{ N}. \tag{Answer}
\end{aligned}
$$

(a)

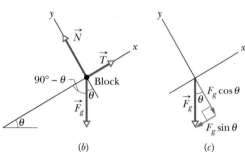

(b)   (c)

Fig. 5-18 Sample Problem 5-7 (a) A block of mass m held stationary by a cord. (b) A free-body diagram for the block. (c) The x and y components of $\vec{F}_g$.

Similarly, for the y axis, Eq. 5-27 yields

$$0 + N - mg \cos \theta = 0$$

or

$$
\begin{aligned}
N &= mg \cos \theta \\
&= (15 \text{ kg})(9.8 \text{ m/s}^2)(\cos 27°) \\
&= 131 \text{ N} \approx 130 \text{ N}. \tag{Answer}
\end{aligned}
$$

(b) We now cut the cord. As the block then slides down the inclined plane, does it accelerate? If so, what is its acceleration?

SOLUTION: Cutting the cord removes force $\vec{T}$ from the block. Along the y axis, the normal force and component F_{gy} are still in equilibrium. However, along the x axis, only force component F_{gx} acts on the block; because it is directed down the plane (along the x axis), that component must cause the block to accelerate down the plane. Our Key Idea here is that we can relate F_{gx} to the acceleration a that it produces with Newton's second law written for x components ($F_{\text{net},x} = ma_x$). We get

$$F_{gx} = ma$$

or

$$-mg \sin \theta = ma,$$

which gives us

$$a = -g \sin \theta. \qquad (5-28)$$

Substituting known data then yields

$$a = -(9.8 \text{ m/s}^2)(\sin 27°) = -4.4 \text{ m/s}^2. \quad \text{(Answer)}$$

The magnitude of this acceleration a is less than the magnitude 9.8 m/s^2 of the free-fall acceleration because only a component of $\vec{F}_g$ (the component that is directed down the plane) is producing acceleration a.

✔CHECKPOINT 7:   In the figure, horizontal force $\vec{F}$ is applied to a block on a ramp. (a) Is the component of $\vec{F}$ that is perpendicular to the ramp $F \cos \theta$ or $F \sin \theta$? (b) Does the presence of $\vec{F}$ increase or decrease the magnitude of the normal force on the block from the ramp?

Sample Problem 5-8

In Fig. 5-19a, a passenger of mass $m = 72.2$ kg stands on a platform scale in an elevator cab. We are concerned with the scale readings when the cab is stationary, and when it is moving up or down.

(a) Find a general solution for the scale reading, whatever the vertical motion of the cab.

SOLUTION: One Key Idea here is that the scale reading is equal to the magnitude of the normal force $\vec{N}$ on the passenger from the scale. The only other force acting on the passenger is the gravitational force $\vec{F}_g$, as shown in the free-body diagram of the passenger in Fig. 5-19b.

A second Key Idea is that we can relate the forces on the passenger to the acceleration $\vec{a}$ of the passenger with Newton's second law ($\vec{F}_{\text{net}} = m\vec{a}$). However, recall that we can use this law only in an inertial frame. If the cab accelerates, then it is *not* an inertial frame. So, we choose the ground to be our inertial frame and make any measure of the passenger's acceleration relative to it.

Because the two forces on the passenger and the passenger's acceleration are all directed vertically, along the y axis shown in

Fig. 5-19b, we can use Newton's second law written for y components ($F_{\text{net},y} = ma_y$) to get

$$N - F_g = ma$$

or

$$N = F_g + ma. \qquad (5-29)$$

This tells us that the scale reading, which is equal to N, depends on the vertical acceleration a of the cab. Substituting mg for F_g gives us

$$N = m(g + a) \qquad \text{(Answer)} \quad (5-30)$$

for any choice of acceleration a.

(b) What does the scale read if the cab is stationary or moving upward at a constant 0.50 m/s?

SOLUTION: The Key Idea here is that for any constant velocity (zero or otherwise), the acceleration a of the passenger is zero. Substituting this and other known values into Eq. 5-30, we find

$$N = (72.2 \text{ kg})(9.8 \text{ m/s}^2 + 0) = 708 \text{ N}. \quad \text{(Answer)}$$

This is the weight of the passenger and is equal to the magnitude F_g of the gravitational force on him.

(c) What does the scale read if the cab accelerates upward at 3.20 m/s^2 and downward at 3.20 m/s^2?

SOLUTION: For $a = 3.20$ m/s^2, Eq. 5-30 gives

$$N = (72.2 \text{ kg})(9.8 \text{ m/s}^2 + 3.20 \text{ m/s}^2)$$
$$= 939 \text{ N}, \qquad \text{(Answer)}$$

and for $a = -3.20$ m/s^2, it gives

$$N = (72.2 \text{ kg})(9.8 \text{ m/s}^2 - 3.20 \text{ m/s}^2)$$
$$= 477 \text{ N}. \qquad \text{(Answer)}$$

So, for an upward acceleration (either the cab's upward speed is increasing or its downward speed is decreasing), the scale reading

Fig. 5-19 Sample Problem 5-8. (a) A passenger stands on a platform scale that indicates his weight or apparent weight. (b) The free-body diagram for the passenger, showing the normal force $\vec{N}$ on him from the scale and the gravitational force $\vec{F}_g$.

(a)   (b)

is greater than the passenger's weight. That reading is a measurement of an apparent weight, because it is made in a noninertial frame. Similarly, for a downward acceleration (either the cab's upward speed is decreasing or its downward speed is increasing), the scale reading is less than the passenger's weight.

(d) During the upward acceleration in part (c), what is the magnitude F_{net} of the net force on the passenger, and what is the magnitude $a_{p,cab}$ of the passenger's acceleration as measured in the frame of the cab? Does $\vec{F}_{net} = m\vec{a}_{p,cab}$?

SOLUTION: One Key Idea here is that the magnitude F_g of the gravitational force on the passenger does not depend on the motion of the passenger or the cab, so, from part (b), F_g is 708 N. From part

(c), the magnitude N of the normal force on the passenger during the upward acceleration is the 939 N reading on the scale. Thus, the net force on the passenger is

$$F_{net} = N - F_g = 939 \text{ N} - 708 \text{ N} = 231 \text{ N}, \text{(Answer)}$$

during the upward acceleration. However, the acceleration $a_{p,cab}$ of the passenger relative to the frame of the cab is zero. Thus, in the noninertial frame of the accelerating cab, F_{net} is not equal to $ma_{p,cab}$, and Newton's second law does not hold.

✔CHECKPOINT 8:   In this sample problem what does the scale read if the elevator cable breaks, so that the cab falls freely; that is, what is the apparent weight of the passenger in free fall?

Sample Problem 5-9

In Fig. 5-20a, a constant horizontal force $\vec{F}_{ap}$ of magnitude 20 N is applied to block A of mass $m_A = 4.0$ kg, which pushes against block B of mass $m_B = 6.0$ kg. The blocks slide over a frictionless surface, along an x axis.

(a) What is the acceleration of the blocks?

SOLUTION: We shall first examine a solution with a serious error, then a dead-end solution, and then a successful solution.

   Serious Error: Because force $\vec{F}_{ap}$ is applied directly to block A, we use Newton's second law to relate that force to the acceleration $\vec{a}$ of block A. Because the motion is along the x axis, we use that law for x components ($F_{net,x} = ma_x$), writing it as

$$F_{ap} = m_A a.$$

However, this is seriously wrong because $\vec{F}_{ap}$ is not the only horizontal force acting on block A. There is also the force $\vec{F}_{AB}$ from block B (as shown in Fig. 5-20b).

   Dead-End Solution: Let us now include force $\vec{F}_{AB}$ by writing, again for the x axis,

$$F_{ap} - F_{AB} = m_A a.$$

(We use the minus sign to include the direction of $\vec{F}_{AB}$.) However, F_{AB} is a second unknown, so we cannot solve this equation for the desired acceleration a.

   Successful Solution: The Key Idea here is that, because of the direction in which force $\vec{F}_{ap}$ is applied, the two blocks form a rigidly connected system. We can relate the net force on the system to the acceleration of the system with Newton's second law. Here, once again for the x axis, we can write that law as

$$F_{ap} = (m_A + m_B)a,$$

Fig. 5-20 Sample Problem 5-9. (a) A constant horizontal force $\vec{F}_{ap}$ is applied to block A, which pushes against block B. (b) Two horizontal forces act on block A: applied force $\vec{F}_{ap}$ and force $\vec{F}_{AB}$ from block B. (c) Only one horizontal force acts on block B: force $\vec{F}_{BA}$ from block A.

where now we properly apply $\vec{F}_{ap}$ to the system with total mass $m_A + m_B$. Solving for a and substituting known values, we find

$$a = \frac{F_{ap}}{m_A + m_B} = \frac{20 \text{ N}}{4.0 \text{ kg} + 6.0 \text{ kg}} = 2.0 \text{ m/s}^2.   \text{(Answer)}$$

Thus, the acceleration of the system and of each block is in the positive direction of the x axis and has the magnitude 2.0 m/s².

(b) What is the force $\vec{F}_{BA}$ on block B from block A (Fig. 5-20c)?

SOLUTION: The Key Idea here is that we can relate the net force on block B to the block's acceleration with Newton's second law. Here we can write that law, still for components along the x axis, as

$$F_{BA} = m_B a,$$

which, with known values, gives

$$F_{BA} = (6.0 \text{ kg})(2.0 \text{ m/s}^2) = 12 \text{ N}.   \text{(Answer)}$$

Thus, force $\vec{F}_{BA}$ is in the positive direction of the x axis and has a magnitude of 12 N.

REVIEW & SUMMARY

Newtonian Mechanics   The velocity of a particle or a particle-like body can change (the particle can accelerate) when the particle is acted on by one or more **forces** (pushes or pulls) from other objects. *Newtonian mechanics* relates accelerations and forces.

Force   Forces are vector quantities. Their magnitudes are defined in terms of the acceleration they would give the standard kilogram. A force that accelerates that standard body by exactly 1 m/s² is defined to have a magnitude of 1 N. The direction of a force is the direction of the acceleration it causes. Forces are combined according to the rules of vector algebra. The **net force** on a body is the vector sum of all the forces acting on it.

Mass The **mass** of a body is the characteristic of that body that relates the body's acceleration to the force (or net force) causing the acceleration. Masses are scalar quantities.

Newton's First Law If there is no net force on a body, the body must remain at rest if it is initially at rest, or move in a straight line at constant speed if it is in motion.

Inertial Reference Frames Reference frames in which Newtonian mechanics holds are called *inertial reference frames* or simply *inertial frames*. We can approximate the ground as an inertial frame if Earth's motions can be neglected. Reference frames in which Newtonian mechanics does not hold are called *noninertial reference frames* or simply *noninertial frames*. An elevator accelerating relative to the ground is a noninertial frame.

Newton's Second Law The net force $\vec{F}_{net}$ on a body with mass m is related to the body's acceleration $\vec{a}$ by

$$\vec{F}_{net} = m\vec{a}, \tag{5-1}$$

which may be written in the component versions

$$F_{net,x} = ma_x \quad F_{net,y} = ma_y \quad \text{and} \quad F_{net,z} = ma_z. \tag{5-2}$$

The second law indicates that in SI units

$$1 \text{ N} = 1 \text{ kg} \cdot \text{m/s}^2. \tag{5-3}$$

A **free-body diagram** is helpful in solving problems with the second law: It is a stripped-down diagram in which only *one* body is considered. That body is represented by a sketch or simply a dot. The external forces on the body are drawn, and a coordinate system is superimposed, oriented so as to simplify the solution.

Some Particular Forces A **gravitational force** $\vec{F}_g$ on a body is a pull by another body. In most situations in this book, the other body is Earth or some other astronomical body. For Earth, the force is directed down toward the ground, which is assumed to be an inertial frame. With that assumption, the magnitude of the force is

$$F_g = mg, \tag{5-8}$$

where m is the body's mass and g is the magnitude of the free-fall acceleration.

The **weight** W of a body is the magnitude of the upward force needed to balance the gravitational force on the body due to Earth (or another astronomical body). It is related to the body's mass by

$$W = mg. \tag{5-12}$$

A **normal force** $\vec{N}$ is the force on a body from a surface against which the body presses. The normal force is always perpendicular to the surface.

A **frictional force** $\vec{f}$ is the force on a body when the body slides or attempts to slide along a surface. The force is always parallel to the surface and directed so as to oppose the motion of the body. On a *frictionless surface*, the frictional force is negligible.

When a cord is under **tension,** it pulls on a body at each of its ends. The pull is directed along the cord, away from the point of attachment to each body. For a *massless* cord (a cord with negligible mass), the pulls at both ends of the cord have the same magnitude T, even if the cord runs around a *massless, frictionless pulley* (a pulley with negligible mass and negligible friction on its axle to oppose its rotation).

Newton's Third Law If a force $\vec{F}_{BC}$ acts on body B due to body C, then there is a force $\vec{F}_{CB}$ on body C due to body B. The forces are equal in magnitude and opposite in direction:

$$\vec{F}_{BC} = -\vec{F}_{CB}.$$

QUESTIONS

1. Two horizontal forces,

$$\vec{F}_1 = (3 \text{ N})\hat{i} - (4 \text{ N})\hat{j} \quad \text{and} \quad \vec{F}_2 = -(1 \text{ N})\hat{i} - (2 \text{ N})\hat{j},$$

pull a banana split across a frictionless lunch counter. Without using a calculator, determine which of the vectors in the free-body diagram of Fig. 5-21 best represent (a) $\vec{F}_1$ and (b) $\vec{F}_2$. What is the net-force component along (c) the x axis and (d) the y axis? Into which quadrants do (e) the net-force vector and (f) the split's acceleration vector point?

2. At time $t = 0$, a single force $\vec{F}$ of constant magnitude begins to act on a rock that is moving along an x axis through deep space. The rock continues to move along that axis. (a) For time

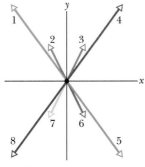

Fig. 5-21 Question 1.

$t > 0$, which of the following is a possible function $x(t)$ for the rock's position: (1) $x = 4t - 3$, (2) $x = -4t^2 + 6t - 3$, (3) $x = 4t^2 + 6t - 3$? (b) For which function is $\vec{F}$ directed opposite the rock's initial direction of motion?

3. Figure 5-22 shows overhead views of four situations in which forces act on a block that lies on a frictionless floor. If the force

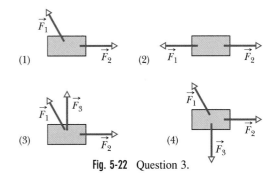

Fig. 5-22 Question 3.

magnitudes are chosen properly, in which situations is it possible that the block is (a) stationary and (b) moving with a constant velocity?

4. In Fig. 5-23, two forces $\vec{F}_1$ and $\vec{F}_2$ act on a "Rocky and Bullwinkle" lunch box as the lunch box slides at constant velocity over a frictionless lunchroom floor. We are to decrease the angle θ of $\vec{F}_1$ without changing the magnitude of $\vec{F}_1$. To keep the lunch box sliding at constant velocity, should we increase, decrease, or maintain the magnitude of $\vec{F}_2$?

Fig. 5-23 Question 4.

5. Figure 5-24 gives the free-body diagram for four situations in which an object is pulled by several forces across a frictionless floor, as seen from overhead. In which situations does the object's acceleration $\vec{a}$ have (a) an x component and (b) a y component? (c) In each situation, give the direction of $\vec{a}$ by naming either a quadrant or a direction along an axis. (This can be done with a few mental calculations.)

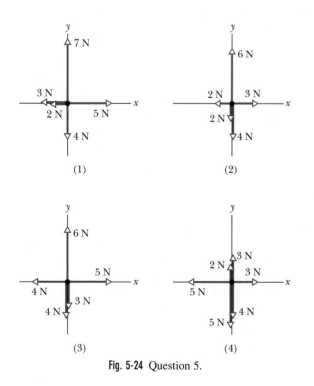

Fig. 5-24 Question 5.

6. Figure 5-25 gives three graphs of velocity component $v_x(t)$ and three graphs of velocity component $v_y(t)$. The graphs are not to scale. Which $v_x(t)$ graph and which $v_y(t)$ graph best correspond to each of the four situations in Question 5 and Fig. 5-24?

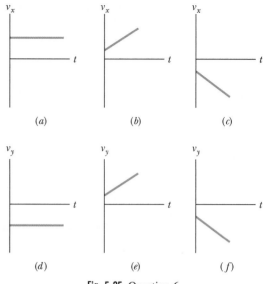

Fig. 5-25 Question 6.

7. The body that is suspended by a rope in Fig. 5-10c has a weight of 75 N. Is T equal to, greater than, or less than 75 N when the body is moving downward at (a) increasing speed and (b) decreasing speed?

8. A vertical force $\vec{F}$ is applied to a block of mass m that lies on a floor. What happens to the magnitude of the normal force $\vec{N}$ on the block from the floor as magnitude F is increased from zero if force $\vec{F}$ is (a) downward and (b) upward?

9. Figure 5-26 shows a train of four blocks being pulled across a frictionless floor by force $\vec{F}$. What total mass is accelerated to the right by (a) force $\vec{F}$, (b) cord 3, and (c) cord 1? (d) Rank the blocks according to their accelerations, greatest first. (e) Rank the cords according to their tension, greatest first. (Warm-up for Problems 34 and 36)

Fig. 5-26 Question 9.

10. Figure 5-27 shows a group of three blocks being pushed across a frictionless floor by horizontal force $\vec{F}$. What total mass is accelerated to the right by (a) force $\vec{F}$, (b) force $\vec{F}_{21}$ on block 2 from block 1, and (c) force $\vec{F}_{32}$ on block 3 from block 2? (d) Rank the blocks according to the magnitudes of their acceleration, greatest first. (e) Rank forces $\vec{F}$, $\vec{F}_{21}$, and $\vec{F}_{32}$ according to their magnitude, greatest first. (Warm-up for Problem 31)

Fig. 5-27 Question 10.

11. In Fig. 5-28a, a toy box is on top of a (heavier) dog house, which sits on a wood floor. In Fig. 5-28b, these objects are represented by dots at the corresponding heights, and six vertical vectors (not to scale) are shown. Which of the vectors best represents

(a)

(b)

Fig. 5-28 Question 11.

(a) the gravitational force on the dog house, (b) the gravitational force on the toy box, (c) the force on the toy box from the dog house, (d) the force on the dog house from the toy box, (e) the force on the dog house from the floor, and (f) the force on the floor from the dog house? (g) Which of the forces are equal in magnitude? Which are (h) greatest and (i) least in magnitude?

12. In Fig. 5-29a, a block is attached by a rope to a bar that is itself rigidly attached to a ramp. Determine whether the magnitudes of the following increase, decrease, or remain the same as the angle θ of the ramp is increased from zero: (a) the component of the

gravitational force $\vec{F}_g$ on the block that is along the ramp, (b) the tension in the cord, (c) the component of $\vec{F}_g$ that is perpendicular to the ramp, and (d) the normal force on the block from the ramp. (e) Which of the curves in Fig. 5-29b corresponds to each of the quantities in parts (a) through (d)?

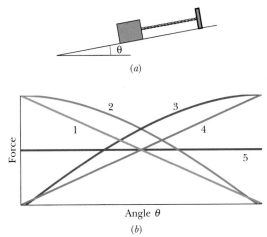

(a)

(b)

Fig. 5-29 Question 12.

EXERCISES & PROBLEMS

SEC. 5-5 Newton's Second Law

1E. If the 1 kg standard body has an acceleration of 2.00 m/s^2 at 20° to the positive direction of the x axis, then what are (a) the x component and (b) the y component of the net force on it, and (c) what is the net force in unit-vector notation?

2E. Two horizontal forces act on a 2.0 kg chopping block that can slide over a frictionless kitchen counter, which lies in an xy plane. One force is $\vec{F}_1 = (3.0 \text{ N})\hat{i} + (4.0 \text{ N})\hat{j}$. Find the acceleration of the chopping block in unit-vector notation when the other force is (a) $\vec{F}_2 = (-3.0 \text{ N})\hat{i} + (-4.0 \text{ N})\hat{j}$, (b) $\vec{F}_2 = (-3.0 \text{ N})\hat{i} + (4.0 \text{ N})\hat{j}$, and (c) $\vec{F}_2 = (3.0 \text{ N})\hat{i} + (-4.0 \text{ N})\hat{j}$.

3E. Only two horizontal forces act on a 3.0 kg body. One force is 9.0 N, acting due east, and the other is 8.0 N, acting 62° north of west. What is the magnitude of the body's acceleration?

4E. While two forces act on it, a particle is to move at the constant velocity $\vec{v} = (3 \text{ m/s})\hat{i} - (4 \text{ m/s})\hat{j}$. One of the forces is $\vec{F}_1 = (2 \text{ N})\hat{i} + (-6 \text{ N})\hat{j}$. What is the other force?

5E. Three forces act on a particle that moves with unchanging velocity $\vec{v} = (2 \text{ m/s})\hat{i} - (7 \text{ m/s})\hat{j}$. Two of the forces are $\vec{F}_1 = (2 \text{ N})\hat{i} + (3 \text{ N})\hat{j} + (-2 \text{ N})\hat{k}$ and $\vec{F}_2 = (-5 \text{ N})\hat{i} + (8 \text{ N})\hat{j} + (-2 \text{ N})\hat{k}$. What is the third force?

6P. Three astronauts, propelled by jet backpacks, push and guide a 120 kg asteroid toward a processing dock, exerting the forces

shown in Fig. 5-30. What is the asteroid's acceleration (a) in unit-vector notation and as (b) a magnitude and (c) a direction?

7P. There are two forces on the 2.0 kg box in the overhead view of Fig. 5-31 but only one is shown. The figure also shows the acceleration of the box. Find the second force (a) in unit-vector notation and as (b) a magnitude and (c) a direction. ssm

8P. Figure 5-32 is an overhead view of a 12 kg tire that is to be pulled by three ropes. One force ($\vec{F}_1$, with magnitude 50 N) is indicated. Orient the other two forces $\vec{F}_2$ and $\vec{F}_3$ so that the magnitude of the resulting acceleration of the tire is least, and find that magnitude if (a) $F_2 = 30 \text{ N}$, $F_3 = 20 \text{ N}$; (b) $F_2 = 30 \text{ N}$, $F_3 = 10 \text{ N}$; and (c) $F_2 = F_3 = 30 \text{ N}$.

SEC. 5-6 Some Particular Forces

9E. (a) An 11.0 kg salami is supported by a cord that runs to a spring scale, which is supported

Fig. 5-30 Problem 6.

Fig. 5-31 Problem 7.

Fig. 5-32 Problem 8.

(b)

(a)

(c)

Fig. 5-33 Exercise 9.

by another cord from the ceiling (Fig. 5-33a). What is the reading on the scale, which is marked in weight units? (b) In Fig. 5-33b the salami is supported by a cord that runs around a pulley and to a scale. The opposite end of the scale is attached by a cord to a wall. What is the reading on the scale? (c) In Fig. 5-33c the wall has been replaced with a second 11.0 kg salami on the left, and the assembly is stationary. What is the reading on the scale now? ssm www

10E. A block with a weight of 3.0 N is at rest on a horizontal surface. A 1.0 N upward force is applied to the block by means of an attached vertical string. What are the magnitude and the direction of the force of the block on the horizontal surface?

11E. A certain particle has a weight of 22 N at a point where $g = 9.8$ m/s². What are its (a) weight and (b) mass at a point where $g = 4.9$ m/s²? What are its (c) weight and (d) mass if it is moved to a point in space where $g = 0$? ssm

12E. Compute the weight of a 75 kg space ranger (a) on Earth, (b) on Mars, where $g = 3.8$ m/s², and (c) in interplanetary space, where $g = 0$. (d) What is the ranger's mass at each of these locations?

SEC. 5-8   Applying Newton's Laws

13E. When a nucleus captures a stray neutron, it must bring the neutron to a stop within the diameter of the nucleus by means of the *strong force*. That force, which "glues" the nucleus together, is approximately zero outside the nucleus. Suppose that a stray neutron with an initial speed of 1.4×10^7 m/s is just barely captured by a nucleus with diameter $d = 1.0 \times 10^{-14}$ m. Assuming that the strong force on the neutron is constant, find the magnitude of that force. The neutron's mass is 1.67×10^{-27} kg. ssm

14E. A 29.0 kg child, with a 4.50 kg backpack on his back, first stands on a sidewalk and then jumps up into the air. Find the magnitude and direction of the force on the sidewalk from the child

when the child is (a) standing still and (b) in the air. Now find the magnitude and direction of the *net* force on Earth due to the child when the child is (c) standing still and (d) in the air.

15E. Refer to Fig. 5-18. Let the mass of the block be 8.5 kg and the angle θ be 30°. Find (a) the tension in the cord and (b) the normal force acting on the block. (c) If the cord is cut, find the magnitude of the block's acceleration. ssm

16E. A 50 kg passenger rides in an elevator that starts from rest on the ground floor of a building at $t = 0$ and rises to the top floor during a 10 s interval. The acceleration of the elevator as a function of the time is shown in Fig. 5-34, where positive values of the acceleration mean that it is directed upward. Give the magnitude and direction of the following forces: (a) the maximum force on the passenger from the floor, (b) the minimum force on the passenger from the floor, and (c) the maximum force on the floor from the passenger.

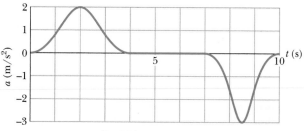

Fig. 5-34 Exercise 16.

17E. *Sunjamming.* A "sun yacht" is a spacecraft with a large sail that is pushed by sunlight. Although such a push is tiny in everyday circumstances, it can be large enough to send the spacecraft outward from the Sun on a cost-free but slow trip. Suppose that the spacecraft has a mass of 900 kg and receives a push of 20 N. (a) What is the magnitude of the resulting acceleration? If the craft starts from rest, (b) how far will it travel in 1 day and (c) how fast will it then be moving?

18E. The tension at which a fishing line snaps is commonly called the line's "strength." What minimum strength is needed for a line that is to stop a salmon of weight 85 N in 11 cm if the fish is initially drifting at 2.8 m/s? Assume a constant deceleration.

19E. An experimental rocket sled can be accelerated at a constant rate from rest to 1600 km/h in 1.8 s. What is the magnitude of the required net force if the sled has a mass of 500 kg? ssm

20E. A car that weighs 1.30×10^4 N is initially moving at a speed of 40 km/h when the brakes are applied and the car is brought to a stop in 15 m. Assuming that the force that stops the car is constant, find (a) the magnitude of that force and (b) the time required for the change in speed. If the initial speed is doubled, and the car experiences the same force during the braking, by what factors are (c) the stopping distance and (d) the stopping time multiplied? (There could be a lesson here about the danger of driving at high speeds.)

21E. An electron with a speed of 1.2×10^7 m/s moves horizontally into a region where a constant vertical force of 4.5×10^{-16} N acts on it. The mass of the electron is 9.11×10^{-31} kg. Determine the vertical distance the electron is deflected during the time it has moved 30 mm horizontally. ssm

22E. A car traveling at 53 km/h hits a bridge abutment. A passenger in the car moves forward a distance of 65 cm (with respect to the road) while being brought to rest by an inflated air bag. What magnitude of force (assumed constant) acts on the passenger's upper torso, which has a mass of 41 kg?

23E. Tarzan, who weighs 820 N, swings from a cliff at the end of a 20 m vine that hangs from a high tree limb and initially makes an angle of 22° with the vertical. Immediately after Tarzan steps off the cliff, the tension in the vine is 760 N. Choose a coordinate system for which the x axis points horizontally away from the edge of the cliff and the y axis points upward. (a) What is the force of the vine on Tarzan in unit-vector notation? (b) What is the net force acting on Tarzan in unit-vector notation? What are (c) the magnitude and (d) the direction of the net force acting on Tarzan? What are (e) the magnitude and (f) the direction of Tarzan's acceleration?

24P. A 50 kg skier is pulled up a frictionless ski slope that makes an angle of 8.0° with the horizontal by holding onto a tow rope that moves parallel to the slope. Determine the magnitude of the force of the rope on the skier at an instant when (a) the rope is moving with a constant speed of 2.0 m/s and (b) the rope is moving with a speed of 2.0 m/s but that speed is increasing at a rate of 0.10 m/s^2.

25P. A 40 kg girl and an 8.4 kg sled are on the frictionless ice of a frozen lake, 15 m apart but connected by a rope of negligible mass. The girl exerts a horizontal 5.2 N force on the rope. (a) What is the acceleration of the sled? (b) What is the acceleration of the girl? (c) How far from the girl's initial position do they meet? ssm

26P. You pull a short refrigerator with a constant force $\vec{F}$ across a greased (frictionless) floor, either with $\vec{F}$ horizontal (case 1) or with $\vec{F}$ tilted upward at an angle θ (case 2). (a) What is the ratio of the refrigerator's speed in case 2 to its speed in case 1 if you pull for a certain time t? (b) What is this ratio if you pull for a certain distance d?

27P. A firefighter with a weight of 712 N slides down a vertical pole with an acceleration of 3.00 m/s^2, directed downward. What are the magnitudes and directions of the vertical forces (a) on the firefighter from the pole and (b) on the pole from the firefighter? ilw

28P. For sport, a 12 kg armadillo runs onto a large pond of level, frictionless ice with an initial velocity of 5.0 m/s along the positive direction of an x axis. Take its initial position on the ice as being the origin. It slips over the ice while being pushed by a wind with a force of 17 N in the positive direction of the y axis. In unit-vector notation, what are the animal's (a) velocity and (b) position vector when it has slid for 3.0 s?

29P. A sphere of mass 3.0×10^{-4} kg is suspended from a cord. A steady horizontal breeze pushes the sphere so that the cord makes a constant angle of 37° with the vertical. Find (a) the magnitude of that push and (b) the tension in the cord. ssm ilw www

30P. A 40 kg skier comes directly down a frictionless ski slope that is inclined at an angle of 10° with the horizontal while a strong wind blows parallel to the slope. Determine the magnitude and direction of the force of the wind on the skier if (a) the magnitude of the skier's velocity is constant, (b) the magnitude of the skier's

velocity is increasing at a rate of 1.0 m/s^2, and (c) the magnitude of the skier's velocity is increasing at a rate of 2.0 m/s^2.

31P. Two blocks are in contact on a frictionless table. A horizontal force is applied to the larger block, as shown in Fig. 5-35. (a) If $m_1 = 2.3$ kg, $m_2 = 1.2$ kg, and $F = 3.2$ N, find the magnitude of the force between the two blocks. (b) Show that if a force of the same magnitude F is applied to the smaller block but in the opposite direction, the magnitude of the force between the blocks is 2.1 N, which is not the same value calculated in (a). (c) Explain the difference. ssm ilw

Fig. 5-35 Problem 31.

32P. A 1400 kg jet engine is fastened to the fuselage of a passenger jet by just three bolts (this is the usual practice). Assume that each bolt supports one-third of the load. (a) Calculate the force on each bolt as the plane waits in line for clearance to take off. (b) During flight, the plane encounters turbulence, which suddenly imparts an upward vertical acceleration of 2.6 m/s^2 to the plane. Calculate the force on each bolt now.

33P. An elevator and its load have a combined mass of 1600 kg. Find the tension in the supporting cable when the elevator, originally moving downward at 12 m/s, is brought to rest with constant acceleration in a distance of 42 m. ssm

34P. Figure 5-36 shows four penguins that are being playfully pulled along very slippery (frictionless) ice by a curator. The masses of three penguins and the tension in two of the cords are given. Find the penguin mass that is not given.

Fig. 5-36 Problem 34.

35P. An 80 kg person is parachuting and experiencing a downward acceleration of 2.5 m/s^2. The mass of the parachute is 5.0 kg. (a) What is the upward force on the open parachute from the air? (b) What is the downward force on the parachute from the person? ssm

36P. In Fig. 5-37, three blocks are connected and pulled to the right on a horizontal frictionless table by a force with a magnitude of $T_3 = 65.0$ N. If $m_1 = 12.0$ kg, $m_2 = 24.0$ kg, and $m_3 = 31.0$ kg, calculate (a) the acceleration of the system and the tensions (b) T_1 and (c) T_2 in the interconnecting cords.

Fig. 5-37 Problem 36.

37P. Imagine a landing craft approaching the surface of Callisto, one of Jupiter's moons. If the engine provides an upward force (thrust) of 3260 N, the craft descends at constant speed; if the engine provides only 2200 N, the craft accelerates downward at 0.39 m/s². (a) What is the weight of the landing craft in the vicinity of Callisto's surface? (b) What is the mass of the craft? (c) What is the magnitude of the free-fall acceleration near the surface of Callisto? ssm

38P. A worker drags a crate across a factory floor by pulling on a rope tied to the crate (Fig. 5-38). The worker exerts a force of 450 N on the rope, which is inclined at 38° to the horizontal, and the floor exerts a horizontal force of 125 N that opposes the motion. Calculate the magnitude of the acceleration of the crate if (a) its mass is 310 kg and (b) its weight is 310 N.

Fig. 5-38 Problem 38.

39P. A motorcycle and 60.0 kg rider accelerate at 3.0 m/s² up a ramp inclined 10° above the horizontal. (a) What is the magnitude of the net force acting on the rider? (b) What is the magnitude of the force on the rider from the motorcycle?

40P. An 85 kg man lowers himself to the ground from a height of 10.0 m by holding onto a rope that runs over a frictionless pulley to a 65 kg sandbag. With what speed does the man hit the ground if he started from rest?

41P. In Fig. 5-39, a chain consisting of five links, each of mass 0.100 kg, is lifted vertically with a constant acceleration of 2.50 m/s². Find the magnitudes of (a) the force on link 1 from link

Fig. 5-39 Problem 41.

2, (b) the force on link 2 from link 3, (c) the force on link 3 from link 4, and (d) the force on link 4 from link 5. Then find the magnitudes of (e) the force $\vec{F}$ on the top link from the person lifting the chain and (f) the *net* force accelerating each link. ssm

42P. A Navy jet (Fig. 5-40) with a weight of 231 kN requires an airspeed of 85 m/s for liftoff. The engine develops a maximum force of 107 kN, but that is insufficient for reaching takeoff speed in the 90 m runway available on an aircraft carrier. What minimum force (assumed constant) is needed from the catapult that is used to help launch the jet? Assume that the catapult and the jet's engine each exert a constant force over the 90 m distance used for takeoff.

Fig. 5-40 Problem 42.

43P. A block of mass $m_1 = 3.70$ kg on a frictionless inclined plane of angle 30.0° is connected by a cord over a massless, frictionless pulley to a second block of mass $m_2 = 2.30$ kg hanging vertically (Fig. 5-41). What are (a) the magnitude of the acceleration of each block and (b) the direction of the acceleration of the hanging block? (c) What is the tension in the cord? ssm ilw www

Fig. 5-41 Problem 43.

44P. In Fig. 5-42, a 1.0 kg pencil box on a 30° frictionless incline is connected to a 3.0 kg pen box on a horizontal frictionless surface. The pulley is frictionless and massless. (a) If the magnitude of $\vec{F}$ is 2.3 N, what is the tension in the connecting cord? (b) What is the largest value that the magnitude of $\vec{F}$ may have without the connecting cord becoming slack?

Fig. 5-42 Problem 44.

45P. A block is projected up a frictionless inclined plane with initial speed $v_0 = 3.50$ m/s. The angle of incline is $\theta = 32.0°$. (a) How far up the plane does it go? (b) How long does it take to get there? (c) What is its speed when it gets back to the bottom? ssm

46P. An interstellar ship has a mass of 1.20×10^6 kg and is initially at rest relative to a star system. (a) What constant acceleration is needed to bring the ship up to a speed of $0.10c$ (where c is the speed of light, 3.0×10^8 m/s) relative to the star system in 3.0 days? (b) What is that acceleration in g units? (c) What force is required for the acceleration? (d) If the engines are shut down when $0.10c$ is reached (the speed then remains constant), how long does the ship take (start to finish) to journey 5.0 light-months, the distance that light travels in 5.0 months?

47P. A 10 kg monkey climbs up a massless rope that runs over a frictionless tree limb and back down to a 15 kg package on the ground (Fig. 5-43). (a) What is the magnitude of the least acceleration the monkey must have if it is to lift the package off the ground? If, after the package has been lifted, the monkey stops its climb and holds onto the rope, what are (b) the magnitude and (c) the direction of the monkey's acceleration, and (d) what is the tension in the rope? ssm

Fig. 5-43 Problem 47.

48P. In earlier days, horses pulled barges down canals in the manner shown in Fig. 5-44. Suppose that the horse pulls on the rope with a force of 7900 N at an angle of 18° to the direction of motion of the barge, which is headed straight along the canal. The mass of the barge is 9500 kg, and its acceleration is 0.12 m/s². What are (a) the magnitude and (b) the direction of the force on the barge from the water?

Fig. 5-44 Problem 48.

49P. In Fig. 5-45, a 5.00 kg block is pulled along a horizontal frictionless floor by a cord that exerts a force of magnitude $F = 12.0$ N at an angle $\theta = 25.0°$ above the horizontal. (a) What is the magnitude of the block's acceleration? (b) The force magnitude F is slowly increased. What is its value just before the block is lifted (completely) off the floor? (c) What is the magnitude of the block's acceleration just before it is lifted (completely) off the floor? ssm

Fig. 5-45 Problem 49.

50P. Figure 5-46 shows a man sitting in a bosun's chair that dangles from a massless rope, which runs over a massless, frictionless pulley and back down to the man's hand. The combined mass of man and chair is 95.0 kg. With what force magnitude must the man pull on the rope if he is to rise (a) with a constant velocity and (b) with an upward acceleration of 1.30 m/s²? (*Hint:* A free-body diagram can really help.) Problem continues, next column.

Fig. 5-46 Problem 50.

Suppose, instead, that the rope on the right extends to the ground, where it is pulled by a co-worker. With what force magnitude must the co-worker pull for the man to rise (c) with a constant velocity and (d) with an upward acceleration of 1.30 m/s²? What is the magnitude of the force on the ceiling from the pulley system in (e) part a (f) part b, (g) part c, and (h) part d?

51P. A block of mass M is pulled along a horizontal frictionless surface by a rope of mass m, as shown in Fig. 5-47. A horizontal force $\vec{F}$ is applied to one end of the rope. (a) Show that the rope *must* sag, even if only by an imperceptible amount. Then, assuming that the sag is negligible, find (b) the acceleration of rope and block, (c) the force on the block from the rope, and (d) the tension in the rope at its midpoint. ssm

Fig. 5-47 Problem 51.

52P. In Fig. 5-48, a 100 kg crate is pushed at constant speed up the frictionless 30.0° ramp by a horizontal force $\vec{F}$. What are the magnitudes of (a) $\vec{F}$ and (b) the force on the crate from the ramp?

Fig. 5-48 Problem 52.

53P. A hot-air balloon of mass M is descending vertically with downward acceleration of magnitude a. How much mass (ballast) must be thrown out to give the balloon an upward acceleration of magnitude a (same magnitude but opposite direction)? Assume that the upward force from the air (the lift) does not change because of the decrease in mass. ssm ilw www

54P. Figure 5-49 shows a section of an alpine cable-car system. The maximum permissible mass of each car with occupants is 2800 kg. The cars, riding on a support cable, are pulled by a second cable attached to each pylon (support tower); assume the cables are straight. What is the difference in tension between adjacent sections of pull cable if the cars are at the maximum permissible mass and are being accelerated up the 35° incline at 0.81 m/s²?

Fig. 5-49 Problem 54.

55P. An elevator with a weight of 27.8 kN is given an upward acceleration of 1.22 m/s² by a cable. (a) Calculate the tension in the cable. (b) What is the tension when the elevator is decelerating at the rate of 1.22 m/s² but is still moving upward?

56P. A lamp hangs vertically from a cord in a descending elevator that decelerates at 2.4 m/s². (a) If the tension in the cord is 89 N, what is the lamp's mass? (b) What is the cord's tension when the elevator ascends with an upward acceleration of 2.4 m/s²?

NEW PROBLEMS

N1. *Blowing off the units.* Throughout your physics course, your instructor will expect you to be careful with the units in your calculations. Yet, some students tend to neglect them and just trust that they always work out properly. Maybe this real-world example will keep you from such a sloppy habit.

   On July 23, 1983, Air Canada Flight 143 was being readied for its long trip from Montreal to Edmonton when the flight crew asked the ground crew to determine how much fuel was already onboard the airplane. The flight crew knew that they needed to begin the trip with 22 300 kg of fuel. They knew that amount in kilograms because Canada had recently switched to the metric system; previously fuel had been measured in pounds. The ground crew could measure the onboard fuel only in liters, which they reported as 7 682 L. Thus, to determine how much fuel was onboard and how much additional fuel must be added, the flight crew asked the ground crew for the conversion factor from liters to kilograms of fuel. The response was 1.77, which the flight crew used (1.77 kg corresponds to 1 L). (a) How many kilograms of fuel did the flight crew think they had? (In this problem, take all the given data as being exact.) (b) How many liters did they ask to be added to the airplane?

   Unfortunately, the response from the ground crew was based on pre-metric habits—the number 1.77 was actually the conversion factor from liters to pounds of fuel (1.77 lb corresponds to 1 L). (c) How many kilograms of fuel were actually onboard? (Except for the given 1.77, use four significant figures for other conversion factors.) (d) How many liters of additional fuel were actually needed? (e) When the airplane left Montreal, what percentage of the required fuel did it actually have?

   On route to Edmonton, at an altitude of 7.9 km, the airplane ran out of fuel and began to fall. Although the airplane then had no power, the pilot somehow managed to put it into a downward glide. However, the nearest working airport was too far to reach by only gliding, so the pilot somehow angled the glide toward an old, nonworking airport.

   Unfortunately, the runway at that airport had been converted to a track for race cars, and a steel barrier had been constructed across it. Fortunately, as the airplane hit the runway, the front landing gear collapsed, dropping the nose of the airplane onto the runway. The skidding slowed the airplane so that it stopped just short of the steel barrier, with stunned race drivers and fans looking on. All onboard the airplane emerged safely. The point here is this: Take care of the units.

N2. In Fig. 5-17 of Sample Problem 5-6, we could change the angle between cord 2 and the ceiling, call it θ_2, by changing the length of cord 2. (The angle between cord 1 and the ceiling is still 28°.) (a) Graph the tension T_1 in cord 1 versus angle θ_2 for θ_2 ranging from 0° to 90°, using the data in the sample problem. Approximately what is the (b) maximum value and (c) minimum value of T_1 and are they physically possible?

N3. A 0.20 kg hockey puck has a velocity of 2.0 m/s toward the east as it slides over the frictionless surface of an ice hockey rink. What constant net force (magnitude and direction) must act on the puck during a 0.40 s time interval to change its velocity to (a) 5.0 m/s toward the west and (b) 5.0 m/s toward the south?

N4. In Fig. 5N-1a, a constant horizontal force $\vec{F}_a$ is applied to block A, which pushes against block B with a 20.0 N force horizontally to the right. In Fig. 5N-1b, the same force $\vec{F}_a$ is applied to block B; now block A pushes on block B with a 10.0 N force horizontally to the left. The blocks have a total mass of 12.0 kg. What are the magnitudes of (a) their acceleration in Fig. 5N-1a and (b) force $\vec{F}_a$?

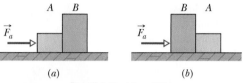

Fig. 5N-1 Problem N4.

N5. The only two forces acting on a body have magnitudes of 20 N and 35 N and directions that differ by 80°. The resulting acceleration has a magnitude of 20 m/s². What is the mass of the body?

N6. An elevator cab is pulled directly upward by a single cable. The elevator cab and its single occupant have a mass of 2000 kg. When that occupant drops a coin, its acceleration relative to the cab is 8.00 m/s² downward. What is the tension in the cable?

N7. A 2.00 kg object is subjected to three forces that give it an acceleration $\vec{a} = -(8.00 \text{ m/s}^2)\hat{i} + (6.00 \text{ m/s}^2)\hat{j}$. If two of the three forces are $\vec{F}_1 = (30.0 \text{ N})\hat{i} + (16.0 \text{ N})\hat{j}$ and $\vec{F}_2 = -(12.0 \text{ N})\hat{i} + (8.00 \text{ N})\hat{j}$, find the third force.

N8. A 2.0 kg particle moves along an x axis, being propelled by a variable force directed along that axis. Its position is given by

$$x = 3.0 \text{ m} + (4.0 \text{ m/s})t + ct^2 - (2.0 \text{ m/s}^3)t^3,$$

with x in meters and t in seconds. The factor c is a constant. At $t = 3.0$ s, the force on the particle has a magnitude of 36 N and is in the negative direction of the axis. What is c?

N9. A dated box of dates, of mass 5.00 kg, is sent sliding up a frictionless ramp at an angle of θ to the horizontal. Figure 5N-2 gives, as a function of time t, the component v_x of the box's velocity along an x axis that extends directly up the ramp. What is the magnitude of the normal force on the box from the ramp?

Fig. 5N-2 Problem N9.

N10. A 2.00 kg FedEx package is pushed across a frictionless floor and over an xy coordinate system by a constant horizontal force $\vec{F}_a$. Figure 5N-3 gives the x and y velocity components of the package versus time t. What are the (a) magnitude and (b) direction of $\vec{F}_a$?

Fig. 5N-3 Problem N10.

N11. Figure 5N-4 shows an overhead view of a lemon half and two of the three horizontal forces that act on it as it is on a frictionless table. Force $\vec{F}_1$ has a magnitude of 6.00 N and is at $\theta_1 = 30°$. Force $\vec{F}_2$ has a magnitude of 7.00 N and is at $\theta_2 = 30°$. The lemon half has mass 0.0250 kg. In unit-vector notation, what is the third force if the lemon half (a) is stationary, (b) has constant velocity $\vec{v} = (13.0\hat{i} - 14.0\hat{j})$ m/s, and (c) has varying velocity $\vec{v} = (13.0t\hat{i} - 14.0t\hat{j})$ m/s^2, where t is time?

Fig. 5N-4 Problem N11.

N12. Figure 5N-5 shows an arrangement in which four disks are suspended by cords. The longer, top cord loops over a frictionless pulley and pulls with a force of magnitude 98 N on the wall to which it is attached. The tensions in the shorter cords are $T_1 = 58.8$ N, $T_2 = 49.0$ N, and $T_3 = 9.8$ N. What are the masses of (a) disk A, (b) disk B, (c) disk C, and (d) disk D?

N13. Suppose that the 1 kg standard body accelerates at 4.00 m/s^2 at 160° from the positive x direction owing to two forces; one is $\vec{F}_1 = (2.50$ N$)\hat{i} + (4.60$ N$)\hat{j}$. What is the other force (a) in unit-vector notation and (b) as a magnitude and direction?

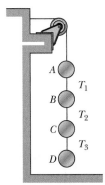

Fig. 5N-5 Problem N12.

N14. In Fig. 5N-6, elevator cabs A and B are connected by a short cable and can be pulled upward or lowered by the cable above cab A. Cab A has mass 1700 kg; cab B has mass 1300 kg. A 12.0 kg box of catnip lies on the floor of cab A. The tension in the cable connecting the cabs is 1.91×10^4 N. What is the magnitude of the normal force on the box from the floor?

Fig. 5N-6 Problem N14.

N15. Starting from rest, an airplane accelerates for takeoff at 2.3 m/s^2. It has two jet engines, each exerting a force (thrust) on the airplane of 1.4×10^5 N. What is the airplane's weight?

N16. In Fig. 5N-7, three ballot boxes are connected by cords, one of which wraps over a pulley with negligible friction on its axle and negligible mass. The masses are the following: $m_A = 30.0$ kg, $m_B = 40.0$ kg, $m_C = 10.0$ kg.

Fig. 5N-7 Problem N16.

When the assembly is released from rest, (a) what is the tension in the cord that connects boxes B and C, and (b) how far does box A move in the first 0.250 s (assuming it does not reach the pulley)?

N17. In Fig. 5N-8, forces act on blocks A and B, which are connected by a string. Force $\vec{F}_A = (12$ N$)\hat{i}$ acts on block A, with mass 4.0 kg. Force $\vec{F}_B = (24$ N$)\hat{i}$ acts on block B, with mass 6.0 kg. What is the tension in the string?

Fig. 5N-8 Problem N17.

N18. Two horizontal forces $\vec{F}_1$ and $\vec{F}_2$ act on a 4.0 kg disk that slides over frictionless ice, on which an xy coordinate system is laid out. Force $\vec{F}_1$ is in the positive direction of the x axis and has a magnitude of 7.0 N. Force $\vec{F}_2$ has a magnitude of 9.0 N. Figure 5N-9 gives the x component v_x of the velocity of the disk as a function of time t during the sliding. What is the angle between the constant directions of forces $\vec{F}_1$ and $\vec{F}_2$?

Fig. 5N-9 Problem N18.

N19. A rocket and its payload have a total mass of 5.0×10^4 kg. How large is the force produced by the engine (the thrust) when (a) the rocket is "hovering" over the launchpad just after ignition, and (b) the rocket is accelerating upward at 20 m/s^2?

N20. Figure 5N-10 shows three blocks attached by cords that loop over frictionless pulleys. Block B lies on a frictionless table; the masses are: $m_A = 6.00$ kg, $m_B = 8.00$ kg, $m_C = 10.0$ kg. When the blocks are released, what is the tension in the cord at the right?

Fig. 5N-10 Problem N20.

N21. An 80 kg man drops to a concrete patio from a window only 0.50 m above the patio. He neglects to bend his knees on landing, taking 2.0 cm to stop. (a) What is his average acceleration from when his feet first touch the patio to when he stops? (b) What is the magnitude of the average stopping force?

N22. Figure 5N-11 gives, as a function of time t, the force component F_x that acts on a 3.00 kg ice block, which can move only along the x axis. At $t = 0$, the block is moving in the positive direction of the axis, with a speed of 3.0 m/s. What are its (a) speed and (b) direction of travel at $t = 11$ s?

Fig. 5N-11 Problem N22.

N23. Two people pull with 90 N and 92 N in opposite directions on a 25 kg sled on frictionless ice. What is the sled's acceleration magnitude?

6 Force and Motion—II

Cats, who enjoy sleeping on window sills, are often kept in apartment buildings. When a cat accidentally falls out of a window and onto a sidewalk, the extent of injury (such as the number of fractured bones or the certainty of death) *decreases* with height if the fall is more than seven or eight floors. (There is even a record of a cat who fell 32 floors and suffered only slight damage to its thorax and one tooth.)

How can the danger possibly decrease with height?

The answer is in this chapter.

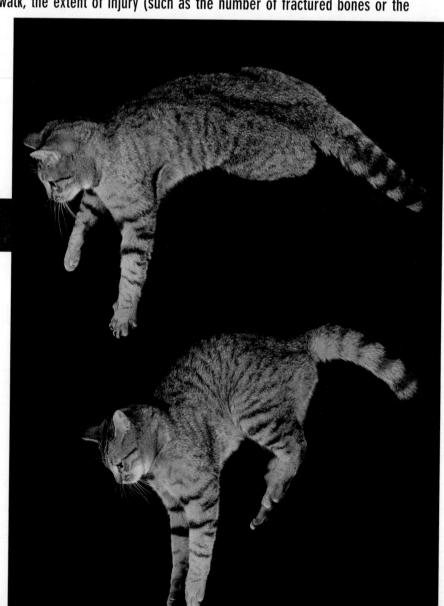

6-1 Friction

Frictional forces are unavoidable in our daily lives. If we were not able to counteract them, they would stop every moving object and bring to a halt every rotating shaft. About 20% of the gasoline used in an automobile is needed to counteract friction in the engine and in the drive train. On the other hand, if friction were totally absent, we could not get an automobile to go anywhere, and we could not walk or ride a bicycle. We could not hold a pencil and, if we could, it would not write. Nails and screws would be useless, woven cloth would fall apart, and knots would untie.

Here we deal with the frictional forces that exist between dry solid surfaces, either stationary relative to each other or moving across each other at slow speeds. Consider three simple thought experiments:

1. Send a book sliding across a long horizontal counter. As expected, the book slows and then stops. This means the book must have an acceleration parallel to the counter surface, in the direction opposite the book's velocity. From Newton's law, then, a force must act on the book parallel to the counter surface, in the direction opposite its velocity. That force is a frictional force.

2. Push horizontally on the book to make it travel at constant velocity along the counter. Can the force from you be the only horizontal force on the book? No, because then the book would accelerate. From Newton's second law, there must be a second force, directed opposite your force but with the same magnitude, so that the two forces balance. That second force is a frictional force, directed parallel to the counter.

3. Push horizontally on a heavy crate. The crate does not move. From Newton's second law, a second force must also act on the crate to counteract your force. Moreover, it must be directed opposite your force and have the same magnitude as your force, so that the two forces balance. That second force is a frictional force. Push even harder. The crate still does not move. Apparently the frictional force can change in magnitude so that the two forces still balance. Now push with all your strength. The crate begins to slide. Evidently, there is a maximum magnitude of the frictional force. When you exceed that maximum magnitude, the crate slides.

Figure 6-1 shows a similar situation in detail. In Fig. 6-1a, a block rests on a tabletop, with the gravitational force $\vec{F}_g$ balanced by a normal force $\vec{N}$. In Fig. 6-1b, you exert a force $\vec{F}$ on the block, attempting to pull it to the left. In response, a frictional force $\vec{f}_s$ is directed to the right, exactly balancing your force. The force $\vec{f}_s$ is called the **static frictional force.** The block does not move.

Figures 6-1c and 6-1d show that as you increase the magnitude of your applied force, the magnitude of the static frictional force $\vec{f}_s$ also increases and the block remains at rest. When the applied force reaches a certain magnitude, however, the block "breaks away" from its intimate contact with the tabletop and accelerates leftward (Fig. 6-1e). The frictional force that then opposes the motion is called the **kinetic frictional force** $\vec{f}_k$.

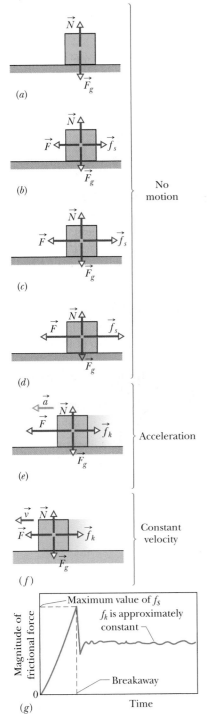

Fig. 6-1 (a) The forces on a stationary block. (b–d) An external force $\vec{F}$, applied to the block, is balanced by a static frictional force $\vec{f}_s$. As F is increased, f_s also increases, until f_s reaches a certain maximum value. (e) The block then "breaks away," accelerating suddenly in the direction of $\vec{F}$. (f) If the block is now to move with constant velocity, F must be reduced from the maximum value it had just before the block broke away. (g) Some experimental results for the sequence (a) through (f).

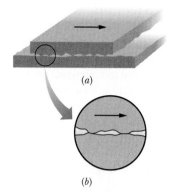

(a)

(b)

Fig. 6-2 The mechanism of sliding friction. (*a*) The upper surface is sliding to the right over the lower surface in this enlarged view. (*b*) A detail, showing two spots where cold-welding has occurred. Force is required to break the welds and maintain the motion.

Usually, the magnitude of the kinetic frictional force, which acts when there is motion, is less than the maximum magnitude of the static frictional force, which acts when there is no motion. Thus, if you wish the block to move across the surface with a constant speed, you must usually decrease the magnitude of the applied force once the block begins to move, as in Fig. 6-1*f*. As an example, Fig. 6-1*g* shows the results of an experiment in which the force on a block was slowly increased until breakaway occurred. Note the reduced force needed to keep the block moving at constant speed after breakaway.

A frictional force is, in essence, the vector sum of many forces acting between surface atoms of one body and those of another body. If two highly polished and carefully cleaned metal surfaces are brought together in a very good vacuum (to keep them clean), they cannot be made to slide over each other. Because the surfaces are so smooth, many atoms of one surface contact many atoms of the other surface, and the surfaces *cold-weld* together instantly, forming a single piece of metal. If a machinist's specially polished gage blocks are brought together in air, there is less atom-to-atom contact but the blocks stick firmly to each other and can be separated only by means of a wrenching motion. Usually, however, this much atom-to-atom contact is not possible. Even a highly polished metal surface is far from being flat on the atomic scale. Moreover, the surfaces of everyday objects have layers of oxides and other contaminants that reduce cold-welding.

When two ordinary surfaces are placed together, only the high points touch each other. (It is like having the Alps of Switzerland turned over and placed down on the Alps of Austria.) The actual *micro*scopic area of contact is much less than the apparent *macro*scopic contact area, perhaps by a factor of 10^4. Nonetheless, many contact points do cold-weld together. These welds produce static friction when an applied force attempts to slide the surfaces relative to each other.

If the applied force is great enough to pull one surface across the other, there is first a tearing of welds (at breakaway) and then a continuous re-forming and tearing apart of welds as movement occurs and chance contacts are made (Fig. 6-2). The kinetic frictional force $\vec{f}_k$ that opposes the motion is the vector sum of the forces at those many chance contacts.

If the two surfaces are pressed together harder, many more points cold-weld. Then, getting the surfaces to slide relative to each other requires a greater applied force: The static frictional force $\vec{f}_s$ has a greater maximum value. Once the surfaces are sliding, there are many more points of momentary cold-welding, so the kinetic frictional force $\vec{f}_k$ also has a greater magnitude.

Often, the sliding motion of one surface on another is "jerky," because the two surfaces alternately stick together and then slip. Such repetitive *stick-and-slip* can produce squeaking or squealing, as when tires skid on dry pavement, fingernails scratch along a chalkboard, or a rusty hinge is opened. It can also produce beautiful sounds, as when a bow is drawn properly across a violin string.

6-2 Properties of Friction

Experiment shows that when a dry and unlubricated body presses against a surface in the same condition, and a force $\vec{F}$ attempts to slide the body along the surface, the resulting frictional force has three properties:

Property 1.   If the body does not move, then the static frictional force $\vec{f}_s$ and the component of $\vec{F}$ that is parallel to the surface balance each other. They are equal in magnitude, and $\vec{f}_s$ is directed opposite that component of $\vec{F}$.

Property 2.   The magnitude of $\vec{f}_s$ has a maximum value $f_{s,\text{max}}$ that is given by

$$f_{s,\text{max}} = \mu_s N, \tag{6-1}$$

where μ_s is the **coefficient of static friction** and N is the magnitude of the normal force on the body from the surface. If the magnitude of the component of $\vec{F}$ that is parallel to the surface exceeds $f_{s,max}$, then the body begins to slide along the surface.

Property 3.   If the body begins to slide along the surface, the magnitude of the frictional force rapidly decreases to a value f_k given by

$$f_k = \mu_k N, \qquad (6\text{-}2)$$

where μ_k is the **coefficient of kinetic friction.** Thereafter, during the sliding, a kinetic frictional force $\vec{f}_k$ with magnitude given by Eq. 6-2 opposes the motion.

The magnitude N of the normal force appears in properties 2 and 3 as a measure of how firmly the body presses against the surface. If the body presses harder, then, by Newton's third law, N is greater. Properties 1 and 2 are worded in terms of a single applied force $\vec{F}$, but they also hold for the net force of several applied forces acting on the body. Equations 6-1 and 6-2 are *not* vector equations; the direction of $\vec{f}_s$ or $\vec{f}_k$ is always parallel to the surface and opposed to the attempted sliding, and the normal force $\vec{N}$ is perpendicular to the surface.

The coefficients μ_s and μ_k are dimensionless and must be determined experimentally. Their values depend on certain properties of both the body and the surface; hence, they are usually referred to with the preposition "between," as in "the value of μ_s *between* an egg and a Teflon-coated skillet is 0.04, but that *between* rock-climbing shoes and rock is as much as 1.2." We assume that the value of μ_k does not depend on the speed at which the body slides along the surface.

✔**CHECKPOINT 1:**   A block lies on a floor. (a) What is the magnitude of the frictional force on it from the floor? (b) If a horizontal force of 5 N is now applied to the block, but the block does not move, what is the magnitude of the frictional force on it? (c) If the maximum value $f_{s,max}$ of the static frictional force on the block is 10 N, will the block move if the magnitude of the horizontally applied force is 8 N? (d) If the magnitude is 12 N? (e) What is the magnitude of the frictional force in part (c)?

Sample Problem 6-1

If a car's wheels are "locked" (kept from rolling) during emergency braking, the car slides along the road. Ripped-off bits of tire and small melted sections of road form the "skid marks" that reveal that cold-welding occurred during the slide. The record for the longest skid marks on a public road was reportedly set in 1960 by a Jaguar on the M1 highway in England (Fig. 6-3a)—the marks were 290 m long! Assuming that $\mu_k = 0.60$ and the car's acceleration was constant during the braking, how fast was the car going when the wheels became locked?

SOLUTION: One **Key Idea** here is that, because the acceleration is assumed constant, we can use the equations of Table 2-1 to find the car's initial speed v_0. Let us try Eq. 2-16,

$$v^2 = v_0^2 + 2a(x - x_0), \qquad (6\text{-}3)$$

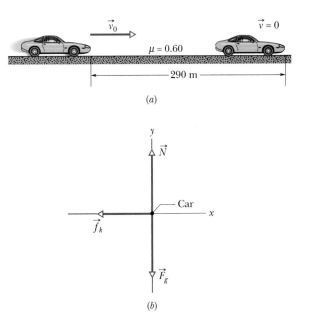

(a)

(b)

Fig. 6-3 Sample Problem 6-1. (a) A car, sliding to the right and finally stopping after a displacement of 290 m. (b) A free-body diagram for the car.

assuming that the car moved in the positive direction of an x axis. We know that the displacement $x - x_0$ was 290 m, we assume that the final speed v was 0, and we want v_0. However, we do not know the car's acceleration a.

To find a, we use another Key Idea: If we neglect the effects of the air on the car, the acceleration a was due only to a kinetic frictional force $\vec{f}_k$ on the car from the road, directed opposite the direction of the car's motion (Fig. 6-3b). We can relate this force to the acceleration by writing Newton's second law for x components ($F_{net,x} = ma_x$) as

$$-f_k = ma, \tag{6-4}$$

where m is the car's mass. The minus sign indicates the direction of the kinetic frictional force.

From Eq. 6-2, that frictional force has the magnitude $f_k = \mu_k N$, where N is the magnitude of the normal force on the car from the road. Because the car is not accelerating vertically, we know

from Fig. 6-3b and Newton's second law that the magnitude of $\vec{N}$ is equal to the magnitude of the gravitational force $\vec{F}_g$ on the car, which is mg. Thus, we have $N = mg$.

Now solving Eq. 6-4 for a and substituting $f_k = \mu_k N = \mu_k mg$ for f_k yield

$$a = -\frac{f_k}{m} = -\frac{\mu_k mg}{m} = -\mu_k g,$$

where the minus sign indicates that the acceleration is in the negative direction of the x axis, opposite the direction of the velocity. Next, substituting this for a in Eq. 6-3 with $v = 0$ and solving for v_0 give

$$v_0 = \sqrt{2\mu_k g(x - x_0)} = \sqrt{(2)(0.60)(9.8 \text{ m/s}^2)(290 \text{ m})}$$
$$= 58 \text{ m/s} = 210 \text{ km/h}. \quad \text{(Answer)}$$

We assumed that $v = 0$ at the far end of the skid marks. Actually, the marks ended only because the Jaguar left the road after 290 m. So v_0 was at least 210 km/h, and possibly much more.

Sample Problem 6-2

In Fig. 6-4a, a woman pulls a loaded sled of mass $m = 75$ kg along a horizontal surface at constant velocity. The coefficient of kinetic friction μ_k between the runners and the snow is 0.10, and the angle ϕ is 42°.

(a) What is the magnitude of the force $\vec{T}$ on the sled from the rope?

SOLUTION: We need three Key Ideas here:

1. Because the sled's velocity is constant, its acceleration is zero in spite of the woman's pull.

2. Acceleration is prevented by a kinetic frictional force $\vec{f}_k$ on the sled from the snow.

3. We can relate the (zero) acceleration of the sled to the forces on the sled, including the desired $\vec{T}$, with Newton's second law ($\vec{F}_{net} = m\vec{a}$).

Figure 6-4b shows the forces on the sled, including the gravitational force $\vec{F}_g$ and the normal force $\vec{N}$ from the snow surface. For these forces, Newton's second law, with $\vec{a} = 0$, gives us

$$\vec{T} + \vec{N} + \vec{F}_g + \vec{f}_k = 0. \tag{6-5}$$

We cannot solve Eq. 6-5 for $\vec{T}$ directly on a vector-capable calculator because we do not know the other vectors, so we rewrite it for components along the x and y axes of Fig. 6-4b. For the x axis, we get

$$T_x + 0 + 0 - f_k = 0$$

or

$$T \cos \phi - \mu_k N = 0, \tag{6-6}$$

where we have used Eq. 6-2 to substitute $\mu_k N$ for f_k. For the y axis, we have

$$T_y + N - F_g + 0 = 0$$

or

$$T \sin \phi + N - mg = 0, \tag{6-7}$$

where we have substituted mg for F_g.

Equations 6-6 and 6-7 are simultaneous equations with unknowns T and N. To solve them for T, we must first solve Eq. 6-6

for N and then substitute that expression into Eq. 6-7 to get

$$T = \frac{\mu_k mg}{\cos \phi + \mu_k \sin \phi}$$
$$= \frac{(0.10)(75 \text{ kg})(9.8 \text{ m/s}^2)}{\cos 42° + (0.10)(\sin 42°)}$$
$$= 91 \text{ N}. \quad \text{(Answer)}$$

(Instead, we could substitute known data into Eqs. 6-6 and 6-7 and then solve them by using a simultaneous-equations solver on a calculator.)

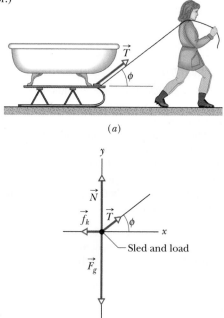

(a)

(b)

Fig. 6-4 Sample Problem 6-2. (a) A woman pulls a loaded sled at constant velocity via a force $\vec{T}$ on the sled from a rope. (b) A free-body diagram for the loaded sled.

$\phi = ?$ FOR $T =$ MINIMUM
FOR WORD DONE $=$ MINIMUM ?

(b) If the woman increases her pull on the rope, so that T is greater than 91 N, is the magnitude f_k of the frictional force greater than, less than, or the same as in (a)?

SOLUTION: The Key Idea here is that, by Eq. 6-2, the magnitude of f_k depends directly on the magnitude N of the normal force. Thus, we can answer if we find a relation between N and T. Equation 6-7 is such a relation. Rewriting it as

$$N = mg - T \sin \phi, \tag{6-8}$$

we note that if T is increased, then N will decrease. (The physical reason is that the upward component of the rope's pull is greater, and thus the force on the sled from the snow is less.) Because $f_k = \mu_k N$, we see that f_k will be less than it was.

✓CHECKPOINT 2:   In the figure, horizontal force $\vec{F}_1$ of magnitude 10 N is applied to a box on a floor, but the box does not slide. Then, as the magnitude of vertically applied force $\vec{F}_2$ is increased from zero but before the box begins to slide, do the following quantities increase, decrease, or stay the same: (a) the magnitude of the frictional force on the box; (b) the magnitude of the normal force on the box from the floor; (c) the maximum value $f_{s,\text{max}}$ of the static frictional force on the box?

Sample Problem 6-3

Figure 6-5a shows a coin of mass m at rest on a book that has been tilted at an angle θ with the horizontal. By experimenting, you find that when θ is increased to 13°, the coin is on the *verge* of sliding down the book, which means that even a slight increase beyond 13° produces sliding. What is the coefficient of static friction μ_s between the coin and the book?

SOLUTION: If the book were frictionless, the coin would surely slide down it for any tilt of the book because of the gravitational force on the coin. Thus, one Key Idea here is that a frictional force $\vec{f}_s$ must be holding the coin in place. A second Key Idea is that, because the coin is *on the verge* of sliding *down* the book, that force is at its *maximum* magnitude $f_{s,\text{max}}$ and is directed *up* the book. Also, from Eq. 6-1, we know that $f_{s,\text{max}} = \mu_s N$, where N is the magnitude of the normal force $\vec{N}$ on the coin from the book. Thus,

$$f_s = f_{s,\text{max}} = \mu_s N,$$

from which
$$\mu_s = \frac{f_s}{N}. \tag{6-9}$$

To evaluate this equation, we need to find the force magnitudes f_s and N. To do that, we use another Key Idea: When the coin is on the verge of sliding, it is stationary and thus its acceleration $\vec{a}$ is zero. We can relate this acceleration to the forces on the coin with Newton's second law ($\vec{F}_{\text{net}} = m\vec{a}$). As shown in the free-body diagram of the coin in Fig. 6-5b, those forces are (1) the frictional force $\vec{f}_s$, (2) the normal force $\vec{N}$, and (3) the gravitational force $\vec{F}_g$ on the coin, with magnitude equal to mg. Then, from Newton's second law with $\vec{a} = 0$, we have

$$\vec{f}_s + \vec{N} + \vec{F}_g = 0. \tag{6-10}$$

To find f_s and N, we rewrite Eq. 6-10 for components along the x and y axes of the tilted coordinate system in Fig. 6-5b. For the x axis and with mg substituted for F_g, we have

$$f_s + 0 - mg \sin \theta = 0,$$

so
$$f_s = mg \sin \theta. \tag{6-11}$$

Similarly, for the y axis we have

$$0 + N - mg \cos \theta = 0,$$

so
$$N = mg \cos \theta. \tag{6-12}$$

Substituting Eqs. 6-11 and 6-12 into Eq. 6-9 produces

$$\mu_s = \frac{mg \sin \theta}{mg \cos \theta} = \tan \theta, \tag{6-13}$$

which here means

$$\mu_s = \tan 13° = 0.23. \qquad \text{(Answer)}$$

Actually you do not need to measure θ to get μ_s. Instead, measure the two lengths shown in Fig. 5-6a and then substitute h/d for $\tan \theta$ in Eq. 6-13.

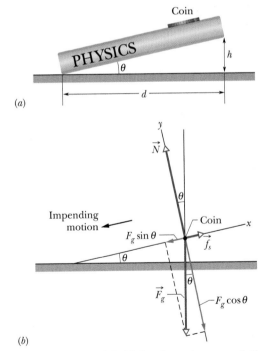

(a)

(b)

Fig. 6-5 Sample Problem 6-3. (a) A coin on the verge of sliding down a book. (b) A free-body diagram for the coin, showing the three forces (drawn to scale) that act on it. The gravitational force $\vec{F}_g$ is shown resolved into its components along the x and the y axes, whose orientations are chosen to simplify the problem. Component $F_g \sin \theta$ tends to slide the coin down the book. Component $F_g \cos \theta$ presses the coin onto the book.

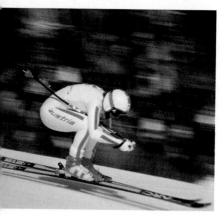

Fig. 6-6 This skier crouches in an "egg position" so as to minimize her effective cross-sectional area and thus the air drag acting on her.

6-3 The Drag Force and Terminal Speed

A **fluid** is anything that can flow—generally either a gas or a liquid. When there is a relative velocity between a fluid and a body (either because the body moves through the fluid or because the fluid moves past the body), the body experiences a **drag force** $\vec{D}$ that opposes the relative motion and points in the direction in which the fluid flows relative to the body.

Here we examine only cases in which air is the fluid, the body is blunt (like a baseball) rather than slender (like a javelin), and the relative motion is fast enough so that the air becomes turbulent (breaks up into swirls) behind the body. In such cases, the magnitude of the drag force $\vec{D}$ is related to the relative speed v by an experimentally determined **drag coefficient** C according to

$$D = \tfrac{1}{2}C\rho A v^2, \tag{6-14}$$

where ρ is the air density (mass per volume) and A is the **effective cross-sectional area** of the body (the area of a cross section taken perpendicular to the velocity $\vec{v}$). The drag coefficient C (typical values range from 0.4 to 1.0) is not truly a constant for a given body, because if v varies significantly, the value of C can vary as well. Here, we ignore such complications.

Downhill speed skiers know well that drag depends on A and v^2. To reach high speeds a skier must reduce D as much as possible by, for example, riding the skis in the "egg position" (Fig. 6-6) to minimize A.

When a blunt body falls from rest through air, the drag force $\vec{D}$ is directed upward; its magnitude gradually increases from zero as the speed of the body increases. This upward force $\vec{D}$ opposes the downward gravitational force $\vec{F}_g$ on the body. We can relate these forces to the body's acceleration by writing Newton's second law for a vertical y axis ($F_{\text{net},y} = ma_y$) as

$$D - F_g = ma, \tag{6-15}$$

where m is the mass of the body. As suggested in Fig. 6-7, if the body falls long enough, D eventually equals F_g. From Eq. 6-15, this means that $a = 0$ and so the body's speed no longer increases. The body then falls at a constant speed, called the **terminal speed** v_t.

To find v_t, we set $a = 0$ in Eq. 6-15 and substitute for D from Eq. 6-14, obtaining

$$\tfrac{1}{2}C\rho A v_t^2 - F_g = 0,$$

which gives

$$v_t = \sqrt{\frac{2F_g}{C\rho A}}. \tag{6-16}$$

Table 6-1 gives values of v_t for some common objects.

According to calculations* based on Eq. 6-14, a cat must fall about six floors to reach terminal speed. Until it does so, $F_g > D$ and the cat accelerates downward because of the net downward force. Recall from Chapter 2 that your body is an accelerometer, not a speedometer. Because the cat also senses the acceleration, it is frightened and keeps its feet underneath its body, its head tucked in, and its spine bent upward, making A small, v_t large, and injury on landing likely.

However, if the cat does reach v_t, the acceleration vanishes and the cat relaxes somewhat, stretching its legs and neck horizontally outward and straightening its spine (it then resembles a flying squirrel). These actions increase area A and thus

---

*W. O. Whitney and C. J. Mehlhaff, "High-Rise Syndrome in Cats." *The Journal of the American Veterinary Medical Association*, 1987, Vol. 191, pp. 1399–1403.

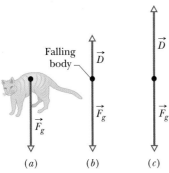

Fig. 6-7 The forces that act on a body falling through air: (a) the body when it has just begun to fall and (b) the free-body diagram a little later, after a drag force has developed. (c) The drag force has increased until it balances the gravitational force on the body. The body now falls at its constant terminal speed.

TABLE 6-1 Some Terminal Speeds in Air

Object	Terminal Speed (m/s)	95% Distance[a] (m)
Shot (from shot put)	145	2500
Sky diver (typical)	60	430
Baseball	42	210
Tennis ball	31	115
Basketball	20	47
Ping-Pong ball	9	10
Raindrop (radius = 1.5 mm)	7	6
Parachutist (typical)	5	3

[a]This is the distance through which the body must fall from rest to reach 95% of its terminal speed.

Source: Adapted from Peter J. Brancazio, *Sport Science*, 1984, Simon & Schuster, New York.

Fig. 6-8 A sky diver in a horizontal "spread eagle" maximizes the air drag.

also, by Eq. 6-14, the drag D. The cat begins to slow because now $D > F_g$ (the net force is upward), until a new, smaller v_t is reached. The decrease in v_t reduces the possibility of serious injury on landing. Just before the end of the fall, when it sees it is nearing the ground, the cat pulls its legs back beneath its body to prepare for the landing.

Humans often fall from great heights for the fun of skydiving. However, in April 1987, during a jump, sky diver Gregory Robertson noticed that fellow sky diver Debbie Williams had been knocked unconscious in a collision with a third sky diver and was unable to open her parachute. Robertson, who was well above Williams at the time and who had not yet opened his parachute for the 4 km plunge, reoriented his body head-down so as to minimize A and maximize his downward speed. Reaching an estimated v_t of 320 km/h, he caught up with Williams and then went into a horizontal "spread eagle" (as in Fig. 6-8) to increase D so that he could grab her. He opened her parachute and then, after releasing her, his own, a scant 10 s before impact. Williams received extensive internal injuries due to her lack of control on landing but survived.

Sample Problem 6-4

If a falling cat reaches a first terminal speed of 97 km/h while it is tucked in and then stretches out, doubling A, how fast is it falling when it reaches a new terminal speed?

SOLUTION: The Key Idea here is that the terminal speeds of the cat depend on (among other things) the effective cross-sectional areas A of the cat, according to Eq. 6-16. Thus, we can use that equation

to set up a ratio of speeds. We let v_{to} and v_{tn} represent the original and new terminal speeds, and A_o and A_n the original and new areas. Then by Eq. 6-16,

$$\frac{v_{tn}}{v_{to}} = \frac{\sqrt{2F_g/C\rho A_n}}{\sqrt{2F_g/C\rho A_o}} = \sqrt{\frac{A_o}{A_n}} = \sqrt{\frac{A_o}{2A_o}} = \sqrt{0.5} \approx 0.7,$$

which means that $v_{tn} \approx 0.7 v_{to}$, or about 68 km/h.

Sample Problem 6-5

A raindrop with radius $R = 1.5$ mm falls from a cloud that is at height $h = 1200$ m above the ground. The drag coefficient C for the drop is 0.60. Assume that the drop is spherical throughout its fall. The density of water ρ_w is 1000 kg/m³, and the density of air ρ_a is 1.2 kg/m³.

(a) What is the terminal speed of the drop?

SOLUTION: The Key Idea here is that the drop reaches a terminal speed v_t when the gravitational force on it is balanced by the air drag force on it, so its acceleration is zero. We could then apply Newton's second law and the drag force equation to find v_t, but Eq. 6-16 does all that for us.

To use Eq. 6-16, we need the drop's effective cross-sectional area A and the magnitude F_g of the gravitational force. Because the drop is spherical, A is the area of a circle (πR^2) with the same radius as the sphere. To find F_g, we use three facts: (1) $F_g = mg$, where m is the drop's mass; (2) the (spherical) drop's volume is $V = \frac{4}{3}\pi R^3$; and (3) the density of the water in the drop is the mass per volume, or $\rho_w = m/V$. Thus, we find

$$F_g = V\rho_w g = \tfrac{4}{3}\pi R^3 \rho_w g.$$

We next substitute this, the expression for A, and the given data into Eq. 6-16. Being careful to distinguish between the air density ρ_a and the water density ρ_w, we obtain

$$v_t = \sqrt{\frac{2F_g}{C\rho_a A}} = \sqrt{\frac{8\pi R^3 \rho_w g}{3C\rho_a \pi R^2}} = \sqrt{\frac{8R\rho_w g}{3C\rho_a}}$$

$$= \sqrt{\frac{(8)(1.5 \times 10^{-3}\ \text{m})(1000\ \text{kg/m}^3)(9.8\ \text{m/s}^2)}{(3)(0.60)(1.2\ \text{kg/m}^3)}}$$

$$= 7.4\ \text{m/s} \approx 27\ \text{km/h.}\qquad\text{(Answer)}$$

Note that the height of the cloud does not enter into the calculation. As Table 6-1 indicates, the raindrop reaches terminal speed after falling just a few meters.

(b) What would be the drop's speed just before impact if there were no drag force?

SOLUTION: The Key Idea here is that, with no drag force to reduce the drop's speed during the fall, the drop would fall with the constant free-fall acceleration g, so the constant-acceleration equations of Table 2-1 apply. Because we know the acceleration is g, the initial velocity v_0 is 0, and the displacement $x - x_0$ is h, we use Eq. 2-16 to find v:

$$v = \sqrt{2gh} = \sqrt{(2)(9.8\ \text{m/s}^2)(1200\ \text{m})}$$

$$= 153\ \text{m/s} \approx 550\ \text{km/h.}\qquad\text{(Answer)}$$

For that speed, Shakespeare would scarcely have written, "it droppeth as the gentle rain from heaven, upon the place beneath."

✔**CHECKPOINT 3:**   Near the ground, is the speed of large raindrops greater than, less than, or the same as the speed of small raindrops, assuming that all raindrops are spherical and have the same drag coefficient?

6-4 Uniform Circular Motion

From Section 4-7, recall that when a body moves in a circle (or a circular arc) at constant speed v, it is said to be in uniform circular motion. Also recall that the body has a centripetal acceleration (directed toward the center of the circle), of constant magnitude given by

$$a = \frac{v^2}{R}\qquad\text{(centripetal acceleration),}\qquad(6\text{-}17)$$

where R is the radius of the circle.

Let us examine two examples of uniform circular motion:

1. *Rounding a curve in a car.* You are sitting in the center of the rear seat of a car moving at a constant high speed along a flat road. When the driver suddenly turns left, rounding a corner in a circular arc, you slide across the seat toward the right and then jam against the car wall for the rest of the turn. What is going on?

   While the car moves in the circular arc, it is in uniform circular motion; that is, it has an acceleration that is directed toward the center of the circle. By Newton's second law, a force must cause this acceleration. Moreover, the force must also be directed toward the center of the circle. Thus, it is a **centripetal force,** where the adjective indicates the direction. In this example, the centripetal force is a frictional force on the tires from the road; it makes the turn possible.

   If you are to move in uniform circular motion along with the car, there must also be a centripetal force on you. However, apparently the frictional force on you from the seat was not great enough to make you go in a circle with the car. Thus, the seat slid beneath you, until the right wall of the car jammed into you. Then, its push on you provided the needed centripetal force on you, and you joined the car's uniform circular motion.

2. *Orbiting Earth.* This time you are a passenger in the space shuttle *Atlantis*. As it and you orbit Earth, you float through your cabin. What is going on?

Both you and the shuttle are in uniform circular motion and have accelerations directed toward the center of the circle. Again by Newton's second law, centripetal forces must cause these accelerations. This time the centripetal forces are gravitational pulls (the pull on you and the pull on the shuttle) by Earth, radially inward, toward the center of Earth.

In both car and shuttle you are in uniform circular motion, acted on by a centripetal force—yet your sensations in the two situations are quite different. In the car, jammed up against the wall, you are aware of being compressed by the wall. In the orbiting shuttle, however, you are floating around with no sensation of any force acting on you. Why this difference?

The difference is due to the nature of the two centripetal forces. In the car, the centripetal force is the push on the part of your body touching the car wall. You can sense the compression on that part of your body. In the shuttle, the centripetal force is Earth's gravitational pull on every atom of your body. Thus, there is no compression (or pull) on any one part of your body and no sensation of a force acting on you. (The sensation is said to be one of "weightlessness," but that description is tricky. The pull on you by Earth has certainly not disappeared and, in fact, is only a little less than it would be with you on the ground.)

Another example of a centripetal force is shown in Fig. 6-9. There a hockey puck moves around in a circle at constant speed v while tied to a string looped around a central peg. This time the centripetal force is the radially inward pull on the puck from the string. Without that force, the puck would slide off in a straight line instead of moving in a circle.

Note again that a centripetal force is not a new kind of force. The name merely indicates the direction of the force. It can, in fact, be a frictional force, a gravitational force, the force from a car wall or a string, or any other force. For any situation:

> A centripetal force accelerates a body by changing the direction of the body's velocity without changing the body's speed.

From Newton's second law and Eq. 6-17 ($a = v^2/R$), we can write the magnitude F of a centripetal force (or a net centripetal force) as

$$F = m\frac{v^2}{R} \qquad \text{(magnitude of centripetal force).} \qquad (6\text{-}18)$$

Because the speed v here is constant, so are also the magnitudes of the acceleration and the force.

However, the directions of the centripetal acceleration and force are not constant; they vary continuously so as to always point toward the center of the circle. For this reason, the force and acceleration vectors are sometimes drawn along a radial axis r that moves with the body and always extends from the center of the circle to the body, as in Fig. 6-9. The positive direction of the axis is radially outward, but the acceleration and force vectors point ı

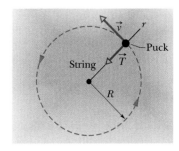

Fig. 6-9 An overhead view of a hockey puck of mass m moving with constant speed v in a circular path of radius R on a horizontal frictionless surface. The centripetal force on the puck is $\vec{T}$, the pull from the string, directed inward along the radial axis r extending through the puck.

Sample Problem 6-6

Igor is a cosmonaut-engineer on the *International Space Station*, in a circular orbit around Earth, at an altitude h of 520 km and with a constant speed v of 7.6 km/s. Igor's mass m is 79 kg.

(a) What is his acceleration?

SOLUTION: The Key Idea here is that Igor is in uniform circular motion and thus has a centripetal acceleration of magnitude given by Eq. 6-17 ($a = v^2/R$). The radius R of Igor's motion is $R_E + h$, where R_E is Earth's radius (6.37×10^6 m, from Appendix C). Thus,

$$a = \frac{v^2}{R} = \frac{v^2}{R_E + h}$$

$$= \frac{(7.6 \times 10^3 \text{ m/s})^2}{6.37 \times 10^6 \text{ m} + 0.52 \times 10^6 \text{ m}}$$

$$= 8.38 \text{ m/s}^2 \approx 8.4 \text{ m/s}^2. \qquad \text{(Answer)}$$

This is the value of the free-fall acceleration at Igor's altitude. If he were lifted to that altitude and released, instead of being put into orbit there, he would fall toward Earth's center, starting out with that value for his acceleration. The difference in the two situations is that when he orbits Earth, he always has a "sideways" motion as well: As he falls, he also moves to the side, so that he ends up moving along a curved path around Earth.

(b) What force does Earth exert on Igor?

SOLUTION: There are two Key Ideas here. First, there must be a centripetal force on Igor if he is to be in uniform circular motion. Second, that force is the gravitational force $\vec{F}_g$ on him from Earth, directed toward his center of rotation (at the center of Earth). From Newton's second law, written along the radial axis r, this force has the magnitude

$$F_g = ma = (79 \text{ kg})(8.38 \text{ m/s}^2)$$

$$= 662 \text{ N} \approx 660 \text{ N}. \qquad \text{(Answer)}$$

If Igor were to stand on a scale placed on the top of a tower with height h = 520 km, the scale would read 660 N. In orbit, the scale (if Igor could "stand" on it) would read zero because he and the scale are in free fall together, and therefore his feet do not actually press against it.

Sample Problem 6-7

In a 1901 circus performance, Allo "Dare Devil" Diavolo introduced the stunt of riding a bicycle in a loop-the-loop (Fig. 6-10a). Assuming that the loop is a circle with radius R = 2.7 m, what is the least speed v Diavolo could have at the top of the loop to remain in contact with it there?

SOLUTION: A Key Idea in analyzing Diavolo's stunt is to assume that he and his bicycle travel through the top of the loop as a single particle in uniform circular motion. Thus, at the top, the acceleration $\vec{a}$ of this particle must have the magnitude $a = v^2/R$ given by Eq. 6-17 and be directed downward, toward the center of the circular loop.

The forces on the particle when it is at the top of the loop are shown in the free-body diagram of Fig 6-10b. The gravitational force $\vec{F}_g$ is directed downward along a y axis. The normal force $\vec{N}$ on the particle from the loop is also directed downward. Thus, Newton's second law for y components ($F_{net,y} = ma_y$) gives us

$$-N - F_g = m(-a).$$

This becomes

$$-N - mg = m\left(-\frac{v^2}{R}\right). \qquad (6\text{-}19)$$

Another Key Idea is that if the particle has the *least speed* v needed to remain in contact, then it is on the *verge of losing contact* with the loop (falling away from the loop), which means that $N = 0$. Substituting this value for N into Eq. 6-19, solving for v, and then substituting known values give us

$$v = \sqrt{gR} = \sqrt{(9.8 \text{ m/s}^2)(2.7 \text{ m})}$$

$$= 5.1 \text{ m/s}. \qquad \text{(Answer)}$$

Diavolo made certain that his speed at the top of the loop was greater than 5.1 m/s so that he did not lose contact with the loop and fall away from it. Note that this speed requirement is indepen-

(a)

Fig. 6-10 Sample Problem 6-7. (a) Contemporary advertisement for Diavolo and (b) free-body diagram for the performer at the top of the loop.

(b)

dent of the mass of Diavolo and his bicycle. Had he feasted on, say, pierogies before his performance, he still would have had to exceed only 5.1 m/s.

✔**CHECKPOINT 4:**   When you ride in a Ferris wheel at constant speed, what are the directions of your acceleration $\vec{a}$ and the normal force $\vec{N}$ on you (from the always upright seat) as you pass through (a) the highest point and (b) the lowest point of the ride?

Sample Problem 6-8

Even some seasoned roller-coaster riders blanch at the thought of riding the Rotor, which is essentially a large, hollow cylinder that is rotated rapidly around its central axis (Fig. 6-11). Before the ride begins, a rider enters the cylinder through a door on the side and stands on a floor, up against a canvas-covered wall. The door is closed, and as the cylinder begins to turn, the rider, wall, and floor move in unison. When the rider's speed reaches some predetermined value, the floor abruptly and alarmingly falls away. The rider does not fall with it but instead is pinned to the wall while the cylinder rotates, as if an unseen (and somewhat unfriendly) agent is pressing the body to the wall. Later, the floor is eased back to the rider's feet, the cylinder slows, and the rider sinks a few centimeters to regain footing on the floor. (Some riders consider all this to be fun.)

Suppose that the coefficient of static friction μ_s between the rider's clothing and the canvas is 0.40 and that the cylinder's radius R is 2.1 m.

(a) What minimum speed v must the cylinder and rider have if the rider is not to fall when the floor drops?

SOLUTION: We start with a question: What force can keep the rider from falling and how is it related to the speed v of her and the cylinder? To answer, we use three Key Ideas:

1. The gravitational force $\vec{F}_g$ on the rider tends to slide her down the wall, but she does not move because a frictional force from the wall acts upward on her (Fig 6-11).

2. If she is to be on the verge of sliding down, that upward force must be a *static* frictional force $\vec{f}_s$ at its maximum value $\mu_s N$, where N is the magnitude of the normal force $\vec{N}$ on her from the cylinder (Fig. 6-11).

3. This normal force is directed horizontally toward the central axis of the cylinder and is the centripetal force that causes the rider

to move in a circular path, with centripetal acceleration of magnitude $a = v^2/R$.

We want speed v in that last expression, for the condition that the rider is on the verge of sliding.

We first place a vertical y axis through the rider, with the positive direction upward. For Key Idea 1, we can then apply Newton's second law to the rider, writing it for y components ($F_{net,y} = ma_y$) as

$$f_s - mg = m(0),$$

where m is the rider's mass and mg is the magnitude of $\vec{F}_g$. For Key Idea 2, we substitute the maximum value $\mu_s N$ for f_s in this equation, getting

$$\mu_s N - mg = 0,$$

or

$$N = \frac{mg}{\mu_s}. \tag{6-20}$$

Next we place a radial r axis through the rider, with the positive direction outward. For Key Idea 3, we can then write Newton's second law for components along that axis as

$$-N = m\left(-\frac{v^2}{R}\right). \tag{6-21}$$

Substituting Eq. 6-20 for N and then solving for v, we find

$$v = \sqrt{\frac{gR}{\mu_s}} = \sqrt{\frac{(9.8 \text{ m/s}^2)(2.1 \text{ m})}{0.40}}$$

$$= 7.17 \text{ m/s} \approx 7.2 \text{ m/s}. \tag{Answer}$$

Note that the result is independent of the rider's mass; it holds for anyone riding the Rotor, from a child to a sumo wrestler, which is why no one has to "weigh in" to ride the Rotor.

(b) If the rider's mass is 49 kg, what is the magnitude of the centripetal force on her?

SOLUTION: According to Eq. 6-21,

$$N = m\frac{v^2}{R} = (49 \text{ kg})\frac{(7.17 \text{ m/s})^2}{2.1 \text{ m}}$$

$$\approx 1200 \text{ N}. \tag{Answer}$$

Although this force is directed toward the central axis, the rider has an overwhelming sensation that the force pinning her against the wall is directed radially outward. Her sensation stems from the fact that she is in a noninertial frame (she and it are accelerating). As measured from such frames, forces can be illusionary. The illusion is part of the Rotor's attraction.

✓CHECKPOINT 5: If the Rotor initially moves at the minimum required speed for the rider not to fall and then its speed is increased in steps, do the following increase, decrease, or remain the same: (a) the magnitude of $\vec{f}_s$; (b) the magnitude of $\vec{N}$; (c) the value of $f_{s,max}$?

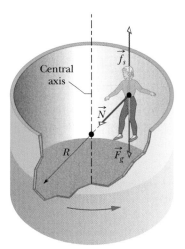

Fig. 6-11 Sample Problem 6-8. A Rotor in an amusement park, showing the forces on a rider. The centripetal force is the normal force with which the wall pushes inward on the rider.

Sample Problem 6-9

Figure 6-12a represents a stock car of mass $m = 1600$ kg traveling at a constant speed $v = 20$ m/s around a flat, circular track of radius $R = 190$ m. For what value of μ_s between the track and the tires of the car will the car be on the verge of sliding off the track?

SOLUTION: We need to relate μ_s to the circular motion of the car. We start with four Key Ideas, all related to one force on the car:

1. A centripetal force must act on the car if the car is moving around a circular path; the force must be directed horizontally toward the center of the circle.

2. The only horizontal force acting on the car is a frictional force on the tires from the road. So, the required centripetal force is a frictional force.

3. Because the car is not sliding, the frictional force must be a *static* frictional force, the $\vec{f}_s$ shown in Fig 6-12a.

4. If the car is just on the verge of sliding, the magnitude f_s of the frictional force is just equal to the maximum value $f_{s,\text{max}} = \mu_s N$, where N is the magnitude of the normal force on the car from the track.

Thus, to find μ_s, we start with the centripetal force on the car. Figure 6-12b is a free-body diagram for the car, shown on a radial axis r that always extends from the center of the circle through the

car as the car moves. The centripetal force $\vec{f}_s$ is directed inward along that axis, in the negative direction of the axis, and so is the car's centripetal acceleration $\vec{a}$ (of magnitude v^2/R). We can relate this force and acceleration by writing Newton's second law for components along the r axis ($F_{\text{net},r} = ma_r$) as

$$-f_s = m\left(-\frac{v^2}{R}\right). \tag{6-22}$$

Substituting $f_{s,\text{max}} = \mu_s N$ for f_s and solving for μ_s give us

$$\mu_s = \frac{mv^2}{NR}. \tag{6-23}$$

Because the car does not accelerate vertically, the two vertical forces acting on it (see Fig. 6-12b) must balance; that is, the magnitude N of the normal force must equal the magnitude mg of the gravitational force. Substituting $N = mg$ into Eq. 6-23, we find

$$\mu_s = \frac{mv^2}{mgR} = \frac{v^2}{gR} \tag{6-24}$$

$$= \frac{(20 \text{ m/s})^2}{(9.8 \text{ m/s}^2)(190 \text{ m})} = 0.21. \quad \text{(Answer)}$$

This means that if $\mu_s = 0.21$, the car is on the verge of sliding off the track; if $\mu_s > 0.21$, the car is in no danger of sliding off; and if $\mu_s < 0.21$, the car will certainly slide off.

Equation 6-24 contains two important lessons for road engineers. First, the value of μ_s (required to prevent sliding) depends on the *square* of v. Much more friction is required when the turning speed is increased. You may have noted this effect if you have ever taken a flat turn too fast and suddenly felt the tires slip. Second, the mass m dropped out in our derivation of Eq. 6-24. Thus, Eq. 6-24 holds for a vehicle of any mass, from a kiddy car to a bicycle to a heavy truck.

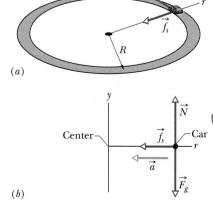

Fig. 6-12 Sample Problem 6-9. (a) A car moves around a flat curved road at constant speed v. The frictional force $\vec{f}_s$ provides the necessary centripetal force along a radial axis r. (b) A free-body diagram (not to scale) for the car, in the vertical plane containing r.

✔**CHECKPOINT 6:** In Fig. 6-12, suppose the car is on the verge of sliding when the radius of the circle is R_1. (a) If we double the car's speed, what is the least radius that would now keep the car from sliding? (b) If we also double the mass of the car (say by adding sandbags), what is the least radius that would now keep the car from sliding?

REVIEW & SUMMARY

Friction When a force $\vec{F}$ tends to slide a body along a surface, a **frictional force** from the surface acts on the body. The frictional force is parallel to the surface and directed so as to oppose the sliding. It is due to bonding between the body and the surface.

If the body does not slide, the frictional force is a **static frictional force** $\vec{f}_s$. If there is sliding, the frictional force is a **kinetic frictional force** $\vec{f}_k$.

Three Properties of Friction

1. If the body does not move, then the static frictional force $\vec{f}_s$ and the component of $\vec{F}$ that is parallel to the surface are equal in

magnitude, and $\vec{f}_s$ is directed opposite that component. If that parallel component increases, magnitude f_s also increases.

2. The magnitude of $\vec{f}_s$ has a maximum value $f_{s,\text{max}}$ that is given by

$$f_{s,\text{max}} = \mu_s N, \tag{6-1}$$

where μ_s is the **coefficient of static friction** and N is the magnitude of the normal force. If the component of $\vec{F}$ that is parallel to the surface exceeds $f_{s,\text{max}}$, then the body slides on the surface.

3. If the body begins to slide on the surface, the magnitude of the frictional force rapidly decreases to a constant value f_k given by

$$f_k = \mu_k N, \tag{6-2}$$

where μ_k is the **coefficient of kinetic friction.**

Drag Force When there is a relative motion between air (or some other fluid) and a body, the body experiences a **drag force** $\vec{D}$ that opposes the relative motion and points in the direction in which the fluid flows relative to the body. The magnitude of $\vec{D}$ is related to the relative speed v by an experimentally determined **drag coefficient** C according to

$$D = \tfrac{1}{2}C\rho Av^2, \qquad (6\text{-}14)$$

where ρ is the fluid density (mass per volume) and A is the **effective cross-sectional area** of the body (the area of a cross section taken perpendicular to the relative velocity $\vec{v}$).

Terminal Speed When a blunt object has fallen far enough through air, the magnitudes of the drag force $\vec{D}$ and the gravitational force $\vec{F}_g$ on the body become equal. The body then falls at a con-stant **terminal speed** v_t given by

$$v_t = \sqrt{\frac{2F_g}{C\rho A}}. \qquad (6\text{-}16)$$

Uniform Circular Motion If a particle moves in a circle or a circular arc with radius R at constant speed v, it is said to be in **uniform circular motion.** It then has a **centripetal acceleration** $\vec{a}$ with magnitude given by

$$a = \frac{v^2}{R}. \qquad (6\text{-}17)$$

This acceleration is due to a net **centripetal force** on the particle, with magnitude given by

$$F = \frac{mv^2}{R}, \qquad (6\text{-}18)$$

where m is the particle's mass. The vector quantities $\vec{a}$ and $\vec{F}$ are directed toward the center of curvature of the particle's path.

QUESTIONS

1. In three experiments, three different horizontal forces are applied to the same block lying on the same countertop. The force mag-nitudes are $F_1 = 12$ N, $F_2 = 8$ N, and $F_3 = 4$ N. In each experiment, the block remains stationary in spite of the applied force. Rank the forces according to (a) the magnitude f_s of the static frictional force on the block from the countertop and (b) the maximum value $f_{s,\text{max}}$ of that force, greatest first.

2. In Fig. 6-13a, a "Batman" thermos is sent sliding leftward across a long plastic tray. What are the directions of the kinetic frictional forces on (a) the thermos and (b) the tray from each other? (c) Does the former increase or decrease the speed of the thermos relative to the floor? In Fig. 6-13b, the tray is now sent sliding leftward beneath the thermos. What now are the directions of the kinetic frictional forces on (d) the thermos and (e) the tray from each other? (f) Does the former increase or decrease the speed of the thermos relative to the floor? (g) *Do kinetic frictional forces always slow objects?*

Fig. 6-13 Question 2.

3. In Fig. 6-14, horizontal force $\vec{F}_1$ of magnitude 10 N is applied to a box on a floor, but the box does not slide. Then, as the mag-nitude of vertical force $\vec{F}_2$ is increased from zero, do the following quantities increase, decrease, or stay the same: (a) the magnitude of the frictional force $\vec{f}_s$ on the box; (b) the magnitude of the nor-mal force $\vec{N}$ on the box from the floor; (c) the maximum value $f_{s,\text{max}}$ of the magnitude of the static frictional force on the box? (d) Does the box eventually slide?

4. If you press an apple crate against a wall so hard that the crate cannot slide down the wall, what is the direction of (a) the

Fig. 6-14 Question 3.

static frictional force $\vec{f}_s$ on the crate from the wall and (b) the normal force $\vec{N}$ on the crate from the wall? If you increase your push, what happens to (c) f_s, (d) N, and (e) $f_{s,\text{max}}$?

5. In Fig. 6-15, if the box is stationary and the angle θ of force $\vec{F}$ is increased, do the following quantities increase, decrease, or re-main the same: (a) F_x; (b) f_s; (c) N; (d) $f_{s,\text{max}}$? (e) If, instead, the box is sliding and θ is increased, does the magnitude of the fric-tional force on the box increase, decrease, or remain the same?

Fig. 6-15 Question 5.

6. Repeat Question 5 for force $\vec{F}$ angled upward instead of downward as drawn.

7. Figure 6-16 shows a block of mass m on a slab of mass M and a horizontal force $\vec{F}$ applied to the block, causing it to slide over the slab. There is friction between the block and the slab (but not between the slab and the floor). (a) What mass determines the mag-nitude of the frictional force between the block and the slab? (b) At the block–slab interface, is the magnitude of the frictional force acting on the block greater than, less than, or equal to that of the frictional force acting on the slab? (c) What are the directions of those two frictional forces? (d) If we write Newton's second law for the slab, what mass should be multiplied by the acceleration of the slab? (Warm-up for Problem 27)

Fig. 6-16 Question 7.

8. *Follow-up to Problem 25.* Suppose the larger of the two blocks in Fig. 6-33 is, instead, fixed to the surface below it. Again, the smaller block is not to slip down the face of the larger block. (a) Is the magnitude of the frictional force between the blocks then greater than, less than, or the same as in the original prob-lem? (b) Is the required minimum magnitude of the horizontal force $\vec{F}$ then greater than, less than, or the same as in the original problem?

9. Figure 6-17 shows the path of a park ride that travels at constant speed through five circular arcs of radii R_0, $2R_0$, and $3R_0$. Rank the arcs according to the magnitude of the centripetal force on a rider traveling in the arcs, greatest first.

Fig. 6-17 Question 9.

10. A person riding a Ferris wheel moves through positions at (1) the top, (2) the bottom, and (3) midheight. If the wheel rotates at a constant rate, rank these three positions according to (a) the magnitude of the person's centripetal acceleration, (b) the magnitude of the net centripetal force on the person, and (c) the magnitude of the normal force on the person, greatest first.

EXERCISES & PROBLEMS

SEC. 6-2   Properties of Friction

1E. A bedroom bureau with a mass of 45 kg, including drawers and clothing, rests on the floor. (a) If the coefficient of static friction between the bureau and the floor is 0.45, what is the magnitude of the minimum horizontal force that a person must apply to start the bureau moving? (b) If the drawers and clothing, with 17 kg mass, are removed before the bureau is pushed, what is the new minimum magnitude?   ssm

2E. The coefficient of static friction between Teflon and scrambled eggs is about 0.04. What is the smallest angle from the horizontal that will cause the eggs to slide across the bottom of a Teflon-coated skillet?

3E. A baseball player with mass $m = 79$ kg, sliding into second base, is retarded by a frictional force of magnitude 470 N. What is the coefficient of kinetic friction μ_k between the player and the ground?   ssm

4E. *The mysterious sliding stones.* Along the remote Racetrack Playa in Death Valley, California, stones sometimes gouge out prominent trails in the desert floor, as if they had been migrating (Fig. 6-18). For years curiosity mounted about why the stones moved. One explanation was that strong winds during the occasional rainstorms would drag the rough stones over ground softened by rain. When the desert dried out, the trails behind the stones were hard-baked in place. According to measurements, the coefficient of kinetic friction between the stones and the wet playa ground is about 0.80. What horizontal force is needed on a stone of typical mass 20 kg to maintain the stone's motion once a gust has started it moving? (Story continues with Exercise 32.)

Fig. 6-18 Exercise 4.

5E. A person pushes horizontally with a force of 220 N on a 55 kg crate to move it across a level floor. The coefficient of kinetic friction is 0.35. (a) What is the magnitude of the frictional force? (b) What is the magnitude of the crate's acceleration?   ssm ilw

6E. A house is built on the top of a hill with a nearby 45° slope (Fig. 6-19). An engineering study indicates that the slope angle should be reduced because the top layers of soil along the slope might slip past the lower layers. If the static coefficient of friction between two such layers is 0.5, what is the least angle ϕ through which the present slope should be reduced to prevent slippage?

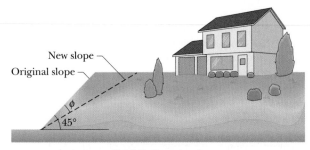

Fig. 6-19 Exercise 6.

7E. A 110 g hockey puck sent sliding over ice is stopped in 15 m by the frictional force on it from the ice. (a) If its initial speed is 6.0 m/s, what is the magnitude of the frictional force? (b) What is the coefficient of friction between the puck and the ice?   ssm

8E. In Fig. 6-20, a 49 kg rock climber is climbing a "chimney" between two rock slabs. The static coefficient of friction between her shoes and the rock is 1.2; between her back and the rock it is 0.80. She has reduced her push against the rock until her back and her shoes are on the verge of slipping. (a) Draw a free-body diagram of the climber. (b) What is her push against the rock? (c) What fraction of her weight is supported by the frictional force on her shoes?

9P. A 12 N horizontal force $\vec{F}$ pushes a block weighing 5.0 N against a vertical wall (Fig. 6-21). The coefficient of static friction between the wall and the block is 0.60, and the coefficient of kinetic friction is 0.40. Assume that the block is not moving initially.

Fig. 6-20 Exercise 8.

(a) Will the block move? (b) In unit-vector notation, what is the force on the block from the wall? ssm www

Fig. 6-21 Problem 9.

10P. A 2.5 kg block is initially at rest on a horizontal surface. A 6.0 N horizontal force and a vertical force $\vec{P}$ are applied to the block as shown in Fig. 6-22. The coefficients of friction for the block and surface are $\mu_s = 0.40$ and $\mu_k = 0.25$. Determine the magnitude and direction of the frictional force acting on the block if the magnitude of $\vec{P}$ is (a) 8.0 N, (b) 10 N, and (c) 12 N.

Fig. 6-22 Problem 10.

11P. A worker wishes to pile a cone of sand onto a circular area in his yard. The radius of the circle is R, and no sand is to spill onto the surrounding area (Fig. 6-23). If μ_s is the static coefficient of friction between each layer of sand along the slope and the sand beneath it (along which it might slip), show that the greatest volume of sand that can be stored in this manner is $\pi\mu_s R^3/3$. (The volume of a cone is $Ah/3$, where A is the base area and h is the cone's height.) ssm

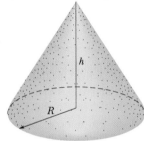

Fig. 6-23 Problem 11.

12P. A worker pushes horizontally on a 35 kg crate with a force of magnitude 110 N. The coefficient of static friction between the crate and the floor is 0.37. (a) What is the frictional force on the crate from the floor? (b) What is the maximum magnitude $f_{s,max}$ of the static frictional force under the circumstances? (c) Does the crate move? (d) Suppose, next, that a second worker pulls directly upward on the crate to help out. What is the least vertical pull that will allow the first worker's 110 N push to move the crate? (e) If, instead, the second worker pulls horizontally to help out, what is the least pull that will get the crate moving?

13P. A 68 kg crate is dragged across a floor by pulling on a rope attached to the crate and inclined 15° above the horizontal. (a) If the coefficient of static friction is 0.50, what minimum force magnitude is required from the rope to start the crate moving? (b) If $\mu_k = 0.35$, what is the magnitude of the initial acceleration of the crate? ssm

Fig. 6-24 Problem 14.

14P. A slide-loving pig slides down a certain 35° slide (Fig. 6-24) in twice the time it would take to slide down a frictionless 35° slide. What is the coefficient of kinetic friction between the pig and the slide?

15P. In Fig. 6-25, blocks A and B have weights of 44 N and 22 N,

respectively. (a) Determine the minimum weight of block C to keep A from sliding if μ_s between A and the table is 0.20. (b) Block C suddenly is lifted off A. What is the acceleration of block A if μ_k between A and the table is 0.15? ssm

Fig. 6-25 Problem 15.

16P. A 3.5 kg block is pushed along a horizontal floor by a force $\vec{F}$ of magnitude 15 N at an angle $\theta = 40°$ with the horizontal (Fig. 6-26). The coefficient of kinetic friction between the block and the floor is 0.25. Calculate the magnitudes of (a) the frictional force on the block from the floor and (b) the acceleration of the block.

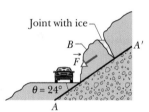

Fig. 6-26 Problem 16.

17P. Figure 6-27 shows the cross section of a road cut into the side of a mountain. The solid line AA' represents a weak bedding plane along which sliding is possible. Block B directly above the highway is separated from uphill rock by a large crack (called a *joint*), so that only friction between the block and the bedding plane prevents sliding. The mass of the block is 1.8×10^7 kg, the *dip angle* θ of the bedding plane is 24°, and the coefficient of static friction between block and plane is 0.63. (a) Show that the block will not slide. (b) Water seeps into the joint and expands upon freezing, exerting on the block a force $\vec{F}$ parallel to AA'. What minimum value of F will trigger a slide?

Fig. 6-27 Problem 17.

18P. A loaded penguin sled weighing 80 N rests on a plane inclined at 20° to the horizontal (Fig. 6-28). Between the sled and the plane, the coefficient of static friction is 0.25, and the coefficient of kinetic friction is 0.15. (a) What is the minimum magnitude of the force $\vec{F}$, parallel to the plane, that will prevent the sled from slipping down the plane? (b) What is the minimum magnitude F that will start the sled moving up the plane? (c) What value of F is required to move the sled up the plane at constant velocity?

Fig. 6-28 Problem 18.

19P. Block B in Fig. 6-29 weighs 711 N. The coefficient of static friction between block and table is 0.25; assume that the cord between B and the knot is horizontal. Find the maximum weight of block A for which the system will be stationary. ssm

20P. A force $\vec{P}$, parallel to a sur-

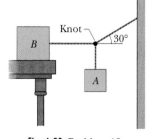

Fig. 6-29 Problem 19.

face inclined 15° above the horizontal, acts on a 45 N block, as shown in Fig. 6-30. The coefficients of friction for the block and surface are $\mu_s = 0.50$ and $\mu_k = 0.34$. If the block is initially at rest, determine the magnitude and direction of the frictional force acting on the block for magnitudes of $\vec{P}$ of (a) 5.0 N, (b) 8.0 N, and (c) 15 N.

Fig. 6-30 Problem 20.

21P. Body A in Fig. 6-31 weighs 102 N, and body B weighs 32 N. The coefficients of friction between A and the incline are $\mu_s = 0.56$ and $\mu_k = 0.25$. Angle θ is 40°. Find the acceleration of A if (a) A is initially at rest, (b) A is initially moving up the incline, and (c) A is initially moving down the incline. ssm www

22P. In Fig. 6-31, two blocks are connected over a pulley. The mass of block A is 10 kg and the coefficient of kinetic friction between A and the incline is 0.20. Angle θ of the incline is 30°. Block A slides down the incline at constant speed. What is the mass of block B?

Frictionless, massless pulley

Fig. 6-31 Problems 21 and 22.

23P. Two blocks, of weights 3.6 N and 7.2 N, are connected by a massless string and slide down a 30° inclined plane. The coefficient of kinetic friction between the lighter block and the plane is 0.10; that between the heavier block and the plane is 0.20. Assuming that the lighter block leads, find (a) the magnitude of the acceleration of the blocks and (b) the tension in the string. (c) Describe the motion if, instead, the heavier block leads. ssm

24P. In Fig. 6-32, a box of Cheerios and a box of Wheaties are accelerated across a horizontal surface by a horizontal force $\vec{F}$ applied to the Cheerios box. The magnitude of the frictional force on the Cheerios box is 2.0 N, and the magnitude of the frictional force on the Wheaties box is 4.0 N. If the magnitude of $\vec{F}$ is 12 N, what is the magnitude of the force on the Wheaties box from the Cheerios box?

1.0 kg $\vec{F}$ 3.0 kg

Fig. 6-32 Problem 24.

25P. The two blocks (with $m = 16$ kg and $M = 88$ kg) shown in Fig. 6-33 are not attached. The coefficient of static friction between the blocks is $\mu_s = 0.38$, but the surface beneath the larger block is frictionless. What is the minimum magnitude of the horizontal force $\vec{F}$ required to keep the smaller block from slipping down the larger block? (See Question 8 for a follow-up.) ilw

26P. In Fig. 6-34, a box of ant aunts (total mass $m_1 = 1.65$ kg) and a box of ant uncles (total mass $m_2 = 3.30$ kg) slide down

m $\vec{F}$ M Frictionless

Fig. 6-33 Problem 25.

m_1 m_2 θ

Fig. 6-34 Problem 26.

an inclined plane while attached by a massless rod parallel to the plane. The angle of incline is $\theta = 30°$. The coefficient of kinetic friction between the aunt box and the incline is $\mu_1 = 0.226$; that between the uncle box and the incline is $\mu_2 = 0.113$. Compute (a) the tension in the rod and (b) the common acceleration of the two boxes. (c) How would the answers to (a) and (b) change if the uncles trailed the aunts?

27P. A 40 kg slab rests on a frictionless floor. A 10 kg block rests on top of the slab (Fig. 6-35). The coefficient of static friction μ_s between the block and the slab is 0.60, whereas their kinetic friction coefficient μ_k is 0.40. The 10 kg block is pulled by a horizontal force with a magnitude of 100 N. What are the resulting accelerations of (a) the block and (b) the slab? ssm www

28P. A locomotive accelerates a 25-car train along a level track. Every car has a mass of 5.0×10^4 kg and is subject to a frictional force $f = 250v$, where the speed v is in meters per second and the force f is in newtons. At the instant when the speed of the train is 30 km/h, the magnitude of its acceleration is 0.20 m/s². (a) What is the tension in the coupling between the first car and the locomotive? (b) If this tension is equal to the maximum force the locomotive can exert on the train, what is the steepest grade up which the locomotive can pull the train at 30 km/h?

100 N   10 kg   Frictionless   40 kg

Fig. 6-35 Problem 27.

29P. In Fig. 6-36, a crate slides down an inclined right-angled trough. The coefficient of kinetic friction between the crate and the trough is μ_k. What is the acceleration of the crate in terms of μ_k, θ, and g? ssm

Fig. 6-36 Problem 29.

90°

30P. An initially stationary box of sand is to be pulled across a floor by means of a cable in which the tension should not exceed 1100 N. The coefficient of static friction between the box and the floor is 0.35. (a) What should be the angle between the cable and the horizontal in order to pull the greatest possible amount of sand, and (b) what is the weight of the sand and box in that situation?

31P. A 1000 kg boat is traveling at 90 km/h when its engine is shut off. The magnitude of the frictional force $\vec{f}_k$ between boat and water is proportional to the speed v of the boat: $f_k = 70v$, where v is in meters per second and f_k is in newtons. Find the time required for the boat to slow to 45 km/h. ssm

SEC. 6-3 The Drag Force and Terminal Speed

32E. *Continuation of Exercise 4.* First reread the explanation of how the wind might drag desert stones across the playa. Now assume that Eq. 6-14 gives the magnitude of the air drag force on the typical 20 kg stone, which presents a vertical cross-sectional area to the wind of 0.040 m² and has a drag coefficient C of 0.80. Take the air density to be 1.21 kg/m³, and the coefficient of kinetic friction to be 0.80. (a) In kilometers per hour, what wind speed V along the ground is needed to maintain the stone's motion once it has started moving? Because winds along the ground are retarded by the ground, the wind speeds reported for storms are often measured at a height of 10 m. Assume wind speeds are 2.00 times those along the ground. (b) For your answer to (a), what wind speed would be reported for the storm and is that value reasonable for a high-speed wind in a storm? (Story continues with Problem 51.)

33E. Calculate the drag force on a missile 53 cm in diameter cruising with a speed of 250 m/s at low altitude, where the density of air is 1.2 kg/m^3. Assume $C = 0.75$. **ssm**

34E. The terminal speed of a sky diver is 160 km/h in the spread-eagle position and 310 km/h in the nosedive position. Assuming that the diver's drag coefficient C does not change from one position to the other, find the ratio of the effective cross-sectional area A in the slower position to that in the faster position.

35P. Calculate the ratio of the drag force on a passenger jet flying with a speed of 1000 km/h at an altitude of 10 km to the drag force on a prop-driven transport flying at half the speed and half the altitude of the jet. At 10 km the density of air is 0.38 kg/m^3, and at 5.0 km it is 0.67 kg/m^3. Assume that the airplanes have the same effective cross-sectional area and the same drag coefficient C.

SEC. 6-4 Uniform Circular Motion

36E. During an Olympic bobsled run, the Jamaican team makes a turn of radius 7.6 m at a speed of 96.6 km/h. What is their acceleration in g-units?

37E. Suppose the coefficient of static friction between the road and the tires on a Formula One car is 0.6 during a Grand Prix auto race. What speed will put the car on the verge of sliding as it rounds a level curve of 30.5 m radius? **ssm**

38E. A roller-coaster car has a mass of 1200 kg when fully loaded with passengers. As the car passes over the top of a circular hill of radius 18 m, its speed is not changing. What are the magnitude and direction of the force of the track on the car at the top of the hill if the car's speed is (a) 11 m/s and (b) 14 m/s?

39E. What is the smallest radius of an unbanked (flat) track around which a bicyclist can travel if her speed is 29 km/h and the coefficient of static friction between tires and track is 0.32? **ilw**

40P. An amusement park ride consists of a car moving in a vertical circle on the end of a rigid boom of negligible mass. The combined weight of the car and riders is 5.0 kN, and the radius of the circle is 10 m. What are the magnitude and direction of the force of the boom on the car at the top of the circle if the car's speed there is (a) 5.0 m/s and (b) 12 m/s?

41P. A puck of mass m slides on a frictionless table while attached to a hanging cylinder of mass M by a cord through a hole in the table (Fig. 6-37). What speed keeps the cylinder at rest? **ssm**

Fig. 6-37 Problem 41.

42P. A bicyclist travels in a circle of radius 25.0 m at a constant speed of 9.00 m/s. The bicycle–rider mass is 85.0 kg. Calculate the magnitudes of (a) the force of friction on the bicycle from the road and (b) the *net* force on the bicycle from the road.

43P. A student of weight 667 N rides a steadily rotating Ferris wheel (the student sits upright). At the highest point, the magnitude of the normal force $\vec{N}$ on the student from the seat is 556 N. (a) Does the student feel "light" or "heavy" there? (b) What is the magnitude of $\vec{N}$ at the lowest point? (c) What is the magnitude N if the wheel's speed is doubled? **ssm ilw www**

44P. An old streetcar rounds a flat corner of radius 9.1 m, at 16 km/h. What angle with the vertical will be made by the loosely hanging hand straps?

45P. An airplane is flying in a horizontal circle at a speed of 480 km/h. If its wings are tilted 40° to the horizontal, what is the radius of the circle in which the plane is flying? (See Fig. 6-38.) Assume that the required force is provided entirely by an "aerodynamic lift" that is perpendicular to the wing surface. **ssm**

Fig. 6-38 Problem 45.

46P. A high-speed railway car goes around a flat, horizontal circle of radius 470 m at a constant speed. The magnitudes of the horizontal and vertical components of the force of the car on a 51.0 kg passenger are 210 N and 500 N, respectively. (a) What is the magnitude of the net force (of *all* the forces) on the passenger? (b) What is the speed of the car?

47P. As shown in Fig. 6-39, a 1.34 kg ball is connected by means of two massless strings to a vertical, rotating rod. The strings are tied to the rod and are taut. The tension in the upper string is 35 N. (a) Draw the free-body diagram for the ball. What are (b) the tension in the lower string, (c) the net force on the ball, and (d) the speed of the ball? **ssm ilw**

Fig. 6-39 Problem 47.

Additional Problem

48. *Continuation of Exercises 4 and 32.* Another explanation is that the stones move only when the water dumped on the playa during a storm freezes into a large, thin sheet. The stones are trapped in place in the sheet. Then, as air flows across the sheet during a wind, the air drag forces on the sheet and on the stones move them, with the stones gouging out the trails. The magnitude of the air drag force on this horizontal "ice sail" is given by $D_{ice} = 4C_{ice}\rho A_{ice}v^2$, where C_{ice} is the drag coefficient (about 2.0×10^{-3}), ρ is the air density (1.21 kg/m^3), A_{ice} is the horizontal area of the ice, and v is the wind speed along the ice.

Assume the following: The sheet measures 400 m by 500 m by 4.0 mm and has a coefficient of kinetic friction of 0.10 with the ground and a density of 917 kg/m^3. Also assume that 100 stones identical to the one in Exercise 4 are trapped in the ice. To maintain the motion of the sheet, what are the required wind speeds (a) near the sheet and (b) at a height of 10 m? (c) Are these reasonable values for high-speed winds in a storm?

NEW PROBLEMS

N1. A filing cabinet with a weight of 556 N rests on the floor. The coefficient of static friction between it and the floor is 0.68, and the coefficient of kinetic friction is 0.56. In four different attempts to move it, it is pushed with horizontal forces of magnitudes (a) 222 N, (b) 334 N, (c) 445 N, and (d) 556 N. For each attempt, determine whether the cabinet moves, and calculate the magnitude of the frictional force on it from the floor. The cabinet is initially at rest for each attempt.

N2. *Engineering a highway curve.* If a car goes through a curve too fast, the car tends to slide out of the curve, as discussed in Sample Problems 6-9 for a flat curve with friction. For a banked curve with friction, a frictional force acts on a fast car to oppose the tendency to slide out of the curve; the force is directed down the bank (in the direction water would drain). Consider a circular curve of radius $R = 200$ m and bank angle θ, where the coefficient of static friction between tires and pavement is μ_s. A car is driven around the curve as shown in Fig. 6N-1. (a) Find an expression for the car speed v_{max} that puts the car on the verge of sliding out. (b) On the same graph, plot v_{max} versus angle θ for the range 0° to 50°, first for $\mu_s = 0.60$ (dry pavement) and then for $\mu_s = 0.050$ (wet or icy pavement). In kilometers per hour, evaluate v_{max} for a bank angle of $\theta = 10°$ and for (c) $\mu_s = 0.60$ and (d) $\mu_s = 0.050$. (Now you can see why accidents occur in highway curves when wet or icy conditions are not obvious to drivers, who tend to drive at normal speeds.)

Fig. 6N-1 Problem N2.

N3. The coefficient of kinetic friction in Fig. 6N-2 is 0.20. What is the block's acceleration (magnitude and direction) if (a) it is sliding down the slope and (b) it has been given an upward shove and is still sliding up the slope?

Fig. 6N-2 Problem N3.

N4. In the early afternoon, a car is parked on a street that runs down a steep hill, at an angle of 35.0° relative to the horizontal. Just then the coefficient of static friction between the tires and the street surface is 0.725. Later, after nightfall, a sleet storm hits the area, and the coefficient decreases due to both the ice and a chemical change in the road surface because of the temperature decrease. By what percentage must the coefficient decrease if the car is to be in danger of sliding down the street?

N5. The floor of a railroad flatcar is loaded with loose crates having a coefficient of static friction of 0.25 with the floor. If the train is initially moving at a speed of 48 km/h, in how short a distance can the train be stopped at constant acceleration without causing the crates to slide over the floor?

N6. A force $\vec{F}$ is applied to a crate of mass m on a floor, at downward angle θ as shown in Fig. 6N-3. The coefficient of static friction between the floor and the bottom of the crate is μ_s. As θ is gradually increased from 0°, the magnitude F of the force is continuously adjusted so that the crate is always on the verge of sliding. For $\mu_s = 0.70$, (a) plot the ratio F/mg of force magnitude to weight versus angle θ and (b) determine the angle θ_{inf} at which the ratio approaches an infinite value. (c) Does lubricating the floor increase or decrease the value of θ_{inf}, or is the value unchanged? (d) What is θ_{inf} for $\mu_s = 0.60$?

Fig. 6N-3 Problem N6.

N7. In Fig. 6N-4 a fastidious worker pushes directly along the handle of a mop with a force $\vec{F}$. The handle is at an angle θ with the vertical, and μ_s and μ_k are the coefficients of static and kinetic friction between the head of the mop and the floor. Ignore the mass of the handle and assume that all the mop's mass m is in its head. (a) If the mop head moves along the floor with a constant velocity, then what is F? (b) Show that if θ is less than a certain value θ_0, then $\vec{F}$ (still directed along the handle) is unable to move the mop head. Find θ_0.

Fig. 6N-4 Problem N7.

N8. A sling-thrower puts a stone (0.250 kg) in the sling's pouch (0.010 kg) and then begins to make the stone and pouch move in a vertical circle of radius 0.650 m. The cord between the pouch and the person's hand has negligible mass and will break when the tension in the cord is 33.0 N or more. Suppose the sling-thrower could gradually increase the speed of the stone. (a) Will the breaking occur at the lowest point of the circle or at the highest point? (b) At what speed of the stone will that breaking occur?

N9. A 4.0 kg block is put on top of a 5.0 kg block. To cause the top block to slip on the bottom one, while the bottom one is held fixed, a horizontal force of at least 12 N must be applied to the top block. The assembly of blocks is now placed on a horizontal, frictionless table (Fig. 6N-5). Find the magnitudes of (a) the maximum horizontal force $\vec{F}$ that can be applied to the lower block so that the blocks will move together and (b) the resulting acceleration of the blocks.

Fig. 6N-5 Problem N9.

N10. You testify as an *expert witness* in a case involving an accident in which car A slid into the rear of car B, which was stopped at a red light along a road headed down a hill (Fig. 6N-6). You find that the slope of the hill is $\theta = 12.0°$, that the cars were separated by distance $d = 24.0$ m when the driver of car A put the car into a slide (it lacked any automatic anti-brake-lock system), and that the speed of car A at the onset of braking was $v_0 = 18.0$ m/s. With what speed did car A hit car B if the coefficient of kinetic friction was (a) 0.60 (dry road surface) and (b) 0.10 (road surface covered with wet leaves)?

Fig. 6N-6 Problem N10.

N11. A 3.0 kg block on a 30° incline is connected by a string to a 2.0 kg block on a horizontal surface, as shown in Fig. 6N-7. The 30° incline is frictionless and the coefficient of kinetic friction between the 2.0 kg block and the horizontal surface is 0.25. The pulley is frictionless and has negligible mass. After the blocks are released, what is the tension in the string that connects the blocks?

Fig. 6N-7 Problem N11.

N12. An 8.00 kg block of steel is at rest on a horizontal table. The coefficient of static friction between the block and the table is 0.45. A force is to be applied to the block. To three significant figures, what is the magnitude of that applied force if it puts the block on the verge of sliding when the force is directed (a) horizontally, (b) upward at 60.0° from the horizontal, and (c) downward at 60.0° from the horizontal?

N13. A box of canned goods slides down a ramp from street level into the basement of a grocery store with acceleration 0.75 m/s² directed down the ramp. The ramp makes an angle of 40° with the horizontal. What is the coefficient of kinetic friction between the box and the ramp?

N14. Figure 6N-8 shows three crates being pushed over a concrete floor by a horizontal force $\vec{F}$ of magnitude 440 N. The masses of the crates are $m_1 = 30.0$ kg, $m_2 = 10.0$ kg, and $m_3 = 20.0$ kg. The coefficient of kinetic friction between the floor and each of the crates is 0.70. (a) What is the magnitude F_{32} of the force on crate 3 from crate 2? (b) If the crates then slide onto a polished floor, where the coefficient of kinetic friction is less than 0.70, is magnitude F_{32} more than, less than, or the same as it was when the coefficient was 0.70?

Fig. 6N-8 Problem N14.

N15. Luggage is transported from one location to another in an airport by a conveyor belt. At a certain location, the belt moves down an incline that makes an angle of 2.5° with the horizontal. Assume that with such a slight angle there is no slipping of the luggage. Determine the magnitude and direction of the frictional force by the belt on a box weighing 69 N when the box is on the inclined portion of the belt for the following situations: (a) The belt is stationary. (b) The belt has a speed of 0.65 m/s that is constant. (c) The belt has a speed of 0.65 m/s that is increasing at a rate of 0.20 m/s². (d) The belt has a speed of 0.65 m/s that is decreasing at a rate of 0.20 m/s². (e) The belt has a speed of 0.65 m/s that is increasing at a rate of 0.57 m/s².

N16. A bolt is threaded onto one end of a thin horizontal rod, and the rod is then rotated horizontally about its other end. An engineer monitors the motion by flashing a strobe lamp onto the rod and bolt, adjusting the strobe rate until the bolt appears to be in the same eight places during each full rotation of the rod (Fig. 6N-9). The strobe rate is 2000 flashes per second; the bolt has mass 30 g and is at radius 3.5 cm. What is the magnitude of the force on the bolt from the rod?

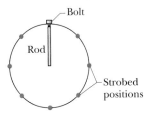

Fig. 6N-9 Problem N16.

N17. A circular-motion addict of mass 80 kg rides a Ferris wheel around in a vertical circle of radius 10 m at a constant speed of 6.1 m/s. (a) What is the period of the motion? What is the magnitude of the normal force on the addict from the seat when both go through (b) the highest point of the circular path and (c) the lowest point?

N18. In Fig. 6N-10, a car is driven at constant speed over a circular hill and then into a circular valley with the same radius. At the top of the hill, the normal force on the driver from the car seat is 0. The driver's mass is 70.0 kg. What is the magnitude of the normal force on the driver from the seat when the car passes through the bottom of the valley?

Fig. 6N-10 Problem N18.

N19. Fig. 6N-11 shows a *conical pendulum*, in which the bob (the small object at the lower end of the cord) moves in a horizontal circle at constant speed. (The cord sweeps out a cone as the bob rotates.) The bob has a mass of 0.040 kg, the string has length $L = 0.90$ m and a mass that is negligible relative to the bob's mass, and the bob follows a circular path of circumference 0.94 m. What are (a) the tension in the string and (b) the period of the motion?

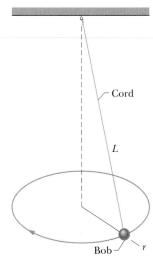

Fig. 6N-11 Problem N19.

N20. A 2.0 kg block and a 1.0 kg block are connected by a string and are pushed across a horizontal surface by a force applied to the 1.0 kg block as shown in Fig. 6N-12. The coefficient of kinetic friction between the blocks and the horizontal surface is 0.20. If the magnitude of $\vec{F}$ is 20 N, what is the tension in the string that connects the blocks?

Fig. 6N-12 Problem N20.

N21. What is the terminal speed of a 6.0 kg spherical ball with a radius of 3.0 cm and a drag coefficient of 1.60? The density of the air through which the ball falls is 1.2 kg/m³.

N22. A police officer in hot pursuit drives her car through a circular turn of radius 300 m with a constant speed of 80.0 km/h. Her mass is 55.0 kg. What are (a) the magnitude and (b) the angle (relative to vertical) of the *net* force of the officer on the car seat? (*Hint:* Consider both horizontal and vertical forces.)

N23. You must push a crate across a floor to a docking bay. The crate weighs 165 N. The coefficient of static friction between the crate and the floor is 0.51 and the coefficient of kinetic friction is 0.32. Your force on the crate is directed horizontally. (a) With what least magnitude must you push to get the crate moving? (b) With what magnitude must you then push to keep the crate moving at a constant velocity? (c) If, instead, you then push with the same magnitude as the answer to (a), what would be the magnitude of the crate's acceleration?

N24. Block *A* in Fig. 6N-13 has mass $m_A = 4.0$ kg and block *B* has mass $m_B = 2.0$ kg. The coefficient of kinetic friction between block *B* and the horizontal plane is 0.50. The inclined plane is frictionless and at angle 30°. Find (a) the tension in the connecting cord (which has a negligible mass) and (b) the magnitude of the acceleration of the blocks.

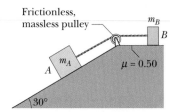

Fig. 6N-13 Problem N24.

N25. A 4.10 kg block is pushed along a floor by a constant applied force that is horizontal and has a magnitude of 40.0 N. Figure 6N-14 gives the block's speed *v* versus time *t* as the block moves along an *x* axis on the floor. What is the coefficient of kinetic friction between the block and the floor?

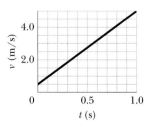

Fig. 6N-14 Problem N25.

N26. A student wants to determine the coefficients of static friction and kinetic friction between a box and a plank. She places the box on the plank and gradually raises one end of the plank. When the angle of inclination with the horizontal reaches 30°, the box starts to slip, and it slides 2.5 m down the plank in 4.0 s. What are (a) the coefficient of static friction and (b) the coefficient of kinetic friction?

N27. A 5.00 kg stone is rubbed across a horizontal ceiling of a cave passageway (Fig. 6N-15). If the coefficient of kinetic friction is 0.65 and the force applied to the stone is angled at $\theta = 70.0°$, what must the magnitude of the force be for the stone to move at constant velocity?

Fig. 6N-15 Problem N27.

N28. The three blocks in Fig. 6N-16 are released from rest and then accelerate with a magnitude of 0.500 m/s^2. What is the coefficient of kinetic friction between the sliding block and the table?

Fig. 6N-16 Problem N28.

N29. A circular curve of highway is designed for traffic moving at 60 km/h. (a) If the radius of the curve is 150 m, what is the correct angle of banking of the road? (b) If the curve were not banked, what would be the minimum coefficient of friction between tires and road that would keep traffic from skidding out of the turn when traveling at 60 km/h?

N30. A block slides down an inclined plane of slope angle θ with constant velocity. It is then projected up the same plane with an initial speed v_0. (a) How far up the incline will it move before coming to rest? (b) Will it slide down again? Give an argument to back your answer.

N31. A cat dozes on a stationary merry-go-round, at a radius of 5.4 m from the center of the ride. Then the operator turns on the ride and brings it up to its proper turning rate of one complete rotation every 6.0 s. What is the least coefficient of static friction

between the cat and the merry-go-round that will allow the cat to stay in place, without sliding?

N32. In a pickup game of dorm shuffleboard, students crazed by final exams use a broom to propel a calculus book along the dorm hallway. If the 3.5 kg book is pushed from rest through a distance of 0.90 m by the horizontal 25 N force from the broom and then has a speed of 1.60 m/s, what is the coefficient of kinetic friction between the book and floor?

N33. A block weighing 22 N is held at rest against a vertical wall by a horizontal force $\vec{F}$ of magnitude 60 N as shown in Fig. 6N-17. The coefficient of static friction between the wall and the block is 0.55 and the coefficient of kinetic friction between them is 0.38. A second force $\vec{P}$ acting parallel to the wall is applied to the block. For the following magnitudes and directions of $\vec{P}$, determine whether the block moves, the direction of motion, and the magnitude and direction of the frictional force acting on the block: (a) 34 N, up, (b) 12 N, up, (c) 48 N, up, (d) 62 N, up, (e) 10 N, down, and (f) 18 N, down.

Fig. 6N-17 Problem N33.

N34. Playing near a road construction site, a child trips over an edge and falls onto a dirt slope angled at 35° to the horizontal. As the child slides down the slope, he has an acceleration of 0.50 m/s^2 up the slope. What is the coefficient of kinetic friction between the child and the slope?

N35. An M&M candy piece is placed on a stationary turntable, at a distance of 6.6 cm from the center. Then the turntable is turned on and its turning rate adjusted so that it makes 2.00 full turns in 2.45 s. The candy rides the turntable without slipping. (a) What is the magnitude of its acceleration? (b) If the turning rate is gradually increased to 2.00 full turns every 1.80 s, the candy piece then begins to slide. What is the coefficient of static friction between it and the turntable?

7 Kinetic Energy and Work

In the weight-lifting competition of the 1996 Olympics, Andrey Chemerkin lifted a record-breaking 260.0 kg from the floor to over his head (about 2 m). In 1957 Paul Anderson stooped beneath a reinforced wood platform, placed his hands on a short stool to brace himself, and then pushed upward on the platform with his back, lifting the platform and its load about a centimeter. On the platform were auto parts and a safe filled with lead; the composite weight of the load was 27 900 N (6270 lb)!

Who did more work on the objects he lifted—Chemerkin or Anderson?

The answer is in this chapter.

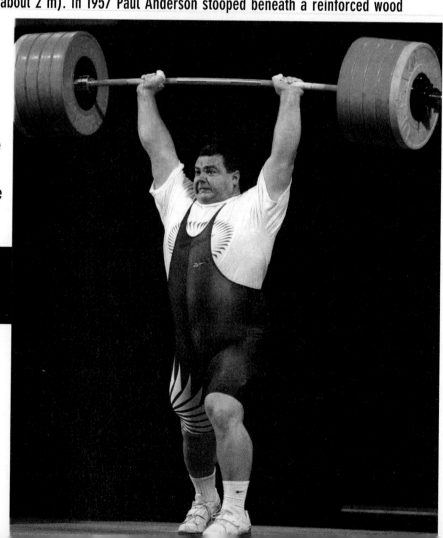

7-1 Energy

Newton's laws of motion allow us to analyze many kinds of motion. However, the analysis is often complicated, requiring details about the motion that we simply do not know. Here is an example: A puck is sent sliding along an inclined frictionless track that includes several ups and downs (hills and valleys) of various shapes. The puck's initial speed is 4.0 m/s, and its initial height is 0.46 m. Using Newton's second law, can you calculate the puck's speed when it reaches the end of the track, at zero height? No, not without the details of how the incline varies all along the track, and then the calculations can be very complicated.

Long ago, scientists and engineers gradually began to realize that there is another, sometimes more powerful technique for analyzing motion. Moreover, this technique could be, and eventually was, extended to other situations, such as chemical reactions, geological processes, and biological functions, that do not involve motion. This other technique involves **energy**, which comes in a great many *forms* (or types). In fact, the term *energy* is so broad that a clear definition for it is difficult to write. Technically, energy is a scalar quantity that is associated with a state (or condition) of one or more objects. However, this definition is too vague to be of help to us now.

A looser definition might at least get us started. Energy is a number that we associate with a system of one or more objects. If a force changes one of the objects by, say, making it move, then the number changes. After countless experiments, scientists and engineers realized that if the scheme by which we assign these energy numbers is planned carefully, then the numbers can be used to predict the outcomes of experiments. (For example, they would allow us to easily find the puck's speed in the previous example.) However, learning how to use the numbering scheme is not easy and the scheme is not obvious. Thus, in this chapter we focus on only one form of energy—kinetic energy. Other forms of energy will show up throughout this book and your work in science or engineering.

Kinetic energy K is energy associated with the *state of motion* of an object. The faster the object moves, the greater is its kinetic energy. When the object is stationary, its kinetic energy is zero.

For an object of mass m whose speed v is well below the speed of light, we define kinetic energy as

$$K = \tfrac{1}{2}mv^2 \qquad \text{(kinetic energy).} \qquad (7\text{-}1)$$

For example, a 3.0 kg duck flying past us at 2.0 m/s has a kinetic energy of 6.0 kg · m²/s²; that is, we associate that number with the duck's motion.

The SI unit of kinetic energy (and every other type of energy) is the **joule** (J), named for James Prescott Joule, an English scientist of the 1800s. It is defined directly from Eq. 7-1 in terms of the units for mass and velocity:

$$1 \text{ joule} = 1 \text{ J} = 1 \text{ kg} \cdot \text{m}^2/\text{s}^2. \qquad (7\text{-}2)$$

Thus, the flying duck has a kinetic energy of 6.0 J.

Sample Problem 7-1

In 1896 in Waco, Texas, William Crush of the "Katy" railroad parked two locomotives at opposite ends of a 6.4-km-long track, fired them up, tied their throttles open, and then allowed them to crash head-on at full speed (Fig. 7-1) in front of 30,000 spectators. Hundreds of people were hurt by flying debris; several were killed. Assuming each locomotive weighed 1.2×10^6 N and its acceleration along the track was a constant 0.26 m/s², what was the total kinetic energy of the two locomotives just before the collision?

Fig. 7-1 Sample Problem 7-1. The aftermath of an 1896 crash of two locomotives.

SOLUTION: One Key Idea here is to find the kinetic energy of each locomotive with Eq. 7-1, but that means we need each locomotive's speed just before the collision and its mass. A second Key Idea is

that, because we can assume each locomotive had constant acceleration, we can use the equations in Table 2-1 to find its speed v just before the collision. We choose Eq. 2-16 because we know values for all the variables except v:

$$v^2 = v_0^2 + 2a(x - x_0).$$

With $v_0 = 0$ and $x - x_0 = 3.2 \times 10^3$ m (half the initial separation), this yields

$$v^2 = 0 + 2(0.26 \text{ m/s}^2)(3.2 \times 10^3 \text{ m}),$$

or $v = 40.8$ m/s

(about 150 km/h).

A third Key Idea is that we can find the mass of each locomotive by dividing its given weight by g:

$$m = \frac{1.2 \times 10^6 \text{ N}}{9.8 \text{ m/s}^2} = 1.22 \times 10^5 \text{ kg}.$$

Now, using Eq. 7-1, we find the total kinetic energy of the two locomotives just before the collision as

$$K = 2(\tfrac{1}{2}mv^2) = (1.22 \times 10^5 \text{ kg})(40.8 \text{ m/s})^2$$

$$= 2.0 \times 10^8 \text{ J.} \qquad \text{(Answer)}$$

Sitting near this collision was like sitting near an exploding bomb.

7-2 Work

If you accelerate an object to a greater speed by applying a force to the object, you increase the kinetic energy $K \ (= \tfrac{1}{2}mv^2)$ of the object. Similarly, if you decelerate the object to a lesser speed by applying a force, you decrease the kinetic energy of the object. We account for these changes in kinetic energy by saying that your force has transferred energy *to* the object from yourself or *from* the object to yourself.

In such a transfer of energy via a force, **work** W is said to be *done on the object by the force*. More formally, we define work as follows:

> Work W is energy transferred to or from an object by means of a force acting on the object. Energy transferred to the object is positive work, and energy transferred from the object is negative work.

"Work," then, is transferred energy; "doing work" is the act of transferring the energy. Work has the same units as energy and is a scalar quantity.

The term *transfer* can be misleading. It does not mean that anything material flows into or out of the object; that is, the transfer is not like a flow of water. Rather it is like the electronic transfer of money between two bank accounts: The number in one account goes up while the number in the other account goes down, with nothing material passing between the two accounts.

Note that we are not concerned here with the common meaning of the word "work," which implies that *any* physical or mental labor is work. For example, if you push hard against a wall, you tire because of the continuously repeated muscle contractions that are required, and you are, in the common sense, working. However, such effort does not cause an energy transfer to or from the wall and thus is not work done on the wall as defined here.

To avoid confusion in this chapter, we shall use the symbol W only for work and shall represent a weight with its equivalent mg.

7-3 Work and Kinetic Energy

Finding an Expression for Work

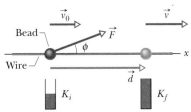

Let us find an expression for work by considering a bead that can slide along a frictionless wire, which is stretched along a horizontal x axis (Fig. 7-2). A constant force $\vec{F}$, directed at an angle ϕ to the wire, accelerates the bead along the wire. We can relate the force and the acceleration with Newton's second law, written for components along the x axis:

$$F_x = ma_x, \tag{7-3}$$

where m is the bead's mass. As the bead moves through a displacement $\vec{d}$, the force changes the bead's velocity from an initial value $\vec{v}_0$ to another value $\vec{v}$. Because the force is constant, we know that the acceleration is also constant. Thus, we can use Eq. 2-16 (one of the basic constant-acceleration equations of Chapter 2) to write, for components along the x axis,

$$v^2 = v_0^2 + 2a_x d. \tag{7-4}$$

Solving this equation for a_x, substituting into Eq. 7-3, and rearranging then give us

$$\tfrac{1}{2}mv^2 - \tfrac{1}{2}mv_0^2 = F_x d. \tag{7-5}$$

The first term on the left side of the equation is the kinetic energy K_f of the bead at the end of the displacement d, and the second term is the kinetic energy K_i of the bead at the start of the displacement. Thus, the left side of Eq. 7-5 tells us the kinetic energy has been changed by the force, and the right side tells us the change is equal to $F_x d$. Therefore, the work W done on the bead by the force (the energy transfer due to the force) is

$$W = F_x d. \tag{7-6}$$

If we know values for F_x and d, we can use this equation to calculate the work W done on the bead by the force.

Fig. 7-2 A constant force $\vec{F}$ directed at angle ϕ to the displacement $\vec{d}$ of a bead on a wire accelerates the bead along the wire, changing the velocity of the bead from $\vec{v}_0$ to $\vec{v}$. A "kinetic energy gauge" indicates the resulting change in the kinetic energy of the bead, from the value K_i to K_f.

> To calculate the work done on an object by a force during a displacement, we use only the force component along the object's displacement. The force component perpendicular to the displacement does zero work.

From Fig. 7-2, we see that we can write F_x as $F \cos \phi$, where ϕ is the angle between the directions of the displacement $\vec{d}$ and the force $\vec{F}$. We can rewrite Eq. 7-6 in a more general form as

$$W = Fd \cos \phi \qquad \text{(work done by a constant force)}. \tag{7-7}$$

This equation is useful for calculating the work if we know values for F, d, and ϕ. Because the right side of this equation is equivalent to the scalar (or dot) product $\vec{F} \cdot \vec{d}$, we can also write

$$W = \vec{F} \cdot \vec{d} \qquad \text{(work done by a constant force)}. \tag{7-8}$$

(You may wish to review the discussion of scalar products in Section 3-7.) Equation 7-8 is especially useful for calculating the work when $\vec{F}$ and $\vec{d}$ are given in unit-vector notation.

Cautions: There are two restrictions to using Eqs. 7-6 through 7-8 to calculate work done on an object by a force. First, the force must be a *constant force*; that is, it must not change in magnitude or direction as the object moves. (Later, we shall discuss what to do with a *variable force* that changes in magnitude.) Second, the object must be *particle-like*. This means that the object must be *rigid*; all parts of the object must move together, in the same direction. In this chapter we consider only particle-like objects, such as the bed and its rider being pushed in Fig. 7-3.

Fig. 7-3 A contestant in a bed race. We can approximate the bed and its rider as being a particle for the purpose of calculating the work done on them by the force applied by the student.

Signs for work. The work done on an object by a force can be either positive work or negative work. For example, if the angle ϕ in Eq. 7-7 is less than 90°, then $\cos \phi$ is positive and thus so is the work. If ϕ is greater than 90° (up to 180°), then $\cos \phi$ is negative and thus so is the work. (Can you see that the work is zero when $\phi = 90°$?) These results lead to a simple rule. To find the sign of the work done by a force, consider the vector component of the force along the displacement:

> ▶ A force does positive work when it has a vector component in the same direction as the displacement, and it does negative work when it has a vector component in the opposite direction. It does zero work when it has no such vector component.

Units for work. Work has the SI unit of the joule, the same as kinetic energy. However, from Eqs. 7-6 and 7-7 we can see that an equivalent unit is the newton-meter (N · m). The corresponding unit in the British system is the foot-pound (ft · lb). Extending Eq. 7-2, we have

$$1 \text{ J} = \text{kg} \cdot \text{m}^2/\text{s}^2 = 1 \text{ N} \cdot \text{m} = 0.738 \text{ ft} \cdot \text{lb}. \tag{7-9}$$

Net work done by several forces. When two or more forces act on an object, the **net work** done on the object is the sum of the works done by the individual forces. We can calculate the net work in two ways. (1) We can find the work done by each force and then sum those works. (2) Alternatively, we can first find the net force $\vec{F}_{net}$ of those forces. Then we can use Eq. 7-7, substituting the magnitude F_{net} for F, and the angle between the directions of $\vec{F}_{net}$ and the displacement for ϕ; or we can use Eq. 7-8 with $\vec{F}_{net}$ substituted for $\vec{F}$.

Work−Kinetic Energy Theorem

Equation 7-5 relates the change in kinetic energy of the bead (from an initial $K_i = \frac{1}{2}mv_0^2$ to a later $K_f = \frac{1}{2}mv^2$) to the work $W \ (= F_x d)$ done on the bead. For such particle-like objects, we can generalize that equation. Let ΔK be the change in the kinetic energy of the object, and W the net work done on it. Then we can write

$$\Delta K = K_f - K_i = W, \tag{7-10}$$

which says that

$$\begin{pmatrix} \text{change in the kinetic} \\ \text{energy of a particle} \end{pmatrix} = \begin{pmatrix} \text{net work done on} \\ \text{the particle} \end{pmatrix}.$$

We can also write

$$K_f = K_i + W, \tag{7-11}$$

which says that

$$\begin{pmatrix} \text{kinetic energy after} \\ \text{the net work is done} \end{pmatrix} = \begin{pmatrix} \text{kinetic energy} \\ \text{before the net work} \end{pmatrix} + \begin{pmatrix} \text{the net} \\ \text{work done} \end{pmatrix}.$$

These statements are known traditionally as the **work–kinetic energy theorem** for particles. They hold for both positive and negative work: If the net work done on a particle is positive, then the particle's kinetic energy increases by the amount of the work. If the net work done is negative, then the particle's kinetic energy decreases by the amount of the work.

For example, if the kinetic energy is initially 5 J and there is a net transfer of 2 J to the particle (positive net work), then the final kinetic energy is 7 J. If, instead, there is a net transfer of 2 J from the particle (negative net work), then the final kinetic energy is 3 J.

✔CHECKPOINT 1:   A particle moves along an x axis. Does the kinetic energy of the particle increase, decrease, or remain the same if the particle's velocity changes (a) from -3 m/s to -2 m/s and (b) from -2 m/s to 2 m/s? (c) In each situation, is the work done on the particle positive, negative, or zero?

Sample Problem 7-2

Figure 7-4a shows two industrial spies sliding an initially stationary 225 kg floor safe a displacement $\vec{d}$ of magnitude 8.50 m, straight toward their truck. The push $\vec{F}_1$ of Spy 001 is 12.0 N, directed at an angle of 30° downward from the horizontal; the pull $\vec{F}_2$ of Spy 002 is 10.0 N, directed at 40° above the horizontal. The magnitudes and directions of these forces do not change as the safe moves, and the floor and safe make frictionless contact.

(a) What is the net work done on the safe by forces $\vec{F}_1$ and $\vec{F}_2$ during the displacement $\vec{d}$?

SOLUTION: We use two Key Ideas here. First, the net work W done on the safe by the two forces is the sum of the works they do individually. Second, because we can treat the safe as a particle and the forces are constant in both magnitude and direction, we can use either Eq. 7-7 ($W = Fd \cos \phi$) or Eq. 7-8 ($W = \vec{F} \cdot \vec{d}$) to calculate those works. Since we know the magnitudes and directions of the forces, we choose Eq. 7-7. From it and the free-body diagram for the safe in Fig. 7-4b, the work done by $\vec{F}_1$ is

$$W_1 = F_1 d \cos \phi_1 = (12.0 \text{ N})(8.50 \text{ m})(\cos 30°)$$
$$= 88.33 \text{ J},$$

and the work done by $\vec{F}_2$ is

$$W_2 = F_2 d \cos \phi_2 = (10.0 \text{ N})(8.50 \text{ m})(\cos 40°)$$
$$= 65.11 \text{ J}.$$

Thus, the net work W is

$$W = W_1 + W_2 = 88.33 \text{ J} + 65.11 \text{ J}$$
$$= 153.4 \text{ J} \approx 153 \text{ J}. \qquad \text{(Answer)}$$

During the 8.50 m displacement, therefore, the spies transfer 153 J of energy to the kinetic energy of the safe.

(b) During the displacement, what is the work W_g done on the safe by the gravitational force $\vec{F}_g$ and what is the work W_N done on the safe by the normal force $\vec{N}$ from the floor?

SOLUTION: The Key Idea is that, because these forces are constant in both magnitude and direction, we can find the work they do with

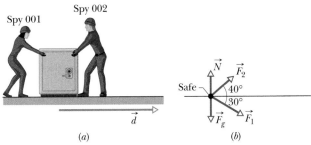

Spy 001   Spy 002

Safe

$\vec{N}$ $\vec{F}_2$

40°
30°

$\vec{F}_g$ $\vec{F}_1$

$\vec{d}$

(a)   (b)

Fig. 7-4 Sample Problem 7-2. (a) Two spies move a floor safe through displacement $\vec{d}$. (b) A free-body diagram for the safe.

Eq. 7-7. Thus, with mg as the magnitude of the gravitational force, we write

$$W_g = mgd \cos 90° = mgd(0) = 0 \qquad \text{(Answer)}$$

and

$$W_N = Nd \cos 90° = Nd(0) = 0. \qquad \text{(Answer)}$$

We should have known this result. Because these forces are perpendicular to the displacement of the safe, they do zero work on the safe and do not transfer any energy to or from it.

(c) The safe is initially stationary. What is its speed v_f at the end of the 8.50 m displacement?

SOLUTION: Here the Key Idea is that the speed of the safe changes because its kinetic energy is changed when energy is transferred to it by $\vec{F}_1$ and $\vec{F}_2$. We relate the speed to the work done by combining Eqs. 7-10 and 7-1:

$$W = K_f - K_i = \tfrac{1}{2}mv_f^2 - \tfrac{1}{2}mv_i^2.$$

The initial speed v_i is zero, and we now know that the work done is 153.4 J. Solving for v_f and then substituting known data, we find that

$$v_f = \sqrt{\frac{2W}{m}} = \sqrt{\frac{2(153.4 \text{ J})}{225 \text{ kg}}}$$
$$= 1.17 \text{ m/s}. \qquad \text{(Answer)}$$

Sample Problem 7-3

During a storm, a crate of crepe is sliding across a slick, oily parking lot through a displacement $\vec{d} = (-3.0 \text{ m})\hat{i}$ while a steady wind pushes against the crate with a force $\vec{F} = (2.0 \text{ N})\hat{i} + (-6.0 \text{ N})\hat{j}$. The situation and coordinate axes are shown in Fig. 7-5.

(a) How much work does this force from the wind do on the crate during the displacement?

y

x

$\vec{d}$ $\vec{F}$

Fig. 7-5 Sample Problem 7-3. Constant force $\vec{F}$ slows a crate during displacement $\vec{d}$.

SOLUTION: The Key Idea here is that, because we can treat the crate as a particle and because the wind force is constant ("steady") in both magnitude and direction during the displacement, we can use either Eq. 7-7 ($W = Fd \cos \phi$) or Eq. 7-8 ($W = \vec{F} \cdot \vec{d}$) to calculate the work. Since we know $\vec{F}$ and $\vec{d}$ in unit-vector notation, we choose Eq. 7-8 and write

$$W = \vec{F} \cdot \vec{d} = [(2.0 \text{ N})\hat{i} + (-6.0 \text{ N})\hat{j}] \cdot [(-3.0 \text{ m})\hat{i}].$$

Of the possible unit-vector dot products, only $\hat{i} \cdot \hat{i}$, $\hat{j} \cdot \hat{j}$, and $\hat{k} \cdot \hat{k}$ are nonzero (see Appendix E). Here we obtain

$$W = (2.0 \text{ N})(-3.0 \text{ m})\hat{i} \cdot \hat{i} + (-6.0 \text{ N})(-3.0 \text{ m})\hat{j} \cdot \hat{i}$$
$$= (-6.0 \text{ J})(1) + 0 = -6.0 \text{ J}. \quad \text{(Answer)}$$

Thus, the force does a negative 6.0 J of work on the crate, transfering 6.0 J of energy from the kinetic energy of the crate.

(b) If the crate has a kinetic energy of 10 J at the beginning of displacement $\vec{d}$, what is its kinetic energy at the end of $\vec{d}$?

SOLUTION: The Key Idea here is that, because the force does negative work on the crate, it reduces the crate's kinetic energy. Using the work–kinetic energy theorem in the form of Eq. 7-11, we have

$$K_f = K_i + W = 10 \text{ J} + (-6.0 \text{ J}) = 4.0 \text{ J}. \quad \text{(Answer)}$$

Because the kinetic energy is decreased to 4.0 J, the crate has been slowed.

✓CHECKPOINT 2:   The figure shows four situations in which a force acts on a box while the box slides rightward a distance d across a frictionless floor. The magnitudes of the forces are identical; their orientations are as shown. Rank the situations according to the work done on the box during the displacement, from most positive to most negative.

(a)   (b)   (c)   (d)

7-4 Work Done by a Gravitational Force

Fig. 7-6 A particle-like tomato of mass m thrown upward slows from velocity $\vec{v}_0$ to velocity $\vec{v}$ during displacement $\vec{d}$ because the gravitational force $\vec{F}_g$ acts on it. A kinetic energy gauge indicates the resulting change in the kinetic energy of the object, from K_i ($= \frac{1}{2}mv_0^2$) to K_f ($= \frac{1}{2}mv^2$).

We next examine the work done on an object by a particular type of force—namely, the gravitational force acting on it. Figure 7-6 shows a particle-like tomato of mass m that is thrown upward with initial speed v_0 and thus with initial kinetic energy $K_i = \frac{1}{2}mv_0^2$. As the tomato rises, it is slowed by a gravitational force $\vec{F}_g$; that is, the tomato's kinetic energy decreases because $\vec{F}_g$ does work on the tomato as it rises.

Because we can treat the tomato as a particle, we can use Eq. 7-7 ($W = Fd \cos \phi$) to express the work done during a displacement $\vec{d}$. For the force magnitude F, we use mg as the magnitude of $\vec{F}_g$. Thus, the work W_g done by the gravitational force $\vec{F}_g$ is

$$W_g = mgd \cos \phi \quad \text{(work done by gravitational force).} \quad (7\text{-}12)$$

For a rising object, force $\vec{F}_g$ is directed opposite the displacement $\vec{d}$, as indicated in Fig. 7-6. Thus, $\phi = 180°$ and

$$W_g = mgd \cos 180° = mgd(-1) = -mgd. \quad (7\text{-}13)$$

The minus sign tells us that during the object's rise, the gravitational force on the object transfers energy in the amount mgd from the kinetic energy of the object. This is consistent with the slowing of the object as it rises.

After the object has reached its maximum height and is falling back down, the angle ϕ between force $\vec{F}_g$ and displacement $\vec{d}$ is zero. Thus,

$$W_g = mgd \cos 0° = mgd(+1) = +mgd. \quad (7\text{-}14)$$

The plus sign tells us that the gravitational force now transfers energy in the amount mgd to the kinetic energy of the object. This is consistent with the speeding up of the object as it falls. (Actually, as we shall see in Chapter 8, energy transfers associated with lifting and lowering an object involve not just the object, but the full object–Earth system. Without Earth, of course, "lifting" would be meaningless.)

Work Done in Lifting and Lowering an Object

Now suppose we lift a particle-like object by applying a vertical force $\vec{F}$ to it. During the upward displacement, our applied force does positive work W_a on the object while the gravitational force does negative work W_g on it. Our force tends to transfer energy to the object while the gravitational force tends to transfer energy from it. By Eq. 7-10, the change ΔK in the kinetic energy of the object due to these two energy transfers is

$$\Delta K = K_f - K_i = W_a + W_g, \tag{7-15}$$

in which K_f is the kinetic energy at the end of the displacement and K_i is that at the start of the displacement. This equation also applies if we lower the object, but then the gravitational force tends to transfer energy *to* the object while our force tends to transfer energy *from* it.

In one common situation the object is stationary before and after the lift—for example, when you lift a book from the floor to a shelf. Then K_f and K_i are both zero, and Eq. 7-15 reduces to

$$W_a + W_g = 0$$

or

$$W_a = -W_g. \tag{7-16}$$

Note that we get the same result if K_f and K_i are not zero but are still equal. Either way, the result means that the work done by the applied force is the negative of the work done by the gravitational force; that is, the applied force transfers the same amount of energy to the object as the gravitational force transfers from the object. Using Eq. 7-12, we can rewrite Eq. 7-16 as

$$W_a = -mgd \cos \phi \quad \text{(work in lifting and lowering; } K_f = K_i\text{)}, \tag{7-17}$$

with ϕ being the angle between $\vec{F}_g$ and $\vec{d}$. If the displacement is vertically upward (Fig. 7-7a), then $\phi = 180°$ and the work done by our force equals mgd. If the displacement is vertically downward (Fig. 7-7b), then $\phi = 0°$ and the work done by the applied force equals $-mgd$.

Equations 7-16 and 7-17 apply to any situation in which an object is lifted or lowered, with the object stationary before and after the lift. They are independent of the magnitude of the force used. For example, when Chemerkin made his record-breaking lift, his force on the object he lifted varied considerably during the lift. Still, because the object was stationary before and after the lift, the work he did is given by Eqs. 7-16 and 7-17, where, in Eq. 7-17, mg is the weight of the object he lifted and d is the distance he lifted it.

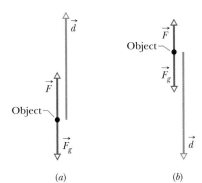

(a)   (b)

Fig. 7-7 (a) An applied force $\vec{F}$ lifts an object. The object's displacement $\vec{d}$ makes an angle $\phi = 180°$ with the gravitational force $\vec{F}_g$ on the object. The applied force does positive work on the object. (b) An applied force $\vec{F}$ lowers an object. The displacement $\vec{d}$ of the object makes an angle $\phi = 0°$ with the gravitational force $\vec{F}_g$. The applied force does negative work on the object.

Sample Problem 7-4

Let us return to the lifting feats of Andrey Chemerkin and Paul Anderson.

(a) Chemerkin made his record-breaking lift with rigidly connected objects (a barbell and disk weights) having a total mass $m = 260.0$ kg; he lifted them a distance of 2.0 m. During the lift, how much work was done on the objects by the gravitational force $\vec{F}_g$ acting on them?

SOLUTION: The Key Idea here is that we can treat the rigidly connected objects as a single particle and thus use Eq. 7-12 ($W_g =$

$mgd \cos \phi$) to find the work W_g done on them by $\vec{F}_g$. The total weight mg was 2548 N, the magnitude d of the displacement was 2.0 m, and the angle ϕ between the directions of the downward gravitational force and the upward displacement was 180°. Thus,

$$W_g = mgd \cos \phi = (2548 \text{ N})(2.0 \text{ m})(\cos 180°)$$
$$= -5100 \text{ J}. \quad \text{(Answer)}$$

(b) How much work was done on the objects by Chemerkin's force during the lift?

SOLUTION: We do not have an expression for Chemerkin's force on the object, and even if we did, his force was certainly not constant, Thus, one Key Idea here is that we *cannot* just substitute his force into Eq. 7-7 to find his work. However, we know that the objects were stationary at the start and end of the lift. Therefore, as a second Key Idea, we know that the work W_{AC} done by Chemerkin's applied force was the negative of the work W_g done by the gravitational force $\vec{F}_g$. Equation 7-16 expresses this fact and gives us

$$W_{AC} = -W_g = +5100 \text{ J}. \qquad \text{(Answer)}$$

(c) While Chemerkin held the objects stationary above his head, how much work was done on them by his force?

SOLUTION: The Key Idea is that when he supported the objects, they were stationary. Thus, their displacement $d = 0$ and, by Eq. 7-7, the work done on them was zero (even though supporting them was a very tiring task).

(d) How much work was done by the force Paul Anderson applied to lift objects with a total weight of 27 900 N, a distance of 1.0 cm?

SOLUTION: Following the argument of parts (a) and (b), but now with

$mg = 27\,900$ N and $d = 1.0$ cm, we find

$$W_{PA} = -W_g = -mgd \cos \phi = -mgd \cos 180°$$
$$= -(27\,900 \text{ N})(0.010 \text{ m})(-1) = 280 \text{ J}. \qquad \text{(Answer)}$$

Anderson's lift required a tremendous upward force but only a small energy transfer of 280 J, because of the short displacement involved. This photo shows another of his lifts.

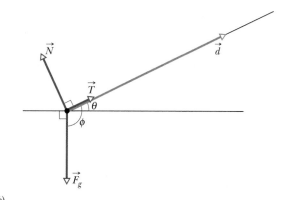

Sample Problem 7-5

An initially stationary 15.0 kg crate of cheese wheels is pulled, via a cable, a distance $L = 5.70$ m up a frictionless ramp, to a height h of 2.50 m, where it stops (Fig. 7-8a).

(a) How much work W_g is done on the crate by the gravitational force $\vec{F}_g$ during the lift?

SOLUTION: A Key Idea is that we can treat the crate as a particle and thus use Eq. 7-12 ($W_g = mgd \cos \phi$) to find the work W_g done by $\vec{F}_g$. However, we do not know the angle ϕ between the directions of $\vec{F}_g$ and displacement $\vec{d}$. From the crate's free-body diagram in Fig. 7-8b, we find that ϕ is $\theta + 90°$, where θ is the (unknown) angle of the ramp. Equation 7-12 then gives us

$$W_g = mgd \cos(\theta + 90°) = -mgd \sin \theta, \qquad (7\text{-}18)$$

where we have used a trigonometic identity to simplify the expression. The result seems to be useless because θ is unknown. But (continuing with physics courage) we see from Fig. 7-8a that $d \sin \theta = h$, where h is a known quantity (2.50 m). With this substitution, Eq. 7-18 then gives us

$$W_g = -mgh \qquad (7\text{-}19)$$
$$= -(15.0 \text{ kg})(9.8 \text{ m/s}^2)(2.50 \text{ m})$$
$$= -368 \text{ J}. \qquad \text{(Answer)}$$

Note that Eq. 7-19 tells us that the work W_g done by the gravitational force depends on the vertical displacement but (surprisingly) not on the horizontal displacement. (We return to this point in Chapter 8.)

(b) How much work W_T is done on the crate by the force $\vec{T}$ from the cable during the lift?

(a)

(b)

Fig. 7-8 Sample Problem 7-5. (*a*) A crate is pulled up a frictionless ramp by a force $\vec{T}$ parallel to the ramp. (*b*) A free-body diagram for the crate, showing also the displacement $\vec{d}$.

SOLUTION: We cannot just substitute the force magnitude T for F in Eq. 7-7 ($W = Fd \cos \phi$) because we do not know the value of T. However, a Key Idea to get us going is that we can treat the crate as a particle and then apply the work–kinetic energy theorem ($\Delta K = W$) to it. Because the crate is stationary before and after the lift, the change ΔK in its kinetic energy is zero. For the net work W done on the crate, we must sum the works done by all three forces acting on the crate. From (a), the work W_g done by the gravitational force $\vec{F}_g$ is -368 J. The work W_N done by the normal force $\vec{N}$ on the crate from the ramp is zero because $\vec{N}$ is perpendicular to the displacement. We want the work W_T done by $\vec{T}$. Thus,

the work–kinetic energy theorem gives us

$$\Delta K = W_T + W_g + W_N$$

or

$$0 = W_T - 368 \text{ J} + 0,$$

and so

$$W_T = 368 \text{ J}. \qquad \text{(Answer)}$$

✔**CHECKPOINT 3:** Suppose we raise the crate by the same height h but with a longer ramp. (a) Is the work done by force $\vec{T}$ now greater, smaller, or the same as before? (b) Is the magnitude of $\vec{T}$ needed to move the crate now greater, smaller, or the same as before?

Sample Problem 7-6

An elevator cab of mass $m = 500$ kg is descending with speed $v_i = 4.0$ m/s when its supporting cable begins to slip, allowing it to fall with constant acceleration $\vec{a} = \vec{g}/5$ (Fig. 7-9a).

(a) During the fall through a distance $d = 12$ m, what is the work W_g done on the cab by the gravitational force $\vec{F}_g$?

SOLUTION: The Key Idea here is that we can treat the cab as a particle and thus use Eq. 7-12 ($W_g = mgd \cos \phi$) to find the work W_g done on it by $\vec{F}_g$. From Fig. 7-9b (a free-body diagram of the cab with the cab's displacement $\vec{d}$ included), we see that the angle between the directions of $\vec{F}_g$ and the cab's displacement $\vec{d}$ is $0°$. Then, from Eq. 7-12, we find

$$W_g = mgd \cos 0° = (500 \text{ kg})(9.8 \text{ m/s}^2)(12 \text{ m})(1)$$
$$= 5.88 \times 10^4 \text{ J} \approx 59 \text{ kJ}. \qquad \text{(Answer)}$$

(b) During the 12 m fall, what is the work W_T done on the cab by the upward pull $\vec{T}$ of the elevator cable?

SOLUTION: A Key Idea here is that we can calculate the work W_T with Eq. 7-7 ($W = Fd \cos \phi$) if we first find an expression for the magnitude T of the cable's pull. A second Key Idea is that we can find that expression by writing Newton's second law for components along the y axis in Fig. 7-9b ($F_{\text{net},y} = ma_y$). We get

$$T - F_g = ma.$$

Solving for T, substituting mg for F_g, and then substituting the result in Eq. 7-7, we obtain

$$W_T = Td \cos \phi = m(a + g)d \cos \phi.$$

Next, substituting $-g/5$ for the (downward) acceleration a and then $180°$ for the angle ϕ between the directions of forces $\vec{T}$ and $m\vec{g}$, we find

$$W_T = m\left(-\frac{g}{5} + g\right)d \cos \phi = \frac{4}{5} mgd \cos \phi$$

$$= \frac{4}{5} (500 \text{ kg})(9.8 \text{ m/s}^2)(12 \text{ m}) \cos 180°$$

$$= -4.70 \times 10^4 \text{ J} \approx -47 \text{ kJ}. \qquad \text{(Answer)}$$

Now note that W_T is not simply the negative of W_g, which we found in (a). The reason is that, because the cab accelerates during the fall, its speed changes during the fall, and thus its kinetic

Fig. 7-9 Sample Problem 7-6. An elevator cab, descending with speed v_i, suddenly begins to accelerate downward. (a) It moves through a displacement $\vec{d}$ with constant acceleration $\vec{a} = \vec{g}/5$. (b) A free-body diagram for the cab, displacement included.

energy also changes. Therefore, Eq. 7-16 (which assumes that the initial and final kinetic energies are equal) does *not* apply here.

(c) What is the net work W done on the cab during the fall?

SOLUTION: The Key Idea here is that the net work is the sum of the works done by the forces acting on the cab:

$$W = W_g + W_T = 5.88 \times 10^4 \text{ J} - 4.70 \times 10^4 \text{ J}$$
$$= 1.18 \times 10^4 \text{ J} \approx 12 \text{ kJ}. \qquad \text{(Answer)}$$

(d) What is the cab's kinetic energy at the end of the 12 m fall?

SOLUTION: The Key Idea here is that the kinetic energy changes *because* of the net work done on the cab, according to Eq. 7-11 ($K_f = K_i + W$). From Eq. 7-1, we can write the kinetic energy at the start of the fall as $K_i = \frac{1}{2}mv_i^2$. We can then write Eq. 7-11 as

$$K_f = K_i + W = \frac{1}{2}mv_i^2 + W$$
$$= \frac{1}{2}(500 \text{ kg})(4.0 \text{ m/s})^2 + 1.18 \times 10^4 \text{ J}$$
$$= 1.58 \times 10^4 \text{ J} \approx 16 \text{ kJ}. \qquad \text{(Answer)}$$

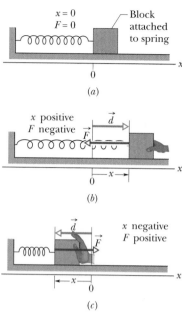

Fig. 7-10 (*a*) A spring in its relaxed state. The origin of an *x* axis has been placed at the end of the spring that is attached to a block. (*b*) The block is displaced by $\vec{d}$, and the spring is stretched by a positive amount *x*. Note the restoring force $\vec{F}$ exerted by the spring. (*c*) The spring is compressed by a negative amount *x*. Again, note the restoring force.

7-5 Work Done by a Spring Force

We next want to examine the work done on a particle-like object by a particular type of *variable force*—namely, a **spring force**, the force from a spring. Many forces in nature have the same mathematical form as the spring force. Thus, by examining this one force, you can gain an understanding of many others.

The Spring Force

Figure 7-10*a* shows a spring in its **relaxed state**—that is, neither compressed nor extended. One end is fixed, and a particle-like object, say, a block, is attached to the other, free end. If we stretch the spring by pulling the block to the right as in Fig. 7-10*b*, the spring pulls on the block toward the left. (Because a spring's force acts to restore the relaxed state, it is sometimes said to be a *restoring force*.) If we compress the spring by pushing the block to the left as in Fig. 7-10*c*, the spring now pushes on the block toward the right.

To a good approximation for many springs, the force $\vec{F}$ from a spring is proportional to the displacement $\vec{d}$ of the free end from its position when the spring is in the relaxed state. The *spring force* is given by

$$\vec{F} = -k\vec{d} \quad \text{(Hooke's law)}, \tag{7-20}$$

which is known as **Hooke's law** after Robert Hooke, an English scientist of the late 1600s. The minus sign in Eq. 7-20 indicates that the spring force is always opposite in direction from the displacement of the free end. The constant *k* is called the **spring constant** (or **force constant**) and is a measure of the stiffness of the spring. The larger *k* is, the stiffer the spring; that is, the stronger will be its pull or push for a given displacement. The SI unit for *k* is the newton per meter.

In Fig. 7-10 an *x* axis has been placed parallel to the length of a spring, with the origin ($x = 0$) at the position of the free end when the spring is in its relaxed state. For this common arrangement, we can write Eq. 7-20 as

$$F = -kx \quad \text{(Hooke's law)}. \tag{7-21}$$

If *x* is positive (the spring is stretched toward the right on the *x* axis), then *F* is negative (it is a pull toward the left). If *x* is negative (the spring is compressed toward the left), then *F* is positive (it is a push toward the right).

Note that a spring force is a *variable force* because its magnitude and direction depend on the position *x* of the free end; *F* can be symbolized as $F(x)$. Also note that Hooke's law is a *linear* relationship between *F* and *x*.

The Work Done by a Spring Force

To find an expression for the work done by the spring force as the block in Fig. 7-10*a* moves, let us make two simplifying assumptions about the spring. (1) It is *massless*; that is, its mass is negligible compared to the block's mass. (2) It is an *ideal spring*; that is, it obeys Hooke's law exactly. Let us also assume that the contact between the block and the floor is frictionless and that the block is particle-like.

We give the block a rightward jerk to get it moving, and then leave it alone. As the block moves rightward, the spring force $\vec{F}$ does work on the block, decreasing the kinetic energy and slowing the block. However, we *cannot* find this work by using Eq. 7-7 ($W = Fd \cos \phi$) because that equation assumes a constant force. The spring force is a variable force.

To find the work done by the spring, we use calculus. Let the block's initial position be x_i and its later position x_f. Then divide the distance between those two positions into many segments, each of tiny length Δx. Label these segments, starting from x_i, as segments 1, 2, and so on. As the block moves through a segment, the spring force hardly varies because the segment is so short that x hardly varies. Thus, we can approximate the force magnitude as being constant within the segment. Label these magnitudes as F_1 in segment 1, F_2 in segment 2, and so on.

With the force now constant in each segment, we *can* find the work done within each segment by using Eq. 7-7 ($W = Fd \cos \phi$). Here $\phi = 0$, so $\cos \phi = 1$. Then the work done is $F_1 \Delta x$ in segment 1, $F_2 \Delta x$ in segment 2, and so on. The net work W_s done by the spring, from x_i to x_f, is the sum of all these works:

$$W_s = \sum F_j \, \Delta x, \tag{7-22}$$

where j labels the segments. In the limit as Δx goes to zero, Eq. 7-22 becomes

$$W_s = \int_{x_i}^{x_f} F \, dx. \tag{7-23}$$

Substituting for F from Eq. 7-21, we find

$$W_s = \int_{x_i}^{x_f} (-kx) \, dx = -k \int_{x_i}^{x_f} x \, dx$$
$$= (-\tfrac{1}{2}k) \, [x^2]_{x_i}^{x_f} = (-\tfrac{1}{2}k)(x_f^2 - x_i^2). \tag{7-24}$$

Multiplied out, this yields

$$W_s = \tfrac{1}{2}kx_i^2 - \tfrac{1}{2}kx_f^2 \qquad \text{(work by a spring force).} \tag{7-25}$$

This work W_s done by the spring force can have a positive or negative value, depending on whether the *net* transfer of energy is to or from the block as the block moves from x_i to x_f. *Caution:* The final position x_f appears in the *second* term on the right side of Eq. 7-25. Therefore, Eq. 7-25 tells us:

> ➤ Work W_s is positive if the block ends up closer to the relaxed position ($x = 0$) than it was initially. It is negative if the block ends up farther away from $x = 0$. It is zero if the block ends up at the same distance from $x = 0$.

If $x_i = 0$ and if we call the final position x, then Eq. 7-25 becomes

$$W_s = -\tfrac{1}{2}kx^2 \qquad \text{(work by a spring force).} \tag{7-26}$$

The Work Done by an Applied Force

Now suppose that we displace the block along the x axis while continuing to apply a force $\vec{F}_a$ to it. During the displacement, our applied force does work W_a on the block while the spring force does work W_s. By Eq. 7-10, the change ΔK in the kinetic energy of the block due to these two energy transfers is

$$\Delta K = K_f - K_i = W_a + W_s, \tag{7-27}$$

in which K_f is the kinetic energy at the end of the displacement and K_i is that at the start of the displacement. If the block is stationary before and after the displacement, then K_f and K_i are both zero and Eq. 7-27 reduces to

$$W_a = -W_s. \tag{7-28}$$

> ➤ If a block that is attached to a spring is stationary before and after a displacement, then the work done on it by the applied force displacing it is the negative of the work done on it by the spring force.

Caution: If the block is not stationary before and after the displacement, then this statement is *not* true.

> ✓ CHECKPOINT 4:   For three situations, the initial and final positions, respectively, along the *x* axis for the block in Fig. 7-10 are (a) −3 cm, 2 cm; (b) 2 cm, 3 cm; and (c) −2 cm, 2 cm. In each situation is the work done by the spring force on the block positive, negative, or zero?

Sample Problem 7-7

A package of spicy Cajun pralines lies on a frictionless floor, attached to the free end of a spring in the arrangement of Fig. 7-10*a*. An applied force of magnitude F_a = 4.9 N would be needed to hold the package stationary at x_1 = 12 mm.

(a) How much work does the spring force do on the package if the package is pulled rightward from x_0 = 0 to x_2 = 17 mm?

SOLUTION: A Key Idea here is that as the package moves from one position to another, the spring force does work on it as given by Eq. 7-25 or Eq. 7-26. We know that the initial position x_i is 0 and the final position x_f is 17 mm, but we do not know the spring constant k.

   We can probably find k with Eq. 7-21 (Hooke's law), but we need a second Key Idea to use it: Were the package held stationary at x_1 = 12 mm, the spring force would have to balance the applied force (according to Newton's second law). Thus, the spring force F would have to be −4.9 N (toward the left in Fig. 7-10*b*), so Eq. 7-21 ($F = -kx$) gives us

$$k = -\frac{F}{x_1} = -\frac{-4.9 \text{ N}}{12 \times 10^{-3} \text{ m}} = 408 \text{ N/m}.$$

Now, with the package at x_2 = 17 mm, Eq. 7-26 yields

$$W_s = -\tfrac{1}{2}kx_2^2 = -\tfrac{1}{2}(408 \text{ N/m})(17 \times 10^{-3} \text{ m})^2$$
$$= -0.059 \text{ J}. \qquad \text{(Answer)}$$

(b) Next, the package is moved leftward to x_3 = −12 mm. How much work does the spring force do on the package during this displacement? Explain the sign of this work.

SOLUTION: The Key Idea here is the first one we noted in part (a). Now x_i = +17 mm and x_f = −12 mm, and Eq. 7-25 yields

$$W_s = \tfrac{1}{2}kx_i^2 - \tfrac{1}{2}kx_f^2 = \tfrac{1}{2}k(x_i^2 - x_f^2)$$
$$= \tfrac{1}{2}(408 \text{ N/m})[(17 \times 10^{-3} \text{ m})^2 - (-12 \times 10^{-3} \text{ m})^2]$$
$$= 0.030 \text{ J} = 30 \text{ mJ}. \qquad \text{(Answer)}$$

This work done on the block by the spring force is positive because the spring force does more positive work as the block moves from x_i = +17 mm to the spring's relaxed position than it does negative work as the block moves from the spring's relaxed position to x_f = −12 mm.

Sample Problem 7-8

In Fig. 7-11, a cumin canister of mass m = 0.40 kg slides across a horizontal frictionless counter with speed v = 0.50 m/s. It then runs into and compresses a spring of spring constant k = 750 N/m. When the canister is momentarily stopped by the spring, by what distance d is the spring compressed?

SOLUTION: There are three Key Ideas here:

1. The work W_s done on the canister by the spring force is related to the requested distance d by Eq. 7-26 ($W_s = -\tfrac{1}{2}kx^2$), with d replacing x.

Fig. 7-11 Sample Problem 7-8. A canister of mass m moves at velocity $\vec{v}$ toward a spring with spring constant k.

2. The work W_s is also related to the kinetic energy of the canister by Eq. 7-10 ($K_f - K_i = W$).

3. The canister's kinetic energy has an initial value of $K = \tfrac{1}{2}mv^2$ and a value of zero when the canister is momentarily at rest.

Putting the first two of these ideas together, we write the work–kinetic energy theorem for the canister as

$$K_f - K_i = -\tfrac{1}{2}kd^2.$$

Substituting according to the third idea makes this

$$0 - \tfrac{1}{2}mv^2 = -\tfrac{1}{2}kd^2.$$

Simplifying, solving for d, and substituting known data then give us

$$d = v\sqrt{\frac{m}{k}} = (0.50 \text{ m/s})\sqrt{\frac{0.40 \text{ kg}}{750 \text{ N/m}}}$$
$$= 1.2 \times 10^{-2} \text{ m} = 1.2 \text{ cm}. \qquad \text{(Answer)}$$

7-6 Work Done by a General Variable Force

One-Dimensional Analysis

(a)

(b)

(c)

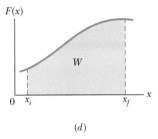

(d)

Fig. 7-12 (a) A one-dimensional force $\vec{F}$ plotted against the displacement x of a particle on which it acts. The particle moves from x_i to x_f. (b) Same as (a) but with the area under the curve divided into narrow strips. (c) Same as (b) but with the area divided into narrower strips. (d) The limiting case. The work done by the force is given by Eq. 7-32 and is represented by the shaded area between the curve and the x axis and between x_i and x_f.

Let us return to the situation of Fig. 7-2 but now consider the force to be directed along the x axis and the force magnitude to vary with position x. Thus, as the bead (particle) moves, the magnitude of the force doing work on it changes. Only the magnitude of this variable force changes, not its direction, and the magnitude at any position does not change with time.

Figure 7-12a shows a plot of such a *one-dimensional variable force*. We want an expression for the work done on the particle by this force as the particle moves from an initial point x_i to a final point x_f. However, we *cannot* use Eq. 7-7 because it applies only for a constant force $\vec{F}$. Here, again, we shall use calculus. We divide the area under the curve of Fig. 7-12a into a number of narrow strips of width Δx (Fig. 7-12b). We choose Δx small enough to permit us to take the force $F(x)$ as being reasonably constant over that interval. We let $F_{j,\text{avg}}$ be the average value of $F(x)$ within the jth interval. Then in Fig. 7-12b, $F_{j,\text{avg}}$ is the height of the jth strip.

With $F_{j,\text{avg}}$ considered constant, the increment (small amount) of work ΔW_j done by the force in the jth interval is now approximately given by Eq. 7-7 and is

$$\Delta W_j = F_{j,\text{avg}} \, \Delta x. \qquad (7\text{-}29)$$

In Fig. 7-12b, ΔW_j is then equal to the area of the jth rectangular, shaded strip.

To approximate the total work W done by the force as the particle moves from x_i to x_f, we add the areas of all the strips between x_i and x_f in Fig. 7-12b:

$$W = \sum \Delta W_j = \sum F_{j,\text{avg}} \, \Delta x. \qquad (7\text{-}30)$$

Equation 7-30 is an approximation because the broken "skyline" formed by the tops of the rectangular strips in Fig. 7-12b only approximates the actual curve of $F(x)$.

We can make the approximation better by reducing the strip width Δx and using more strips, as in Fig. 7-12c. In the limit, we let the strip width approach zero; the number of strips then becomes infinitely large and we have, as an exact result,

$$W = \lim_{\Delta x \to 0} \sum F_{j,\text{avg}} \, \Delta x. \qquad (7\text{-}31)$$

This limit is exactly what we mean by the integral of the function $F(x)$ between the limits x_i and x_f. Thus, Eq. 7-31 becomes

$$W = \int_{x_i}^{x_f} F(x) \, dx \qquad \text{(work: variable force).} \qquad (7\text{-}32)$$

If we know the function $F(x)$, we can substitute it into Eq. 7-32, introduce the proper limits of integration, carry out the integration, and thus find the work. (Appendix E contains a list of common integrals.) Geometrically, the work is equal to the area between the $F(x)$ curve and the x axis, between the limits x_i and x_f (shaded in Fig. 7-12d).

Three-Dimensional Analysis

Consider now a particle that is acted on by a three-dimensional force

$$\vec{F} = F_x \hat{i} + F_y \hat{j} + F_z \hat{k}, \qquad (7\text{-}33)$$

in which the components F_x, F_y, and F_z can depend on the position of the particle; that is, they can be functions of that position. However, we make three simplifications: F_x may depend on x but not on y or z, F_y may depend on y but not on x or z, and F_z may depend on z but not on x or y. Now let the particle move through an incremental displacement

$$d\vec{r} = dx\hat{i} + dy\hat{j} + dz\hat{k}. \tag{7-34}$$

The increment of work dW done on the particle by $\vec{F}$ during the displacement $d\vec{r}$ is, by Eq. 7-8,

$$dW = \vec{F} \cdot d\vec{r} = F_x\, dx + F_y\, dy + F_z\, dz. \tag{7-35}$$

The work W done by $\vec{F}$ while the particle moves from an initial position r_i with coordinates (x_i, y_i, z_i) to a final position r_f with coordinates (x_f, y_f, z_f) is then

$$W = \int_{r_i}^{r_f} dW = \int_{x_i}^{x_f} F_x\, dx + \int_{y_i}^{y_f} F_y\, dy + \int_{z_i}^{z_f} F_z\, dz. \tag{7-36}$$

If $\vec{F}$ has only an x component, then the y and z terms in Eq. 7-36 are zero and the equation reduces to Eq. 7-32.

Work–Kinetic Energy Theorem with a Variable Force

Equation 7-32 gives the work done by a variable force on a particle in a one-dimensional situation. Let us now make certain that the work calculated with Eq. 7-32 is indeed equal to the change in kinetic energy of the particle, as the work–kinetic energy theorem states.

Consider a particle of mass m, moving along the x axis and acted on by a net force $F(x)$ that is directed along that axis. The work done on the particle by this force as the particle moves from an initial position x_i to a final position x_f is given by Eq. 7-32 as

$$W = \int_{x_i}^{x_f} F(x)\, dx = \int_{x_i}^{x_f} ma\, dx, \tag{7-37}$$

in which we use Newton's second law to replace $F(x)$ with ma. We can write the quantity $ma\, dx$ in Eq. 7-37 as

$$ma\, dx = m\frac{dv}{dt}\, dx. \tag{7-38}$$

From the "chain rule" of calculus, we have

$$\frac{dv}{dt} = \frac{dv}{dx}\frac{dx}{dt} = \frac{dv}{dx}v, \tag{7-39}$$

and Eq. 7-38 becomes

$$ma\, dx = m\frac{dv}{dx}v\, dx = mv\, dv. \tag{7-40}$$

Substituting Eq. 7-40 into Eq. 7-37 yields

$$W = \int_{v_i}^{v_f} mv\, dv = m\int_{v_i}^{v_f} v\, dv$$
$$= \tfrac{1}{2}mv_f^2 - \tfrac{1}{2}mv_i^2. \tag{7-41}$$

Note that when we change the variable from x to v we are required to express the limits on the integral in terms of the new variable. Note also that because the mass m is a constant, we are able to move it outside the integral.

Recognizing the terms on the right side of Eq. 7-41 as kinetic energies allows us to write this equation as

$$W = K_f - K_i = \Delta K,$$

which is the work–kinetic energy theorem.

Sample Problem 7-9

Force $\vec{F} = (3x^2 \text{ N})\hat{i} + (4 \text{ N})\hat{j}$, with x in meters, acts on a particle, changing only the kinetic energy of the particle. How much work is done on the particle as it moves from coordinates (2 m, 3 m) to (3 m, 0 m)? Does the speed of the particle increase, decrease, or remain the same?

SOLUTION: The Key Idea here is that the force is a variable force because its x component depends on the value of x. Thus, we cannot use Eqs. 7-7 and 7-8 to find the work done. Instead, we must use

Eq. 7-36 to integrate the force:

$$W = \int_2^3 3x^2 \, dx + \int_3^0 4 \, dy = 3\int_2^3 x^2 \, dx + 4\int_3^0 dy$$
$$= 3[\tfrac{1}{3}x^3]_2^3 + 4[y]_3^0 = [3^3 - 2^3] + 4[0 - 3]$$
$$= 7.0 \text{ J.} \qquad\qquad\qquad\qquad \text{(Answer)}$$

The positive result means that energy is transferred to the particle by force $\vec{F}$. Thus, the kinetic energy of the particle increases, and so must its speed.

7-7 Power

A contractor wishes to lift a load of bricks from the sidewalk to the top of a building by means of a winch. We can now calculate how much work the force applied by the winch must do on the load to make the lift. The contractor, however, is much more interested in the *rate* at which that work is done. Will the job take 5 minutes (acceptable) or a week (unacceptable)?

The time rate at which work is done by a force is said to be the **power** due to the force. If an amount of work W is done in an amount of time Δt by a force, the **average power** due to the force during that time interval is

$$P_{\text{avg}} = \frac{W}{\Delta t} \qquad \text{(average power).} \qquad (7\text{-}42)$$

The **instantaneous power** P is the instantaneous time rate of doing work, which we can write as

$$P = \frac{dW}{dt} \qquad \text{(instantaneous power).} \qquad (7\text{-}43)$$

Suppose we know the work $W(t)$ done by a force as a function of time. Then to get the instantaneous power P at, say, time $t = 3.0$ s during the work, we would first take the time derivative of $W(t)$, and then evaluate the result for $t = 3.0$ s.

The SI unit of power is the joule per second. This unit is used so often that it has a special name, the **watt** (W), after James Watt, who greatly improved the rate at which steam engines could do work. In the British system, the unit of power is the foot-pound per second. Often the horsepower is used. Some relations among these units are

$$1 \text{ watt} = 1 \text{ W} = 1 \text{ J/s} = 0.738 \text{ ft} \cdot \text{lb/s} \qquad (7\text{-}44)$$

and $\qquad 1 \text{ horsepower} = 1 \text{ hp} = 550 \text{ ft} \cdot \text{lb/s} = 746 \text{ W.} \qquad (7\text{-}45)$

Inspection of Eq. 7-42 shows that work can be expressed as power multiplied by time, as in the common unit, the kilowatt-hour. Thus,

$$1 \text{ kilowatt-hour} = 1 \text{ kW} \cdot \text{h} = (10^3 \text{ W})(3600 \text{ s}) \qquad (7\text{-}46)$$
$$= 3.60 \times 10^6 \text{ J} = 3.60 \text{ MJ.}$$

Perhaps because they appear on our utility bills, the watt and the kilowatt-hour have become identified as electrical units. They can be used equally well as units for other examples of power and work or energy. Thus, if you pick up this book from the floor and put it on a tabletop, you are free to report the work that you have done as $4 \times 10^{-6} \text{ kW} \cdot \text{h}$ (or more conveniently as 4 mW $\cdot$ h).

We can also express the rate at which a force does work on a particle (or particle-like object) in terms of that force and the particle's velocity. For a particle that is moving along a straight line (say, the x axis) and is acted on by a constant force $\vec{F}$ directed at some angle ϕ to that line, Eq. 7-43 becomes

$$P = \frac{dW}{dt} = \frac{F \cos \phi \, dx}{dt} = F \cos \phi \left(\frac{dx}{dt} \right),$$

or

$$P = Fv \cos \phi. \tag{7-47}$$

Reorganizing the right side of Eq. 7-47 as the dot product $\vec{F} \cdot \vec{v}$, we may also write Eq. 7-47 as

$$P = \vec{F} \cdot \vec{v} \quad \text{(instantaneous power)}. \tag{7-48}$$

For example, the truck in Fig. 7-13 exerts a force $\vec{F}$ on the trailing load, which has velocity $\vec{v}$ at some instant. The instantaneous power due to $\vec{F}$ is the rate at which $\vec{F}$ does work on the load at that instant and is given by Eqs. 7-47 and 7-48. Saying that this power is "the power of the truck" is often acceptable, but we should keep in mind what is meant: Power is the rate at which the applied *force* does work.

✓**CHECKPOINT 5:**   A block moves with uniform circular motion because a cord tied to the block is anchored at the center of a circle. Is the power due to the force on the block from the cord positive, negative, or zero?

Fig. 7-13 The power due to the truck's applied force on the trailing load is the rate at which that force does work on the load.

Sample Problem 7-10

Figure 7-14 shows constant forces $\vec{F}_1$ and $\vec{F}_2$ acting on a box as the box slides rightward across a frictionless floor. Force $\vec{F}_1$ is horizontal, with magnitude 2.0 N; force $\vec{F}_2$ is angled upward by 60° to the floor and has magnitude 4.0 N. The speed v of the box at a certain instant is 3.0 m/s.

Fig. 7-14 Sample Problem 7-10. Two forces $\vec{F}_1$ and $\vec{F}_2$ act on a box that slides rightward across a frictionless floor. The velocity of the box is $\vec{v}$.

(a) What is the power due to each force acting on the box at that instant, and what is the net power? Is the net power changing at that instant?

SOLUTION: A Key Idea here is that we want an instantaneous power, not an average power over a time period. Also, we know the particle's velocity (rather than the work done on it). Therefore, we use Eq. 7-47 for each force. For force $\vec{F}_1$, at angle $\phi_1 = 180°$ to velocity

$\vec{v}$, we have

$$P_1 = F_1 v \cos \phi_1 = (2.0 \text{ N})(3.0 \text{ m/s}) \cos 180°$$
$$= -6.0 \text{ W}. \qquad \text{(Answer)}$$

This result tells us that force $\vec{F}_1$ is transferring energy *from* the box at the rate of 6.0 J/s.

For force $\vec{F}_2$, at angle $\phi_2 = 60°$ to velocity $\vec{v}$, we have

$$P_2 = F_2 v \cos \phi_2 = (4.0 \text{ N})(3.0 \text{ m/s}) \cos 60°$$
$$= 6.0 \text{ W}. \qquad \text{(Answer)}$$

This result tells us that force $\vec{F}_2$ is transferring energy *to* the box at the rate of 6.0 J/s.

A second **Key Idea** is that the net power is the sum of the individual powers:

$$P_{\text{net}} = P_1 + P_2$$
$$= -6.0 \text{ W} + 6.0 \text{ W} = 0, \qquad \text{(Answer)}$$

which tells us that the net rate of transfer of energy to or from the box is zero. Thus, the kinetic energy ($K = \frac{1}{2}mv^2$) of the box is not changing, and so the speed of the box will remain at 3.0 m/s. With

neither the forces $\vec{F}_1$ and $\vec{F}_2$ nor the velocity $\vec{v}$ changing, we see from Eq. 7-48 that P_1 and P_2 are constant and thus so is P_{net}.

(b) If the magnitude of $\vec{F}_2$ is, instead, 6.0 N, what now is the net power, and is it changing?

SOLUTION: The same **Key Ideas** as above give us, for the power now due to $\vec{F}_2$,

$$P_2 = F_2 v \cos \phi_2 = (6.0 \text{ N})(3.0 \text{ m/s}) \cos 60°$$
$$= 9.0 \text{ W}.$$

The power of force $\vec{F}_1$ is still $P_1 = -6.0 \text{ W}$, so the net power is now

$$P_{\text{net}} = P_1 + P_2 = -6.0 \text{ W} + 9.0 \text{ W}$$
$$= 3.0 \text{ W}, \qquad \text{(Answer)}$$

which tells us that the net rate of transfer of energy to the box has a positive value. Thus, the kinetic energy of the box is increasing, and so also is the speed of the box. With the speed increasing, we see from Eq. 7-48 that the values of P_1 and P_2, and thus also of P_{net}, will be changing. Hence, this net power of 3.0 W is the net power only at the instant the speed is the given 3.0 m/s.

REVIEW & SUMMARY

Kinetic Energy   The **kinetic energy** K associated with the motion of a particle of mass m and speed v, where v is well below the speed of light, is

$$K = \tfrac{1}{2}mv^2 \qquad \text{(kinetic energy).} \qquad (7\text{-}1)$$

Work   Work W is energy transferred to or from an object via a force acting on the object. Energy transferred to the object is positive work, and energy transferred from the object is negative work.

Work Done by a Constant Force   The work done on a particle by a constant force $\vec{F}$ during displacement $\vec{d}$ of the particle is

$$W = Fd \cos \phi = \vec{F} \cdot \vec{d} \qquad \text{(work, constant force),} \qquad (7\text{-}7, 7\text{-}8)$$

in which ϕ is the constant angle between the directions of $\vec{F}$ and $\vec{d}$. Only the component of $\vec{F}$ that is along the displacement $\vec{d}$ can do work on the object. When two or more forces act on an object, their **net work** is the sum of the individual works by the forces, which is also equal to the work that would be done on the object by the net force $\vec{F}_{\text{net}}$ of those forces.

Work and Kinetic Energy   We can relate a change ΔK in kinetic energy of a particle to the net work W done on the particle with

$$\Delta K = K_f - K_i = W \qquad \text{(work–kinetic energy theorem),} \qquad (7\text{-}10)$$

in which K_i is the initial kinetic energy of the object and K_f is the kinetic energy after the work is done. Equation 7-10 rearranged gives us

$$K_f = K_i + W. \qquad (7\text{-}11)$$

Work Done by the Gravitational Force   The work W_g done by the gravitational force $\vec{F}_g$ on a particle-like object of mass m during a displacement $\vec{d}$ of the object is given by

$$W_g = mgd \cos \phi, \qquad (7\text{-}12)$$

in which ϕ is the angle between $\vec{F}_g$ and $\vec{d}$.

Work Done in Lifting and Lowering an Object   The work W_a done by an applied force during a lifting or lowering of a particle-like object is related to the work W_g done by the gravitational force and the change ΔK in the object's kinetic energy by

$$\Delta K = K_f - K_i = W_a + W_g. \qquad (7\text{-}15)$$

If the kinetic energy at the beginning of a lift equals that at the end of the lift, then Eq. 7-15 reduces to

$$W_a = -W_g, \qquad (7\text{-}16)$$

which tells us that the applied force transfers as much energy to the object as the gravitational force transfers from the object.

Spring Force   The force $\vec{F}$ from a spring is

$$\vec{F} = -k\vec{d} \qquad \text{(Hooke's law),} \qquad (7\text{-}20)$$

where $\vec{d}$ is the displacement of its free end from the position when the spring is in its **relaxed state** (neither compressed nor extended), and k is the **spring constant** (a measure of the spring's stiffness). If an x axis lies along the spring, with the origin at the location of the spring's free end when the spring is in its relaxed state, Eq. 7-20 can be written as

$$F = -kx \qquad \text{(Hooke's law).} \qquad (7\text{-}21)$$

A spring force is thus a variable force: It varies with the displacement of the spring's free end.

Work Done by a Spring Force

If an object is attached to the spring's free end, the work W_s done on the object by the spring force when the object is moved from an initial position x_i to a final position x_f is

$$W_s = \tfrac{1}{2}kx_i^2 - \tfrac{1}{2}kx_f^2. \qquad (7\text{-}25)$$

If $x_i = 0$ and $x_f = x$, then Eq. 7-25 becomes

$$W_s = -\tfrac{1}{2}kx^2. \qquad (7\text{-}26)$$

Work Done by a Variable Force

When the force $\vec{F}$ on a particle-like object depends on the position of the object, the work done by $\vec{F}$ on the object while the object moves from an initial position r_i with coordinates (x_i, y_i, z_i) to a final position r_f with coordinates (x_f, y_f, z_f) must be found by integrating the force. If we assume that component F_x may depend on x but not on y or z, component F_y may depend on y but not on x or z, and component F_z may depend on z but not on x or y, then the work is

$$W = \int_{x_i}^{x_f} F_x\, dx + \int_{y_i}^{y_f} F_y\, dy + \int_{z_i}^{z_f} F_z\, dz. \qquad (7\text{-}36)$$

If $\vec{F}$ has only an x component, then Eq. 7-36 reduces to

$$W = \int_{x_i}^{x_f} F(x)\, dx. \qquad (7\text{-}32)$$

Power

The **power** due to a force is the *rate* at which that force does work on an object. If the force does work W during a time interval Δt, the *average power* due to the force over that time interval is

$$P_{\text{avg}} = \frac{W}{\Delta t}. \qquad (7\text{-}42)$$

Instantaneous power is the instantaneous rate of doing work:

$$P = \frac{dW}{dt}. \qquad (7\text{-}43)$$

If the direction of a force $\vec{F}$ is at an angle ϕ to the direction of travel of the object, the instantaneous power is

$$P = Fv\cos\phi = \vec{F}\cdot\vec{v}, \qquad (7\text{-}47, 7\text{-}48)$$

in which $\vec{v}$ is the instantaneous velocity of the object.

QUESTIONS

1. Rank the following velocities according to the kinetic energy a particle will have with each velocity, greatest first: (a) $\vec{v} = 4\hat{i} + 3\hat{j}$, (b) $\vec{v} = -4\hat{i} + 3\hat{j}$, (c) $\vec{v} = -3\hat{i} + 4\hat{j}$, (d) $\vec{v} = 3\hat{i} - 4\hat{j}$, (e) $\vec{v} = 5\hat{i}$, and (f) $v = 5$ m/s at 30° to the horizontal.

2. Is positive or negative work done by a constant force $\vec{F}$ on a particle during a straight-line displacement $\vec{d}$ if (a) the angle between $\vec{F}$ and $\vec{d}$ is 30°; (b) the angle is 100°; (c) $\vec{F} = 2\hat{i} - 3\hat{j}$ and $\vec{d} = -4\hat{i}$?

3. Figure 7-15 shows six situations in which two forces act simultaneously on a box after the box has been sent sliding over a frictionless surface, either to the left or to the right. The forces are either 1 N or 2 N in magnitude, as indicated by the vector lengths.

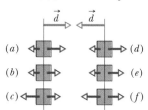

Fig. 7-15 Question 3.

For each situation, is the work done on the box by the net force during the indicated displacement $\vec{d}$ positive, negative, or zero?

4. Figure 7-16 shows the values of a force $\vec{F}$, directed along an x

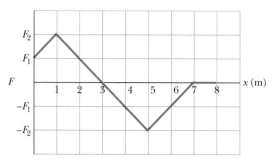

Fig. 7-16 Question 4.

axis, that will act on a particle at the corresponding values of x. If the particle begins at rest at $x = 0$, what is the particle's coordinate when it has (a) its greatest kinetic energy, (b) its greatest speed, and (c) zero speed? (d) What is the particle's direction of travel after it reaches $x = 6$ m?

5. Figure 7-17 gives, for three situations, the position versus time of a box of contraband pulled by applied forces directed along an x axis on a frictionless surface. Line B is straight; the others are curved. Rank the situations according to the kinetic energy of the box at (a) time t_1 and (b) time t_2, greatest first. (c) Now rank them according to the net work done on the box by the applied forces during the time period t_1 to t_2, greatest first.

(d) For each situation, which of the following best describes the net work done by the applied forces during the period t_1 to t_2?

(1) Energy is transferred to the box.

(2) Energy is transferred from the box.

(3) The net work is zero.

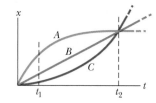

Fig. 7-17 Question 5.

6. Figure 7-18 shows four graphs (drawn to the same scale) of the x component of a variable force $\vec{F}$ (directed along an x axis) versus the position x of a particle on which the force acts. Rank the graphs according to the work done by $\vec{F}$ on the particle from $x = 0$ to x_1, from most positive work to most negative work.

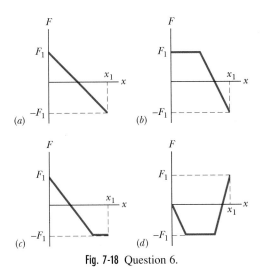

(a) ... (b) ... (c) ... (d)

Fig. 7-18 Question 6.

7. In Fig. 7-19, a greased pig has a choice of three frictionless slides along which to slide to the ground. Rank the slides according to how much work the gravitational force does on the pig during the descent, greatest first.

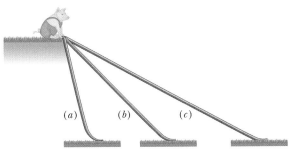

(a)   (b)   (c)

Fig. 7-19 Question 7.

8. You lift an armadillo to a shelf. Does the work done by your force on the armadillo depend on (a) the mass of the armadillo, (b) the weight of the armadillo, (c) the height of the shelf, (d) the time you take, or (e) whether you move the armadillo sideways or directly upward?

9. Figure 7-20 shows a bundle of magazines that is lifted by a cord through distance d. The table shows six pairs of values of the initial speed v_0 and final speed v (in meters per second) of the bundle at the beginning and end of distance d. Rank the pairs according to the work done by the cord's force on the bundle over distance d, most positive first, most negative last.

	a	b	c	d	e	f
$v=$	0	2	2	2	0	1
$v_0=$	0	2	0	1	2	2

Fig. 7-20 Question 9.

10. Spring A is stiffer than spring B; that is, $k_A > k_B$. The spring force of which spring does more work if the springs are compressed (a) the same distance and (b) by the same applied force?

11. A block is attached to a relaxed spring as in Fig. 7-21a. The spring constant k of the spring is such that for a certain rightward displacement $\vec{d}$ of the block, the spring force acting on the block has magnitude F_1 and has done work W_1 on the block. A second, identical spring is then attached on the opposite side of the block, as shown in Fig. 7-21b; both springs are in their relaxed state in the figure. If the block is again displaced by $\vec{d}$, (a) what is the magnitude of the net force on it from both springs and (b) how much work has been done on it by the spring forces?

(a)   (b)

Fig. 7-21 Question 11.

12. Figure 7-22 gives the velocity versus time of a scooter car being moved along an axis by a varying applied force. The time axis shows six periods: $\Delta t_1 = \Delta t_2 = \Delta t_3 = \Delta t_6 = 2\Delta t_4 = \frac{2}{3}\Delta t_5$. (a) During which of the time periods is energy transferred *from* the scooter car by the applied force? (b) Rank the time periods according to the work done on the scooter car by the applied force during the period, most positive work first, most negative work last. (c) Rank the periods according to the rate at which the applied force transfers energy, greatest rate of transfer *to* the scooter car first, greatest rate of transfer *from* it last.

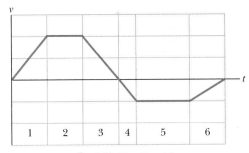

Fig. 7-22 Question 12.

13. In three situations, an initially stationary can of crayons is sent sliding over a frictionless floor by a different applied force. Plots of the resulting acceleration versus time for the three situations are given in Fig. 7-23. Rank the plots according to the work done by the applied force during the acceleration period, greatest first.

Fig. 7-23 Question 13.

EXERCISES & PROBLEMS

SEC. 7-1   Energy

1E. If an electron (mass $m = 9.11 \times 10^{-31}$ kg) in copper near the lowest possible temperature has a kinetic energy of 6.7×10^{-19} J, what is the speed of the electron? ssm

2E. On August 10, 1972, a large meteorite skipped across the atmosphere above western United States and Canada, much like a stone skipped across water. The accompanying fireball was so bright that it could be seen in the daytime sky (Fig. 7-24). The meteorite's mass was about 4×10^6 kg; its speed was about 15 km/s. Had it entered the atmosphere vertically, it would have hit Earth's surface with about the same speed. (a) Calculate the meteorite's loss of kinetic energy (in joules) that would have been associated with the vertical impact. (b) Express the energy as a multiple of the explosive energy of 1 megaton of TNT, which is 4.2×10^{15} J. (c) The energy associated with the atomic bomb explosion over Hiroshima was equivalent to 13 kilotons of TNT. To how many Hiroshima bombs would the meteorite impact have been equivalent?

Fig. 7-24 Exercise 2. A large meteorite skips across the atmosphere in the sky above the mountains (upper right).

3E. Calculate the kinetic energies of the following objects moving at the given speeds: (a) a 110 kg football linebacker running at 8.1 m/s; (b) a 4.2 g bullet at 950 m/s; (c) the aircraft carrier *Nimitz*, 91,400 tons at 32 knots.

4P. A father racing his son has half the kinetic energy of the son, who has half the mass of the father. The father speeds up by 1.0 m/s and then has the same kinetic energy as the son. What are the original speeds of (a) the father and (b) the son?

5P. A proton (mass $m = 1.67 \times 10^{-27}$ kg) is being accelerated along a straight line at 3.6×10^{15} m/s^2 in a machine. If the proton has an initial speed of 2.4×10^7 m/s and travels 3.5 cm, what then is (a) its speed and (b) the increase in its kinetic energy? ssm

SEC. 7-3   Work and Kinetic Energy

6E. A floating ice block is pushed through a displacement $\vec{d} = (15 \text{ m})\hat{i} - (12 \text{ m})\hat{j}$ along a straight embankment by rushing water, which exerts a force $\vec{F} = (210 \text{ N})\hat{i} - (150 \text{ N})\hat{j}$ on the block. How much work does the force do on the block during the displacement?

7E. To pull a 50 kg crate across a horizontal frictionless floor, a worker applies a force of 210 N, directed 20° above the horizontal. As the crate moves 3.0 m, what work is done on the crate by (a) the worker's force, (b) the gravitational force on the crate, and (c) the normal force on the crate from the floor? (d) What is the total work done on the crate? ssm

8E. A 1.0 kg standard body is at rest on a frictionless horizontal air track when a constant horizontal force $\vec{F}$ acting in the positive direction of an x axis along the track is applied to the body. A stroboscopic graph of the position of the body as it slides to the right is shown in Fig. 7-25. The force $\vec{F}$ is applied to the body at $t = 0$, and the graph records the position of the body at 0.50 s intervals. How much work is done on the body by the applied force $\vec{F}$ between $t = 0$ and $t = 2.0$ s?

Fig. 7-25 Exercise 8.

9E. A luge and its rider, with a total mass of 85 kg, emerges from a downhill track onto a horizontal straight track with an initial speed of 37 m/s. If they stop at a constant deceleration of 2.0 m/s^2, (a) what magnitude F is required for the decelerating force, (b) what distance d do they travel while decelerating, and (c) what work W is done on them by the decelerating force? What are (d) F, (e) d, and (f) W for a deceleration of 4.0 m/s^2? ilw

10P. A force acts on a 3.0 kg particle-like object in such a way that the position of the object as a function of time is given by $x = 3.0t - 4.0t^2 + 1.0t^3$, with x in meters and t in seconds. Find the work done on the object by the force from $t = 0$ to $t = 4.0$ s. (*Hint:* What are the speeds at those times?)

11P. Figure 7-26 shows three forces applied to a trunk that moves leftward by 3.00 m over a frictionless floor. The force magnitudes are $F_1 = 5.00$ N, $F_2 = 9.00$ N, and $F_3 = 3.00$ N. During the displacement, (a) what is the net work done on the trunk by the three forces and (b) does the kinetic energy of the trunk increase or decrease? ssm

Fig. 7-26 Problem 11.

12P. The only force acting on a 2.0 kg canister that is moving in an xy plane has a magnitude of 5.0 N. The canister initially has a velocity of 4.0 m/s in the positive x direction, and some time later has a velocity of 6.0 m/s in the positive y direction. How much work is done on the canister by the 5.0 N force during this time?

13P. Figure 7-27 shows an overhead view of three horizontal forces acting on a cargo canister that was initially stationary but that now moves across a frictionless floor. The force magnitudes are $F_1 = 3.00$ N, $F_2 = 4.00$ N, and $F_3 = 10.0$ N. What is the net work done on the canister by the three forces during the first 4.00 m of displacement? ssm

Fig. 7-27 Problem 13.

SEC. 7-4 Work Done by a Gravitational Force

14E. (a) In 1975 the roof of Montreal's Velodrome, with a weight of 360 kN, was lifted by 10 cm so that it could be centered. How much work was done on the roof by the forces making the lift? (b) In 1960 Mrs. Maxwell Rogers of Tampa, Florida, reportedly raised one end of a car that had fallen onto her son when a jack failed. If her panic lift effectively raised 4000 N (about $\frac{1}{4}$ of the car's weight) by 5.0 cm, how much work did her force do on the car?

15E. In Fig. 7-28, a cord runs around two massless, frictionless pulleys; a canister with mass $m = 20$ kg hangs from one pulley; and you exert a force $\vec{F}$ on the free end of the cord. (a) What must be the magnitude of $\vec{F}$ if you are to lift the canister at a constant speed? (b) To lift the canister by 2.0 cm, how far must you pull the free end of the cord? During that lift, what is the work done on the canister by (c) your force (via the cord) and (d) the gravitational force on the canister? (*Hint:* When a cord loops around a pulley as shown, it pulls on the pulley with a net force that is twice the tension in the cord.)

Fig. 7-28 Exercise 15.

16E. A 45 kg block of ice slides down a frictionless incline 1.5 m long and 0.91 m high. A worker pushes up against the ice, parallel to the incline, so that the block slides down at constant speed. (a) Find the magnitude of the worker's force. How much work is done on the block by (b) the worker's force, (c) the gravitational force on the block, (d) the normal force on the block from the surface of the incline, and (e) the net force on the block?

17P. A helicopter lifts a 72 kg astronaut 15 m vertically from the ocean by means of a cable. The acceleration of the astronaut is $g/10$. How much work is done on the astronaut by (a) the force from the helicopter and (b) the gravitational force on her? What are (c) the kinetic energy and (d) the speed of the astronaut just before she reaches the helicopter? ssm www

18P. A cave rescue team lifts an injured spelunker directly upward and out of a sinkhole by means of a motor-driven cable. The lift is performed in three stages, each requiring a vertical distance of 10.0 m: (a) the initially stationary spelunker is accelerated to a speed of 5.00 m/s; (b) he is then lifted at the constant speed of 5.00 m/s; (c) finally he is decelerated to zero speed. How much work is done on the 80.0 kg rescuee by the force lifting him during each stage?

19P. A cord is used to vertically lower an initially stationary block of mass M at a constant downward acceleration of $g/4$. When the block has fallen a distance d, find (a) the work done by the cord's force on the block, (b) the work done by the gravitational force on the block, (c) the kinetic energy of the block, and (d) the speed of the block. ssm

SEC. 7-5 Work Done by a Spring Force

20E. During spring semester at MIT, residents of the parallel buildings of the East Campus dorms battle one another with large catapults that are made with surgical hose mounted on a window frame. A balloon filled with dyed water is placed in a pouch attached to the hose, which is then stretched through the width of the room. Assume that the stretching of the hose obeys Hooke's law with a spring constant of 100 N/m. If the hose is stretched by 5.00 m and then released, how much work does the force from the hose do on the balloon in the pouch by the time the hose reaches its relaxed length?

21E. A spring with a spring constant of 15 N/cm has a cage attached to one end (Fig. 7-29). (a) How much work does the spring force do on the cage when the spring is stretched from its relaxed length by 7.6 mm? (b) How much additional work is done by the spring force when the spring is stretched by an additional 7.6 mm? ssm ilw www

Fig. 7-29 Exercise 21.

22P. A 250 g block is dropped onto a relaxed vertical spring that has a spring constant of $k = 2.5$ N/cm (Fig. 7-30). The block becomes attached to the spring and compresses the spring 12 cm before momentarily stopping. While the spring is being compressed, what work is done on the block by (a) the gravitational force on it and (b) the spring force? (c) What is the speed of the block just before it hits the spring? (Assume that friction is negligible.) (d) If the speed at impact is doubled, what is the maximum compression of the spring?

Fig. 7-30 Problem 22.

23P. The only force acting on a 2.0 kg body as it moves along the positive x axis has an x component $F_x = -6x$ N, where x is in meters. The velocity of the body at $x = 3.0$ m is 8.0 m/s. (a) What is the velocity of the body at $x = 4.0$ m? (b) At what positive value of x will the body have a velocity of 5.0 m/s? ssm

SEC. 7-6   Work Done by a General Variable Force

24E. A 5.0 kg block moves in a straight line on a horizontal frictionless surface under the influence of a force that varies with position as shown in Fig. 7-31. How much work is done by the force as the block moves from the origin to $x = 8.0$ m?

Fig. 7-31 Exercise 24.

25E. A 10 kg brick moves along an x axis. Its acceleration as a function of its position is shown in Fig. 7-32. What is the net work performed on the brick by the force causing the acceleration as the brick moves from $x = 0$ to $x = 8.0$ m?   ssm   ilw

26P. The only force acting on a 2.0 kg body as the body moves along the x axis varies as shown in Fig. 7-33. The velocity of the body at $x = 0$ is 4.0 m/s. (a) What is the kinetic energy of the body at $x = 3.0$ m? (b) At what value of x will the body have a kinetic energy of 8.0 J? (c) What is the maximum kinetic energy attained by the body between $x = 0$ and $x = 5.0$ m?

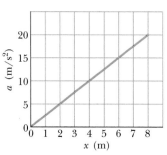

Fig. 7-32 Exercise 25.

27P. The force on a particle is directed along an x axis and given by $F = F_0(x/x_0 - 1)$. Find the work done by the force in moving the particle from $x = 0$ to $x = 2x_0$ by (a) plotting $F(x)$ and measuring the work from the graph and (b) integrating $F(x)$.   ssm

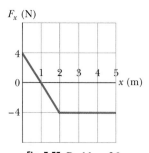

Fig. 7-33 Problem 26.

28P. A 1.5 kg block is initially at rest on a horizontal frictionless surface when a horizontal force in the positive direction of an x axis is applied to the block. The force is given by $\vec{F}(x) = (2.5 - x^2)\hat{i}$ N, where x is in meters and the initial position of the block is $x = 0$. (a) What is the kinetic energy of the block as it passes through $x = 2.0$ m? (b) What is the maximum kinetic energy of the block between $x = 0$ and $x = 2.0$ m?

29P. What work is done by a force $\vec{F} = (2x \text{ N})\hat{i} + (3 \text{ N})\hat{j}$, with x in meters, that moves a particle from a position $\vec{r}_i = (2 \text{ m})\hat{i} + (3 \text{ m})\hat{j}$ to a position $\vec{r}_f = -(4 \text{ m})\hat{i} - (3 \text{ m})\hat{j}$?   ssm

SEC. 7-7   Power

30E. The loaded cab of an elevator has a mass of 3.0×10^3 kg and moves 210 m up the shaft in 23 s at constant speed. At what average rate does the force from the cable do work on the cab?

31E. A 100 kg block is pulled at a constant speed of 5.0 m/s across a horizontal floor by an applied force of 122 N directed 37° above the horizontal. What is the rate at which the force does work on the block?   ssm   ilw

32E. (a) At a certain instant, a particle-like object is acted on by a force $\vec{F} = (4.0 \text{ N})\hat{i} - (2.0 \text{ N})\hat{j} + (9.0 \text{ N})\hat{k}$ while having a velocity $\vec{v} = -(2.0 \text{ m/s})\hat{i} + (4.0 \text{ m/s})\hat{k}$. What is the instantaneous rate at which the force does work on the object? (b) At some other time, the velocity consists of only a y component. If the force is unchanged, and the instantaneous power is -12 W, what is the velocity of the object just then?

33P. A force of 5.0 N acts on a 15 kg body initially at rest. Compute the work done by the force in (a) the first, (b) the second, and (c) the third seconds and (d) the instantaneous power due to the force at the end of the third second.   ssm

34P. A skier is pulled by a tow rope up a frictionless ski slope that makes an angle of 12° with the horizontal. The rope moves parallel to the slope with a constant speed of 1.0 m/s. The force of the rope does 900 J of work on the skier as the skier moves a distance of 8.0 m up the incline. (a) If the rope moved with a constant speed of 2.0 m/s, how much work would the force of the rope do on the skier as the skier moved a distance of 8.0 m up the incline? At what rate is the force of the rope doing work on the skier when the rope moves with a speed of (b) 1.0 m/s and (c) 2.0 m/s?

35P. A fully loaded, slow-moving freight elevator has a cab with a total mass of 1200 kg, which is required to travel upward 54 m in 3.0 min, starting and ending at rest. The elevator's counterweight has a mass of only 950 kg, so the elevator motor must help pull the cab upward. What average power is required of the force the motor exerts on the cab via the cable?   ssm   www

36P. A 0.30 kg ladle sliding on a horizontal frictionless surface is attached to one end of a horizontal spring (with $k = 500$ N/m) whose other end is fixed. The ladle has a kinetic energy of 10 J as it passes through its equilibrium position (the point at which the spring force is zero). (a) At what rate is the spring doing work on the ladle as the ladle passes through its equilibrium position? (b) At what rate is the spring doing work on the ladle when the spring is compressed 0.10 m and the ladle is moving away from the equilibrium position?

37P. The force (but not the power) required to tow a boat at constant velocity is proportional to the speed. If a speed of 4.0 km/h requires 7.5 kW, how much power does a speed of 12 km/h require?   ssm

38P. Boxes are transported from one location to another in a warehouse by means of a conveyor belt that moves with a constant speed of 0.50 m/s. At a certain location the conveyor belt moves for 2.0 m up an incline that makes an angle of 10° with the horizontal, then for 2.0 m horizontally, and finally for 2.0 m down an incline that makes an angle of 10° with the horizontal. Assume that a 2.0 kg box rides on the belt without slipping. At what rate is the force of the conveyor belt doing work on the box (a) as the box moves up the 10° incline, (b) as the box moves horizontally, and (c) as the box moves down the 10° incline?

39P. A horse pulls a cart with a force of 40 lb at an angle of 30° above the horizontal and moves along at a speed of 6.0 mi/h. (a) How much work does the force do in 10 min? (b) What is the average power (in horsepower) of the force?

40P. An initially stationary 2.0 kg object accelerates horizontally and uniformly to a speed of 10 m/s in 3.0 s. (a) In that 3.0 s interval, how much work is done on the object by the force accelerating it? What is the instantaneous power due to that force (b) at the end of the interval and (c) at the end of the first half of the interval?

NEW PROBLEMS

N1. Figure 7N-1 gives the acceleration of a 2.00 kg particle as it moves from rest along an x axis by an applied force $\vec{F}_a$, from x = 0 to x = 9 m. How much work has the force done on the particle when the particle reaches (a) x = 4 m, (b) x = 7 m, and (c) x = 9 m? What is the particle's speed and direction of travel when it reaches (d) x = 4 m, (e) x = 7 m, and (f) x = 9 m?

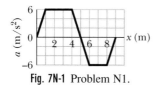

Fig. 7N-1 Problem N1.

N2. In Fig. 7N-2a, a 2.0 N force is applied to a 4.0 kg block at a downward angle θ as the block moves rightward through 1.0 m across a frictionless floor. Find an expression for the speed v_f of the block at the end of that distance if the block's initial velocity is (a) 0 and (b) 1.0 m/s to the right. (c) The situation in Fig. 7N-2b is similar in that the block is initially moving at 1.0 m/s to the right, but now the 2.0 N force is directed downward to the left. Find an expression for the speed v_f of the block at the end of the 1.0 m distance. (d) Graph all three expressions for v_f versus downward angle θ, for θ = 0° to θ = 90°. Interpret the graphs.

(a)   (b)

Fig. 7N-2 Problem N2.

N3. *Numerical integration.* A bread box is made to move along an x axis from x = 0.15 m to x = 1.20 m by a force with a magnitude given by F = exp($-2x^2$), with x in meters and F in newtons. (Here exp is the exponential function.) How much work is done on the bread box by the force?

N4. The block in Fig. 7-10a lies on a horizontal frictionless surface and is attached to the free end of the spring, with a spring constant of 50 N/m. Initially, the spring is at its relaxed length and the block is stationary at position x = 0. Then an applied force with a constant magnitude of 3.0 N pulls the block in the positive direction of the x axis, stretching the spring until the block stops. When that stopping point is reached, what are (a) the position of the block, (b) the work that has been done on the block by the applied force, and (c) the work that has been done on the block by the spring force? During the block's displacement, what are (d) the block's position when its kinetic energy is maximum and (e) the value of that maximum kinetic energy?

N5. A coin slides over a frictionless plane and across an xy coordinate system from the origin to a point with xy coordinates (3.0 m, 4.0 m) while a constant force acts on it. The force has magnitude 2.0 N and is directed at a counterclockwise angle of 100° from the positive direction of the x axis. How much work is done by the force on the coin during the displacement?

N6. A 2.0 kg lunchbox is sent sliding over a frictionless surface, in the positive direction of an x axis along the surface. Beginning at time t = 0, a steady wind pushes on the lunchbox in the negative direction of the x axis. Figure 7N-3 shows the position x of the lunchbox as a function of time t as the wind pushes on the lunchbox. From the graph, estimate the kinetic energy of the lunchbox at (a) t = 1.0 s and (b) t = 5.0 s. (c) How much work does the force from the wind do on the lunchbox from t = 1.0 s to t = 5.0 s?

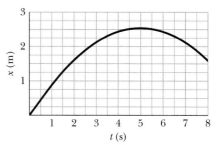

Fig. 7N-3 Problem N6.

N7. A force $\vec{F}$ = (4.0 N)$\hat{i}$ + c$\hat{j}$ acts on a particle as the particle goes through displacement $\vec{d}$ = (3.0 m)$\hat{i}$ − (2.0 m)$\hat{j}$. (Other forces also act on the particle.) What is the value of c if the work done on the particle by force $\vec{F}$ is (a) zero, (b) 17 J, and (c) −18 J?

N8. If a vehicle with a mass of 1200 kg has a speed of 120 km/h, what is the vehicle's kinetic energy as determined by someone at rest alongside the vehicle's road?

N9. A machine carries a 4.0 kg package from an initial position of $\vec{d}_i$ = (0.50 m)$\hat{i}$ + (0.75 m)$\hat{j}$ + (0.20 m)$\hat{k}$ at t = 0 to a final position of $\vec{d}_f$ = (7.50 m)$\hat{i}$ + (12.0 m)$\hat{j}$ + (7.20 m)$\hat{k}$ at t = 12 s. The constant force applied by the machine on the package is $\vec{F}$ = (2.00 N)$\hat{i}$ + (4.00 N)$\hat{j}$ + (6.00 N)$\hat{k}$. For that displacement, find (a) the work done on the package by the machine's force and (b) the power of the machine's force on the package.

N10. In Fig. 7N-4, a block of ice slides down a frictionless ramp at angle θ = 50°, while an ice worker pulls up the ramp (via a rope) with a force of magnitude F_r = 50 N. As the block slides through distance d = 0.50 m along the ramp, its kinetic energy increases by 80 J. How much greater would its kinetic energy have been if the rope had not been attached to the block?

Fig. 7N-4 Problem N10.

N11. Figure 7N-5 shows a cold package of hot dogs sliding rightward across a frictionless floor through a distance d = 20.0 cm while three forces are applied to it. Two of the forces are horizontal and have the magnitudes F_1 = 5.00 N and F_2 = 1.00 N; the third force is angled down by θ = 60.0° and has the magnitude F_3 = 4.00 N. (a) For the 20.0 cm displacement, what is the *net* work done on the package by the three applied forces, the gravitational force on the package, and the normal force on the package? (b) If the package has a mass of 2.0 kg and an initial kinetic energy of 0, what is its speed at the end of the displacement?

Fig. 7N-5 Problem N11.

N12. In Fig. 7N-6, a horizontal force $\vec{F}_a$ of magnitude 20.0 N is applied to a 3.00 kg psychology book, as the book slides a distance $d = 0.500$ m up a frictionless ramp. (a) During that displacement, what is the *net* work done on the book by $\vec{F}_a$, the gravitational force on the book, and the normal force on the book? (b) If the book has zero kinetic energy at the start of the displacement, what is its speed at the end of the displacement?

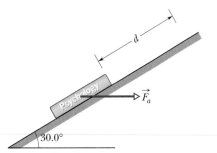

Fig. 7N-6 Problem N12.

N13. A CD case slides along a floor in the positive direction of an x axis while an applied force $\vec{F}_a$ acts on it. The force is directed along the x axis and has the x component $F_{ax} = 9x - 3x^2$, with x in meters and F_{ax} in newtons. The case starts at rest at the position $x = 0$, and it moves until it is again at rest. (a) Plot the work done on the case by $\vec{F}_a$ versus position x. (b) At what position is the work maximum, and (c) what is that maximum value? (d) At what position has the work decreased to zero? (e) At what position is the case again at rest?

N14. A can of bolts and nuts is pushed 2.00 m along an x axis by a broom along the greasy (frictionless) floor of a car repair shop in a version of shuffleboard. Figure 7N-7 gives the work W done on the can by the constant horizontal force from the broom, versus the can's position x. (a) What is the magnitude of that force? (b) If the can had an initial kinetic energy of 3.00 J, moving in the positive direction of the x axis, what is its kinetic energy at the end of the 2.00 m displacement?

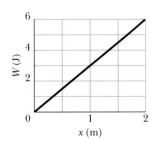

Fig. 7N-7 Problem N14.

N15. (a) Estimate the work done by the force represented by the graph of Fig. 7N-8 in displacing a particle from $x = 1$ m to $x = 3$ m. (b) The curve is given by $F = a/x^2$, with $a = 9$ N·m². Calculate the work using integration.

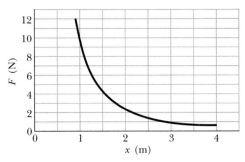

Fig. 7N-8 Problem N15.

N16. A force $\vec{F}$ in the direction of increasing x acts on an object moving along the x axis. If the magnitude of the force is $F = 10e^{-x/2.0}$ N, where x is in meters, find the work done by $\vec{F}$ as the object moves from $x = 0$ to $x = 2.0$ m by (a) plotting $F(x)$ and estimating the area under the curve and (b) integrating to find the work analytically.

N17. A single force acts on a body that moves along an x axis. Figure 7N-9 shows the velocity component v versus time t for the body. For each of the intervals AB, BC, CD, and DE, give the sign (plus or minus) of the work done by the force on the body, or state that the work is zero.

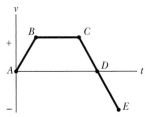

Fig. 7N-9 Problem N17.

N18. To push a 25.0 kg crate up a frictionless incline, angled at 25.0° to the horizontal, a worker exerts a force of 209 N, parallel to the incline. As the crate slides 1.50 m, how much work is done on the crate by (a) the worker's applied force, (b) the gravitational force on the crate, and (c) the normal force exerted by the incline on the crate? (d) What is the total work done on the crate?

N19. An elevator cab in the New York Marriott Marquis has a mass of 4500 kg and can carry a maximum load of 1800 kg. If the cab is moving upward at full load at 3.80 m/s, what power is required of the force moving the cab to maintain that speed?

N20. A force $\vec{F} = (3.0\hat{i} + 7.0\hat{j} + 7.0\hat{k})$ N acts on a 2.00 kg object that moves from an initial position of $\vec{d}_i = (3.0\hat{i} - 2.0\hat{j} + 5.0\hat{k})$ m to a final position of $\vec{d}_f = (-5.0\hat{i} + 4.0\hat{j} + 7.0\hat{k})$ m in 4.0 s. Find (a) the work done on the object by the force in that time interval, (b) the average power due to the force during that time interval, and (c) the angle between vectors $\vec{d}_i$ and $\vec{d}_f$.

8 Potential Energy and Conservation of Energy

The prehistoric people of Easter Island carved hundreds of giant stone statues in their quarry and then moved them to sites all over the island. How they managed to move them by as much as 10 km without the use of sophisticated machines has been a hotly debated subject, with many fanciful theories about the source of the required energy.

How much energy *was* required to move one of the statues using only primitive means?

The answer is in this chapter.

8-1 Potential Energy

Fig. 8-1 When Chemerkin lifted the weights above his head, he increased the separation between the weights and Earth and thus changed the configuration of the weights–Earth system from that in (a) to that in (b).

In this chapter we continue the discussion of energy that we began in Chapter 7. To do so, we define a second form of energy: **potential energy** U is energy that can be associated with the configuration (or arrangement) of a system of objects that exert forces on one another. If the configuration of the system changes, then the potential energy of the system can also change.

One type of potential energy is the **gravitational potential energy** that is associated with the state of separation between objects, which attract one another via the gravitational force. For example, when Andrey Chemerkin lifted the record-breaking weights above his head in the 1996 Olympics, he increased the separation between the weights and Earth. The work his force did changed the gravitational potential energy of the weights–Earth system because it changed the configuration of the system—that is, the force shifted the relative locations of the weights and Earth (Fig. 8-1).

Another type of potential energy is **elastic potential energy,** which is associated with the state of compression or extension of an elastic (springlike) object. If you compress or extend a spring, you do work to change the relative locations of the coils within the spring. The result of the work done by your force is an increase in the elastic potential energy of the spring.

The idea of potential energy can be an enormously powerful tool in understanding situations involving the motion of an object. With it we can, in fact, easily solve problems that would require careful computer programming if we used only the ideas of earlier chapters.

Work and Potential Energy

In Chapter 7 we discussed the relation between work and a change in kinetic energy. Here we discuss the relation between work and a change in potential energy.

Let us throw a tomato upward (Fig. 8-2). We already know that as the tomato rises, the work W_g done on the tomato by the gravitational force is negative because the force transfers energy *from* the kinetic energy of the tomato. We can now finish the story by saying that this energy is transferred by the gravitational force *to* the gravitational potential energy of the tomato–Earth system.

The tomato slows, stops, and then begins to fall back down because of the gravitational force. During the fall, the transfer is reversed: The work W_g done on the tomato by the gravitational force is now positive—that force transfers energy *from* the gravitational potential energy of the tomato–Earth system *to* the kinetic energy of the tomato.

For either rise or fall, the change ΔU in gravitational potential energy is defined to equal the negative of the work done on the tomato by the gravitational force. Using the general symbol W for work, we write this as

$$\Delta U = -W. \tag{8-1}$$

This equation also applies to a block–spring system, as in Fig. 8-3. If we abruptly shove the block to send it moving rightward, the spring force acts leftward and thus does negative work on the block, transferring energy from the kinetic energy of the block to the elastic potential energy of the spring. The block slows and eventually stops, and then begins to move leftward because the spring force is still leftward. The transfer of energy is then reversed—it is from potential energy of the spring to kinetic energy of the block.

Negative work done by the gravitational force

Positive work done by the gravitational force

Fig. 8-2 A tomato is thrown upward. As it rises, the gravitational force does negative work on it, decreasing its kinetic energy. As the tomato descends, the gravitational force does positive work on it, increasing its kinetic energy.

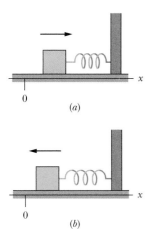

Fig. 8-3 A block, attached to a spring and initially at rest at $x = 0$, is set in motion toward the right. (*a*) As the block moves rightward (as indicated by the arrow), the spring force does negative work on it. (*b*) Then, as the block moves back toward $x = 0$, the spring force does positive work on it.

Conservative and Nonconservative Forces

Let us list the key elements of the two situations we just discussed:

1. The *system* consists of two or more objects.

2. A *force* acts between a particle-like object (tomato or block) in the system and the rest of the system.

3. When the system configuration changes, the force does *work* (call it W_1) on the particle-like object, transferring energy between the kinetic energy K of the object and some other form of energy of the system.

4. When the configuration change is reversed, the force reverses the energy transfer, doing work W_2 in the process.

In a situation in which $W_1 = -W_2$ is always true, the other form of energy is a potential energy, and the force is said to be a **conservative force.** As you might suspect, the gravitational force and the spring force are both conservative (since otherwise we could not have spoken of gravitational potential energy and elastic potential energy, as we did previously).

A force that is not conservative is called a **nonconservative force.** The kinetic frictional force and drag force are nonconservative. For an example, let us send a block sliding across a floor that is not frictionless. During the sliding, a kinetic frictional force from the floor does negative work on the block, slowing the block by transferring energy from its kinetic energy to a form of energy called *thermal energy* (which has to do with the random motions of atoms and molecules). We know from experiment that this energy transfer cannot be reversed (thermal energy cannot be transferred back to kinetic energy of the block by the kinetic frictional force). Thus, although we have a system (made up of the block and the floor), a force that acts between parts of the system, and a transfer of energy by the force, the force is not conservative. Therefore, thermal energy is not a potential energy.

When only conservative forces act on a particle-like object, we can greatly simplify otherwise difficult problems involving motion of the object. The next section, in which we develop a test for identifying conservative forces, provides one means for simplifying such problems.

8-2 Path Independence of Conservative Forces

The primary test for determining whether a force is conservative or nonconservative is this: Let the force act on a particle that moves along any *closed path,* beginning at some initial position and eventually returning to that position (so that the particle makes a *round trip* beginning and ending at the initial position). The force is conservative only if the total energy it transfers to and from the particle during the round trip along this and any other closed path is zero. In other words:

▶ The net work done by a conservative force on a particle moving around every closed path is zero.

We know from experiment that the gravitational force passes this *closed-path test.* An example is the tossed tomato of Fig. 8-2. The tomato leaves the launch point with speed v_0 and kinetic energy $\frac{1}{2}mv_0^2$. The gravitational force acting on the tomato slows it, stops it, and then causes it to fall back down. When the tomato returns to the launch point, it again has speed v_0 and kinetic energy $\frac{1}{2}mv_0^2$. Thus, the gravita-

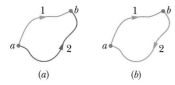

Fig. 8-4 (*a*) As a conservative force acts on it, a particle can move from point *a* to point *b* along either path 1 or path 2. (*b*) The particle moves in a round trip, from point *a* to point *b* along path 1 and then back to point *a* along path 2.

tional force transfers as much energy *from* the tomato during the ascent as it transfers *to* the tomato during the descent back to the launch point. The net work done on the tomato by the gravitational force during the round trip is zero.

An important result of the closed-path test is that

> ▶ The work done by a conservative force on a particle moving between two points does not depend on the path taken by the particle.

For example, suppose that a particle moves from point *a* to point *b* in Fig. 8-4*a* along either path 1 or path 2. If only a conservative force acts on the particle, then the work done on the particle is the same along the two paths. In symbols, we can write this result as

$$W_{ab,1} = W_{ab,2}, \qquad (8\text{-}2)$$

where the subscript *ab* indicates the initial and final points, respectively, and the subscripts 1 and 2 indicate the path.

This result is powerful, because it allows us to simplify difficult problems when only a conservative force is involved. Suppose you need to calculate the work done by a conservative force along a given path between two points, and the calculation is difficult or even impossible without additional information. You can find the work by substituting some other path between those two points for which the calculation is easier and possible. Sample Problem 8-1 gives an example, but first we need to prove Eq. 8-2.

Proof of Equation 8-2

Figure 8-4*b* shows an arbitrary round trip for a particle that is acted upon by a single force. The particle moves from an initial point *a* to point *b* along path 1, and then back to point *a* along path 2. The force does work on the particle as the particle moves along each path. Without worrying about where positive work is done and where negative work is done, let us just represent the work done from *a* to *b* along path 1 as $W_{ab,1}$ and the work done from *b* back to *a* along path 2 as $W_{ba,2}$. If the force is conservative, then the net work done during the round trip must be zero:

$$W_{ab,1} + W_{ba,2} = 0,$$

and thus

$$W_{ab,1} = -W_{ba,2}. \qquad (8\text{-}3)$$

In words, the work done along the outward path must be the negative of the work done along the path back.

Let us now consider the work $W_{ab,2}$ done on the particle by the force when the particle moves from *a* to *b* along path 2, as indicated in Fig. 8-4*a*. If the force is conservative, that work is the negative of $W_{ba,2}$:

$$W_{ab,2} = -W_{ba,2}. \qquad (8\text{-}4)$$

Substituting $W_{ab,2}$ for $-W_{ba,2}$ in Eq. 8-3, we obtain

$$W_{ab,1} = W_{ab,2},$$

which is what we set out to prove.

✔**CHECKPOINT 1:**   The figure shows three paths connecting points *a* and *b*. A single force $\vec{F}$ does the indicated work on a particle moving along each path in the indicated direction. On the basis of this information, is force $\vec{F}$ conservative?

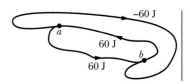

Sample Problem 8-1

Figure 8-5a shows a 2.0 kg block of slippery cheese that slides along a frictionless track from point a to point b. The cheese travels through a total distance of 2.0 m along the track, and a net vertical distance of 0.80 m. How much work is done on the cheese by the gravitational force during the slide?

SOLUTION: A Key Idea here is that we *cannot* use Eq. 7-12 ($W_g = mgd \cos \phi$) to calculate the work done by the gravitational force $\vec{F}_g$ as the cheese moves along the track. The reason is that the angle ϕ between the directions of $\vec{F}_g$ and the displacement $\vec{d}$ varies along the track in an unknown way. (Even if we did know the shape of the track and could calculate ϕ along it, the calculation could be very difficult.)

A second Key Idea is that because $\vec{F}_g$ is a conservative force, we can find the work by choosing some other path between a and b—one that makes the calculation easy. Let us choose the dashed path in Fig. 8-5b; it consists of two straight segments. Along the horizontal segment, the angle ϕ is a constant 90°. Even though we do not know the displacement along that horizontal segment, Eq. 7-12 tells us that the work W_h done there is

$$W_h = mgd \cos 90° = 0.$$

Along the vertical segment, the displacement d is 0.80 m and, with $\vec{F}_g$ and $\vec{d}$ both downward, the angle ϕ is a constant 0°. Thus, Eq. 7-12 gives us, for the work W_v done along the vertical part of

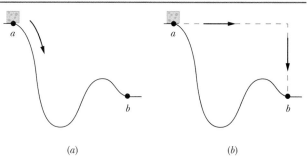

Fig. 8-5 Sample Problem 8-1. (a) A block of cheese slides along a frictionless track from point a to point b. (b) Finding the work done on the cheese by the gravitational force is easier along the dashed path than along the actual path taken by the cheese; the result is the same for both paths.

the dashed path,

$$W_v = mgd \cos 0°$$
$$= (2.0 \text{ kg})(9.8 \text{ m/s}^2)(0.80 \text{ m})(1) = 15.7 \text{ J}.$$

The total work done on the cheese by $\vec{F}_g$ as the cheese moves from point a to point b along the dashed path is then

$$W = W_h + W_v = 0 + 15.7 \text{ J} \approx 16 \text{ J}. \qquad \text{(Answer)}$$

This is also the work done as the cheese moves along the track from a to b.

8-3 Determining Potential Energy Values

Here we find equations that give the value of the two types of potential energy discussed in this chapter: gravitational potential energy and elastic potential energy. However, first we must find a general relation between a conservative force and the associated potential energy.

Consider a particle-like object that is part of a system in which a conservative force $\vec{F}$ acts. When that force does work W on the object, the change ΔU in the potential energy associated with the system is the negative of the work done. We wrote this fact as Eq. 8-1 ($\Delta U = -W$). For the most general case, in which the force may vary with position, we may write the work W as in Eq. 7-32:

$$W = \int_{x_i}^{x_f} F(x) \, dx. \qquad (8\text{-}5)$$

This equation gives the work done by the force when the object moves from point x_i to point x_f, changing the configuration of the system. (Because the force is conservative, the work is the same for all paths between those two points.)

Substituting Eq. 8-5 into Eq. 8-1, we find that the change in potential energy due to the change in configuration is

$$\Delta U = -\int_{x_i}^{x_f} F(x) \, dx. \qquad (8\text{-}6)$$

This is the general relation we sought. Let's put it to use.

Gravitational Potential Energy

We first consider a particle with mass m moving vertically along a y axis (the positive direction is upward). As the particle moves from point y_i to point y_f, the gravitational force $\vec{F}_g$ does work on it. To find the corresponding change in the gravitational potential energy of the particle–Earth system, we use Eq. 8-6 with two changes: (1) We integrate along the y axis instead of the x axis, because the gravitational force acts vertically. (2) We substitute $-mg$ for the force symbol F, because $\vec{F}_g$ has the magnitude mg and is directed down the y axis. We then have

$$\Delta U = -\int_{y_i}^{y_f} (-mg)\, dy = mg \int_{y_i}^{y_f} dy = mg \left[y \right]_{y_i}^{y_f},$$

which yields

$$\Delta U = mg(y_f - y_i) = mg\,\Delta y. \qquad (8\text{-}7)$$

Only *changes* ΔU in gravitational potential energy (or any other type of potential energy) are physically meaningful. However, to simplify a calculation or a discussion, we sometimes would like to say that a certain gravitational potential value U is associated with a certain particle–Earth system when the particle is at a certain height y. To do so, we rewrite Eq. 8-7 as

$$U - U_i = mg(y - y_i). \qquad (8\text{-}8)$$

Then we take U_i to be the gravitational potential energy of the system when it is in a **reference configuration** in which the particle is at a **reference point** y_i. Usually we take $U_i = 0$ and $y_i = 0$. Doing this changes Eq. 8-8 to

$$U(y) = mgy \qquad \text{(gravitational potential energy).} \qquad (8\text{-}9)$$

This equation tells us:

> ▶ The gravitational potential energy associated with a particle–Earth system depends only on the vertical position y (or height) of the particle relative to the reference position $y = 0$, not on the horizontal position.

Elastic Potential Energy

We next consider the block–spring system shown in Fig. 8-3, with the block moving on the end of a spring of spring constant k. As the block moves from point x_i to point x_f, the spring force $F = -kx$ does work on the block. To find the corresponding change in the elastic potential energy of the block–spring system, we substitute $-kx$ for $F(x)$ in Eq. 8-6. We then have

$$\Delta U = -\int_{x_i}^{x_f} (-kx)\, dx = k \int_{x_i}^{x_f} x\, dx = \tfrac{1}{2} k \left[x^2 \right]_{x_i}^{x_f},$$

or

$$\Delta U = \tfrac{1}{2} k x_f^2 - \tfrac{1}{2} k x_i^2. \qquad (8\text{-}10)$$

To associate a potential energy value U with the block at position x, we choose the reference configuration to be when the spring is at its relaxed length and the block is at $x_i = 0$. Then the elastic potential energy U_i is 0, and Eq. 8-10 becomes

$$U - 0 = \tfrac{1}{2} k x^2 - 0,$$

which gives us

$$U(x) = \tfrac{1}{2} k x^2 \qquad \text{(elastic potential energy).} \qquad (8\text{-}11)$$

✔CHECKPOINT 2:   A particle is to move along the x axis from $x = 0$ to x_1 while a conservative force, directed along the x axis, acts on the particle. The figure shows three situations in which the x component of that force varies with x. The force has the same maximum magnitude F_1 in all three situations. Rank the situations according to the change in the associated potential energy during the particle's motion, most positive first.

(1)   (2)   (3)

PROBLEM-SOLVING TACTICS

Tactic 1: *Using the Term "Potential Energy"*

A potential energy is associated with a system as a whole. However, you might see statements that associate it with only part of the system. For example, you might read, "An apple hanging in a tree has a gravitational potential energy of 30 J." Such statements are often acceptable, but you should always keep in mind that the potential energy is actually associated with a system—here the apple–Earth system. Also keep in mind that assigning a particular potential energy value, such as 30 J here, to an object or even a system makes sense *only* if the reference potential energy value is known, as explored in Sample Problem 8-2.

Sample Problem 8-2

A 2.0 kg sloth hangs 5.0 m above the ground (Fig. 8-6).

(a) What is the gravitational potential energy U of the sloth–Earth system if we take the reference point $y = 0$ to be (1) at the ground, (2) at a balcony floor that is 3.0 m above the ground, (3) at the limb, and (4) 1.0 m above the limb? Take the gravitational potential energy to be zero at $y = 0$.

SOLUTION: The Key Idea here is that once we have chosen the reference point for $y = 0$, we can calculate the gravitational potential energy U of the system *relative to that reference point* with Eq. 8-9. For example, for choice (1) the sloth is at $y = 5.0$ m, and

$$U = mgy = (2.0 \text{ kg})(9.8 \text{ m/s}^2)(5.0 \text{ m})$$
$$= 98 \text{ J}. \quad \text{(Answer)}$$

For the other choices, the values of U are

(2) $U = mgy = mg(2.0 \text{ m}) = 39 \text{ J}$,
(3) $U = mgy = mg(0) = 0 \text{ J}$,
(4) $U = mgy = mg(-1.0 \text{ m}) = -19.6 \text{ J} \approx -20 \text{ J}$.   (Answer)

(b) The sloth drops to the ground. For each choice of reference point, what is the change ΔU in the potential energy of the sloth–Earth system due to the fall?

SOLUTION: The Key Idea here is that the *change* in potential energy does not depend on the choice of the reference point for $y = 0$; instead, it depends on the change in height Δy. For all four situations, we have the same $\Delta y = -5.0$ m. Thus, for (1) to (4), Eq. 8-7 tells us that

$$\Delta U = mg\, \Delta y = (2.0 \text{ kg})(9.8 \text{ m/s}^2)(-5.0 \text{ m})$$
$$= -98 \text{ J}. \quad \text{(Answer)}$$

Fig. 8-6 Sample Problem 8-2. Four choices of reference point $y = 0$. Each y axis is marked in units of meters. The choice affects the value of the potential energy U of the sloth–Earth system. However, it does not affect the change ΔU in potential energy of the system if the sloth moves by, say, falling.

8-4 Conservation of Mechanical Energy

The **mechanical energy** E_{mec} of a system is the sum of its potential energy U and the kinetic energy K of the objects within it:

$$E_{mec} = K + U \quad \text{(mechanical energy)}. \quad (8\text{-}12)$$

In this section, we examine what happens to this mechanical energy when only conservative forces cause energy transfers within the system—that is, when frictional and drag forces do not act on the objects in the system. Also, we shall assume that the system is *isolated* from its environment; that is, no *external force* from an object outside the system causes energy changes inside the system.

When a conservative force does work W on an object within the system, it transfers energy between kinetic energy K of the object and potential energy U of the system. From Eq. 7-10, the change ΔK in kinetic energy is

$$\Delta K = W \quad (8\text{-}13)$$

and from Eq. 8-1, the change ΔU in potential energy is

$$\Delta U = -W. \quad (8\text{-}14)$$

Combining Eqs. 8-13 and 8-14, we find that

$$\Delta K = -\Delta U. \quad (8\text{-}15)$$

In words, one of these energies increases exactly as much as the other decreases.

We can rewrite Eq. 8-15 as

$$K_2 - K_1 = -(U_2 - U_1), \quad (8\text{-}16)$$

where the subscripts refer to two different instants and thus to two different arrangements of the objects in the system. Rearranging Eq. 8-16 yields

$$K_2 + U_2 = K_1 + U_1 \quad \text{(conservation of mechanical energy)}. \quad (8\text{-}17)$$

In words, this equation says:

$$\begin{pmatrix} \text{the sum of } K \text{ and } U \text{ for} \\ \text{any state of a system} \end{pmatrix} = \begin{pmatrix} \text{the sum of } K \text{ and } U \text{ for} \\ \text{any other state of the system} \end{pmatrix},$$

when the system is isolated and only conservative forces act on the objects in the system. In other words:

> In an isolated system where only conservative forces cause energy changes, the kinetic energy and potential energy can change, but their sum, the mechanical energy E_{mec} of the system, cannot change.

This result is called the **principle of conservation of mechanical energy.** (Now you can see where *conservative* forces got their name.) With the aid of Eq. 8-15, we can write this principle in one more form, as

$$\Delta E_{mec} = \Delta K + \Delta U = 0. \quad (8\text{-}18)$$

The principle of conservation of mechanical energy allows us to solve problems that would be quite difficult to solve using only Newton's laws:

> When the mechanical energy of a system is conserved, we can relate the sum of kinetic energy and potential energy at one instant to that at another instant *without considering the intermediate motion* and *without finding the work done by the forces involved.*

In olden days, a native Alaskan would be tossed via a blanket to be able to see farther over the flat terrain. Nowadays, it is done just for fun. During the ascent of the child in the photograph, energy is transferred from kinetic energy to gravitational potential energy. The maximum height is reached when that transfer is complete. Then the transfer is reversed during the fall.

Fig. 8-7 A pendulum, with its mass concentrated in a bob at the lower end, swings back and forth. One full cycle of the motion is shown. During the cycle the values of the potential and kinetic energies of the pendulum–Earth system vary as the bob rises and falls, but the mechanical energy E_{mec} of the system remains constant. The energy E_{mec} can be described as continuously shifting between the kinetic and potential forms. In stages (a) and (e), all the energy is kinetic energy. The bob then has its greatest speed and is at its lowest point. In stages (c) and (g), all the energy is potential energy. The bob then has zero speed and is at its highest point. In stages (b), (d), (f), and (h), half the energy is kinetic energy and half is potential energy. If the swinging involved a frictional force at the point where the pendulum is attached to the ceiling, or a drag force due to the air, then E_{mec} would not be conserved, and eventually the pendulum would stop.

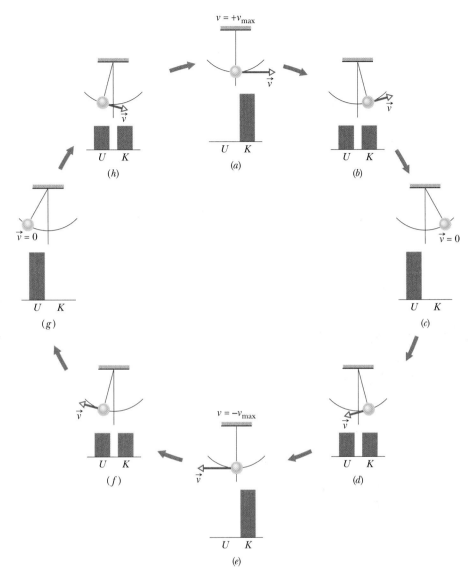

Figure 8-7 shows an example in which the principle of conservation of mechanical energy can be applied: As a pendulum swings, the energy of the pendulum–Earth system is transferred back and forth between kinetic energy K and gravitational potential energy U, with the sum $K + U$ being constant. If we know the gravitational potential energy when the pendulum bob is at its highest point (Fig. 8-7c), Eq. 8-17 gives us the kinetic energy of the bob at the lowest point (Fig. 8-7e).

For example, let us choose the lowest point as the reference point, with the gravitational potential energy $U_2 = 0$. Suppose then that the potential energy at the highest point is $U_1 = 20$ J relative to the reference point. Because the bob momentarily stops at its highest point, the kinetic energy there is $K_1 = 0$. Substituting these values into Eq. 8-17 gives us the kinetic energy K_2 at the lowest point:

$$K_2 + 0 = 0 + 20 \text{ J} \quad \text{or} \quad K_2 = 20 \text{ J}.$$

Note that we get this result without considering the motion between the highest and lowest points (such as in Fig. 8-7d) and without finding the work done by any forces involved in the motion.

✓CHECKPOINT 3:   The figure shows four situations—one in which an initially stationary block is dropped and three in which the block is allowed to slide down frictionless ramps. (a) Rank the situations according to the kinetic energy of the block at point B, greatest first. (b) Rank them according to the speed of the block at point B, greatest first.

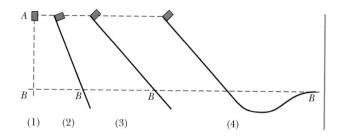

(1)   (2)   (3)   (4)

Sample Problem 8-3

In Fig. 8-8, a child of mass m is released from rest at the top of a water slide, at height $h = 8.5$ m above the bottom of the slide. Assuming that the slide is frictionless because of the water on it, find the child's speed at the bottom of the slide.

SOLUTION: A Key Idea here is that we cannot find her speed at the bottom by using her acceleration along the slide as we might have in earlier chapters because we do not know the slope (angle) of the slide. However, because that speed is related to her kinetic energy, perhaps we can use the principle of conservation of mechanical energy to get the speed. Then we would not need to know the slope or anything about the slide's shape. A second Key Idea is that mechanical energy is conserved in an isolated system when only conservative forces cause energy transfers. Let's check.

   Forces: Two forces act on the child. The *gravitational force,* a conservative force, does work on her. The *normal force* on her from the slide does no work, because its direction at any point during the descent is always perpendicular to the direction in which the child moves.

   System: Because the only force doing work on the child is the gravitational force, we choose the child–Earth system as our system, which we take to be isolated.

   Thus, we have only a conservative force doing work in an isolated system, so we *can* use the principle of conservation of mechanical energy. Let the mechanical energy be $E_{\text{mec},t}$ when the child is at the top of the slide and $E_{\text{mec},b}$ when she is at the bottom. Then the conservation principle tells us

$$E_{\text{mec},b} = E_{\text{mec},t}.$$

Expanding this to show both kinds of mechanical energy, we have

$$K_b + U_b = K_t + U_t,$$

or $\frac{1}{2}mv_b^2 + mgy_b = \frac{1}{2}mv_t^2 + mgy_t.$

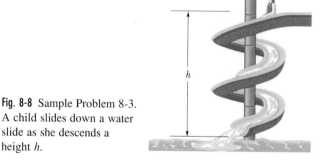

Fig. 8-8 Sample Problem 8-3. A child slides down a water slide as she descends a height h.

Dividing by m and rearranging yield

$$v_b^2 = v_t^2 + 2g(y_t - y_b).$$

Putting $v_t = 0$ and $y_t - y_b = h$ leads to

$$v_b = \sqrt{2gh} = \sqrt{(2)(9.8 \text{ m/s}^2)(8.5 \text{ m})}$$
$$= 13 \text{ m/s}. \qquad \text{(Answer)}$$

This is the same speed that the child would reach if she fell 8.5 m. On an actual slide, some frictional forces would act and the child would not be moving quite so fast.

   Although this problem is hard to solve directly with Newton's laws, using conservation of mechanical energy makes the solution much easier. However, if we were asked to find the time taken for the child to reach the bottom of the slide, energy methods would be of no use; we would need to know the shape of the slide, and we would have a difficult problem.

   Now that we have worked this sample problem, return to the puck example in the first paragraph of Chapter 7. See if you can show that the puck's speed at the end of the track is 5.0 m/s.

Sample Problem 8-4

A 61.0 kg bungee-cord jumper is on a bridge 45.0 m above a river. The elastic bungee cord has a relaxed length of $L = 25.0$ m. Assume that the cord obeys Hooke's law, with a spring constant of 160 N/m. If the jumper stops before reaching the water, what is the height h of her feet above the water at her lowest point?

SOLUTION: Figure 8-9 shows the jumper at the lowest point, with her feet at height h and with the cord stretched by distance d from its relaxed length. If we knew d, we could find h. One Key Idea is that perhaps we can solve for d by applying the principle of conserva-

tion of mechanical energy, between her initial point (on the bridge) and her lowest point. In that case, a second Key Idea is that mechanical energy is conserved in an isolated system when only conservative forces cause energy transfers. Let's check.

   Forces: The gravitational force does work on the jumper throughout her fall. Once the bungee cord becomes taut, the spring-like force from it does work on her, transferring energy to elastic potential energy of the cord. The force from the cord also pulls on the bridge, which is attached to Earth. The gravitational force and the spring-like force are conservative.

Fig. 8-9 Sample Problem 8-4. A bungee-cord jumper at the lowest point of the jump.

change in the elastic potential energy of the bungee cord, and ΔU_g is the change in the jumper's gravitational potential energy. All these changes must be computed between her initial point and her lowest point. Because she is stationary (at least momentarily) both initially and at her lowest point, $\Delta K = 0$. From Fig. 8-9, we see that the change Δy in her height is $-(L + d)$, so we have

$$\Delta U_g = mg\,\Delta y = -mg(L + d),$$

where m is her mass. Also from Fig. 8-9, we see that the bungee cord is stretched by distance d. Thus, we also have

$$\Delta U_e = \tfrac{1}{2}kd^2.$$

Inserting these expressions and the given data into Eq. 8-19, we obtain

$$0 + \tfrac{1}{2}kd^2 - mg(L + d) = 0$$

or $$\tfrac{1}{2}kd^2 - mgL - mgd = 0$$

and then

$$\tfrac{1}{2}(160 \text{ N/m})d^2 - (61.0 \text{ kg})(9.8 \text{ m/s}^2)(25.0 \text{ m})$$
$$- (61.0 \text{ kg})(9.8 \text{ m/s}^2)d = 0.$$

Solving this quadratic equation yields

$$d = 17.9 \text{ m.}$$

The jumper's feet are then a distance of $(L + d) = 42.9$ m below their initial height. Thus,

$$h = 45.0 \text{ m} - 42.9 \text{ m} = 2.1 \text{ m.} \qquad \text{(Answer)}$$

System: The jumper–Earth–cord system includes all these forces and energy transfers, and we can take it to be isolated. Thus, we *can* apply the principle of conservation of mechanical energy to the system. From Eq. 8-18, we can write the principle as

$$\Delta K + \Delta U_e + \Delta U_g = 0, \qquad (8\text{-}19)$$

where ΔK is the change in the jumper's kinetic energy, ΔU_e is the

---

PROBLEM-SOLVING TACTICS

Tactic 2: *Conservation of Mechanical Energy*
Asking the following questions will help you to solve problems involving the conservation of mechanical energy.

For what system is mechanical energy conserved? You should be able to separate your system from its environment. Imagine drawing a closed surface such that whatever is inside is your system and whatever is outside is the environment of that system. In Sample Problem 8-3 the system is the *child + Earth* and in Sample Problem 8-4, it is the *jumper + Earth + cord.*

Is friction or drag present? If friction or drag is present, mechanical energy is not conserved.

Is your system isolated? Conservation of mechanical energy applies only to isolated systems. That means that no *external forces* (forces exerted by objects outside the system) should do work on the objects in the system.

What are the initial and final states of your system? The system changes from some initial state (or configuration) to some final state. You apply the principle of conservation of mechanical energy by saying that E_{mec} has the same value in both these states. Be very clear about what these two states are.

8-5 Reading a Potential Energy Curve

Once again we consider a particle that is part of a system in which a conservative force acts. This time suppose that the particle is constrained to move along an x axis while the conservative force does work on it. We can learn a lot about the motion of the particle from a plot of the system's potential energy $U(x)$. However, before we discuss such plots, we need one more relationship.

Finding the Force Analytically

Equation 8-6 tells us how to find the change ΔU in potential energy between two points in a one-dimensional situation if we know the force $F(x)$. Now we want to go the other way; that is, we know the potential energy function $U(x)$ and want to find the force.

For one-dimensional motion, the work W done by a force that acts on a particle as the particle moves through a distance Δx is $F(x) \Delta x$. We can then write Eq. 8-1 as

$$\Delta U(x) = -W = -F(x) \Delta x.$$

Solving for $F(x)$ and passing to the differential limit yield

$$F(x) = -\frac{dU(x)}{dx} \quad \text{(one-dimensional motion)}, \quad (8\text{-}20)$$

which is the relation we sought.

We can check this result by putting $U(x) = \frac{1}{2}kx^2$, which is the elastic potential energy function for a spring force. Equation 8-20 then yields, as expected, $F(x) = -kx$, which is Hooke's law. Similarly, we can substitute $U(x) = mgx$, which is the gravitational potential energy function for a particle–Earth system, with a particle of mass m at height x above Earth's surface. Equation 8-20 then yields $F = -mg$, which is the gravitational force on the particle.

The Potential Energy Curve

Figure 8-10a is a plot of a potential energy function $U(x)$ for a system in which a particle is in one-dimensional motion while a conservative force $F(x)$ does work on it. We can easily find $F(x)$ by (graphically) taking the slope of the $U(x)$ curve at various points. (Equation 8-20 tells us that $F(x)$ is negative the slope of the $U(x)$ curve.) Figure 8-10b is a plot of $F(x)$ found in this way.

(a)

(b)

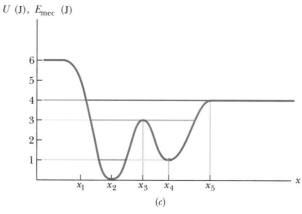

(c)

Fig. 8-10 (a) A plot of $U(x)$, the potential energy function of a system containing a particle confined to move along the x axis. There is no friction, so mechanical energy is conserved. (b) A plot of the force $F(x)$ acting on the particle, derived from the potential energy plot by taking its slope at various points. (c) The $U(x)$ plot of (a) with three different possible values of E_{mec} shown.

Turning Points

In the absence of a nonconservative force, the mechanical energy E of the system has a constant value given by

$$U(x) + K(x) = E_{mec}. \qquad (8\text{-}21)$$

Here $K(x)$ is the *kinetic energy function* of the particle (this $K(x)$ gives the kinetic energy as a function of the particle's location x). We may rewrite Eq. 8-21 as

$$K(x) = E_{mec} - U(x). \qquad (8\text{-}22)$$

Suppose that E_{mec} (which has a constant value, remember) happens to be 5.0 J. It would be represented in Fig. 8-10a by a horizontal line that runs through the value 5.0 J on the energy axis. (It is, in fact, shown there.)

Equation 8-22 tells us how to determine the kinetic energy K for any location x of the particle: On the $U(x)$ curve, find U for that location x and then subtract U from E_{mec}. For example, if the particle is at any point to the right of x_5, then $K = 1.0$ J. The value of K is greatest (5.0 J) when the particle is at x_2, and least (0 J) when the particle is at x_1.

Since K can never be negative (because v^2 is always positive), the particle can never move to the left of x_1, where $E_{mec} - U$ is negative. Instead, as the particle moves toward x_1 from x_2, K decreases (the particle slows) until $K = 0$ at x_1 (the particle stops there).

Note that when the particle reaches x_1, the force on the particle, given by Eq. 8-20, is positive (because the slope dU/dx is negative). This means that the particle does not remain at x_1 but instead begins to move to the right, opposite its earlier motion. Hence x_1 is a **turning point**, a place where $K = 0$ (because $U = E$) and the particle changes direction. There is no turning point (where $K = 0$) on the right side of the graph. When the particle heads to the right, it will continue indefinitely.

Equilibrium Points

Figure 8-10c shows three different values for E_{mec} superposed on the plot of the same potential energy function $U(x)$. Let us see how they would change the situation. If $E_{mec} = 4.0$ J (purple line), the turning point shifts from x_1 to a point between x_1 and x_2. Also, at any point to the right of x_5, the system's mechanical energy is equal to its potential energy; thus, the particle has no kinetic energy and (by Eq. 8-20) no force acts on it, and so it must be stationary. A particle at such a position is said to be in **neutral equilibrium**. (A marble placed on a horizontal tabletop is in that state.)

If $E_{mec} = 3.0$ J (pink line), there are two turning points: One is between x_1 and x_2, and the other is between x_4 and x_5. In addition, x_3 is a point at which $K = 0$. If the particle is located exactly there, the force on it is also zero, and the particle remains stationary. However, if it is displaced even slightly in either direction, a nonzero force pushes it farther in the same direction, and the particle continues to move. A particle at such a position is said to be in **unstable equilibrium**. (A marble balanced on top of a bowling ball is an example.)

Next consider the particle's behavior if $E_{mec} = 1.0$ J (green line). If we place it at x_4, it is stuck there. It cannot move left or right on its own because to do so would require a negative kinetic energy. If we push it slightly left or right, a restoring force appears that moves it back to x_4. A particle at such a position is said to be in **stable equilibrium**. (A marble placed at the bottom of a hemispherical bowl is an example.) If we place the particle in the cuplike *potential well* centered at x_2, it is between two turning points. It can still move somewhat, but only partway to x_1 or x_3.

✔CHECKPOINT 4:   The figure gives the potential energy function $U(x)$ for a system in which a particle is in one-dimensional motion. (a) Rank regions AB, BC, and CD according to the magnitude of the force on the particle, greatest first. (b) What is the direction of the force when the particle is in region AB?

8-6 Work Done on a System by an External Force

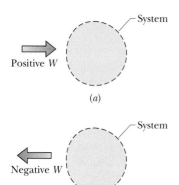

Fig. 8-11 (a) Positive work W on an arbitrary system means a transfer of energy to the system. (b) Negative work W means a transfer of energy from the system.

In Chapter 7, we defined work as being energy transferred to or from an object by means of a force acting on the object. We can now extend that definition to an external force acting on a system of objects.

> Work is energy transferred to or from a system by means of an external force acting on that system.

Figure 8-11a represents positive work (a transfer of energy *to* a system), and Fig. 8-11b represents negative work (a transfer of energy *from* a system). When more than one force acts on a system, their *net work* is the energy transferred to or from the system.

These transfers are like transfers of money to and from a bank account. If a system consists of a single particle or particle-like object, as in Chapter 7, the work done on the system by a force can change only the kinetic energy of the system. The energy statement for such transfers is the work–kinetic energy theorem of Eq. 7-10 ($\Delta K = W$); that is, a single particle has only one energy account, called kinetic energy. External forces can transfer energy into or out of that account. If a system is more complicated, however, an external force can change other forms of energy (such as potential energy); that is, a more complicated system can have multiple energy accounts.

Let us find energy statements for such systems by examining two basic situations, one that does not involve friction and one that does.

No Friction Involved

To compete in a bowling-ball hurling contest, you first squat and cup your hands under the ball on the floor. Then you rapidly straighten up while also pulling your hands up sharply, launching the ball upward at about face level. During your upward motion, your applied force on the ball obviously does work; that is, it is an external force that transfers energy, but to what system?

To answer, we check to see which energies change. There is a change ΔK in the ball's kinetic energy and, because the ball and Earth become more separated, there is a change ΔU in the gravitational potential energy of the ball–Earth system. To include both changes, we need to consider the ball–Earth system. Then your force is an external force doing work on that system, and the work is

$$W = \Delta K + \Delta U, \tag{8-23}$$

or $$W = \Delta E_{\text{mec}} \quad \text{(work done on system, no friction involved)}, \tag{8-24}$$

Fig. 8-12 Positive work W is done on a system of a bowling ball and Earth, causing a change ΔE_{mec} in the mechanical energy of the system, a change ΔK in the ball's kinetic energy, and a change ΔU in the system's gravitational potential energy.

where ΔE_{mec} is the change in the mechanical energy of the system. These two equations, which are represented in Fig. 8-12, are equivalent energy statements for work done on a system by an external force when friction is not involved.

Friction Involved

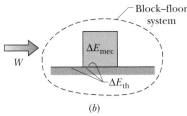

Fig. 8-13 (*a*) A block is pulled across a floor by force $\vec{F}$ while a kinetic frictional force opposes the motion. The block has velocity $\vec{v}_0$ at the start of a displacement $\vec{d}$ and a velocity $\vec{v}$ at the end of the displacement. (*b*) Positive work W is done on the block–floor system by force $\vec{F}$, resulting in a change ΔE_{mec} in the block's mechanical energy and a change ΔE_{th} in the thermal energy of the block and floor.

We next consider the example in Fig. 8-13*a*. A constant horizontal force $\vec{F}$ pulls a block along an x axis and through a displacement of magnitude d, increasing the block's velocity from $\vec{v}_0$ to $\vec{v}$. During the motion, a constant kinetic frictional force $\vec{f}_k$ from the floor acts on the block. Let us first choose the block as our system and apply Newton's second law to it. We can write that law for components along the x axis ($F_{\text{net},x} = ma_x$) as

$$F - f_k = ma. \tag{8-25}$$

Because the forces are constant, the acceleration $\vec{a}$ is also constant. Thus, we can use Eq. 2-16 to write

$$v^2 = v_0^2 + 2ad.$$

Solving this equation for a, substituting the result into Eq. 8-25, and rearranging then give us

$$Fd = \tfrac{1}{2}mv^2 - \tfrac{1}{2}mv_0^2 + f_k d \tag{8-26}$$

or, because $\tfrac{1}{2}mv^2 - \tfrac{1}{2}mv_0^2 = \Delta K$ for the block,

$$Fd = \Delta K + f_k d. \tag{8-27}$$

In a more general situation (say, one in which the block is moving up a ramp), there can be a change in potential energy. To include such a possible change, we generalize Eq. 8-27 by writing

$$Fd = \Delta E_{\text{mec}} + f_k d. \tag{8-28}$$

By experiment we find that the block and the portion of the floor along which it slides become warmer as the block slides. As we shall discuss in Chapter 19, the temperature of an object is related to the object's thermal energy E_{th} (the energy associated with the random motion of the atoms and molecules in the object). Here, the thermal energy of the block and floor increase because (1) there is friction between them and (2) there is sliding. Recall that friction is due to the cold-welding between two surfaces. As the block slides over the floor, the sliding causes repeated tearing and reforming of the welds between the block and the floor, which makes the block and floor warmer. Thus, the sliding increases their thermal energy E_{th}.

Through experiment, we find that the increase ΔE_{th} in thermal energy is equal to the product of the magnitudes f_k and d:

$$\Delta E_{\text{th}} = f_k d \qquad \text{(increase in thermal energy by sliding)}. \tag{8-29}$$

Thus, we can rewrite Eq. 8-28 as

$$Fd = \Delta E_{\text{mec}} + \Delta E_{\text{th}}. \tag{8-30}$$

Fd is the work W done by the external force $\vec{F}$ (the energy transferred by the force), but on which system is the work done (where are the energy transfers made)? To answer, we check to see which energies change. The block's mechanical energy changes, and the thermal energies of the block and floor also change. Therefore, the work done by force $\vec{F}$ is done on the block–floor system. That work is

$$W = \Delta E_{\text{mec}} + \Delta E_{\text{th}} \qquad \text{(work done on system, friction involved)}. \tag{8-31}$$

This equation, which is represented in Fig. 8-13*b*, is the energy statement for the work done on a system by an external force when friction is involved.

✔CHECKPOINT 5:   In three trials, a block is pushed by a horizontal applied force across a floor that is not frictionless, as in Fig. 8-13a. The magnitudes F of the applied force and the result of the pushing on the block's speed are given in the table. In all three trials, the block is pushed through the same distance d. Rank the three trials according to the change in the thermal energy of the block and floor that occurs in that distance d, greatest first.

Trial	F	Result on Block's Speed
a	5.0 N	decreases
b	7.0 N	remains constant
c	8.0 N	increases

Sample Problem 8-5

The giant stone statues of Easter Island were most likely moved by the prehistoric islanders by cradling each statue in a wooden sled and then pulling the sled over a "runway" consisting of almost identical logs acting as rollers. In a modern reenactment of this technique, 25 men were able to move a 9000 kg Easter Island-type statue 45 m over level ground in 2 min.

(a) Estimate the work the net force $\vec{F}$ from the men did during the 45 m displacement of the statue, and determine the system on which that force did the work.

SOLUTION: One Key Idea is that we can calculate the work done with Eq. 7-7 ($W = Fd \cos \phi$). Here d is the distance 45 m, F is the magnitude of the net force on the statue from the 25 men, and $\phi = 0°$. Let us estimate that each man pulled with a force magnitude equal to twice his weight, which we take to be the same value mg for all the men. Thus, the magnitude of the net force was $F = (25)(2mg) = 50mg$. Estimating a man's mass as 80 kg, we can then write Eq. 7-7 as

$$W = Fd \cos \phi = 50mgd \cos \phi$$
$$= (50)(80 \text{ kg})(9.8 \text{ m/s}^2)(45 \text{ m}) \cos 0°$$
$$= 1.8 \times 10^6 \text{ J} \approx 2 \text{ MJ}. \qquad \text{(Answer)}$$

The Key Idea in determining the system on which the work is done is to see which energies change. Because the statue moved, there was certainly a change ΔK in its kinetic energy during the motion. We can easily guess that there must have been considerable kinetic friction between the sled, logs, and ground, resulting in a change ΔE_{th} in their thermal energies. Thus, the system on which the work was done consisted of the statue, sled, logs, and ground.

(b) What was the increase ΔE_{th} in the thermal energy of the system during the 45 m displacement?

SOLUTION: The Key Idea here is that we can relate ΔE_{th} to the work W done by $\vec{F}$ with the energy statement of Eq. 8-31 for a system that involves friction:

$$W = \Delta E_{mec} + \Delta E_{th}.$$

We know the value of W from (a). The change ΔE_{mec} in the crate's mechanical energy was zero because the statue was stationary at the beginning and the end of the move and did not change in elevation. Thus, we find

$$\Delta E_{th} = W = 1.8 \times 10^6 \text{ J} \approx 2 \text{ MJ}. \qquad \text{(Answer)}$$

(c) Estimate the work that would have been done by the 25 men if they had moved the statue 10 km across level ground on Easter Island. Also estimate the total change ΔE_{th} that would have occurred in the statue–sled–logs–ground system.

SOLUTION: The Key Ideas here are the same as in (a) and (b). Thus we calculate W as in (a), but with 1×10^4 m now substituted for d. Also, we again equate ΔE_{th} to W. We get

$$W = \Delta E_{th} = 3.9 \times 10^8 \text{ J} \approx 400 \text{ MJ}. \qquad \text{(Answer)}$$

This would have been a staggering amount of energy for the men to have transferred during the movement of a statue. Still, the 25 men *could* have moved the statue 10 km, and the required energy does not suggest some mysterious source.

Sample Problem 8-6

A food shipper pushes a wood crate of cabbage heads (total mass $m = 14$ kg) across a concrete floor with a constant horizontal force $\vec{F}$ of magnitude 40 N. In a straight-line displacement of magnitude $d = 0.50$ m, the speed of the crate decreases from $v_0 = 0.60$ m/s to $v = 0.20$ m/s.

(a) How much work is done by force $\vec{F}$, and on what system does it do the work?

SOLUTION: One Key Idea is that Eq. 7-7 holds here: The work W done by $\vec{F}$ can be calculated as

$$W = Fd \cos \phi = (40 \text{ N})(0.50 \text{ m}) \cos 0°$$
$$= 20 \text{ J}. \qquad \text{(Answer)}$$

The Key Idea in determining the system on which the work is done is to see which energies change. Because the crate's speed changes, there is certainly a change ΔK in the crate's kinetic energy. Is there friction between the floor and the crate, and thus a change in thermal energy? Note that $\vec{F}$ and the crate's velocity have the same direction. Thus, a Key Idea here is that if there is no friction, then $\vec{F}$ should be accelerating the crate to a *greater* speed. However, the crate is *slowing*, so there must be friction and a change ΔE_{th} in thermal energy of the crate and the floor. Therefore, the system on which the work is done is the crate–floor system, because both energy changes occur in that system.

(b) What is the increase ΔE_{th} in the thermal energy of the crate and floor?

SOLUTION: The Key Idea here is that we can relate ΔE_{th} to the work W done by $\vec{F}$ with the energy statement of Eq. 8-31 for a system

that involves friction:

$$W = \Delta E_{mec} + \Delta E_{th}. \tag{8-32}$$

We know the value of W from (a). The change ΔE_{mec} in the crate's mechanical energy is just the change in its kinetic energy because no potential energy changes occur, so we have

$$\Delta E_{mec} = \Delta K = \tfrac{1}{2}mv^2 - \tfrac{1}{2}mv_0^2.$$

Substituting this into Eq. 8-32 and solving for ΔE_{th}, we find

$$\Delta E_{th} = W - (\tfrac{1}{2}mv^2 - \tfrac{1}{2}mv_0^2) = W - \tfrac{1}{2}m(v^2 - v_0^2)$$

$$= 20 \text{ J} - \tfrac{1}{2}(14 \text{ kg})[(0.20 \text{ m/s})^2 - (0.60 \text{ m/s})^2]$$

$$= 22.2 \text{ J} \approx 22 \text{ J}. \tag{Answer}$$

8-7 Conservation of Energy

We now have discussed several situations in which energy is transferred to or from objects and systems, much like money is transferred between accounts. In each situation we assume that the energy that was involved could always be accounted for; that is, energy could not magically appear or disappear. In more formal language, we assumed (correctly) that energy obeys a law called the **law of conservation of energy,** which is concerned with the **total energy** E of a system. That total is the sum of the system's mechanical energy, thermal energy, and any form of *internal energy* in addition to thermal energy. (We have not yet discussed other forms of internal energy.) The law states that

> The total energy E of a system can change only by amounts of energy that are transferred to or from the system.

The only type of energy transfer that we have considered is work W done on a system. Thus, for us at this point, this law states that

$$W = \Delta E = \Delta E_{mec} + \Delta E_{th} + \Delta E_{int}, \tag{8-33}$$

where ΔE_{mec} is any change in the mechanical energy of the system, ΔE_{th} is any change in the thermal energy of the system, and ΔE_{int} is any change in any other form of internal energy of the system. Included in ΔE_{mec} are changes ΔK in kinetic energy and changes ΔU in potential energy (elastic, gravitational, or any other form we might find).

This law of conservation of energy is *not* something we have derived from basic physics principles. Rather, it is a law based on countless experiments. Scientists and engineers have never found an exception to it.

Isolated System

If a system is isolated from its environment, then there can be no energy transfers to or from it. For that case, the law of conservation of energy states:

> The total energy E of an isolated system cannot change.

Fig. 8-14 To descend, the rock climber must transfer energy from the gravitational potential energy of a system consisting of her, her gear, and Earth. She has wrapped the rope around metal rings so that the rope rubs against the rings. This allows most of the transferred energy to go to the thermal energy of the rope and rings rather than to her kinetic energy.

Many energy transfers may be going on *within* an isolated system, between, say, kinetic energy and a potential energy or kinetic energy and thermal energy. However, the total of all the forms of energy in the system cannot change.

We can use the rock climber in Fig. 8-14 as an example, approximating her, her gear, and Earth as an isolated system. As she rappels down the rock face, changing the configuration of the system, she needs to control the transfer of energy from the gravitational potential energy of the system. (That energy cannot just disappear.) Some of it is transferred to her kinetic energy. However, she obviously does not want very much transferred to that form or she will be moving too quickly, so she has wrapped the rope around metal rings to produce friction between the rope and the rings as she moves down. The sliding of the rings on the rope then transfers the gravitational potential energy of the system to thermal energy of the rings and rope in a way that she can control. The total energy of the climber–gear–Earth system (the total of its gravitational potential energy, kinetic energy, and thermal energy) does not change during her descent.

For an isolated system, the law of conservation of energy can be written in two ways. First, by setting $W = 0$ in Eq. 8-33, we get

$$\Delta E_{mec} + \Delta E_{th} + \Delta E_{int} = 0 \qquad \text{(isolated system).} \qquad (8\text{-}34)$$

We can also let $\Delta E_{mec} = E_{mec,2} - E_{mec,1}$, where the subscripts 1 and 2 refer to two different instants, say before and after a certain process has occurred. Then Eq. 8-34 becomes

$$E_{mec,2} = E_{mec,1} - \Delta E_{th} - \Delta E_{int}. \qquad (8\text{-}35)$$

Equation 8-35 tells us:

> In an isolated system, we can relate the total energy at one instant to the total energy at another instant *without considering the energies at intermediate times*.

This fact can be a very powerful tool in solving problems about isolated systems when you need to relate energies of a system before and after a certain process occurs in the system.

In Section 8-4, we discussed a special situation for isolated systems—namely, the situation in which nonconservative forces (such as a kinetic frictional force) do not act within them. In that special situation, ΔE_{th} and ΔE_{int} are both zero, and so Eq. 8-35 reduces to Eq. 8-18. In other words, the mechanical energy of an isolated system is conserved when nonconservative forces do not act within it.

Power

Now that you have seen how energy can be transferred from one form to another, we can expand the definition of power given in Section 7-7. There it is the rate at which work is done by a force. In a more general sense, power P is the rate at which energy is transferred by a force from one form to another. If an amount of energy ΔE is transferred in an amount of time Δt, the **average power** due to the force is

$$P_{avg} = \frac{\Delta E}{\Delta t}. \qquad (8\text{-}36)$$

Similarly, the **instantaneous power** due to the force is

$$P = \frac{dE}{dt}. \qquad (8\text{-}37)$$

Sample Problem 8-7

In Fig. 8-15, a 2.0 kg package of tamale slides along a floor with speed $v_1 = 4.0$ m/s. It then runs into and compresses a spring, until the package momentarily stops. Its path to the initially relaxed spring is frictionless, but as it compresses the spring, a kinetic frictional force from the floor, of magnitude 15 N, acts on it. The spring constant is 10,000 N/m. By what distance d is the spring compressed when the package stops?

SOLUTION: A starting Key Idea is to examine all the forces acting on the package, and then to determine whether we have an isolated system or a system on which an external force is doing work.

Forces: The normal force on the package from the floor does no work on the package, because its direction is always perpendicular to that of the package's displacement. For the same reason, the gravitational force on the package does no work. As the spring is compressed, however, a spring force does work on the package, transferring energy to elastic potential energy of the spring. The spring force also pushes against a rigid wall. Because there is friction between the package and the floor, the sliding of the package across the floor increases their thermal energies.

System: The package–spring–floor–wall system includes all these forces and energy transfers in one isolated system. Therefore, a second Key Idea is that, because the system is isolated, its total energy cannot change. We can then apply the law of conservation of energy in the form of Eq. 8-35 to the system:

$$E_{mec,2} = E_{mec,1} - \Delta E_{th}. \qquad (8\text{-}38)$$

Let subscript 1 correspond to the initial state of the sliding package, and subscript 2 correspond to the state in which the package is momentarily stopped and the spring is compressed by dis-

Fig. 8-15 Sample Problem 8-7. A package slides across a frictionless floor with velocity $\vec{v}_1$ toward a spring of spring constant k. When the package reaches the spring, a frictional force from the floor acts on it.

tance d. For both states the mechanical energy of the system is the sum of the package's kinetic energy ($K = \frac{1}{2}mv^2$) and the spring's potential energy ($U = \frac{1}{2}kx^2$). For state 1, $U = 0$ (because the spring is not compressed), and the package's speed is v_1. Thus, we have

$$E_{mec,1} = K_1 + U_1 = \tfrac{1}{2}mv_1^2 + 0.$$

For state 2, $K = 0$ (because the package is stopped), and the compression distance is d. Therefore, we have

$$E_{mec,2} = K_2 + U_2 = 0 + \tfrac{1}{2}kd^2.$$

Finally, by Eq. 8-29, we can substitute $f_k d$ for the change ΔE_{th} in the thermal energy of the package and the floor. We can now rewrite Eq. 8-38 as

$$\tfrac{1}{2}kd^2 = \tfrac{1}{2}mv_1^2 - f_k d.$$

Rearranging and substituting known data give us

$$5000d^2 + 15d - 16 = 0.$$

Solving this quadratic equation yields

$$d = 0.055 \text{ m} = 5.5 \text{ cm}. \qquad \text{(Answer)}$$

Sample Problem 8-8

In Fig. 8-16, a circus beagle of mass $m = 6.0$ kg runs onto the left end of a curved ramp with speed $v_0 = 7.8$ m/s at height $y_0 = 8.5$ m above the floor. It then slides to the right and comes to a momentary stop when it reaches a height $y = 11.1$ m from the floor. The ramp is not frictionless. What is the increase ΔE_{th} in the thermal energy of the beagle and ramp because of the sliding?

SOLUTION: A Key Idea to get us started is to examine all the forces on the beagle, and then see if we have an isolated system or a system on which an external force is doing work.

Forces: The normal force on the beagle from the ramp does no work on the beagle, because its direction is always perpendicular to that of the beagle's displacement. The gravitational force on the beagle does do work as the beagle's elevation changes. Because there is friction between the beagle and the ramp, the sliding increases their thermal energy.

System: The beagle–ramp–Earth system includes all these forces and energy transfers in one isolated system. Then a second Key Idea is that, because the system is isolated, its total energy cannot change. We can apply the law of conservation of energy in the form of Eq. 8-34 to this system:

$$\Delta E_{mec} + \Delta E_{th} = 0, \qquad (8\text{-}39)$$

Fig. 8-16 Sample Problem 8-8. A beagle slides along a curved ramp, starting with speed v_0 at height y_0, and reaching a height y at which it momentarily stops.

where the energy changes occur between the initial state and the state when the beagle stops momentarily. Also, the change ΔE_{mec} is the sum of the change ΔK in the kinetic energy of the beagle and the change ΔU in the gravitational potential energy of the system, where

$$\Delta K = 0 - \tfrac{1}{2}mv_0^2$$

and

$$\Delta U = mgy - mgy_0.$$

Substituting these expressions into Eq. 8-39 and solving for ΔE_{th} yield

$$\begin{aligned}
\Delta E_{th} &= \tfrac{1}{2}mv_0^2 - mg(y - y_0) \\
&= \tfrac{1}{2}(6.0 \text{ kg})(7.8 \text{ m/s})^2 \\
&\quad - (6.0 \text{ kg})(9.8 \text{ m/s}^2)(11.1 \text{ m} - 8.5 \text{ m}) \\
&\approx 30 \text{ J}. \qquad\qquad\qquad\qquad\text{(Answer)}
\end{aligned}$$

REVIEW & SUMMARY

Conservative Forces   A force is a **conservative force** if the net work it does on a particle moving around every closed path, from an initial point and then back to that point, is zero. Equivalently, it is conservative if the net work it does on a particle moving between two points does not depend on the path taken by the particle. The gravitational force and the spring force are conservative forces; the kinetic frictional force is a **nonconservative force.**

Potential Energy   A **potential energy** is energy that is associated with the configuration of a system in which a conservative force acts. When the conservative force does work W on a particle within the system, the change ΔU in the potential energy of the system is

$$\Delta U = -W. \qquad (8\text{-}1)$$

If the particle moves from point x_i to point x_f, the change in the potential energy of the system is

$$\Delta U = -\int_{x_i}^{x_f} F(x)\, dx. \qquad (8\text{-}6)$$

Gravitational Potential Energy   The potential energy associated with a system consisting of Earth and a nearby particle is **gravitational potential energy.** If the particle moves from height y_i to height y_f, the change in the gravitational potential energy of the particle–Earth system is

$$\Delta U = mg(y_f - y_i) = mg\,\Delta y. \qquad (8\text{-}7)$$

If the **reference position** of the particle is set as $y_i = 0$ and the corresponding gravitational potential energy of the system is set as $U_i = 0$, then the gravitational potential energy U when the particle is at any height y is

$$U(y) = mgy. \qquad (8\text{-}9)$$

Elastic Potential Energy   **Elastic potential energy** is the energy associated with the state of compression or extension of an elastic object. For a spring that exerts a spring force $F = -kx$ when its free end has displacement x, the elastic potential energy is

$$U(x) = \tfrac{1}{2}kx^2. \qquad (8\text{-}11)$$

The reference configuration has the spring at its relaxed length, at which $x = 0$ and $U = 0$.

Mechanical Energy   The **mechanical energy** E_{mec} of a system is the sum of its kinetic energy K and its potential energy U:

$$E_{mec} = K + U. \qquad (8\text{-}12)$$

An *isolated system* is one in which no *external force* causes energy changes. If only conservative forces do work within an isolated system, then the mechanical energy E_{mec} of the system cannot change. This **principle of conservation of mechanical energy** is written as

$$K_2 + U_2 = K_1 + U_1, \qquad (8\text{-}17)$$

in which the subscripts refer to different instants during an energy transfer process. This conservation principle can also be written as

$$\Delta E_{mec} = \Delta K + \Delta U = 0. \qquad (8\text{-}18)$$

Potential Energy Curves   If we know the **potential energy function** $U(x)$ for a system in which a one-dimensional force F acts on a particle, we can find the force as

$$F(x) = -\frac{dU(x)}{dx}. \qquad (8\text{-}20)$$

If $U(x)$ is given on a graph, then at any value of x, the force F is the negative of the slope of the curve there and the kinetic energy of the particle is given by

$$K(x) = E_{mec} - U(x), \qquad (8\text{-}22)$$

where E_{mec} is the mechanical energy of the system. A **turning point** is a point x at which the particle reverses its motion (there, $K = 0$). The particle is in **equilibrium** at points where the slope of the $U(x)$ curve is zero (there, $F(x) = 0$).

Work Done on a System by an External Force   Work W is energy transferred to or from a system by means of an external force acting on the system. When more than one force acts on a system, their *net work* is the transferred energy. When friction is not involved, the work done on the system and the change ΔE_{mec} in the mechanical energy of the system are equal:

$$W = \Delta E_{mec} = \Delta K + \Delta U. \qquad (8\text{-}24,\ 8\text{-}23)$$

When a kinetic frictional force acts within the system, then the thermal energy E_{th} of the system changes. (This energy is associated with the random motion of atoms and molecules in the system.) The work done on the system is then

$$W = \Delta E_{mec} + \Delta E_{th}. \qquad (8\text{-}31)$$

The change ΔE_{th} is related to the magnitude f_k of the frictional force and the magnitude d of the displacement caused by the external force by

$$\Delta E_{th} = f_k d. \qquad (8\text{-}29)$$

Conservation of Energy   The total energy E of a system (the sum of its mechanical energy and its internal energies, including thermal energy) can change only by amounts of energy that are transferred to or from the system. This experimental fact is known as the **law of conservation of energy.** If work W is done on the system, then

$$W = \Delta E = \Delta E_{mec} + \Delta E_{th} + \Delta E_{int}. \qquad (8\text{-}33)$$

If the system is isolated ($W = 0$), this gives

$$\Delta E_{mec} + \Delta E_{th} + \Delta E_{int} = 0 \qquad (8\text{-}34)$$

and

$$E_{mec,2} = E_{mec,1} - \Delta E_{th} - \Delta E_{int}, \qquad (8\text{-}35)$$

where the subscripts 1 and 2 refer to two different instants.

Power   The **power** due to a force is the *rate* at which that force transfers energy. If an amount of energy ΔE is transferred by a force in an amount of time Δt, the **average power** of the force is

$$P_{avg} = \frac{\Delta E}{\Delta t}. \qquad (8\text{-}36)$$

The **instantaneous power** due to a force is

$$P = \frac{dE}{dt}. \qquad (8\text{-}37)$$

QUESTIONS

1. Figure 8-17 shows one direct path and four indirect paths from point i to point f. Along the direct path and three of the indirect paths, only a conservative force F_c acts on a certain object. Along the fourth indirect path, both F_c and a nonconservative force F_{nc} act on the object. The change ΔE_{mec} in the object's mechanical energy (in joules) in going from i to f is indicated along each straight line segment of the indirect paths. (a) What is ΔE_{mec} in moving from i to f along the direct path? (b) What is ΔE_{mec} due to F_{nc} along the one path where it acts?

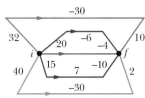

Fig. 8-17 Question 1.

2. A spring is initially stretched by 3.0 cm from its relaxed length. Here are four choices for changing the initial stretch: (a) to a stretch of 2.0 cm, (b) to a compression of 2.0 cm, (c) to a compression of 4.0 cm, and (d) to a stretch of 4.0 cm. Rank the choices according to the changes they make in the elastic potential energy of the spring, most positive first, most negative last.

3. A coconut is thrown from a cliff edge toward a wide, flat valley, with initial speed $v_0 = 8$ m/s. Rank the following choices for the launch direction according to (a) the initial kinetic energy of the coconut and (b) its kinetic energy just before hitting the valley bottom, greatest first: (1) $\vec{v}_0$ almost vertically upward, (2) $\vec{v}_0$ angled upward by 45°, (3) $\vec{v}_0$ horizontal, (4) $\vec{v}_0$ angled downward by 45°, and (5) $\vec{v}_0$ almost vertically downward.

4. In Fig. 8-18, a brave skater slides down three slopes of frictionless ice whose vertical heights d are identical. Rank the slopes according to (a) the work done on the skater by the gravitational force during the descent on each slope and (b) the change in her kinetic energy produced along the slope, greatest first.

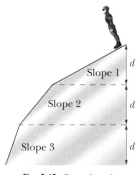

Fig. 8-18 Question 4.

5. In Fig. 8-19, a small, initially stationary block is released on a frictionless ramp at a height of 3.0 m. Hill heights along the ramp are as shown. The hills have identical circular tops (assume that the block does not fly off any hill). (a) Which hill is the first the block cannot cross? (b) What does it do after failing to cross that hill? On which hilltop is (c) the centripetal acceleration of the block greatest and (d) the normal force on the block least?

Fig. 8-19 Question 5.

6. Figure 8-20 gives the potential energy function of a particle. (a) Rank regions AB, BC, CD, and DE according to the magnitude of the force on the particle, greatest first. What value must the mechanical energy E_{mec} of the particle not exceed if the particle is to be (b) trapped in the potential well at the left, (c) trapped in the

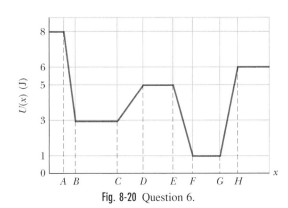

Fig. 8-20 Question 6.

potential well at the right, and (d) able to move between the two potential wells but not to the right of point H? For the situation of (d), in which of regions BC, DE, and FG will the particle have (e) the greatest kinetic energy and (f) the least speed?

7. In Fig. 8-21, a block is released from rest on a track with an initial gravitational potential energy U_i. The curved portions on the track are frictionless, but the horizontal portion, of length L, produces a frictional force f on the block. (a) How much energy is transferred to thermal energy if the block passes once through length L? How many times does the block pass through that length if the initial potential energy U_i is equal to (b) $0.50fL$, (c) $1.25fL$, and (d) $2.25fL$? (e) For each of those three situations, does the block come to a stop at the center of the horizontal portion, to the left of center, or to the right of center? (Warm-up for Problem 63)

Fig. 8-21 Question 7.

8. In Fig. 8-22, a block slides along a track that descends through distance h. The track is frictionless except for the lower section. There the block slides to a stop in a certain distance D because of friction. (a) If we decrease h, will the block now slide to a stop in a distance that is greater than, less than, or equal to D? (b) If, instead, we increase the mass of the block, will the stopping distance now be greater than, less than, or equal to D?

Fig. 8-22 Question 8.

9. In Fig. 8-23, a block slides from A to C along a frictionless ramp, and then it passes through horizontal region CD, where a frictional force acts on it. Is the block's kinetic energy increasing, decreasing, or constant in (a) region AB, (b) region BC, and (c) region CD? (d) Is the block's mechanical energy increasing, decreasing, or constant in those regions?

Fig. 8-23 Question 9.

EXERCISES & PROBLEMS

SEC. 8-3   Determining Potential Energy Values

1E. What is the spring constant of a spring that stores 25 J of elastic potential energy when compressed by 7.5 cm from its relaxed length?   ssm

2E. You drop a 2.00 kg textbook to a friend who stands on the ground 10.0 m below the textbook with outstretched hands 1.50 m above the ground (Fig. 8-24). (a) How much work W_g is done on the textbook by the gravitational force as it drops to your friend's hands? (b) What is the change ΔU in the gravitational potential energy of the textbook–Earth system during the drop? If the gravitational potential energy U of that system is taken to be zero at ground level, what is U

Fig. 8-24 Exercises 2 and 10.

when the textbook (c) is released and (d) reaches the hands? Now take U to be 100 J at ground level and again find (e) W_g, (f) ΔU, (g) U at the release point, and (h) U at the hands.

3E. In Fig. 8-25, a 2.00 g ice flake is released from the edge of a hemispherical bowl whose radius r is 22.0 cm. The flake–bowl contact is frictionless. (a) How much work is done on the flake by the gravitational force during the flake's descent to the bottom of the bowl? (b) What is the change in the potential energy of the flake–Earth system during that descent? (c) If that potential energy is taken to be zero at the bottom of the bowl, what is its

Fig. 8-25 Exercises 3 and 9.

value when the flake is released? (d) If, instead, the potential energy is taken to be zero at the release point, what is its value when the flake reaches the bottom of the bowl? (e) If the mass of the flake were doubled, would the magnitudes of the answers to (a) through (d) increase, decrease, or remain the same?   ssm

4E. In Fig. 8-26, a frictionless roller coaster of mass m tops the first hill with speed v_0. How much work does the gravitational force do on it from that point to (a) point A, (b) point B, and (c) point C? If the gravitational potential energy of the coaster–Earth system is taken to be zero at point C, what is its value when the coaster is at (d) point B and (e) point A? (f) If mass m were doubled, would the change in the gravitational potential energy of the system between points A and B increase, decrease, or remain the same?

Fig. 8-26 Exercises 4 and 12.

5E. Figure 8-27 shows a ball with mass m attached to the end of a thin rod with length L and negligible mass. The other end of the rod is pivoted so that the ball can move in a vertical circle. The rod is held in the horizontal position as shown and then given enough of a downward push to cause the ball to swing down and around and just reach the vertically upward position, with zero speed there. How much work is done on the ball by the gravitational force from the initial point to (a) the lowest point, (b) the highest point, and (c) the point on the right at which the ball is level with the initial point? If the gravitational potential energy of the ball–Earth system is taken to be zero at the initial point, what is its value when the ball reaches (d) the lowest point, (e) the highest point, and (f) the point on the right that is level with the initial point? (g) Suppose the rod were pushed harder so that the ball passed through the highest point with a nonzero speed. Would the change in the gravitational potential energy from the lowest point to the highest point then be greater, less, or the same? ssm

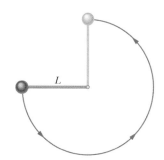

Fig. 8-27 Exercises 5 and 11.

6P. In Fig. 8-28, a small block of mass m can slide along the frictionless loop-the-loop. The block is released from rest at point P, at height $h = 5R$ above the bottom of the loop. How much work does the gravitational force do on the block as the block travels from point P to (a) point Q and (b) the top of the loop? If the gravitational potential energy of the block–Earth system is taken to be zero at the bottom of the loop, what is that potential energy when the block is (c) at point P, (d) at point Q, and (e) at the top of the loop? (f) If, instead of being released, the block is given some initial speed downward along the track, do the answers to (a) through (e) increase, decrease, or remain the same?

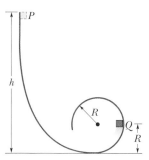

Fig. 8-28 Problems 6 and 20.

7P. A 1.50 kg snowball is fired from a cliff 12.5 m high with an initial velocity of 14.0 m/s, directed 41.0° above the horizontal.

(a) How much work is done on the snowball by the gravitational force during its flight to the flat ground below the cliff? (b) What is the change in the gravitational potential energy of the snowball–Earth system during the flight? (c) If that gravitational potential energy is taken to be zero at the height of the cliff, what is its value when the snowball reaches the ground? ssm

8P. Figure 8-29 shows a thin rod, of length L and negligible mass, that can pivot about one end to rotate in a vertical circle. A heavy ball of mass m is attached to the other end. The rod is pulled aside through an angle θ and released. As the ball descends to its lowest point, (a) how much work does the gravitational force do on it and (b) what is the change in the gravitational potential energy of the ball–Earth system? (c) If the gravitational potential energy is taken to be zero at the lowest point, what is its value just as the ball is released? (d) Do the magnitudes of the answers to (a) through (c) increase, decrease, or remain the same if angle θ is increased?

Fig. 8-29 Problems 8 and 14.

SEC. 8-4 Conservation of Mechanical Energy

9E. (a) In Exercise 3, what is the speed of the flake when it reaches the bottom of the bowl? (b) If we substituted a second flake with twice the mass, what would its speed be? (c) If, instead, we gave the flake an initial downward speed along the bowl, would the answer to (a) increase, decrease, or remain the same? ssm www

10E. (a) In Exercise 2, what is the speed of the textbook when it reaches the hands? (b) If we substituted a second textbook with twice the mass, what would its speed be? (c) If, instead, the textbook were thrown down, would the answer to (a) increase, decrease, or remain the same?

11E. (a) In Exercise 5, what initial speed must be given the ball so that it reaches the vertically upward position with zero speed? What then is its speed at (b) the lowest point and (c) the point on the right at which the ball is level with the initial point? (d) If the ball's mass were doubled, would the answers to (a) through (c) increase, decrease, or remain the same? ssm

12E. In Exercise 4, what is the speed of the coaster at (a) point A, (b) point B, and (c) point C? (d) How high will it go on the last hill, which is too high for it to cross? (e) If we substitute a second coaster with twice the mass, what then are the answers to (a) through (d)?

13E. In Fig. 8-30, a runaway truck with failed brakes is moving downgrade at 130 km/h just before the driver steers the truck up a

Fig. 8-30 Exercise 13.

frictionless emergency escape ramp with an inclination of 15°. The truck's mass is 5000 kg. (a) What minimum length L must the ramp have if the truck is to stop (momentarily) along it? (Assume the truck is a particle, and justify that assumption.) Does the minimum length L increase, decrease, or remain the same if (b) the truck's mass is decreased and (c) its speed is decreased? ssm

14P. (a) In Problem 8, what is the speed of the ball at the lowest point if $L = 2.00$ m, $\theta = 30.0°$, and $m = 5.00$ kg? (b) Does the speed increase, decrease, or remain the same if the mass is increased?

15P. (a) In Problem 7, using energy techniques rather than the techniques of Chapter 4, find the speed of the snowball as it reaches the ground below the cliff. What is that speed (b) if the launch angle is changed to 41.0° *below* the horizontal and (c) if the mass is changed to 2.50 kg? ssm

16P. Figure 8-31 shows an 8.00 kg stone at rest on a spring. The spring is compressed 10.0 cm by the stone. (a) What is the spring constant? (b) The stone is pushed down an additional 30.0 cm and released. What is the elastic potential energy of the compressed spring just before that release?
(c) What is the change in the gravitational potential energy of the stone–Earth system when the stone moves from the release point to its maximum height? (d) What is that maximum height, measured from the release point?

Fig. 8-31 Problem 16.

17P. A 5.0 g marble is fired vertically upward using a spring gun. The spring must be compressed 8.0 cm if the marble is to just reach a target 20 m above the marble's position on the compressed spring. (a) What is the change ΔU_g in the gravitational potential energy of the marble–Earth system during the 20 m ascent? (b) What is the change ΔU_s in the elastic potential energy of the spring during its launch of the marble? (c) What is the spring constant of the spring? ssm www

18P. Figure 8-32 shows a pendulum of length L. Its bob (which effectively has all the mass) has speed v_0 when the cord makes an angle θ_0 with the vertical. (a) Derive an expression for the speed of the bob when it is in its lowest position. What is the least value that v_0 can have if the pendulum is to swing down and then up (b) to a horizontal position, and (c) to a vertical position with the cord remaining straight? (d) Do the answers to (b) and (c) increase, decrease, or remain the same if θ_0 is increased by a few degrees?

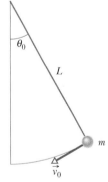

Fig. 8-32 Problem 18.

19P. A 2.00 kg block is placed against a spring on a frictionless 30.0° incline (Fig. 8-33). (The block is not attached to the spring.) The spring, whose spring constant is 19.6 N/cm, is compressed 20.0 cm and then released. (a) What is the elastic potential energy of the compressed spring? (b) What is the change in the gravitational potential energy of the block–Earth system as the block

moves from the release point to its highest point on the incline? (c) How far along the incline is the highest point from the release point? ilw

20P. In Problem 6, what are (a) the horizontal component and (b) the vertical component of the *net* force acting on the block at point Q? (c) At what height h should the block be released from rest so that it is on the verge of losing contact with the track at the top of the loop? (*On the verge of losing contact* means that the normal force on the block from the track has just then become zero.) (d) Graph the magnitude of the normal force on the block at the top of the loop versus initial height h, for the range $h = 0$ to $h = 6R$.

Fig. 8-33 Problem 19.

21P. In Fig. 8-34, a 12 kg block is released from rest on a 30° frictionless incline. Below the block is a spring that can be compressed 2.0 cm by a force of 270 N. The block momentarily stops when it compresses the spring by 5.5 cm. (a) How far does the block move down the incline from its rest position to this stopping point? (b) What is the speed of the block just as it touches the spring? ssm

Fig. 8-34 Problem 21.

22P. At $t = 0$ a 1.0 kg ball is thrown from the top of a tall tower with velocity $\vec{v} = (18$ m/s$)\hat{i} + (24$ m/s$)\hat{j}$. What is the change in the potential energy of the ball–Earth system between $t = 0$ and $t = 6.0$ s?

23P. The string in Fig. 8-35 is $L = 120$ cm long, has a ball attached to one end, and is fixed at its other end. The distance d to the fixed peg at point P is 75.0 cm. When the initially stationary ball is released with the string horizontal as shown, it will swing along the dashed arc. What is its speed when it reaches (a) its lowest point and (b) its highest point after the string catches on the peg? ilw

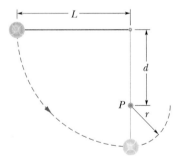

Fig. 8-35 Problems 23 and 29.

24P. A 60 kg skier starts from rest at a height of 20 m above the end of a ski-jump ramp as shown in Fig. 8-36. As the skier leaves the ramp, his velocity makes an angle of 28° with the horizontal.

Fig. 8-36 Problem 24.

Neglect the effects of air resistance and assume the ramp is frictionless. (a) What is the maximum height h of his jump above the end of the ramp? (b) If he increased his weight by putting on a backpack, would h then be greater, less, or the same?

25P. A 2.0 kg block is dropped from a height of 40 cm onto a spring of spring constant $k = 1960$ N/m (Fig. 8-37). Find the maximum distance the spring is compressed. ssm

Fig. 8-37 Problem 25.

26P. Tarzan, who weighs 688 N, swings from a cliff at the end of a convenient vine that is 18 m long (Fig. 8-38). From the top of the cliff to the bottom of the swing, he descends by 3.2 m. The vine will break if the force on it exceeds 950 N. (a) Does the vine break? (b) If no, what is the greatest force on it during the swing? If yes, at what angle with the vertical does it break?

27P. Two children are playing a game in which they try to hit a small box on the floor with a marble fired from a spring-loaded gun that is mounted on a table. The target box is 2.20 m horizontally from the edge of the table; see Fig. 8-39. Bobby compresses the spring 1.10 cm, but the center of the marble falls 27.0 cm short of the center of the box. How far should Rhoda compress the spring to score a direct hit? Assume that neither the spring nor the ball encounters friction in the gun. ssm

Fig. 8-38 Problem 26.

Fig. 8-39 Problem 27

— 2.20 m —

28P. A 700 g block is released from rest at height h_0 above a vertical spring with spring constant $k = 400$ N/m and negligible mass. The block sticks to the spring and momentarily stops after compressing the spring 19.0 cm. How much work is done (a) by the block on the spring and (b) by the spring on the block? (c) What is the value of h_0? (d) If the block were released from height $2h_0$ above the spring, what would be the maximum compression of the spring?

29P. In Fig. 8-35 show that, if the ball is to swing completely around the fixed peg, then $d > 3L/5$. (Hint: The ball must still be moving at the top of its swing. Do you see why?) ssm www

30P. To make a pendulum, a 300 g ball is attached to one end of a string that has a length of 1.4 m and negligible mass. (The other end of the string is fixed.) The ball is pulled to one side until the string makes an angle of 30.0° with the vertical; then (with the string taut) the ball is released from rest. Find (a) the speed of the ball when the string makes an angle of 20.0° with the vertical and (b) the maximum speed of the ball. (c) What is the angle between the string and the vertical when the speed of the ball is one-third its maximum value?

31P. A rigid rod of length L and negligible mass has a ball with mass m attached to one end and its other end fixed, to form a pendulum. The pendulum is inverted, with the rod straight up, and then released. At the lowest point, what are (a) the ball's speed and (b) the tension in the rod? (c) The pendulum is next released at rest from a horizontal position. At what angle from the vertical does the tension in the rod equal the weight of the ball? ssm

32P. In Fig. 8-40, a spring with spring constant $k = 170$ N/m is at the top of a 37.0° frictionless incline. The lower end of the incline is 1.00 m from the end of the spring, which is at its relaxed length. A 2.00 kg canister is pushed against the spring until the spring is compressed 0.200 m and released from rest. (a) What is the speed of the canister at the instant the spring returns to its relaxed length (which is when the canister loses contact with the spring)? (b) What is the speed of the canister when it reaches the lower end of the incline?

Fig. 8-40 Problem 32.

33P*. In Fig. 8-41, a chain is held on a frictionless table with one-fourth of its length hanging over the edge. If the chain has length L and mass m, how much work is required to pull the hanging part back onto the table? ssm

Fig. 8-41 Problem 33.

34P. A spring with spring constant $k = 400$ N/m is placed in a vertical orientation with its lower end supported by a horizontal surface. The upper end is depressed 25.0 cm, and a block with a weight of 40.0 N is placed (unattached) on the depressed spring. The system is then released from rest. Assume the gravitational potential energy U_g of the block is zero at the release point ($y = 0$) and calculate the gravitational potential energy, the elastic potential energy U_e, and the kinetic energy K of the block for y equal to (a) 0, (b) 5.00 cm, (c) 10.0 cm, (d) 15.0 cm, (e) 20.0 cm, (f) 25.0 cm, and (g) 30.0 cm. Also, (h) how far above its point of release does the block rise?

35P*. A boy is seated on the top of a hemispherical mound of ice (Fig. 8-42). He is given a very small push and starts sliding down the ice. Show that he leaves the ice at a point whose height is $2R/3$ if the ice is frictionless. (Hint: The normal force vanishes as he leaves the ice.) ssm

Fig. 8-42 Problem 35.

SEC. 8-5   Reading a Potential Energy Curve

36E. A conservative force $F(x)$ acts on a 2.0 kg particle that moves along the x axis. The potential energy $U(x)$ associated with $F(x)$ is graphed in Fig. 8-43. When the particle is at $x = 2.0$ m, its velocity is -1.5 m/s. (a) What are the magnitude and direction of $F(x)$ at this position? (b) Between what limits of x does the particle move? (c) What is its speed at $x = 7.0$ m?

Fig. 8-43   Exercise 36.

37P. The potential energy of a diatomic molecule (a two-atom system like H_2 or O_2) is given by

$$U = \frac{A}{r^{12}} - \frac{B}{r^6},$$

where r is the separation of the two atoms of the molecule and A and B are positive constants. This potential energy is associated with the force that binds the two atoms together. (a) Find the *equilibrium separation,* that is, the distance between the atoms at which the force on each atom is zero. Is the force repulsive (the atoms are pushed apart) or attractive (they are pulled together) if their separation is (b) smaller and (c) larger than the equilibrium separation?   ssm

38P. A single conservative force $F(x)$ acts on a 1.0 kg particle that moves along an x axis. The potential energy $U(x)$ associated with $F(x)$ is given by

$$U(x) = -4x\, e^{-x/4}\ \text{J},$$

where x is in meters. At $x = 5.0$ m the particle has a kinetic energy of 2.0 J. (a) What is the mechanical energy of the system? (b) Make a plot of $U(x)$ as a function of x for $0 \le x \le 10$ m, and on the same graph draw the line that represents the mechanical energy of the system. Use part (b) to determine (c) the least value of x and (d) the greatest value of x between which the particle can move. Use part (b) to determine (e) the maximum kinetic energy of the particle and (f) the value of x at which it occurs. (g) Determine the equation for $F(x)$ as a function of x. (h) For what (finite) value of x does $F(x) = 0$?

SEC. 8-6   Work Done on a System by an External Force

39E. A collie drags its bed box across a floor by applying a horizontal force of 8.0 N. The kinetic frictional force acting on the box has magnitude 5.0 N. As the box is dragged through 0.70 m along the way, what are (a) the work done by the collie's applied force and (b) the increase in thermal energy of the bed and floor?

40E. The temperature of a plastic cube is monitored while the cube is pushed 3.0 m across a floor at constant speed by a horizontal force of 15 N. The monitoring reveals that the thermal energy of the cube increases by 20 J. What is the increase in the thermal energy of the floor along which the cube slides?

41P. A 3.57 kg block is drawn at constant speed 4.06 m along a horizontal floor by a rope. The force on the block from the rope has a magnitude of 7.68 N and is directed 15.0° above the horizontal. What are (a) the work done by the rope's force, (b) the increase in thermal energy of the block–floor system, and (c) the coefficient of kinetic friction between the block and floor?   ssm

42P. A worker pushed a 27 kg block 9.2 m along a level floor at constant speed with a force directed 32° below the horizontal. If the coefficient of kinetic friction between block and floor was 0.20, what were (a) the work done by the worker's force and (b) the increase in thermal energy of the block–floor system?

SEC. 8-7   Conservation of Energy

43E. A 25 kg bear slides, from rest, 12 m down a lodgepole pine tree, moving with a speed of 5.6 m/s just before hitting the ground. (a) What change occurs in the gravitational potential energy of the bear–Earth system during the slide? (b) What is the kinetic energy of the bear just before hitting the ground? (c) What is the average frictional force that acts on the sliding bear?   ssm   ilw

44E. A 30 g bullet, with a horizontal velocity of 500 m/s, comes to a stop 12 cm within a solid wall. (a) What is the change in its mechanical energy? (b) What is the magnitude of the average force from the wall stopping it?

45E. A 60 kg skier leaves the end of a ski-jump ramp with a velocity of 24 m/s directed 25° above the horizontal. Suppose that as a result of air drag the skier returns to the ground with a speed of 22 m/s, landing 14 m vertically below the end of the ramp. From the launch to the return to the ground, by how much is the mechanical energy of the skier–Earth system reduced because of air drag?

46E. A 75 g Frisbee is thrown from a point 1.1 m above the ground with a speed of 12 m/s. When it has reached a height of 2.1 m, its speed is 10.5 m/s. What was the reduction in the mechanical energy of the Frisbee–Earth system because of air drag?

47E. An outfielder throws a baseball with an initial speed of 81.8 mi/h. Just before an infielder catches the ball at the same level, the ball's speed is 110 ft/s. In foot-pounds, by how much is the mechanical energy of the ball–Earth system reduced because of air drag? (The weight of a baseball is 9.0 oz.)

48E. Approximately 5.5×10^6 kg of water fall 50 m over Niagara Falls each second. (a) What is the decrease in the gravitational potential energy of the water–Earth system each second? (b) If all this energy could be converted to electrical energy (it cannot be), at what rate would electrical energy be supplied? (The mass of 1 m^3 of water is 1000 kg.) (c) If the electrical energy were sold at 1 cent/kW·h, what would be the yearly cost?

49E. During a rockslide, a 520 kg rock slides from rest down a hillside that is 500 m long and 300 m high. The coefficient of kinetic friction between the rock and the hill surface is 0.25. (a) If the gravitational potential energy U of the rock–Earth system is zero at the bottom of the hill, what is the value of U just before the slide? (b) How much energy is transferred to thermal energy during the slide? (c) What is the kinetic energy of the rock as it reaches the bottom of the hill? (d) What is its speed then?

50P. You push a 2.0 kg block against a horizontal spring, compressing the spring by 15 cm. Then you release the block, and the spring sends it sliding across a tabletop. It stops 75 cm from where you released it. The spring constant is 200 N/m. What is the coefficient of kinetic friction between the block and the table?

51P. As Fig. 8-44 shows, a 3.5 kg block is accelerated by a compressed spring whose spring constant is 640 N/m. After leaving the spring at the spring's relaxed length, the block travels over a horizontal surface, with a coefficient of kinetic friction of 0.25, for a distance of 7.8 m before stopping. (a) What is the increase in the thermal energy of the block–floor system? (b) What is the maximum kinetic energy of the block? (c) Through what distance is the spring compressed before the block begins to move? **ssm** **www**

No friction →|← 7.8 m →|
$(\mu_k = 0.25)$

Fig. 8-44 Problem 51.

52P. In Fig. 8-45, a block is moved down an incline a distance of 5.0 m from point A to point B by a force $\vec{F}$ that is parallel to the incline and has magnitude 2.0 N. The magnitude of the frictional force acting on the block is 10 N. If the kinetic energy of the block increases by 35 J between A and B, how much work is done on the block by the gravitational force as the block moves from A to B?

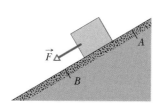

Fig. 8-45 Problem 52.

53P. A certain spring is found *not* to conform to Hooke's law. The force (in newtons) it exerts when stretched a distance x (in meters) is found to have magnitude $52.8x + 38.4x^2$ in the direction opposing the stretch. (a) Compute the work required to stretch the spring from $x = 0.500$ m to $x = 1.00$ m. (b) With one end of the spring fixed, a particle of mass 2.17 kg is attached to the other end of the spring when it is extended by an amount $x = 1.00$ m. If the particle is then released from rest, what is its speed at the instant the spring has returned to the configuration in which the extension is $x = 0.500$ m? (c) Is the force exerted by the spring conservative or nonconservative? Explain. **ssm**

54P. A 4.0 kg bundle starts up a 30° incline with 128 J of kinetic energy. How far will it slide up the incline if the coefficient of kinetic friction between bundle and incline is 0.30?

55P. Two snowy peaks are 850 m and 750 m above the valley between them. A ski run extends down from the top of the higher peak and then back up to the top of the lower one, with a total length of 3.2 km and an average slope of 30° (Fig. 8-46). (a) A

skier starts from rest at the top of the higher peak. At what speed will he arrive at the top of the lower peak if he coasts without using ski poles? Ignore friction. (b) Approximately what coefficient of kinetic friction between snow and skis would make him stop just at the top of the lower peak? **ssm**

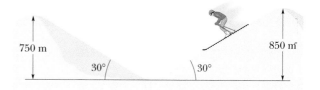

750 m   850 m

30°   30°

Fig. 8-46 Problem 55.

56P. A girl whose weight is 267 N slides down a 6.1 m playground slide that makes an angle of 20° with the horizontal. The coefficient of kinetic friction between slide and child is 0.10. (a) How much energy is transferred to thermal energy? (b) If the girl starts at the top with a speed of 0.457 m/s, what is her speed at the bottom?

57P. In Fig. 8-47, a 2.5 kg block slides head on into a spring with a spring constant of 320 N/m. When the block stops, it has compressed the spring by 7.5 cm. The coefficient of kinetic friction between the block and the horizontal surface is 0.25. While the

2.5 kg   320 N/m

Fig. 8-47 Problem 57.

block is in contact with the spring and being brought to rest, what are (a) the work done by the spring force and (b) the increase in thermal energy of the block–floor system? (c) What is the block's speed just as the block reaches the spring? **ilw**

58P. A factory worker accidentally releases a 180 kg crate that was being held at rest at the top of a 3.7 m-long-ramp inclined at 39° to the horizontal. The coefficient of kinetic friction between the crate and the ramp, and between the crate and the horizontal factory floor, is 0.28. (a) How fast is the crate moving as it reaches the bottom of the ramp? (b) How far will it subsequently slide across the factory floor? (Assume that the crate's kinetic energy does not change as it moves from the ramp onto the floor.) (c) Do the answers to (a) and (b) increase, decrease, or remain the same if we halve the mass of the crate?

59P. In Fig. 8-48, a block slides along a track from one level to a higher level, by moving through an intermediate valley. The track is frictionless until the block reaches the higher level. There a frictional force stops the block in a distance d. The block's initial speed v_0 is 6.0 m/s; the height difference h is 1.1 m; and the coefficient of kinetic friction μ is 0.60. Find d.

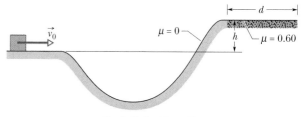

$\vec{v_0}$ $\mu = 0$ d

h $\mu = 0.60$

Fig. 8-48 Problem 59.

60P. A cookie jar is moving up a 40° incline. At a point 55 cm from the bottom of the incline (measured along the incline), it has a speed of 1.4 m/s. The coefficient of kinetic friction between jar and incline is 0.15. (a) How much farther up the incline will the jar move? (b) How fast will it be going when it has slid back to the bottom of the incline? (c) Do the answers to (a) and (b) increase, decrease, or remain the same if we decrease the coefficient of kinetic friction (but do not change the given speed or location)?

61P. A stone with weight w is thrown vertically upward into the air from ground level with initial speed v_0. If a constant force f due to air drag acts on the stone throughout its flight, (a) show that the maximum height reached by the stone is

$$h = \frac{v_0^2}{2g(1 + f/w)}.$$

(b) Show that the stone's speed is

$$v = v_0 \left(\frac{w - f}{w + f}\right)^{1/2}$$

just before impact with the ground. ssm

62P. A playground slide is in the form of an arc of a circle with a maximum height of 4.0 m, with a radius of 12 m, and with the ground tangent to the circle (Fig. 8-49). A 25 kg child starts from rest at the top of the slide and has a speed of 6.2 m/s at the bottom. (a) What is the length of the slide? (b) What average frictional force acts on the child over this distance? If, instead of the ground, a vertical line through the *top of the slide* is tangent to the circle, what are (c) the length of the slide and (d) the average frictional force on the child?

4.0 m

Fig. 8-49 Problem 62.

63P. A particle can slide along a track with elevated ends and a flat central part, as shown in Fig. 8-50. The flat part has length L. The curved portions of the track are frictionless, but for the flat part the coefficient of kinetic friction is $\mu_k = 0.20$. The particle is released from rest at point A, which is a height $h = L/2$ above the flat part of the track. Where does the particle finally stop?

Fig. 8-50 Problem 63.

64P. The cable of the 1800 kg elevator cab in Fig. 8-51 snaps when the cab is at rest at the first floor, where the cab bottom is a distance $d = 3.7$ m above a cushioning spring whose spring constant is $k = 0.15$ MN/m. A safety device clamps the cab against guide rails so that a constant frictional force of 4.4 kN opposes the cab's mo-

tion. (a) Find the speed of the cab just before it hits the spring. (b) Find the maximum distance x that the spring is compressed (the frictional force still acts during this compression). (c) Find the distance that the cab will bounce back up the shaft. (d) Using conservation of energy, find the approximate total distance that the cab will move before coming to rest. (Assume that the frictional force on the cab is negligible when the cab is stationary.)

Fig. 8-51 Problem 64.

65P. At a certain factory, 300 kg crates are dropped vertically from a packing machine onto a conveyor belt moving at 1.20 m/s (Fig. 8-52). (A motor maintains the belt's constant speed.) The coefficient of kinetic friction between the belt and each crate is 0.400. After a short time, slipping between the belt and the crate ceases, and the crate then moves along with the belt. For the period of time during which the crate is being brought to rest relative to the belt, calculate, for a coordinate system at rest in the factory, (a) the kinetic energy supplied to the crate, (b) the magnitude of the kinetic frictional force acting on the crate, and (c) the energy supplied by the motor. (d) Explain why the answers to (a) and (c) are different.

Fig. 8-52 Problem 65.

Additional Problems

66. The maximum force that you can exert on an object with one of your back teeth is about 750 N. Suppose that as you gradually bite on a clump of elastic licorice, the licorice resists its compression by one of the teeth with a spring-like force of spring constant 2.5×10^5 N/m. Find (a) the distance the licorice is compressed by your tooth and (b) the work your tooth does on the licorice during the compression. (c) Plot the magnitude of your force versus the compression distance. (d) If there is a potential energy associated with this compression, plot it versus compression distance.

In the 1990s the pelvis of a particular *Triceratops* dinosaur was found to have deep bite marks. The shape of the marks suggested that they were made by a *Tyrannosaurus rex* dinosaur. To test the idea, researchers made a replica of a *T. rex* tooth from bronze and aluminum and then used a hydraulic press to gradually drive the replica into cow bone to the depth seen in the *Triceratops* bone. A graph of the force required for the penetration versus the depth of penetration is given in Fig. 8-53 for one trial; the required force increased with depth because, as the nearly conical tooth penetrated the bone, more of the tooth came in contact with the bone. (e) How much work was done by the hydraulic press—and thus, presumably, by the *T. rex*—in such a penetration? (f) Is there a potential energy associated with this penetration? The large biting force and energy expenditure attributed to the *T. rex* by this research (and unsurpassed by modern animals) suggest that the animal was a predator and not a scavenger (as had been argued by some researchers).

Fig. 8-53 Problem 66.

67. *Fly-Fishing and Speed Amplification.* If you throw a loose fishing fly, it will travel horizontally only about 1 m. However, if you throw that fly attached to fishing line by casting the line with a rod, the fly will easily travel horizontally to the full length of the line, say, 20 m.

The cast is depicted in Fig. 8-54: Initially (Fig. 8-54a) the line of length L is extended horizontally leftward and moving rightward with a speed v_0. As the fly at the end of the line moves forward, the line doubles over, with the upper section still moving and the lower section stationary (Fig. 8-54b). The upper section decreases in length as the lower section increases in length (Fig. 8-54c), until the line is extended horizontally rightward and there is only a lower section (Fig. 8-54d). If air drag is neglected, the initial kinetic energy of the line in Fig. 8-54a becomes progressively concentrated in the fly and the decreasing portion of the line that is still moving, resulting in an amplification (increase) in the speed of the fly and that portion.

(a) Using the x axis indicated, show that when the fly position is x, the length of the still-moving (upper) section of line is $(L - x)/2$. (b) Assuming that the line is uniform with a linear density ρ (mass per unit length), what is the mass of the still-moving section? Next, let m_f represent the mass of the fly, and assume that the kinetic energy of the moving section does not change from its initial value (when the moving section had length L and speed v_0) even though the length of the moving section is decreasing during the cast. (c) Find an expression for the speed of the still-moving section and the fly.

Assume that initial speed $v_0 = 6.0$ m/s, line length $L = 20$ m, fly mass $m_f = 0.80$ g, and linear density $\rho = 1.3$ g/m. (d) Plot the fly's speed v versus its position x. (e) What is the fly's speed just as the line approaches its final horizontal orientation and the fly is about to flip over and stop? (In more realistic calculations, air drag reduces this final speed.)

Speed amplification can also be produced with a bullwhip and even a rolled-up, wet towel that is popped against a victim in a common locker-room prank. (Adapted from "The Mechanics of Flycasting: The Flyline," by Graig A. Spolek, *American Journal of Physics,* Sept. 1986, Vol. 54, No. 9, pp. 832–836.)

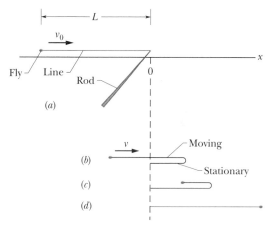

Fig. 8-54 Problem 67.

NEW PROBLEMS

N1. In Fig. 8N-1, a block is released from rest at height d and slides down a frictionless ramp and onto a first plateau, which has length d and where the coefficient of kinetic friction is 0.50. If the block is still moving, it then slides down a second frictionless ramp through a height of $d/2$ and onto a lower plateau, which has length $d/2$ and where the coefficient of kinetic friction is again 0.50. If the block is still moving, it then slides up a frictionless ramp until it (momentarily) stops. Give the stopping point of the block, either as a distance along the first or second plateau (this would be a final stop) or as a height on the ramp at the right (this would be a momentary stop because the block will slide back down the ramp).

Fig. 8N-1 Problem N1.

N2. A large fake cookie sliding on a horizontal surface is attached to one end of a horizontal spring with spring constant $k = 400$ N/m; the other end of the spring is fixed in place. The cookie has a kinetic energy of 20.0 J as it passes through the position where the spring is unstretched. As the cookie slides, a frictional force of magnitude 10.0 N acts on it. (a) How far will the cookie slide from the position where the spring is unstretched before coming momentarily to rest? (b) What will be the kinetic energy of the cookie as it slides back through the position where the spring is unstretched?

N3. The spring in the muzzle of a child's spring gun has a spring constant of 700 N/m. To shoot a ball from the gun, first the spring is compressed and then the ball is placed on it. The gun's trigger then releases the spring, which pushes the ball through the muzzle. The ball leaves the spring just as it leaves the outer end of the muzzle. When the gun is inclined upward by 30° to the horizontal, a 57 g ball is shot to a maximum height of 1.83 m above the gun's muzzle. Assume air drag on the ball is negligible. (a) At what speed does the spring launch the ball? (b) Assuming that friction on the ball within the gun can be neglected, find the initial compression distance of the spring.

N4. A pendulum consists of a 2.0 kg stone swinging on a 4.0 m string of negligible mass. The stone has a speed of 8.0 m/s when it passes its lowest point. (a) What is the speed when the string is at 60° to the vertical? (b) What is the greatest angle with the vertical that the string will reach during the stone's motion? (c) If the potential energy of the pendulum–Earth system is taken to be zero at the stone's lowest point, what is the total mechanical energy of the system?

N5. A spring with a spring constant of 3200 N/m is initially stretched until the elastic potential energy of the spring is 1.44 J. ($U = 0$ for no stretch.) What is the change in the elastic potential energy if the initial stretch is next changed to (a) a stretch of 2.0 cm, (b) a compression of 2.0 cm, and (c) a compression of 4.0 cm?

N6. In Fig. 8N-2a, a block with a kinetic energy of 30 J is about to collide with a spring at its relaxed length. As the block compresses the spring, a frictional force between the block and floor acts on the block. Figure 8N-2b gives the kinetic energy $K(x)$ of the block and the potential energy $U(x)$ of the spring as functions of position x of the block, as the spring is compressed. What is the increase in thermal energy of the block and the floor when (a) the block reaches position $x = 0.10$ m and (b) the spring reaches its maximum compression?

(a)

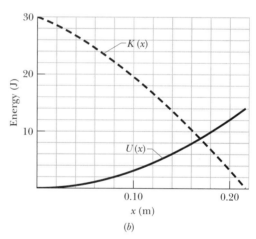

(b)

Fig. 8N-2 Problem N6.

N7. A metal tool is sharpened by being held against the rim of a wheel on a grinding machine by a force of 180 N. The frictional forces between the rim and the tool grind off small pieces of the tool. The wheel has a radius of 20.0 cm and rotates at 2.50 rev/s. The coefficient of kinetic friction between the wheel and the tool is 0.320. At what rate is energy being transferred from the motor driving the wheel to the thermal energy of the wheel and tool and to the kinetic energy of the material thrown from the tool?

N8. A particle can move along only an x axis, where conservative forces act on it in the ranges indicated in Fig. 8N-3 and given in the following table. Because the forces are conservative, a potential energy can be associated with each of them. Assume that the particle is released at $x = 5.00$ m with a kinetic energy of 14.0 J, and take its potential energy U there to be zero. If it is initially moving in the negative direction of the x axis (when it is released), what is its potential energy U and kinetic energy K at (a) $x = 2.00$ m and (b) $x = 0$? If, instead, it is initially moving in the positive direction of the x axis, what is its potential energy U and kinetic energy K at (c) $x = 11.0$ m, (d) $x = 12.0$ m, and (e) $x = 13.0$ m? (f) Plot its potential energy versus x for the range $x = 0$ to $x = 13.0$ m.

Next, let the particle be released from rest at $x = 0$. (g) Can it reach $x = 5.0$ m? If so, what is its kinetic energy there? (h) To what maximum position x_{max} in the positive direction of the x axis can the particle move? (i) What does the particle do after it reaches x_{max}?

Range	Force
0 to 2.00 m	$\vec{F}_1 = +(3.00 \text{ N})\hat{i}$
2.00 m to 3.00 m	$\vec{F}_2 = +(5.00 \text{ N})\hat{i}$
3.00 m to 8.00 m	$F = 0$
8.00 m to 11.0 m	$\vec{F}_3 = -(4.00 \text{ N})\hat{i}$
11.0 m to 12.0 m	$\vec{F}_4 = -(1.00 \text{ N})\hat{i}$
12.0 m to 15.0 m	$F = 0$

Fig. 8N-3 Problems N8 and N9.

N9. For the arrangement of forces in Problem N8 and Fig. 8N-3, suppose that a 2.00 kg particle is released at $x = 5.0$ m with an initial speed of 3.45 m/s. (a) If it is initially headed in the negative direction of the x axis, will it be turned back before it reaches $x = 0$? If so, at what position will it be turned back and if not, what is its speed at $x = 0$? (b) If, instead, it is initially headed in the positive direction of the x axis, will it be turned back before it reaches $x = 13$ m? If so, at what position will it be turned back and if not, what is its speed at $x = 13$ m?

N10. A 3.2 kg sloth hangs 3.0 m above the ground. (a) What is the gravitational potential energy of the sloth–Earth system if we take the reference point $y = 0$ to be at the ground? If the sloth drops to the ground and air drag on it is assumed to be negligible, what are (b) the kinetic energy and (c) the speed of the sloth just before it reaches the ground?

N11. A 2.50 kg beverage can is thrown directly downward from a height of 4.00 m, with an initial speed of 3.00 m/s. The air drag on the can is negligible. What is the kinetic energy of the can (a) as it reaches the ground at the end of its fall and (b) when it is half way to the ground? What are (c) the kinetic energy of the can and (d) the gravitational potential energy of the can–Earth system 0.200 s before the can reaches the ground? For the latter, take the reference point $y = 0$ to be at the ground.

N12. A machine pulls a 40 kg trunk 2.0 m up along a 40° ramp at constant velocity. Its force on the trunk is directed parallel to the ramp. The coefficient of kinetic friction between the trunk and the ramp is 0.40. What are (a) the work done on the trunk by the machine's force and (b) the increase in thermal energy of the trunk and the ramp?

N13. A spring ($k = 200$ N/m) is fixed at the top of a frictionless plane that makes an angle of 40° with the horizontal, as shown in Fig. 8N-4. A 1.0 kg block is projected up the plane, from an initial position that is 0.60 m from the end of the uncompressed spring, with an initial kinetic energy of 16 J. (a) What is the kinetic energy of the block at the instant it has compressed the spring 0.20 m? (b) With what kinetic energy must the block be projected up the plane if it is to stop momentarily when it has compressed the spring by 0.40 m?

Fig. 8N-4 Problem N13.

N14. A conservative force $F(x)$ acts on a particle that moves along the x axis. Figure 8N-5 shows how the potential energy $U(x)$ associated with force $F(x)$ varies with the x position of the particle. (a) Plot $F(x)$ for the range $0 < x < 6$ m. (b) The mechanical energy E of the system is 4.0 J. Plot the kinetic energy $K(x)$ of the particle directly on Fig. 8N-5.

Fig. 8N-5 Problem N14.

N15. If a 70 kg baseball player steals home by sliding into the plate with an initial speed of 10 m/s just as he hits the ground, (a) what is the decrease in the player's kinetic energy and (b) what is the increase in the thermal energy of his body and the ground along which he slides?

N16. A horizontal force of magnitude 35.0 N pushes a block of mass 4.00 kg across a floor where the coefficient of kinetic friction is 0.600. (a) How much work is done by that applied force on the block–floor system when the block slides through a displacement of 3.00 m across the floor? (b) During that displacement, the thermal energy of the block increases by 40.0 J. What is the increase in thermal energy of the floor? (c) What is the increase in the kinetic energy of the block?

N17*. A 3.20 kg block starts at rest and slides a distance d down a frictionless 30.0° incline, where it runs into a spring (Fig. 8N-6). The block slides an additional 21.0 cm before it is brought to rest momentarily by compressing the spring, whose spring constant k is 431 N/m. (a) What is the value of d? (b) What is the distance between the point of first contact and the point where the block's speed is greatest?

Fig. 8N-6 Problem N17.

N18. We move a particle along an x axis, first outward from $x = 1.0$ m to $x = 4.0$ m and then back to $x = 1.0$ m, while an external force acts on it. That force is directed along the x axis, and its x component can have different values for the outward trip and for the return trip. Here are the values (in newtons) for four situations, where x is in meters:

	Outward	Inward
(a)	+3.0	−3.0
(b)	+5.0	+5.0
(c)	+2.0x	−2.0x
(d)	+3.0x^2	+3.0x^2

Find the net work done on the particle by the external force *for the round trip* for each of the four situations. (e) For which, if any, is the external force conservative?

N19. A 60.0 kg circus performer slides 4.00 m down a pole to the circus floor, starting from rest. What is the kinetic energy of the performer as she reaches the floor if the frictional force on her from the pole (a) is negligible (she will be hurt) and (b) has a magnitude of 500 N?

N20. A 0.50 kg banana is thrown directly upward with an initial speed of 4.00 m/s and reaches a maximum height of 0.80 m. What is the change in the mechanical energy of the banana–Earth system during the ascent of the banana to that maximum height, due to air drag on the banana?

N21. A volcanic ash flow is moving across horizontal ground when it encounters a 10° upslope. The front of the flow then travels 920 m on the upslope before stopping. Assume that the gases entrapped in the flow lift the flow and thus make the frictional force from the ground negligible; assume also that mechanical energy of the front of the flow is conserved. What was the initial speed of the front of the flow?

N22. A massless rigid rod of length L has a ball of mass m attached to one end (Fig. 8N-7). The other end is pivoted in such a way that the ball will move in a vertical circle. First, assume that there is no

friction at the pivot. The system is launched downward from the horizontal position A with initial speed v_0. The ball just barely reaches point D and then stops. (a) Derive an expression for v_0 in terms of L, m, and g. (b) What is the tension in the rod when the ball passes through B? (c) A little grit is placed on the pivot to increase the friction there. Then the ball just barely reaches C when launched from A with the same speed as before. What is the decrease in the mechanical energy during this motion? (d) What is the decrease in the mechanical energy by the time the ball finally comes to rest at B after several oscillations?

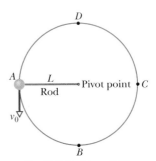

Fig. 8N-7 Problem N22.

N23. Two blocks, of masses M and $2M$ where $M = 2.0$ kg, are connected to a spring of spring constant $k = 200$ N/m that has one end fixed, as shown in Fig. 8N-8. The horizontal surface and the pulley's axle are frictionless, and the pulley has negligible mass. The blocks are released from rest with the spring unstretched. (a) What is the combined kinetic energy of the two blocks when the hanging block has fallen a distance of 0.090 m? (b) What is the kinetic energy of the hanging block when it has fallen that 0.090 m? (c) What maximum distance does the hanging block fall before momentarily stopping?

Fig. 8N-8 Problem N23.

N24. A 9.40 kg projectile is fired vertically upward. Air drag decreases the mechanical energy of the projectile–Earth system by 68.0 kJ during the projectile's ascent. How much higher would the projectile have gone were air drag negligible?

N25. A river descends 15 m through rapids. The speed of the water is 3.2 m/s upon entering the rapids and 13 m/s upon leaving. What percentage of the gravitational potential energy of the water–Earth system is transferred to kinetic energy during the descent? (*Hint:* Consider the descent of, say, 10 kg of water.)

N26. In Fig. 8N-9, a small block is sent through point A with a speed of 7.0 m/s. Its path is without friction until it reaches the section of length $L = 12$ m, where the coefficient of kinetic friction is 0.70. The indicated heights are $h_1 = 6.0$ m and $h_2 = 2.0$ m. What are the speeds of the block at (a) point B and (b) point C? (c) Does the block reach point D? If so, what is its speed there and if not, how far through the section of friction does it travel?

Fig. 8N-9 Problem N26.

N27. The magnitude of the gravitational force between a particle of mass m_1 and one of mass m_2 is given by

$$F(x) = G \frac{m_1 m_2}{x^2},$$

where G is a constant and x is the distance between the particles. (a) What is the corresponding potential energy function $U(x)$? Assume that $U(x) \to 0$ as $x \to \infty$ and that x is positive. (b) How much work is required to increase the separation of the particles from $x = x_1$ to $x = x_1 + d$?

N28. In Fig. 8N-10, a block is sent sliding down a frictionless ramp. Its speeds at points A and B are 2.00 m/s and 2.60 m/s, respectively. Next, it is again sent sliding down the ramp, but this time its speed at point A is 4.00 m/s. What then is its speed at point B?

Fig. 8N-10 Problem N28.

N29. In Fig. 8N-11, a block slides along a path that is without friction until the block reaches the section of length $L = 0.75$ m, which begins at height $h = 2.0$ m. In that section, the coefficient of kinetic friction is 0.40. The block passes through point A with a speed of 8.0 m/s. Does it reach point B (where the section of friction ends)? If so, what is its speed there and if not, what greatest height above point A does it reach?

Fig. 8N-11 Problem N29.

N30. Two blocks are connected by a string, as shown in Fig. 8N-12. They are released from rest. Show that, after they have moved a distance L, their common speed is given by

$$v = \sqrt{\frac{2(m_2 - \mu m_1)gL}{m_1 + m_2}},$$

in which μ is the coefficient of kinetic friction between the upper block and the surface. Assume that the pulley is massless and frictionless.

Fig. 8N-12 Problem N30.

9 Systems of Particles

If you leap forward, chances are that your head and torso will follow a parabolic path, like a baseball thrown in from the outfield. However, when a skilled ballet dancer leaps across the stage in a *grand jeté*, the path taken by her head and torso is nearly horizontal during much of the jump. She seems to be floating across the stage. The audience may not know much about projectile motion, but they still sense that something unusual has happened.

How does the ballerina seemingly "turn off" the gravitational force?

The answer is in this chapter.

(a)

(b)

Fig. 9-1 (a) A ball tossed into the air follows a parabolic path. (b) The center of mass (the black dot) of a baseball bat that is flipped into the air does also, but all other points of the bat follow more complicated curved paths.

9-1 A Special Point

Physicists love to look at something complicated and find in it something simple and familiar. Here is an example. If you flip a baseball bat into the air, its motion as it turns is clearly more complicated than that of, say, a nonspinning tossed ball (Fig. 9-1a), which moves like a particle. Every part of the bat moves in a different way from every other part, so you cannot represent the bat as a tossed particle; instead, it is a system of particles.

However, if you look closely, you will find that one special point of the bat moves in a simple parabolic path, just as a particle would if tossed into the air (Fig. 9-1b). In fact, that special point moves as though (1) the bat's total mass were concentrated there and (2) the gravitational force on the bat acted only there. That special point is said to be the **center of mass** of the bat. In general:

> The center of mass of a body or a system of bodies is the point that moves as though all of the mass were concentrated there and all external forces were applied there.

The center of mass of a baseball bat lies along the bat's central axis. You can locate it by balancing the bat horizontally on an outstretched finger: The center of mass is on the bat's axis just above your finger.

9-2 The Center of Mass

We shall now spend some time determining how to find the center of mass in various systems. We start with a system of a few particles, and then we consider a system of a great many particles (as in a baseball bat).

Systems of Particles

Figure 9-2a shows two particles of masses m_1 and m_2 separated by a distance d. We have arbitrarily chosen the origin of the x axis to coincide with the particle of mass m_1. We *define* the position of the center of mass (com) of this two-particle system to be

$$x_{com} = \frac{m_2}{m_1 + m_2} d. \qquad (9\text{-}1)$$

Suppose, as an example, that $m_2 = 0$. Then there is only one particle, of mass m_1, and the center of mass must lie at the position of that particle; Eq. 9-1 dutifully

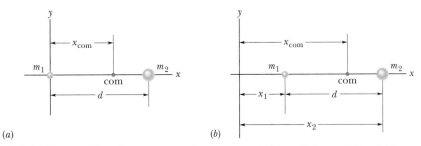

(a)   (b)

Fig. 9-2 (a) Two particles of masses m_1 and m_2 are separated by a distance d. The dot labeled com shows the position of the center of mass, calculated from Eq. 9-1. (b) The same as (a) except that the origin is located farther from the particles. The position of the center of mass is calculated from Eq. 9-2. The location of the center of mass (with respect to the particles) is the same in both cases.

reduces to $x_{com} = 0$. If $m_1 = 0$, there is again only one particle (of mass m_2), and we have, as we expect, $x_{com} = d$. If $m_1 = m_2$, the masses of the particles are equal and the center of mass should be halfway between them; Eq. 9-1 reduces to $x_{com} = \frac{1}{2}d$, again as we expect. Finally, Eq. 9-1 tells us that if neither m_1 nor m_2 is zero, x_{com} can have only values that lie between zero and d; that is, the center of mass must lie somewhere between the two particles.

Figure 9-2b shows a more generalized situation, in which the coordinate system has been shifted leftward. The position of the center of mass is now defined as

$$x_{com} = \frac{m_1 x_1 + m_2 x_2}{m_1 + m_2}. \tag{9-2}$$

Note that if we put $x_1 = 0$, then x_2 becomes d and Eq. 9-2 reduces to Eq. 9-1, as it must. Note also that in spite of the shift of the coordinate system, the center of mass is still the same distance from each particle.

We can rewrite Eq. 9-2 as

$$x_{com} = \frac{m_1 x_1 + m_2 x_2}{M}, \tag{9-3}$$

in which M is the total mass of the system. (Here, $M = m_1 + m_2$.) We can extend this equation to a more general situation in which n particles are strung out along the x axis. Then the total mass is $M = m_1 + m_2 + \cdots + m_n$, and the location of the center of mass is

$$x_{com} = \frac{m_1 x_1 + m_2 x_2 + m_3 x_3 + \cdots + m_n x_n}{M}$$

$$= \frac{1}{M} \sum_{i=1}^{n} m_i x_i. \tag{9-4}$$

Here the subscript i is a running number, or index, that takes on all integer values from 1 to n. It identifies the various particles, their masses, and their x coordinates.

If the particles are distributed in three dimensions, the center of mass must be identified by three coordinates. By extension of Eq. 9-4, they are

$$x_{com} = \frac{1}{M} \sum_{i=1}^{n} m_i x_i, \qquad y_{com} = \frac{1}{M} \sum_{i=1}^{n} m_i y_i, \qquad z_{com} = \frac{1}{M} \sum_{i=1}^{n} m_i z_i. \tag{9-5}$$

We can also define the center of mass with the language of vectors. First recall that the position of a particle at coordinates x_i, y_i, and z_i is given by a position vector:

$$\vec{r}_i = x_i \hat{i} + y_i \hat{j} + z_i \hat{k}. \tag{9-6}$$

Here the index identifies the particle, and $\hat{i}$, $\hat{j}$, and $\hat{k}$ are unit vectors pointing, respectively, in the positive direction of the x, y, and z axes. Similarly, the position of the center of mass of a system of particles is given by a position vector:

$$\vec{r}_{com} = x_{com} \hat{i} + y_{com} \hat{j} + z_{com} \hat{k}. \tag{9-7}$$

The three scalar equations of Eq. 9-5 can now be replaced by a single vector equation,

$$\vec{r}_{com} = \frac{1}{M} \sum_{i=1}^{n} m_i \vec{r}_i, \tag{9-8}$$

where again M is the total mass of the system. You can check that this equation is correct by substituting Eqs. 9-6 and 9-7 into it, and then separating out the x, y, and z components. The scalar relations of Eq. 9-5 result.

Solid Bodies

An ordinary object, such as a baseball bat, contains so many particles (atoms) that we can best treat it as a continuous distribution of matter. The "particles" then become differential mass elements dm, the sums of Eq. 9-5 become integrals, and the coordinates of the center of mass are defined as

$$x_{com} = \frac{1}{M} \int x \, dm, \qquad y_{com} = \frac{1}{M} \int y \, dm, \qquad z_{com} = \frac{1}{M} \int z \, dm, \quad (9\text{-}9)$$

where M is now the mass of the object.

Evaluating these integrals for most common objects (like a television set or a moose) would be difficult, so here we shall consider only *uniform* objects. Such an object has *uniform density*, or mass per unit volume; that is, the density ρ (Greek letter rho) is the same for any given element of the object as for the whole object:

$$\rho = \frac{dm}{dV} = \frac{M}{V}, \qquad (9\text{-}10)$$

where dV is the volume occupied by a mass element dm, and V is the total volume of the object. If we substitute $dm = M/V \, dV$ from Eq. 9-10 into Eq. 9-9, we find that

$$x_{com} = \frac{1}{V} \int x \, dV, \qquad y_{com} = \frac{1}{V} \int y \, dV, \qquad z_{com} = \frac{1}{V} \int z \, dV. \quad (9\text{-}11)$$

You can bypass one or more of these integrals if an object has a point, a line, or a plane of symmetry. The center of mass of such an object then lies at that point, on that line, or in that plane. For example, the center of mass of a uniform sphere (which has a point of symmetry) is at the center of the sphere (which is the point of symmetry). The center of mass of a uniform cone (whose axis is a line of symmetry) lies on the axis of the cone. The center of mass of a banana (which has a plane of symmetry that splits it into two equal parts) lies somewhere in that plane.

The center of mass of an object need not lie within the object. There is no dough at the center of mass of a doughnut, and no iron at the center of mass of a horseshoe.

✔CHECKPOINT 1:   The figure shows a uniform square plate from which four identical squares at the corners will be removed. (a) Where is the center of mass of the plate originally? Where is it after the removal of (b) square 1; (c) squares 1 and 2; (d) squares 1 and 3; (e) squares 1, 2, and 3; (f) all four squares? Answer in terms of quadrants, axes, or points (without calculation, of course).

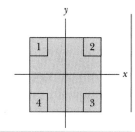

Sample Problem 9-1

Three particles of masses $m_1 = 1.2$ kg, $m_2 = 2.5$ kg, and $m_3 = 3.4$ kg form an equilateral triangle of edge length $a = 140$ cm. Where is the center of mass of this three-particle system?

SOLUTION: A Key Idea to get us started is that we are dealing with particles instead of an extended solid body, so we can use Eq. 9-5 to locate their center of mass. The particles are in the plane of the equilateral triangle, so we need only the first two equations. A second Key Idea is that we can simplify the calculations by choosing the x and y axes so that one of the particles is located at the origin

and the x axis coincides with one of the triangle's sides (Fig. 9-3). The three particles then have the following coordinates:

Particle	Mass (kg)	x (cm)	y (cm)
1	1.2	0	0
2	2.5	140	0
3	3.4	70	121

The total mass M of the system is 7.1 kg.

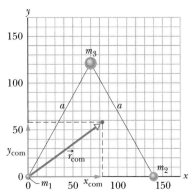

Fig. 9-3 Sample Problem 9-1. Three particles form an equilateral triangle of edge length a. The center of mass is located by the position vector $\vec{r}_{com}$.

From Eq. 9-5, the coordinates of the center of mass are

$$x_{com} = \frac{1}{M} \sum_{i=1}^{3} m_i x_i = \frac{m_1 x_1 + m_2 x_2 + m_3 x_3}{M}$$

$$= \frac{(1.2 \text{ kg})(0) + (2.5 \text{ kg})(140 \text{ cm}) + (3.4 \text{ kg})(70 \text{ cm})}{7.1 \text{ kg}}$$

$$= 83 \text{ cm} \qquad\qquad\qquad\qquad\qquad \text{(Answer)}$$

and $y_{com} = \frac{1}{M} \sum_{i=1}^{3} m_i y_i = \frac{m_1 y_1 + m_2 y_2 + m_3 y_3}{M}$

$$= \frac{(1.2 \text{ kg})(0) + (2.5 \text{ kg})(0) + (3.4 \text{ kg})(121 \text{ cm})}{7.1 \text{ kg}}$$

$$= 58 \text{ cm}. \qquad\qquad\qquad\qquad \text{(Answer)}$$

In Fig. 9-3, the center of mass is located by the position vector $\vec{r}_{com}$, which has components x_{com} and y_{com}.

Sample Problem 9-2

Figure 9-4a shows a uniform metal plate P of radius $2R$ from which a disk of radius R has been stamped out (removed) in an assembly line. Using the xy coordinate system shown, locate the center of mass com$_P$ of the plate.

SOLUTION: First, let us roughly locate the center of plate P by using the Key Idea of symmetry. We note that the plate is symmetric about the x axis (we get the portion below that axis by rotating the upper portion about the axis). Thus, com$_P$ must be on the x axis. The plate (with the disk removed) is not symmetric about the y axis. However, because there is somewhat more mass on the right of the y axis, com$_P$ must be somewhat to the right of that axis. Thus, the location of com$_P$ should be roughly as indicated in Fig. 9-4a.

Another Key Idea here is that plate P is an extended solid body, so we can use Eqs. 9-11 to find the actual coordinates of com$_P$. However, that procedure is difficult. A much easier way is to use this Key Idea: In working with centers of mass, we can assume that the mass of a *uniform* object is concentrated in a particle at the object's center of mass. Here is how we do so:

First, put the stamped-out disk (call it disk S) back into place (Fig. 9-4b) to form the original composite plate (call it plate C). Because of its circular symmetry, the center of mass com$_S$ for disk S is at the center of S, at $x = -R$ (as shown). Similarly, the center of mass com$_C$ for composite plate C is at the center of C, at the origin (as shown). We then have the following:

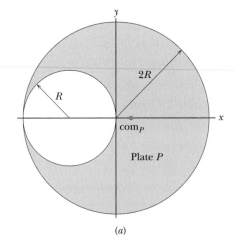

Fig. 9-4 Sample Problem 9-2. (a) Plate P is a metal plate of radius $2R$, with a circular hole of radius R. The center of mass of P is at point com$_P$. (b) Disk S has been put back into place to form a composite plate C. The center of mass com$_S$ of disk S and the center of mass com$_C$ of plate C are shown. (c) The center of mass com$_{S+P}$ of the combination of S and P coincides with com$_C$, which is at $x = 0$.

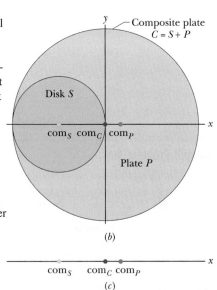

Plate	Center of Mass	Location of com	Mass
P	com$_P$	$x_P = ?$	m_P
S	com$_S$	$x_S = -R$	m_S
C	com$_C$	$x_C = 0$	$m_C = m_S + m_P$

Now we use the Key Idea of concentrated mass: Assume that mass m_S of disk S is concentrated in a particle at $x_S = -R$, and mass m_P is concentrated in a particle at x_P (Fig. 9-4c). Next treat these two particles as a two-particle system, using Eq. 9-2 to find their center of mass x_{S+P}. We get

$$x_{S+P} = \frac{m_S x_S + m_P x_P}{m_S + m_P}. \qquad (9\text{-}12)$$

Next note that the combination of disk S and plate P is composite plate C. Thus, the position x_{S+P} of com$_{S+P}$ must coincide with the position x_C of com$_C$, which is at the origin; so $x_{S+P} = x_C = 0$. Substituting this into Eq. 9-12 and solving for x_P, we get

$$x_P = -x_S \frac{m_S}{m_P}. \qquad (9\text{-}13)$$

Now we seem to have a problem, because we do not know the masses in Eq. 9-13. However, we can relate the masses to the face areas of S and P by noting that

$$\text{mass} = \text{density} \times \text{volume}$$
$$= \text{density} \times \text{thickness} \times \text{area}.$$

Then $\dfrac{m_S}{m_P} = \dfrac{\text{density}_S}{\text{density}_P} \times \dfrac{\text{thickness}_S}{\text{thickness}_P} \times \dfrac{\text{area}_S}{\text{area}_P}.$

Because the plate is uniform, the densities and thicknesses are equal; we are left with

$$\frac{m_S}{m_P} = \frac{\text{area}_S}{\text{area}_P} = \frac{\text{area}_S}{\text{area}_C - \text{area}_S} = \frac{\pi R^2}{\pi(2R)^2 - \pi R^2} = \frac{1}{3}.$$

Substituting this and $x_S = -R$ into Eq. 9-13, we have

$$x_P = \tfrac{1}{3}R. \qquad \text{(Answer)}$$

PROBLEM-SOLVING TACTICS

Tactic 1: *Center-of-Mass Problems*
Sample Problems 9-1 and 9-2 provide three strategies for simplifying center-of-mass problems. (1) Make full use of the symmetry of the object, be it about a point, a line, or a plane. (2) If the object can be divided into several parts, treat each of these parts as a particle, located at its own center of mass. (3) Choose your axes wisely: If your system is a group of particles, choose one of the particles as your origin. If your system is a body with a line of symmetry, let that be your x or y axis. The choice of origin is completely arbitrary; the location of the center of mass is the same regardless of the origin from which it is measured.

9-3 Newton's Second Law for a System of Particles

If you roll a cue ball at a second billiard ball that is at rest, you expect that the two-ball system will continue to have some forward motion after impact. You would be surprised, for example, if both balls came back toward you or if both moved to the right or to the left.

What continues to move forward, its steady motion completely unaffected by the collision, is the center of mass of the two-ball system. If you focus on this point — which is always halfway between these bodies because they have identical masses — you can easily convince yourself by trial at a billiard table that this is so. No matter whether the collision is glancing, head on, or somewhere in between, the center of mass continues to move forward, as if the collision had never occurred. Let us look into this center-of-mass motion in more detail.

To do so, we replace the pair of billiard balls with an assemblage of n particles of (possibly) different masses. We are interested not in the individual motions of these particles but *only* in the motion of their center of mass. Although the center of mass is just a point, it moves like a particle whose mass is equal to the total mass of the system; we can assign a position, a velocity, and an acceleration to it. We state (and shall prove next) that the (vector) equation that governs the motion of the center of mass of such a system of particles is

$$\vec{F}_{\text{net}} = M\vec{a}_{\text{com}} \qquad \text{(system of particles).} \qquad (9\text{-}14)$$

This equation is Newton's second law for the motion of the center of mass of a system of particles. Note that it has the same form ($\vec{F}_{\text{net}} = m\vec{a}$) that holds for the motion of a single particle. However, the three quantities that appear in Eq. 9-14

Fig. 9-5 A fireworks rocket explodes in flight. In the absence of air drag, the center of mass of the fragments would continue to follow the original parabolic path, until fragments began to hit the ground.

must be evaluated with some care:

1. $\vec{F}_{net}$ is the net force of *all external forces* that act on the system. Forces on one part of the system from another (*internal forces*) are not included in Eq. 9-14.

2. M is the *total mass* of the system. We assume that no mass enters or leaves the system as it moves, so that M remains constant. The system is said to be **closed.**

3. $\vec{a}_{com}$ is the acceleration of the *center of mass* of the system. Equation 9-14 gives no information about the acceleration of any other point of the system.

Equation 9-14 is equivalent to three equations involving the components of $\vec{F}_{net}$ and $\vec{a}_{com}$ along the three coordinate axes. These equations are

$$F_{net,x} = Ma_{com,x} \qquad F_{net,y} = Ma_{com,y} \qquad F_{net,z} = Ma_{com,z}. \qquad (9\text{-}15)$$

Now we can go back and examine the behavior of the billiard balls. Once the cue ball has begun to roll, no net external force acts on the (two-ball) system. Thus, because $\vec{F}_{net} = 0$, Eq. 9-14 tells us that $\vec{a}_{com} = 0$ also. Because acceleration is the rate of change of velocity, we conclude that the velocity of the center of mass of the system of two balls does not change. When the two balls collide, the forces that come into play are *internal* forces, on one ball from the other. Such forces do not contribute to the net force $\vec{F}_{net}$, which remains zero. Thus, the center of mass of the system, which was moving forward before the collision, must continue to move forward after the collision, with the same speed and in the same direction.

Equation 9-14 applies not only to a system of particles but also to a solid body, such as the bat of Fig. 9-1*b*. In that case, M in Eq. 9-14 is the mass of the bat and $\vec{F}_{net}$ is the gravitational force on the bat. Equation 9-14 then tells us that $\vec{a}_{com} = \vec{g}$. In other words, the center of mass of the bat moves as if the bat were a single particle of mass M, with force $\vec{F}_g$ acting on it.

Figure 9-5 shows another interesting case. Suppose that at a fireworks display, a rocket is launched on a parabolic path. At a certain point, it explodes into fragments. If the explosion had not occurred, the rocket would have continued along the trajectory shown in the figure. The forces of the explosion are *internal* to the system (first the rocket and then its fragments); that is, they are forces on parts of the system from other parts. If we ignore air drag, the net *external* force $\vec{F}_{net}$ acting on the system is the gravitational force on the system, regardless of whether the rocket explodes. Thus, from Eq. 9-14, the acceleration $\vec{a}_{com}$ of the center of mass of the fragments (while they are in flight) remains equal to $\vec{g}$. This means that the center of mass of the fragments follows the same parabolic trajectory that the rocket would have followed had it not exploded.

When a ballet dancer leaps across the stage in a grand jeté, she raises her arms and stretches her legs out horizontally as soon as her feet leave the stage (Fig. 9-6). These actions shift her center of mass upward through her body. Although the shifting center of mass faithfully follows a parabolic path across the stage, its movement relative to the body decreases the height that is attained by her head and torso, relative to that of a normal jump. The result is that the head and torso follow a nearly horizontal path, giving an illusion that the dancer is floating.

Proof of Equation 9-14

Now let us prove this important equation. From Eq. 9-8 we have, for a system of n particles,

$$M\vec{r}_{com} = m_1\vec{r}_1 + m_2\vec{r}_2 + m_3\vec{r}_3 + \cdots + m_n\vec{r}_n, \qquad (9\text{-}16)$$

Path of head

Path of center of mass

Fig. 9-6 A grand jeté. (Adapted from *The Physics of Dance,* by Kenneth Laws, Schirmer Books, 1984.)

in which M is the system's total mass and $\vec{r}_{com}$ is the vector locating the position of the system's center of mass.

Differentiating Eq. 9-16 with respect to time gives

$$M\vec{v}_{com} = m_1\vec{v}_1 + m_2\vec{v}_2 + m_3\vec{v}_3 + \cdots + m_n\vec{v}_n. \qquad (9\text{-}17)$$

Here $\vec{v}_i \ (= d\vec{r}_i/dt)$ is the velocity of the ith particle, and $\vec{v}_{com} \ (= d\vec{r}_{com}/dt)$ is the velocity of the center of mass.

Differentiating Eq. 9-17 with respect to time leads to

$$M\vec{a}_{com} = m_1\vec{a}_1 + m_2\vec{a}_2 + m_3\vec{a}_3 + \cdots + m_n\vec{a}_n. \qquad (9\text{-}18)$$

Here $\vec{a}_i \ (= d\vec{v}_i/dt)$ is the acceleration of the ith particle, and $\vec{a}_{com} \ (= d\vec{v}_{com}/dt)$ is the acceleration of the center of mass. Although the center of mass is just a geometrical point, it has a position, a velocity, and an acceleration, as if it were a particle.

From Newton's second law, $m_i\vec{a}_i$ is equal to the resultant force $\vec{F}_i$ that acts on the ith particle. Thus, we can rewrite Eq. 9-18 as

$$M\vec{a}_{com} = \vec{F}_1 + \vec{F}_2 + \vec{F}_3 + \cdots + \vec{F}_n. \qquad (9\text{-}19)$$

Among the forces that contribute to the right side of Eq. 9-19 will be forces that the particles of the system exert on each other (internal forces) and forces exerted on the particles from outside the system (external forces). By Newton's third law, the internal forces form third-law force pairs and cancel out in the sum that appears on the right side of Eq. 9-19. What remains is the vector sum of all the *external* forces that act on the system. Equation 9-19 then reduces to Eq. 9-14, the relation that we set out to prove.

✔CHECKPOINT 2:   Two skaters on frictionless ice hold opposite ends of a pole of negligible mass. An axis runs along the pole, and the origin of the axis is at the center of mass of the two-skater system. One skater, Fred, weighs twice as much as the other skater, Ethel. Where do the skaters meet if (a) Fred pulls hand over hand along the pole so as to draw himself to Ethel, (b) Ethel pulls hand over hand to draw herself to Fred, and (c) both skaters pull hand over hand?

Sample Problem 9-3

The three particles in Fig. 9-7a are initially at rest. Each experiences an *external* force due to bodies outside the three-particle system. The directions are indicated, and the magnitudes are $F_1 = 6.0$ N, $F_2 = 12$ N, and $F_3 = 14$ N. What is the acceleration of the center of mass of the system, and in what direction does it move?

SOLUTION: The position of the center of mass, calculated by the method of Sample Problem 9-1, is marked by a dot in the figure. One Key Idea here is that we can treat the center of mass as if it

were a real particle, with a mass equal to the system's total mass $M = 16$ kg. We can also treat the three external forces as if they act at the center of mass (Fig. 9-7b).

A second Key Idea is that we can now apply Newton's second law ($\vec{F}_{net} = m\vec{a}$) to the center of mass, writing

$$\vec{F}_{net} = M\vec{a}_{com} \qquad (9\text{-}20)$$

or

$$\vec{F}_1 + \vec{F}_2 + \vec{F}_3 = M\vec{a}_{com}$$

so

$$\vec{a}_{com} = \frac{\vec{F}_1 + \vec{F}_2 + \vec{F}_3}{M}. \qquad (9\text{-}21)$$

Equation 9-20 tells us that the acceleration $\vec{a}_{com}$ of the center of mass is in the same direction as the net external force $\vec{F}_{net}$ on the system (Fig. 9-7b). Because the particles are initially at rest, the center of mass must also be at rest. As the center of mass then begins to accelerate, it must move off in the common direction of $\vec{a}_{com}$ and $\vec{F}_{net}$.

We can evaluate the right side of Eq. 9-21 directly on a vector-capable calculator, or we can rewrite Eq. 9-21 in component form, find the components of $\vec{a}_{com}$, and then find $\vec{a}_{com}$. Along the x axis, we have

$$a_{com,x} = \frac{F_{1x} + F_{2x} + F_{3x}}{M}$$

$$= \frac{-6.0 \text{ N} + (12 \text{ N}) \cos 45° + 14 \text{ N}}{16 \text{ kg}} = 1.03 \text{ m/s}^2.$$

Along the y axis, we have

$$a_{com,y} = \frac{F_{1y} + F_{2y} + F_{3y}}{M}$$

$$= \frac{0 + (12 \text{ N}) \sin 45° + 0}{16 \text{ kg}} = 0.530 \text{ m/s}^2.$$

From these components, we find that $\vec{a}_{com}$ has the magnitude

$$a_{com} = \sqrt{(a_{com,x})^2 + (a_{com,y})^2}$$

$$= 1.16 \text{ m/s}^2 \approx 1.2 \text{ m/s}^2 \qquad \text{(Answer)}$$

and the angle (from the positive direction of the x axis)

$$\theta = \tan^{-1} \frac{a_{com,y}}{a_{com,x}} = 27°. \qquad \text{(Answer)}$$

(a)

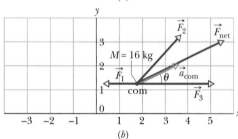

(b)

Fig. 9-7 Sample Problem 9-3. (a) Three particles, initially at rest in the positions shown, are acted on by the external forces shown. The center of mass (com) of the system is marked. (b) The forces are now transferred to the center of mass of the system, which behaves like a particle with a mass M equal to the total mass of the system. The net external force $\vec{F}_{net}$ and the acceleration $\vec{a}_{com}$ of the center of mass are shown.

9-4　Linear Momentum

Momentum is a word that has several meanings in everyday language but only a single precise meaning in physics. The **linear momentum** of a particle is a vector $\vec{p}$, defined as

$$\vec{p} = m\vec{v} \qquad \text{(linear momentum of a particle),} \qquad (9\text{-}22)$$

in which m is the mass of the particle and $\vec{v}$ is its velocity. (The adjective *linear* is often dropped, but it serves to distinguish $\vec{p}$ from *angular* momentum, which is introduced in Chapter 12 and which is associated with rotation.) Since m is always a positive scalar quantity, Eq. 9-22 tells us that $\vec{p}$ and $\vec{v}$ have the same direction. From Eq. 9-22, the SI unit for momentum is the kilogram-meter per second.

Newton actually expressed his second law of motion in terms of momentum:

▶ The time rate of change of the momentum of a particle is equal to the net force acting on the particle and is in the direction of that force.

In equation form this becomes

$$\vec{F}_{net} = \frac{d\vec{p}}{dt}. \tag{9-23}$$

Substituting for $\vec{p}$ from Eq. 9-22 gives

$$\vec{F}_{net} = \frac{d\vec{p}}{dt} = \frac{d}{dt}(m\vec{v}) = m\frac{d\vec{v}}{dt} = m\vec{a}.$$

Thus, the relations $\vec{F}_{net} = d\vec{p}/dt$ and $\vec{F}_{net} = m\vec{a}$ are equivalent expressions of Newton's second law of motion for a particle.

✔CHECKPOINT 3:   The figure gives the linear momentum versus time for a particle moving along an axis. A force directed along the axis acts on the particle. (a) Rank the four regions indicated according to the magnitude of the force, greatest first. (b) In which region is the particle slowing?

9-5 The Linear Momentum of a System of Particles

Now consider a system of n particles, each with its own mass, velocity, and linear momentum. The particles may interact with each other, and external forces may act on them as well. The system as a whole has a total linear momentum $\vec{P}$, which is defined to be the vector sum of the individual particles' linear momenta. Thus,

$$\vec{P} = \vec{p}_1 + \vec{p}_2 + \vec{p}_3 + \cdots + \vec{p}_n$$
$$= m_1\vec{v}_1 + m_2\vec{v}_2 + m_3\vec{v}_3 + \cdots + m_n\vec{v}_n. \tag{9-24}$$

If we compare this equation with Eq. 9-17, we see that

$$\vec{P} = M\vec{v}_{com} \qquad \text{(linear momentum, system of particles)}, \tag{9-25}$$

which gives us another way to define the linear momentum of a system of particles:

▶ The linear momentum of a system of particles is equal to the product of the total mass M of the system and the velocity of the center of mass.

If we take the time derivative of Eq. 9-25, we find

$$\frac{d\vec{P}}{dt} = M\frac{d\vec{v}_{com}}{dt} = M\vec{a}_{com}. \tag{9-26}$$

Comparing Eqs. 9-14 and 9-26 allows us to write Newton's second law for a system

of particles in the equivalent form

$$\vec{F}_{net} = \frac{d\vec{P}}{dt} \quad \text{(system of particles)}, \tag{9-27}$$

where $\vec{F}_{net}$ is the net external force acting on the system. This equation is the generalization of the single-particle equation $\vec{F}_{net} = d\vec{p}/dt$ to a system of many particles.

Sample Problem 9-4

Figure 9-8a shows a 2.0 kg toy race car before and after taking a turn on a track. Its speed is 0.50 m/s before the turn and 0.40 m/s after the turn. What is the change $\Delta\vec{P}$ in the linear momentum of the car due to the turn?

SOLUTION: We treat the car as a system of particles. Then a Key Idea is that to get $\Delta\vec{P}$, we need the car's linear momenta before and after the turn. However, that means we first need its velocity $\vec{v}_i$ before the turn and its velocity $\vec{v}_f$ after the turn. Using the coordinate system of Fig. 9-8a, we write $\vec{v}_i$ and $\vec{v}_f$ as

$$\vec{v}_i = -(0.50 \text{ m/s})\hat{j} \quad \text{and} \quad \vec{v}_f = (0.40 \text{ m/s})\hat{i}.$$

Equation 9-25 then gives us the linear momentum $\vec{P}_i$ before the turn and the linear momentum $\vec{P}_f$ after the turn:

$$\vec{P}_i = M\vec{v}_i = (2.0 \text{ kg})(-0.50 \text{ m/s})\hat{j} = (-1.0 \text{ kg} \cdot \text{m/s})\hat{j}$$

and $\vec{P}_f = M\vec{v}_f = (2.0 \text{ kg})(0.40 \text{ m/s})\hat{i} = (0.80 \text{ kg} \cdot \text{m/s})\hat{i}.$

A second Key Idea is that, because these linear momenta are not directed along the same axis, we *cannot* find the change in linear momentum $\Delta\vec{P}$ by merely subtracting the magnitude of $\vec{P}_i$ from the magnitude of $\vec{P}_f$. Instead, we must write the change in linear momentum as the vector equation

$$\Delta\vec{P} = \vec{P}_f - \vec{P}_i \tag{9-28}$$

(a)   (b)

Fig. 9-8 Sample Problem 9-4. (a) A toy car takes a turn. (b) The change $\Delta\vec{P}$ in the car's linear momentum is the vector sum of its final linear momentum $\vec{P}_f$ and the negative of its initial linear momentum $\vec{P}_i$.

and then as

$$\Delta\vec{P} = (0.80 \text{ kg} \cdot \text{m/s})\hat{i} - (-1.0 \text{ kg} \cdot \text{m/s})\hat{j}$$
$$= (0.8\hat{i} + 1.0\hat{j}) \text{ kg} \cdot \text{m/s}. \quad \text{(Answer)}$$

Figure 9-8b shows $\Delta\vec{P}$, $\vec{P}_f$, and $-\vec{P}_i$. Note that we subtract $\vec{P}_i$ from $\vec{P}_f$ in Eq. 9-28 by adding $-\vec{P}_i$ to $\vec{P}_f$.

9-6 Conservation of Linear Momentum

Suppose that the net external force acting on a system of particles is zero (the system is isolated) and that no particles leave or enter the system (the system is closed). Putting $\vec{F}_{net} = 0$ in Eq. 9-27 then yields $d\vec{P}/dt = 0$, or

$$\vec{P} = \text{constant} \quad \text{(closed, isolated system)}. \tag{9-29}$$

In words,

> If no net external force acts on a system of particles, the total linear momentum $\vec{P}$ of the system cannot change.

This result is called the **law of conservation of linear momentum.** It can also be written as

$$\vec{P}_i = \vec{P}_f \quad \text{(closed, isolated system)}. \tag{9-30}$$

In words, this equation says that, for a closed, isolated system,

$$\begin{pmatrix} \text{total linear momentum} \\ \text{at some initial time } t_i \end{pmatrix} = \begin{pmatrix} \text{total linear momentum} \\ \text{at some later time } t_f \end{pmatrix}.$$

Equations 9-29 and 9-30 are vector equations and, as such, each is equivalent to three equations corresponding to the conservation of linear momentum in three mutually perpendicular directions as in, say, the *xyz* coordinate system. Depending on the forces acting on a system, linear momentum might be conserved in one or two directions but not in all directions. However,

> If the component of the net *external* force on a closed system is zero along an axis, then the component of the linear momentum of the system along that axis cannot change.

As an example, suppose that you toss a grapefruit across a room. During its flight, the only external force acting on the grapefruit (which we take as the system) is the gravitational force $\vec{F}_g$, which is directed vertically downward. Thus, the vertical component of the linear momentum of the grapefruit changes, but since no horizontal external force acts on the grapefruit, the horizontal component of the linear momentum cannot change.

Note that we focus on the external forces acting on a closed system. Although internal forces can change the linear momentum of portions of the system, they cannot change the total linear momentum of the entire system.

✔CHECKPOINT 4:   An initially stationary device lying on a frictionless floor explodes into two pieces, which then slide across the floor. One piece slides in the positive direction of an *x* axis. (a) What is the sum of the momenta of the two pieces after the explosion? (b) Can the second piece move at an angle to the *x* axis? (c) What is the direction of the momentum of the second piece?

Sample Problem 9-5

A ballot box with mass $m = 6.0$ kg slides with speed $v = 4.0$ m/s across a frictionless floor in the positive direction of an *x* axis. It suddenly explodes into two pieces. One piece, with mass $m_1 = 2.0$ kg, moves in the positive direction of the *x* axis with speed $v_1 = 8.0$ m/s. What is the velocity of the second piece, with mass m_2?

SOLUTION: There are two Key Ideas here. First, we could get the velocity of the second piece if we knew its momentum, because we already know its mass is $m_2 = m - m_1 = 4.0$ kg. Second, we can relate the momenta of the two pieces to the original momentum of the box if momentum is conserved. Let's check.

Our reference frame will be that of the floor. Our system, which consists initially of the box and then of the two pieces, is closed but is not isolated, because the box and pieces each experience a normal force from the floor and a gravitational force. However, those forces are both vertical and thus cannot change the horizontal component of the momentum of the system. Neither can the forces produced by the explosion, because those forces are internal to the system. Thus, the horizontal component of the momentum of the system is conserved, and we can apply Eq. 9-30 along the *x* axis.

The initial momentum of the system is that of the box:

$$\vec{P}_i = m\vec{v}.$$

Similarly, we can write the final momenta of the two pieces as

$$\vec{P}_{f1} = m_1\vec{v}_1 \quad \text{and} \quad \vec{P}_{f2} = m_2\vec{v}_2.$$

The final total momentum $\vec{P}_f$ of the system is the vector sum of the momenta of the two pieces:

$$\vec{P}_f = \vec{P}_{f1} + \vec{P}_{f2} = m_1\vec{v}_1 + m_2\vec{v}_2.$$

Since all the velocities and momenta in this problem are vectors along the *x* axis, we can write them in terms of their *x* components. Doing so while applying Eq. 9-30, we now obtain

$$P_i = P_f$$

or

$$mv = m_1v_1 + m_2v_2.$$

Inserting known data, we find

$$(6.0 \text{ kg})(4.0 \text{ m/s}) = (2.0 \text{ kg})(8.0 \text{ m/s}) + (4.0 \text{ kg})v_2$$

and thus $v_2 = 2.0$ m/s.   (Answer)

Since the result is positive, the second piece moves in the positive direction of the *x* axis.

Sample Problem 9-6

Figure 9-9a shows a space hauler and cargo module, of total mass M traveling along an x axis in deep space. They have an initial velocity $\vec{v}_i$ of magnitude 2100 km/h relative to the Sun. With a small explosion, the hauler ejects the cargo module, of mass $0.20M$ (Fig. 9-9b). The hauler then travels 500 km/h faster than the module along the x axis; that is, the relative speed v_{rel} between the hauler and the module is 500 km/h. What then is the velocity $\vec{v}_{HS}$ of the hauler relative to the Sun?

SOLUTION: The Key Idea here is that, because the hauler–module system is closed and isolated, its total linear momentum is conserved; that is,

$$\vec{P}_i = \vec{P}_f, \tag{9-31}$$

where the subscripts i and f refer to values before and after the ejection, respectively. Because the motion is along a single axis, we can write momenta and velocities in terms of their x components. Before the ejection, we have

$$P_i = Mv_i. \tag{9-32}$$

Let v_{MS} be the velocity of the ejected module relative to the Sun. The total linear momentum of the system after the ejection is then

$$P_f = (0.20M)v_{MS} + (0.80M)v_{HS}, \tag{9-33}$$

where the first term on the right is the linear momentum of the module and the second term is that of the hauler.

We do not know the velocity v_{MS} of the module relative to the Sun, but we can relate it to the known velocities with

$$\begin{pmatrix} \text{velocity of} \\ \text{hauler relative} \\ \text{to Sun} \end{pmatrix} = \begin{pmatrix} \text{velocity of} \\ \text{hauler relative} \\ \text{to module} \end{pmatrix} + \begin{pmatrix} \text{velocity of} \\ \text{module relative} \\ \text{to Sun} \end{pmatrix}.$$

In symbols, this gives us

$$v_{HS} = v_{rel} + v_{MS} \tag{9-34}$$

Fig. 9-9 Sample Problem 9-6. (a) A space hauler, with a cargo module, moving at initial velocity $\vec{v}_i$. (b) The hauler has ejected the cargo module. Now the module has velocity $\vec{v}_{MS}$ and the hauler has velocity $\vec{v}_{HS}$.

or

$$v_{MS} = v_{HS} - v_{rel}.$$

Substituting this expression for v_{MS} into Eq. 9-33, and then substituting Eqs. 9-32 and 9-33 into Eq. 9-31, we find

$$Mv_i = 0.20M(v_{HS} - v_{rel}) + 0.80Mv_{HS},$$

which gives us

$$v_{HS} = v_i + 0.20v_{rel},$$

or

$$\begin{aligned} v_{HS} &= 2100 \text{ km/h} + (0.20)(500 \text{ km/h}) \\ &= 2200 \text{ km/h}. \end{aligned} \tag{Answer}$$

✓**CHECKPOINT 5:**   The table gives velocities of the hauler and module (after ejection and relative to the Sun), and the relative speed between the hauler and the module for three situations. What are the missing values?

	Velocities (km/h)		Relative Speed (km/h)
	Module	Hauler	
(a)	1500	2000	
(b)		3000	400
(c)	1000		600

Sample Problem 9-7

A firecracker placed inside a coconut of mass M, initially at rest on a frictionless floor, blows the coconut into three pieces that slide across the floor. An overhead view is shown in Fig. 9-10a. Piece C, with mass $0.30M$, has final speed $v_{fC} = 5.0$ m/s.

(a) What is the speed of piece B, with mass $0.20M$?

SOLUTION: A Key Idea here is to see whether linear momentum is conserved. We note that (1) the coconut and its pieces form a closed system, (2) the explosion forces are internal to that system, and (3) no net external force acts on the system. Therefore, the linear momentum of the system is conserved.

To get started, we superimpose an xy coordinate system as shown in Fig. 9-10b, with the negative direction of the x axis coinciding with the direction of $\vec{v}_{fA}$. The x axis is at 80° with the direction of $\vec{v}_{fC}$ and 50° with the direction of $\vec{v}_{fB}$.

A second Key Idea is that linear momentum is conserved separately along both x and y axes. Let's use the y axis and write

$$P_{iy} = P_{fy}, \tag{9-35}$$

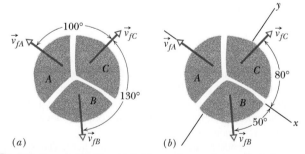

Fig. 9-10 Sample Problem 9-7. Three pieces of an exploded coconut move off in three directions along a frictionless floor. (a) An overhead view of the event. (b) The same with a two-dimensional axis system imposed.

where subscript i refers to the initial value (before the explosion), and subscript y refers to the y component of $\vec{P}_i$ or $\vec{P}_f$.

The component P_{iy} of the initial linear momentum is zero, because the coconut is initially at rest. To get an expression for P_{fy}, we find the y component of the final linear momentum of each piece, using the y-component version of Eq. 9-22 ($p_y = mv_y$):

$$p_{fA,y} = 0,$$

$$p_{fB,y} = -0.20Mv_{fB,y} = -0.20Mv_{fB}\sin 50°,$$

$$p_{fC,y} = 0.30Mv_{fC,y} = 0.30Mv_{fC}\sin 80°.$$

(Note that $p_{fA,y} = 0$ because of our choice of axes.) Equation 9-35 can now be written as

$$P_{iy} = P_{fy} = p_{fA,y} + p_{fB,y} + p_{fC,y}.$$

Then, with $v_{fC} = 5.0$ m/s, we have

$$0 = 0 - 0.20Mv_{fB}\sin 50° + (0.30M)(5.0 \text{ m/s})\sin 80°,$$

from which we find

$$v_{fB} = 9.64 \text{ m/s} \approx 9.6 \text{ m/s.} \qquad \text{(Answer)}$$

(b) What is the speed of piece A?

SOLUTION: Because linear momentum is also conserved along the x axis, we have

$$P_{ix} = P_{fx}, \qquad (9\text{-}36)$$

where $P_{ix} = 0$ because the coconut is initially at rest. To get P_{fx},

we find the x components of the final momenta, using the fact that piece A must have a mass of $0.50M$ ($= M - 0.20M - 0.30M$):

$$p_{fA,x} = -0.50Mv_{fA},$$

$$p_{fB,x} = 0.20Mv_{fB,x} = 0.20Mv_{fB}\cos 50°,$$

$$p_{fC,x} = 0.30Mv_{fC,x} = 0.30Mv_{fC}\cos 80°.$$

Equation 9-36 can now be written as

$$P_{ix} = P_{fx} = p_{fA,x} + p_{fB,x} + p_{fC,x}.$$

Then, with $v_{fC} = 5.0$ m/s and $v_{fB} = 9.64$ m/s, we have

$$0 = -0.50Mv_{fA} + 0.20M(9.64 \text{ m/s})\cos 50°$$
$$+ 0.30M(5.0 \text{ m/s})\cos 80°,$$

from which we find

$$v_{fA} = 3.0 \text{ m/s.} \qquad \text{(Answer)}$$

✔CHECKPOINT 6:   Suppose that the exploding coconut is accelerating in the negative direction of y in Fig. 9-10 (say it is on a ramp that slants downward in that direction). Is linear momentum conserved along (a) the x axis (as stated in Eq. 9-36) and (b) the y axis (as stated in Eq. 9-35)?

PROBLEM-SOLVING TACTICS

Tactic 2: *Conservation of Linear Momentum*
For problems involving the conservation of linear momentum, first make sure that you have chosen a closed, isolated system. *Closed* means that no matter (no particles) passes through the system boundary in any direction. *Isolated* means that the net external force acting on the system is zero. If it is not isolated, then remember that each component of linear momentum is conserved separately if the corresponding component of the net external force is zero. So, you might conserve one component and not another.

Next, select two appropriate states of the system (which you may choose to call the initial state and the final state) and write expressions for the linear momentum of the system in each of these two states. In writing these expressions, make sure that you know what inertial reference frame you are using, and make sure also that you include the entire system, not missing any part of it and not including objects that do not belong to your system.
Finally, set your expressions for $\vec{P}_i$ and $\vec{P}_f$ equal to each other and solve for what is requested.

9-7 Systems with Varying Mass: A Rocket

In the systems we have dealt with so far, we have assumed that the total mass of the system remains constant. Sometimes, as in a rocket (Fig. 9-11), it does not. Most of the mass of a rocket on its launching pad is fuel, all of which will eventually be burned and ejected from the nozzle of the rocket engine.

We handle the variation of the mass of the rocket as the rocket accelerates by applying Newton's second law, not to the rocket alone but to the rocket and its ejected combustion products taken together. The mass of *this* system does *not* change as the rocket accelerates.

Finding the Acceleration

Assume that we are at rest relative to an inertial reference frame, watching a rocket accelerate through deep space with no gravitational or atmospheric drag forces acting

Fig. 9-11 Liftoff of Project Mercury spacecraft.

Fig. 9-12 (*a*) An accelerating rocket of mass *M* at time *t*, as seen from an inertial reference frame. (*b*) The same but at time *t* + *dt*. The exhaust products released during interval *dt* are shown.

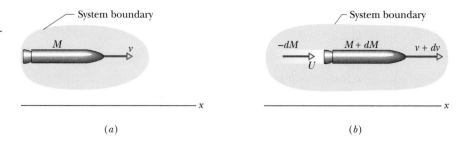

(*a*)   (*b*)

on it. For this one-dimensional motion, let *M* be the mass of the rocket and *v* its velocity at an arbitrary time *t* (see Fig. 9-12*a*).

Figure 9-12*b* shows how things stand a time interval *dt* later. The rocket now has velocity *v* + *dv* and mass *M* + *dM*, where the change in mass *dM* is a *negative quantity*. The exhaust products released by the rocket during interval *dt* have mass −*dM* and velocity *U* relative to our inertial reference frame.

Our system consists of the rocket and the exhaust products released during interval *dt*. The system is closed and isolated, so the linear momentum of the system must be conserved during *dt*; that is,

$$P_i = P_f, \tag{9-37}$$

where the subscripts *i* and *f* indicate the values at the beginning and end of time interval *dt*. We can rewrite Eq. 9-37 as

$$Mv = -dM\,U + (M + dM)(v + dv), \tag{9-38}$$

where the first term on the right is the linear momentum of the exhaust products released during interval *dt* and the second term is the linear momentum of the rocket at the end of interval *dt*.

We can simplify Eq. 9-38 by using the relative speed v_{rel} between the rocket and the exhaust products, which is related to the velocities relative to the frame with

$$\left(\begin{array}{c}\text{velocity of rocket}\\\text{relative to frame}\end{array}\right) = \left(\begin{array}{c}\text{velocity of rocket}\\\text{relative to products}\end{array}\right) + \left(\begin{array}{c}\text{velocity of products}\\\text{relative to frame}\end{array}\right).$$

In symbols, this means

$$(v + dv) = v_{\text{rel}} + U,$$

or

$$U = v + dv - v_{\text{rel}}. \tag{9-39}$$

Substituting this result for *U* into Eq. 9-38 yields, with a little algebra,

$$-dM\,v_{\text{rel}} = M\,dv. \tag{9-40}$$

Dividing each side by *dt* gives us

$$-\frac{dM}{dt}\,v_{\text{rel}} = M\,\frac{dv}{dt}. \tag{9-41}$$

We replace *dM/dt* (the rate at which the rocket loses mass) by −*R*, where *R* is the (positive) mass rate of fuel consumption, and we recognize that *dv/dt* is the acceleration of the rocket. With these changes, Eq. 9-41 becomes

$$Rv_{\text{rel}} = Ma \qquad \text{(first rocket equation)}. \tag{9-42}$$

Equation 9-42 holds at any instant, with the mass *M*, the fuel consumption rate *R*, and the acceleration *a* evaluated at that instant.

The left side of Eq. 9-42 has the dimensions of a force (kg · m/s² = N) and depends only on design characteristics of the rocket engine, namely, the rate *R* at which it consumes fuel mass and the speed v_{rel} with which that mass is ejected relative to the rocket. We call this term Rv_{rel} the **thrust** of the rocket engine and represent it with T. Newton's second law emerges clearly if we write Eq. 9-42 as $T = Ma$, in which a is the acceleration of the rocket at the time that its mass is M.

Finding the Velocity

How will the velocity of a rocket change as it consumes its fuel? From Eq. 9-40 we have

$$dv = -v_{rel} \frac{dM}{M}.$$

Integrating leads to

$$\int_{v_i}^{v_f} dv = -v_{rel} \int_{M_i}^{M_f} \frac{dM}{M},$$

in which M_i is the initial mass of the rocket and M_f its final mass. Evaluating the integrals then gives

$$v_f - v_i = v_{rel} \ln \frac{M_i}{M_f} \qquad \text{(second rocket equation)} \qquad (9\text{-}43)$$

for the increase in the speed of the rocket during the change in mass from M_i to M_f. (The symbol "ln" in Eq. 9-43 means the *natural logarithm*.) We see here the advantage of multistage rockets, in which M_f is reduced by discarding successive stages when their fuel is depleted. An ideal rocket would reach its destination with only its payload remaining.

Sample Problem 9-8

A rocket whose initial mass M_i is 850 kg consumes fuel at the rate $R = 2.3$ kg/s. The speed v_{rel} of the exhaust gases relative to the rocket engine is 2800 m/s.

(a) What thrust does the rocket engine provide?

SOLUTION: The Key Idea here is that the thrust T is equal to the product of the fuel consumption rate R and the relative speed v_{rel} at which exhaust gases are expelled:

$$T = Rv_{rel} = (2.3 \text{ kg/s})(2800 \text{ m/s})$$
$$= 6440 \text{ N} \approx 6400 \text{ N.} \qquad \text{(Answer)}$$

(b) What is the initial acceleration of the rocket?

SOLUTION: We can relate the thrust T of a rocket to the magnitude a of the resulting acceleration with $T = Ma$, where M is the rocket's mass. The Key Idea, however, is that M decreases and a increases as fuel is consumed. Because we want the initial value of a here, we must use the initial value M_i of the mass, finding that

$$a = \frac{T}{M_i} = \frac{6440 \text{ N}}{850 \text{ kg}} = 7.6 \text{ m/s}^2. \qquad \text{(Answer)}$$

To be launched from Earth's surface, a rocket must have an initial acceleration greater than $g = 9.8$ m/s^2. Put another way, the thrust T of the rocket engine must exceed the initial gravitational force

on the rocket, which here has the magnitude $M_i g$, which gives us (850 kg)(9.8 m/s^2), or 8330 N. Because the acceleration or thrust requirement is not met (here $T = 6400$ N), our rocket could not be launched from Earth's surface by itself; it would require another, more powerful, rocket.

(c) Suppose, instead, that the rocket is launched from a spacecraft already in deep space, where we can neglect any gravitational force acting on it. The mass M_f of the rocket when its fuel is exhausted is 180 kg. What is its speed relative to the spacecraft at that time? Assume that the spacecraft is so massive that the launch does not alter its speed.

SOLUTION: The Key Idea here is that the rocket's final speed v_f (when the fuel is exhausted) depends on the ratio M_i/M_f of its initial mass to its final mass, as given by Eq. 9-43. With the initial speed $v_i = 0$, we have

$$v_f = v_{rel} \ln \frac{M_i}{M_f}$$

$$= (2800 \text{ m/s}) \ln \frac{850 \text{ kg}}{180 \text{ kg}}$$

$$= (2800 \text{ m/s}) \ln 4.72 \approx 4300 \text{ m/s.} \qquad \text{(Answer)}$$

Note that the ultimate speed of the rocket can exceed the exhaust speed v_{rel}.

9-8　External Forces and Internal Energy Changes

As the ice skater in Fig. 9-13a pushes herself away from a railing, there is a force $\vec{F}$ on her from the railing at angle ϕ to the horizontal. This force accelerates her, increasing her speed until she leaves the railing (Fig. 9-13b). Thus, her kinetic energy is increased via the force. This example differs in two ways from earlier examples in which an object's kinetic energy is changed via a force:

1. Previously, each part of an object moved rigidly in the same direction. Here the skater's arm does not move like the rest of her body.

2. Previously, energy was transferred between the object (or system) and its environment via an external force; that is, the force did work. Here the energy is transferred internally (from one part of the system to another) via the external force $\vec{F}$. In particular, the energy is transferred from internal biochemical energy of the skater's muscles to kinetic energy of her body as a whole.

We want to relate the external force $\vec{F}$ to that internal energy transfer.

In spite of the two differences listed here, we can relate $\vec{F}$ to the change in kinetic energy much as we did for a particle in Section 7-3. To do so, we first mentally concentrate the mass M of the skater at her center of mass so that we can treat her as a particle at the center of mass. Then we treat the external force $\vec{F}$ as acting on the particle (Fig. 9-13c). The horizontal force component $F \cos \phi$ accelerates that particle, resulting in a change ΔK in its kinetic energy during a displacement of magnitude d. As we prove later, we can relate these quantities with

$$\Delta K = Fd \cos \phi. \tag{9-44}$$

We can imagine cases in which the external force would also change the height of the skater's center of mass and thus cause a change ΔU in the gravitational potential energy of the skater–Earth system. To include such a change ΔU, we can rewrite Eq. 9-44 as

$$\Delta K + \Delta U = Fd \cos \phi. \tag{9-45}$$

The left side here is ΔE_{mec}, the change in the mechanical energy of the system. Thus, in the general case, we have

$$\Delta E_{mec} = Fd \cos \phi \qquad \text{(external force, change in } E_{mec}). \tag{9-46}$$

Next let us consider the energy transfers in the ice skater, which we now take as the system. Although an external force acts on the system, the force does not transfer energy to or from the system. Thus, the total energy E of the system cannot change: $\Delta E = 0$. We know that when the ice skater pushes off from the rail, there is a change not only in the mechanical energy of her center of mass, but also in the energy of her muscles. Without going into the details of the change in muscular energy, let us just represent it with ΔE_{int} (for a change in internal energy). Then we can write $\Delta E = 0$ as

$$\Delta E_{int} + \Delta E_{mec} = 0, \tag{9-47}$$

or

$$\Delta E_{int} = -\Delta E_{mec}. \tag{9-48}$$

This equation means that as E_{mec} increases for the ice skater, E_{int} decreases by just as much. Substituting for ΔE_{mec} in Eq. 9-48 from Eq. 9-46, we have

$$\Delta E_{int} = -Fd \cos \phi \qquad \text{(external force, internal energy change).} \tag{9-49}$$

(a)

(b)

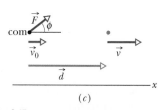

(c)

Fig. 9-13 (a) As a skater pushes herself away from a railing, the force on her from the railing is $\vec{F}$. (b) After the skater leaves the railing, her center of mass has velocity $\vec{v}$. (c) External force $\vec{F}$ is taken to act at the skater's center of mass, at angle ϕ with a horizontal x axis. When the center of mass goes through displacement $\vec{d}$, its velocity is changed from $\vec{v}_0$ to $\vec{v}$ by the horizontal component of $\vec{F}$.

Fig. 9-14 A vehicle accelerates to the right using four-wheel drive. The road exerts four frictional forces (two of them shown) on the bottom surfaces of the tires. Taken together, these four forces make up the net external force $\vec{F}$ acting on the car.

This equation relates the change ΔE_{int} in internal energy that is made via the external force $\vec{F}$. If $\vec{F}$ is not constant in magnitude, we can replace the symbol F in Eqs. 9-46 and 9-49 with the symbol F_{avg} for the average magnitude of $\vec{F}$.

Although we have used an ice skater to derive Eqs. 9-46 and 9-49, the equations hold for other objects in which a change ΔE_{int} in the internal energy of a system is made via an external force. For example, consider a vehicle with four-wheel drive (all four wheels are made to turn by the engine) as its speed is increased. During the acceleration, the engine causes the tires to push backward on the road's surface. This push produces frictional forces that act on each tire in the forward direction (Fig. 9-14). The net external force $\vec{F}$, which is the sum of these frictional forces, gives the car's center of mass an acceleration $\vec{a}$. Thus, there is a transfer of energy from the internal energy stored in the fuel to the kinetic energy of the car. If $\vec{F}$ is constant, then for a given displacement $\vec{d}$ of the car's center of mass along level road, we can relate the change ΔK in the car's kinetic energy to the external force $\vec{F}$ with Eq. 9-45, with $\Delta U = 0$ and $\phi = 0$.

If the driver applies the brakes, Eq. 9-45 still holds. Now $\vec{F}$ due to the frictional forces is toward the rear, and $\phi = 180°$. Energy is now transferred from kinetic energy of the car's center of mass to thermal energy of the brakes.

Proof of Equation 9-44

Let us return to the ice skater and Fig. 9-13. Suppose that during displacement $\vec{d}$ of her center of mass, the velocity of her center of mass changes from $\vec{v}_0$ to $\vec{v}$. Then the magnitude of $\vec{v}$ is, by Eq. 2-16,

$$v^2 = v_0^2 + 2a_x d, \tag{9-50}$$

where a_x is her acceleration. Multiplying both sides of Eq. 9-50 by the skater's mass M and rearranging yield

$$\tfrac{1}{2}Mv^2 - \tfrac{1}{2}Mv_0^2 = Ma_x d. \tag{9-51}$$

The left side of Eq. 9-51 is the difference between the final kinetic energy K_f of the center of mass and the initial kinetic energy K_i. This difference is the change ΔK in the kinetic energy of the center of mass due to the force $\vec{F}$. Making that substitution, and substituting the product $F \cos \phi$ for the product Ma_x according to Newton's second law, we get

$$\Delta K = Fd \cos \phi,$$

as we intended to prove.

Sample Problem 9-9

When a click beetle is upside down on its back, it jumps upward by suddenly arching its back, transferring energy stored in a muscle to mechanical energy. This launching mechanism produces an audible click, giving the beetle its name. Videotape of a certain click-beetle jump shows that the center of mass of a click beetle of mass $m = 4.0 \times 10^{-6}$ kg moved directly upward by 0.77 mm during the launch and then to a maximum height of $h = 0.30$ m. What was the average magnitude of the external force $\vec{F}$ on the beetle's back from the floor during the launch?

SOLUTION: The Key Idea here is that during the launch, energy is transferred from an internal energy of the beetle to the mechanical energy of the beetle–Earth system, in the amount ΔE_{mec}. The transfer is made via the external force $\vec{F}$. To find the magnitude of the force with Eq. 9-46, we first need an expression for ΔE_{mec}.

Let the system's mechanical energy be $E_{mec,0}$ just before the launch and $E_{mec,1}$ at the end of the launch. Then the change ΔE_{mec} is

$$\Delta E_{mec} = E_{mec,1} - E_{mec,0}. \tag{9-52}$$

We must now find expressions for $E_{mec,0}$ and $E_{mec,1}$. Let the gravitational potential energy of the beetle–Earth system be $U_0 = 0$ when the beetle is on the floor. Also, just before the launch, the kinetic energy of the beetle's center of mass is $K_0 = 0$, so the beginning mechanical energy $E_{mec,0}$ is 0. Unfortunately, when we try to find $E_{mec,1}$, we get stuck because we do not know the beetle's kinetic energy K_1 or speed v_1 at the end of the launch.

We get unstuck with a second Key Idea: The mechanical energy of the system does not change from the time the beetle is launched until it reaches its maximum height. We can find this mechanical energy $E_{mec,1}$ at the maximum height because we know the beetle's speed ($v = 0$) and height ($y = h$) there. Thus, we have

$$E_{mec,1} = K + U = \tfrac{1}{2}mv^2 + mgy = 0 + mgh = mgh.$$

Substituting this and $E_{mec,0} = 0$ into Eq. 9-52 now gives us

$$\Delta E_{mec} = mgh - 0 = mgh. \tag{9-53}$$

We can now use Eq. 9-46 to relate this change in mechanical energy to the external force. We write

$$\Delta E_{mec} = F_{avg}d \cos \phi. \tag{9-54}$$

F_{avg} is the average magnitude of the external force on the beetle, d is the magnitude (0.77 mm) of the displacement of the beetle's center of mass during the launch (when the external force acts on it), and ϕ ($= 0°$) is the angle between the directions of the external force and the displacement.

Solving Eq. 9-54 for F_{avg} and using Eq. 9-53 yield

$$
\begin{aligned}
F_{avg} &= \frac{\Delta E_{mec}}{d \cos \phi} = \frac{mgh}{d \cos \phi} \\
&= \frac{(4.0 \times 10^{-6}\ \text{kg})(9.8\ \text{m/s}^2)(0.30\ \text{m})}{(7.7 \times 10^{-4}\ \text{m})(\cos 0°)} \\
&= 1.5 \times 10^{-2}\ \text{N}. \qquad \text{(Answer)}
\end{aligned}
$$

This force magnitude may seem small, but to the click beetle it is enormous because, as you can show, it gives the beetle an acceleration of over $380g$ during the launch.

REVIEW & SUMMARY

Center of Mass   The **center of mass** of a system of n particles is defined to be the point whose coordinates are given by

$$x_{com} = \frac{1}{M}\sum_{i=1}^{n} m_i x_i, \quad y_{com} = \frac{1}{M}\sum_{i=1}^{n} m_i y_i, \quad z_{com} = \frac{1}{M}\sum_{i=1}^{n} m_i z_i, \tag{9-5}$$

or

$$\vec{r}_{com} = \frac{1}{M}\sum_{i=1}^{n} m_i \vec{r}_i, \tag{9-8}$$

where M is the total mass of the system. If the mass is continuously distributed, the center of mass is given by

$$x_{com} = \frac{1}{M}\int x\, dm, \quad y_{com} = \frac{1}{M}\int y\, dm, \quad z_{com} = \frac{1}{M}\int z\, dm. \tag{9-9}$$

If the density (mass per unit volume) is uniform, then Eq. 9-9 can be rewritten as

$$x_{com} = \frac{1}{V}\int x\, dV, \quad y_{com} = \frac{1}{V}\int y\, dV, \quad z_{com} = \frac{1}{V}\int z\, dV, \tag{9-11}$$

where V is the volume occupied by M.

Newton's Second Law for a System of Particles   The motion of the center of mass of any system of particles is governed by **Newton's second law for a system of particles**, which is

$$\vec{F}_{net} = M\vec{a}_{com}. \tag{9-14}$$

Here $\vec{F}_{net}$ is the net force of all the *external* forces acting on the system, M is the total mass of the system, and $\vec{a}_{com}$ is the acceleration of the system's center of mass.

Linear Momentum and Newton's Second Law   For a single particle, we define a quantity $\vec{p}$ called its **linear momentum** as

$$\vec{p} = m\vec{v}, \tag{9-22}$$

and can write Newton's second law in terms of this momentum:

$$\vec{F}_{net} = \frac{d\vec{p}}{dt}. \tag{9-23}$$

For a system of particles these relations become

$$\vec{P} = M\vec{v}_{com} \quad \text{and} \quad \vec{F}_{net} = \frac{d\vec{P}}{dt}. \tag{9-25, 9-27}$$

Conservation of Linear Momentum   If a system is isolated so that no net *external* force acts on the system, the linear momentum $\vec{P}$ of the system remains constant:

$$\vec{P} = \text{constant} \qquad \text{(closed, isolated system).} \tag{9-29}$$

This can also be written as

$$\vec{P}_i = \vec{P}_f \qquad \text{(closed, isolated system),} \tag{9-30}$$

where the subscripts refer to the values of $\vec{P}$ at some initial time and at a later time. Equations 9-29 and 9-30 are equivalent statements of the **law of conservation of linear momentum.**

Variable-Mass Systems   If a system has varying mass, we redefine the system, enlarging its boundaries until they encompass a larger system whose mass *does* remain constant; then we apply the law of conservation of linear momentum. For a rocket, this means that the system includes both the rocket and its exhaust gases. Analysis of such a system shows that in the absence of external forces a rocket accelerates at an instantaneous rate given by

$$Rv_{rel} = Ma \qquad \text{(first rocket equation),} \tag{9-42}$$

in which M is the rocket's instantaneous mass (including unexpended fuel), R is the fuel consumption rate, and v_{rel} is the fuel's exhaust speed relative to the rocket. The term Rv_{rel} is the **thrust** of the rocket engine. For a rocket with constant R and v_{rel}, whose speed changes from v_i to v_f when its mass changes from M_i to M_f,

$$v_f - v_i = v_{rel} \ln \frac{M_i}{M_f} \qquad \text{(second rocket equation).} \qquad (9\text{-}43)$$

External Forces and Internal Energy Changes Energy can be transferred inside a system between an internal energy and mechanical energy via an external force $\vec{F}$. The change ΔE_{int} in the internal energy is related by

$$\Delta E_{int} = -Fd \cos \phi \qquad (9\text{-}49)$$

to the external force, the displacement $\vec{d}$ of the system's center of mass, and the angle ϕ between the directions of $\vec{F}$ and $\vec{d}$. The change in the mechanical energy is

$$\Delta E_{mec} = \Delta K + \Delta U = Fd \cos \phi \qquad (9\text{-}46, 9\text{-}45)$$

QUESTIONS

1. Figure 9-15 shows four uniform square metal plates that have had a section removed. The origin of the x and y axes is at the center of the original plate, and in each case the center of mass of the removed section was at the origin. In each case, where is the center of mass of the remaining section of the plate? Answer in terms of quadrants, lines, or points.

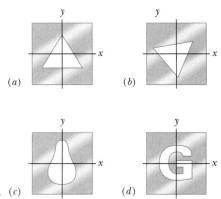

Fig. 9-15 Question 1. (*c*)

2. Some skilled basketball players seem to hang in midair during a jump at the basket, allowing them more time to shift the ball from hand to hand and then into the basket. If a player raises arms or legs during a jump, does the player's time in the air increase, decrease, or stay the same?

3. In Fig. 9-16, a penguin stands at the left edge of a uniform sled of length L, which lies on frictionless ice. The sled and penguin have equal masses. (a) Where is the center of mass of the sled? (b) How far and in what direction is the center of the sled from the center of mass of the sled–penguin system?

Fig. 9-16 Question 3.

The penguin then waddles to the right edge of the sled, and the sled slides on the ice. (c) Does the center of mass of the sled–penguin system move leftward, rightward, or not at all? (d) Now how far and in what direction is the center of the sled from the center of mass of the sled–penguin system? (e) How far does the penguin move relative to the sled? Relative to the center of mass of the sled–penguin system, how far does (f) the center of the sled move and (g) the penguin move? (Warm-up for Problem 19)

4. In Question 3 and Fig. 9-16, suppose that the sled and penguin are initially moving rightward at speed v_0. (a) As the penguin waddles to the right edge of the sled, is the speed v of the sled less than, greater than, or equal to v_0? (b) If the penguin then waddles back to the left edge, during that motion is the speed v of the sled less than, greater than, or equal to v_0?

5. Figure 9-17 shows an overhead view of four particles of equal mass sliding over a frictionless surface at constant velocity. The directions of the velocities are indicated; their magnitudes are equal. Consider pairing the particles. Which pairs form a system with a center of mass that (a) is stationary, (b) is stationary and at the origin, and (c) passes through the origin?

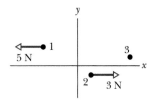

Fig. 9-17 Question 5.

6. Figure 9-18 shows an overhead view of three particles on which external forces act. The magnitudes and directions of the forces on two of the particles are indicated. What are the magnitude and direction of the force acting on the third particle if the center of mass of the three-particle system is (a) stationary, (b) moving at a constant velocity rightward, and (c) accelerating rightward?

Fig. 9-18 Question 6.

7. A container sliding along an x axis on a frictionless surface explodes into three pieces. The pieces then move along the x axis in the directions indicated in Fig. 9-19. The following table gives four sets of magnitudes (in kg · m/s) for the momenta $\vec{p}_1$, $\vec{p}_2$, and $\vec{p}_3$ of the pieces. Rank the sets according to the initial speed of the container, greatest first.

	p_1	p_2	p_3			p_1	p_2	p_3
(a)	10	2	6		(b)	10	6	2
(c)	2	10	6		(d)	6	2	10

Fig. 9-19 Question 7.

8. A spaceship that is moving along an x axis separates into two parts, like the hauler in Fig. 9-9. (a) Which of the graphs in

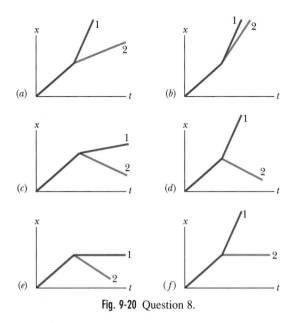

Fig. 9-20 Question 8.

Fig. 9-20 could possibly give the position versus time for the ship and the two parts? (b) Which of the numbered lines pertains to the trailing part? (c) Rank the possible graphs according to the relative speed between the parts, greatest first.

9. Consider a box, like that in Sample Problem 9-5, which explodes into two pieces while moving with a constant positive velocity along an x axis. If one piece, with mass m_1, ends up with positive velocity $\vec{v}_1$, then the second piece, with mass m_2, could end up with (a) a positive velocity $\vec{v}_2$ (Fig. 9-21a), (b) a negative velocity $\vec{v}_2$ (Fig. 9-21b), or (c) zero velocity (Fig. 9-21c). Rank those three possible results for the second piece according to the corresponding magnitude of $\vec{v}_1$, greatest first.

Fig. 9-21 Question 9.

EXERCISES & PROBLEMS

SEC. 9-2   The Center of Mass

1E. (a) How far is the center of mass of the Earth–Moon system from the center of Earth? (Appendix C gives the masses of Earth and the Moon and the distance between the two.) (b) Express the answer to (a) as a fraction of Earth's radius R_e.   ssm

2E. A distance of 1.131×10^{-10} m lies between the centers of the carbon and oxygen atoms in a carbon monoxide (CO) gas molecule. Locate the center of mass of a CO molecule relative to the carbon atom. (Find the masses of C and O in Appendix F.)

3E. What are (a) the x coordinate and (b) the y coordinate of the center of mass of the three-particle system shown in Fig. 9-22? (c) What happens to the center of mass as the mass of the topmost particle is gradually increased?   ssm

4E. Three thin rods, each of length L, are arranged in an inverted U, as shown in Fig. 9-23. The two rods on the arms of the U each have mass M; the third rod has mass $3M$. Where is the center of mass of the assembly?

Fig. 9-22 Exercise 3.

Fig. 9-23 Exercise 4.

5E. A uniform square plate 6 m on a side has had a square piece 2 m on a side cut out of it (Fig. 9-24). The center of that piece is at $x = 2$ m, $y = 0$. The center of the square plate is at $x = y = 0$. Find (a) the x coordinate and (b) the y coordinate of the center of mass of the remaining piece.

6P. Figure 9-25 shows the dimensions of a composite slab; half the slab is made of aluminum (density $= 2.70$ g/cm³) and half is made of iron (density $= 7.85$ g/cm³). Where is the center of mass of the slab?

Fig. 9-24 Exercise 5.

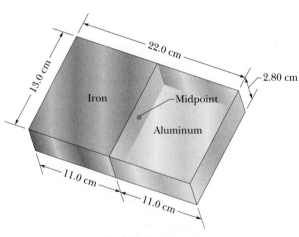

Fig. 9-25 Problem 6.

7P. In the ammonia (NH_3) molecule (see Fig. 9-26), the three hydrogen (H) atoms form an equilateral triangle; the center of the triangle is 9.40×10^{-11} m from each hydrogen atom. The nitrogen (N) atom is at the apex of a pyramid, with the three hydrogen atoms forming the base. The nitrogen-to-hydrogen atomic mass ratio is 13.9, and the nitrogen-to-hydrogen distance is 10.14×10^{-11} m. Locate the center of mass of the molecule relative to the nitrogen atom. ilw

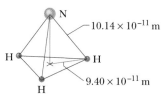

10.14×10^{-11} m

9.40×10^{-11} m

Fig. 9-26 Problem 7.

8P. Figure 9-27 shows a cubical box that has been constructed from metal plate of uniform density and negligible thickness. The box is open at the top and has edge length 40 cm. Find (a) the x coordinate, (b) the y coordinate, and (c) the z coordinate of the center of mass of the box.

Fig. 9-27 Problem 8.

9P*. A right cylindrical can with mass M, height H, and uniform density is initially filled with soda of mass m (Fig. 9-28). We punch small holes in the top and bottom to drain the soda; we then consider the height h of the center of mass of the can and any soda within it. What is h (a) initially and (b) when all the soda has drained? (c) What happens to h during the draining of the soda? (d) If x is the height of the remaining soda at any given instant, find x (in terms of M, H, and m) when the center of mass reaches its lowest point. ssm

Fig. 9-28 Problem 9.

SEC. 9-3 Newton's Second Law for a System of Particles

10E. Two skaters, one with mass 65 kg and the other with mass 40 kg, stand on an ice rink holding a pole of length 10 m and negligible mass. Starting from the ends of the pole, the skaters pull themselves along the pole until they meet. How far does the 40 kg skater move?

11E. An old Chrysler with mass 2400 kg is moving along a straight stretch of road at 80 km/h. It is followed by a Ford with mass 1600 kg moving at 60 km/h. How fast is the center of mass of the two cars moving? ssm

12E. A man of mass m clings to a rope ladder suspended below a balloon of mass M; see Fig. 9-29. The balloon is stationary with respect to the ground. (a) If the man begins to climb the ladder at speed v (with respect to the ladder), in what direction and with what speed (with respect to the ground) will the balloon move?

(b) What is the state of the motion after the man stops climbing?

13P. A stone is dropped at $t = 0$. A second stone, with twice the mass of the first, is dropped from the same point at $t = 100$ ms. (a) How far below the release point is the center of mass of the two stones at $t = 300$ ms? (Neither stone has yet reached the ground.) (b) How fast is the center of mass of the two-stone system moving at that time? ilw

14P. A 1000 kg automobile is at rest at a traffic signal. At the instant the light turns green, the automobile starts to move with a constant acceleration of 4.0 m/s^2.

Fig. 9-29 Exercise 12.

At the same instant a 2000 kg truck, traveling at a constant speed of 8.0 m/s, overtakes and passes the automobile. (a) How far is the center of mass of the automobile–truck system from the traffic light at $t = 3.0$ s? (b) What is the speed of the center of mass of the automobile–truck system then?

15P. A shell is shot with an initial velocity $\vec{v}_0$ of 20 m/s, at an angle of 60° with the horizontal. At the top of the trajectory, the shell explodes into two fragments of equal mass (Fig. 9-30). One fragment, whose speed immediately after the explosion is zero, falls vertically. How far from the gun does the other fragment land, assuming that the terrain is level and that air drag is negligible? ssm

Explosion

$\vec{v}_0$

60°

Fig. 9-30 Problem 15.

16P. A big olive ($m = 0.50$ kg) lies at the origin and a big Brazil nut ($M = 1.5$ kg) lies at the point (1.0, 2.0) m in an xy plane. At $t = 0$, a force $\vec{F}_o = (2\hat{i} + 3\hat{j})$ N begins to act on the olive, and a force $\vec{F}_n = (-3\hat{i} - 2\hat{j})$ N begins to act on the nut. In unit-vector notation, what is the displacement of the center of mass of the olive–nut system at $t = 4.0$ s, with respect to its position at $t = 0$?

17P. Two identical containers of sugar are connected by a massless cord that passes over a massless, frictionless pulley with a diameter of 50 mm (Fig. 9-31). The two containers are at the same level. Each originally has a mass of 500 g. (a) What is the horizontal position of their center of mass? (b) Now 20 g of sugar is transferred from one container to the other, but the containers are prevented from moving. What is the

Fig. 9-31 Problem 17.

new horizontal position of their center of mass, relative to the central axis through the lighter container? (c) The two containers are now released. In what direction does the center of mass move? (d) What is its acceleration? ssm

18P. Ricardo, of mass 80 kg, and Carmelita, who is lighter, are enjoying Lake Merced at dusk in a 30 kg canoe. When the canoe is at rest in the placid water, they exchange seats, which are 3.0 m apart and symmetrically located with respect to the canoe's center. Ricardo notices that the canoe moves 40 cm relative to a submerged log during the exchange and calculates Carmelita's mass, which she has not told him. What is it?

19P. In Fig. 9-32a, a 4.5 kg dog stands on an 18 kg flatboat and is 6.1 m from the shore. He walks 2.4 m along the boat toward shore and then stops. Assuming there is no friction between the boat and the water, find how far the dog is then from the shore. (*Hint:* See Fig. 9-32b. The dog moves leftward and the boat moves rightward, but does the center of mass of the *boat + dog* system move?) ssm www

(a)

Dog's displacement $\vec{d}_d$

Boat's displacement $\vec{d}_b$

(b)

Fig. 9-32 Problem 19.

SEC. 9-5   The Linear Momentum of a System of Particles

20E. How fast must an 816 kg VW Beetle travel (a) to have the same linear momentum as a 2650 kg Cadillac going 16 km/h and (b) to have the same kinetic energy?

21E. Suppose that your mass is 80 kg. How fast would you have to run to have the same linear momentum as a 1600 kg car moving at 1.2 km/h?

22E. A 0.70 kg ball is moving horizontally with a speed of 5.0 m/s when it strikes a vertical wall. The ball rebounds with a speed of 2.0 m/s. What is the magnitude of the change in linear momentum of the ball?

23P. A 2100 kg truck traveling north at 41 km/h turns east and accelerates to 51 km/h. (a) What is the change in the kinetic energy of the truck? What are (b) the magnitude and (c) the direction of the change in the linear momentum of the truck? ilw

24P. A 0.165 kg cue ball with an initial speed of 2.00 m/s bounces off the rail in a game of pool, as shown from an overhead view in Fig. 9-33. For *x* and *y* axes located as shown, the bounce reverses the *y* component of the ball's velocity but does not alter the *x* component. (a) What is θ in Fig. 9-33? (b) What is the change in the ball's linear momentum in unit-vector notation? (The fact that the ball rolls is not relevant to either question.)

Fig. 9-33 Problem 24.

25P. An object is tracked by a radar station and found to have a position vector given by $\vec{r} = (3500 - 160t)\hat{i} + 2700\hat{j} + 300\hat{k}$,

with $\vec{r}$ in meters and *t* in seconds. The radar station's *x* axis points east, its *y* axis north, and its *z* axis vertically up. If the object is a 250 kg meteorological missile, what are (a) its linear momentum, (b) its direction of motion, and (c) the net force on it?

26P. A 0.30 kg softball has a velocity of 15 m/s at an angle of 35° below the horizontal just before making contact with the bat. What is the magnitude of the change in momentum of the ball while it is in contact with the bat if the ball leaves the bat with a velocity of (a) 20 m/s, vertically downward and (b) 20 m/s, horizontally away from the batter and back toward the pitcher?

SEC. 9-6   Conservation of Linear Momentum

27E. A 91 kg man lying on a surface of negligible friction shoves a 68 g stone away from him, giving it a speed of 4.0 m/s. What velocity does the man acquire as a result? ssm

28E. Two blocks of masses 1.0 kg and 3.0 kg are connected by a spring and rest on a frictionless surface. They are given velocities toward each other such that the 1.0 kg block travels initially at 1.7 m/s toward the center of mass, which remains at rest. What is the initial velocity of the other block?

29E. A 75 kg man is riding on a 39 kg cart traveling at a speed of 2.3 m/s. He jumps off with zero horizontal speed relative to the ground. What is the resulting change in the speed of the cart? ssm

30E. A mechanical toy slides along an *x* axis on a frictionless surface with a velocity of $(-0.40 \text{ m/s})\hat{i}$ when two internal springs separate the toy into three parts, as given in the table. What is the velocity of part *A*?

Part	Mass (kg)	Velocity (m/s)
A	0.50	?
B	0.60	$0.20\hat{i}$
C	0.20	$0.30\hat{i}$

31E. A space vehicle is traveling at 4300 km/h relative to Earth when the exhausted rocket motor is disengaged and sent backward with a speed of 82 km/h relative to the command module. The mass of the motor is four times the mass of the module. What is the speed of the command module relative to Earth just after the separation?

32E. A railroad flatcar of weight *W* can roll without friction along a straight horizontal track. Initially, a man of weight *w* is standing on the car, which is moving to the right with speed v_0 (see Fig. 9-34). What is the change in velocity of the car if the man runs to the left (in the figure) so that his speed relative to the car is v_{rel}?

Fig. 9-34 Exercise 32.

33P. The last stage of a rocket, which is traveling at a speed of 7600 m/s, consists of two parts that are clamped together: a rocket case with a mass of 290.0 kg and a payload capsule with a mass of 150.0 kg. When the clamp is released, a compressed spring causes the two parts to separate with a relative speed of 910.0 m/s. What are the speeds of (a) the rocket case and (b) the payload after they have separated? Assume that all velocities are along the same line. Find the total kinetic energy of the two parts (c) before and (d) after they separate; account for any difference. **ssm**

34P. A 4.0 kg mess kit sliding on a frictionless surface explodes into two 2.0 kg parts, one moving at 3.0 m/s, due north, and the other at 5.0 m/s, 30° north of east. What is the original speed of the mess kit?

35P. A certain radioactive nucleus can transform to another nucleus by emitting an electron and a neutrino. (The *neutrino* is one of the fundamental particles of physics.) Suppose that in such a transformation, the initial nucleus is stationary, the electron and neutrino are emitted along perpendicular paths, and the magnitudes of the linear momenta are 1.2×10^{-22} kg·m/s for the electron and 6.4×10^{-23} kg·m/s for the neutrino. As a result of the emissions, the new nucleus moves (recoils). (a) What is the magnitude of its linear momentum? What is the angle between its path and the path of (b) the electron and (c) the neutrino? (d) What is its kinetic energy if its mass is 5.8×10^{-26} kg? **ilw**

36P. Particle *A* and particle *B* are held together with a compressed spring between them. When they are released, the spring pushes them apart and they then fly off in opposite directions, free of the spring. The mass of *A* is 2.00 times the mass of *B*, and the energy stored in the spring was 60 J. Assume that the spring has negligible mass and that all its stored energy is transferred to the particles. Once that transfer is complete, what are the kinetic energies of (a) particle *A* and (b) particle *B*?

37P. A 20.0 kg body is moving in the positive *x* direction with a speed of 200 m/s when, owing to an internal explosion, it breaks into three parts. One part, with a mass of 10.0 kg, moves away from the point of explosion with a speed of 100 m/s in the positive *y* direction. A second fragment, with a mass of 4.00 kg, moves in the negative *x* direction with a speed of 500 m/s. (a) What is the velocity of the third (6.00 kg) fragment? (b) How much energy is released in the explosion? Ignore effects due to the gravitational force. **ssm   ilw   www**

38P. An object, with mass *m* and speed *v* relative to an observer, explodes into two pieces, one three times as massive as the other; the explosion takes place in deep space. The less massive piece stops relative to the observer. How much kinetic energy is added to the system in the explosion, as measured in the observer's reference frame?

39P. A vessel at rest explodes, breaking into three pieces. Two pieces, having equal mass, fly off perpendicular to one another with the same speed of 30 m/s. The third piece has three times the mass of each other piece. What are the magnitude and direction of its velocity immediately after the explosion? **ssm**

40P. An 8.0 kg body is traveling at 2.0 m/s with no external force acting on it. At a certain instant an internal explosion occurs, splitting the body into two chunks of 4.0 kg mass each. The explosion gives the chunks an additional 16 J of kinetic energy. Neither chunk leaves the line of original motion. Determine the speed and direction of motion of each of the chunks after the explosion.

SEC. 9-7 Systems with Varying Mass: A Rocket

41E. A 6090 kg space probe, moving nose-first toward Jupiter at 105 m/s relative to the Sun, fires its rocket engine, ejecting 80.0 kg of exhaust at a speed of 253 m/s relative to the space probe. What is the final velocity of the probe? **ssm**

42E. A rocket is moving away from the solar system at a speed of 6.0×10^3 m/s. It fires its engine, which ejects exhaust with a speed of 3.0×10^3 m/s relative to the rocket. The mass of the rocket at this time is 4.0×10^4 kg, and its acceleration is 2.0 m/s². (a) What is the thrust of the engine? (b) At what rate, in kilograms per second, is exhaust ejected during the firing?

43E. A rocket, which is in deep space and initially at rest relative to an inertial reference frame, has a mass of 2.55×10^5 kg, of which 1.81×10^5 kg is fuel. The rocket engine is then fired for 250 s, during which fuel is consumed at the rate of 480 kg/s. The speed of the exhaust products relative to the rocket is 3.27 km/s. (a) What is the rocket's thrust? After the 250 s firing, what are (b) the mass and (c) the speed of the rocket? **ssm   ilw**

44E. Consider a rocket that is in deep space and at rest relative to an inertial reference frame. The rocket's engine is to be fired for a certain interval. What must be the rocket's *mass ratio* (ratio of initial to final mass) over that interval if the rocket's original speed relative to the inertial frame is to be equal to (a) the exhaust speed (speed of the exhaust products relative to the rocket) and (b) 2.0 times the exhaust speed?

45E. During a lunar mission, it is necessary to increase the speed of a spacecraft by 2.2 m/s when it is moving at 400 m/s relative to the Moon. The speed of the exhaust products from the rocket engine is 1000 m/s relative to the spacecraft. What fraction of the initial mass of the spacecraft must be burned and ejected to accomplish the speed increase?

46E. A railroad car moves at a constant speed of 3.20 m/s under a grain elevator. Grain drops into it at the rate of 540 kg/min. What is the magnitude of the force needed to keep the car moving at constant speed if friction is negligible?

47P. In Fig. 9-35, two long barges are moving in the same direction in still water, one with a speed of 10 km/h and the other with a speed of 20 km/h. While they are passing each other, coal is shov-

Fig. 9-35 Problem 47.

eled from the slower to the faster one at a rate of 1000 kg/min. How much additional force must be provided by the driving engines of (a) the fast barge and (b) the slow barge if neither is to change speed? Assume that the shoveling is always perfectly sideways and that the frictional forces between the barges and the water do not depend on the mass of the barges. ssm

48P. A 6100 kg rocket is set for vertical firing from the ground. If the exhaust speed is 1200 m/s, how much gas must be ejected each second if the thrust (a) is to equal the magnitude of the gravitational force on the rocket and (b) is to give the rocket an initial upward acceleration of 21 m/s²?

SEC. 9-8 External Forces and Internal Energy Changes

49E. In 1981, Daniel Goodwin climbed 443 m up the *exterior* of the Sears Building in Chicago using suction cups and metal clips. (a) Approximate his mass and then compute how much energy he had to transfer from biomechanical (internal) energy to the gravitational potential energy of the Earth–Goodwin system to lift his center of mass to that height. (b) How much energy would he have had to transfer if he had, instead, taken the stairs inside the building (to the same height)?

50E. The summit of Mount Everest is 8850 m above sea level. (a) How much energy would a 90 kg climber expend against the gravitational force on him in climbing to the summit from sea level? (b) How many candy bars, at 1.25 MJ per bar, would supply an energy equivalent to this? Your answer should suggest that work done against the gravitational force is a very small part of the energy expended in climbing a mountain.

51E. A sprinter who weighs 670 N runs the first 7.0 m of a race in 1.6 s, starting from rest and accelerating uniformly. What are the sprinter's (a) speed and (b) kinetic energy at the end of the 1.6 s? (c) What average power does the sprinter generate during the 1.6 s interval?

52E. The luxury liner *Queen Elizabeth 2* has a diesel-electric powerplant with a maximum power of 92 MW at a cruising speed of 32.5 knots. What forward force is exerted on the ship at this speed? (1 knot = 1.852 km/h.)

53E. A swimmer moves through the water at an average speed of 0.22 m/s. The average drag force opposing this motion is 110 N. What average power is required of the swimmer?

54E. An automobile with passengers has weight 16,400 N and is moving at 113 km/h when the driver brakes to a stop. The frictional force on the wheels from the road has a magnitude of 8230 N. Find the stopping distance.

55E. A 55 kg woman leaps vertically from a crouching position in which her center of mass is 40 cm above the ground. As her feet leave the floor her center of mass is 90 cm above the ground; it rises to 120 cm at the top of her leap. (a) As she is pressing down on the ground during the leap, what is the average magnitude of the force on her from the ground? (b) What maximum speed does she attain? ssm www

56P. A 1500 kg automobile starts from rest on a horizontal road and gains a speed of 72 km/h in 30 s. (a) What is the kinetic energy of the auto at the end of the 30 s? (b) What is the average power required of the car during the 30 s interval? (c) What is the instantaneous power at the end of the 30 s interval, assuming that the acceleration is constant?

57P. A locomotive with a power capability of 1.5 MW can accelerate a train from a speed of 10 m/s to 25 m/s in 6.0 min. (a) Calculate the mass of the train. Find (b) the speed of the train and (c) the force accelerating the train as functions of time (in seconds) during the 6.0 min interval. (d) Find the distance moved by the train during the interval.

58P. Resistance to the motion of an automobile consists of road friction, which is almost independent of speed, and air drag, which is proportional to speed-squared. For a certain car with a weight of 12,000 N, the total resistant force F is given by $F = 300 + 1.8v^2$, where F is in newtons and v is in meters per second. Calculate the power (in horsepower) required to accelerate the car at 0.92 m/s² when the speed is 80 km/h.

Additional Problem

59. Some solidified lava contains a pattern of horizontal bubble layers separated vertically with few intermediate bubbles. (Researchers must slice open solidified lava to see these bubbles.) Apparently, as the lava was cooling, bubbles rising from the bottom of the lava separated into these layers and then were locked into place when the lava solidified. Similar layering of bubbles has been studied in certain creamy stouts poured fresh from a tap into a clear glass. The rising bubbles quickly become sorted into layers (Fig. 9-36). The bubbles trapped within a layer rise at speed v_i; the free bubbles between the layers rise at a greater speed v_f. Bubbles breaking free from the top of one layer rise to join the bottom of the next layer. Assume that the rate at which a layer loses height at its top is $dy/dt = v_f$ and the rate at which it gains height at its bottom is $dy/dt = v_f$. Also assume that $v_f = 2.0v_i = 1.0$ cm/s. What are the speed and direction of motion of the layer's center of mass?

Fig. 9-36 Problem 59.

NEW PROBLEMS

N1. In Fig. 9N-1, a stationary block explodes into two pieces L and R that slide across a frictionless floor and then into regions with friction, where they stop. Piece L, with a mass of 2.0 kg, encounters a coefficient of kinetic friction $\mu_L = 0.40$ and slides to a stop in distance $d_L = 0.15$ m. Piece R encounters a coefficient of kinetic friction $\mu_R = 0.50$ and slides to a stop in distance $d_R = 0.25$ m. What was the mass of the block?

Fig. 9N-1 Problem N1.

N2. *"Relative" is an important word.* In Fig. 9N-2, block L of mass $m_L = 1.00$ kg and block R of mass $m_R = 0.500$ kg are held in place with a compressed spring between them. When the blocks are released, the spring sends them sliding across a frictionless

Fig. 9N-2 Problem N2.

floor. (The spring has negligible mass and falls to the floor after the blocks leave it.) (a) If the spring gives block L a release speed of 1.20 m/s *relative* to the floor, how far does block R travel in the next 0.800 s? (b) If, instead, the spring gives block L a release speed of 1.20 m/s *relative* to the velocity that the spring gives block R, how far does block R travel in the next 0.800 s?

N3. At a certain instant, four particles have the xy coordinates and velocities given in the following table. At that instant, what are (a) the coordinates of their center of mass and (b) the velocity of their center of mass?

Particle	Mass (kg)	Position (m)	Velocity (m/s)
1	2.0	0, 3.0	$-9.0\hat{j}$
2	4.0	3.0, 0	$6.0\hat{i}$
3	3.0	0, -2.0	$6.0\hat{j}$
4	12	-1.0, 0	$-2.0\hat{i}$

N4. Figure 9N-3 shows an arrangement with an air track, in which a cart is connected by a cord to a hanging block. The cart has mass $m_1 = 0.600$ kg and its center is initially at xy coordinates $(-0.500$ m, 0 m); the block has mass $m_2 = 0.400$ kg and its center is initially at xy coordinates $(0, -0.100$ m). The mass of the cord and pulley are negligible. The cart is released from rest, and both cart and block move until the cart hits the pulley. The friction between the cart and the air track and between the pulley and its axle is negligible. (a) In unit-vector notation what is the acceleration of the center of mass of the cart–block system? (b) What is the velocity of the center of mass as a function of time t? (c) Sketch the path taken by the system's center of mass. (d) If the path is curved, does it bulge upward to the right or downward to the left? If, instead, it is straight, give the angle between it and the x axis.

Fig. 9N-3 Problem N4.

N5. Figure 9N-4 shows a rocket of mass M moving along an x axis at the constant speed $v_i = 40$ m/s. A small explosion separates the rocket into a rear section (of mass m_1) and a front section; both sec-

Fig. 9N-4 Problem N5.

tions move along the x axis. The relative speed between the rear and front sections is 20 m/s. Approximately what are (a) the minimum possible and (b) the maximum possible values of the final speed v_f of the front section and for what limiting values of m_1 do they occur?

N6. The script for an action movie calls for a small race car (of mass 1500 kg and length 3.0 m) to accelerate along a flattop boat (of mass 4000 kg and length 14 m), from one end to the other, where the car will then

Fig. 9N-5 Problem N6.

jump the gap between the boat and a somewhat lower dock. You are the technical advisor for the movie. The boat will initially touch the dock, as in Fig. 9N-5; the boat can slide through the water without significant resistance; both the car and the boat can be approximated as uniform in their mass distribution. Determine what the width of the gap will be just as the car is about to make the jump.

N7. Show that the ratio of the distances of two particles from their center of mass is the inverse ratio of their masses.

N8. Two particles P and Q are initially at rest 1.0 m apart. P has a mass of 0.10 kg and Q a mass of 0.30 kg. P and Q attract each other with a constant force of 1.0×10^{-2} N. No external forces act on the system. (a) Describe the motion of the center of mass. (b) At what distance from P's original position do the particles collide?

N9. A 110 kg ice hockey player skates at 3.00 m/s toward a railing at the edge of the ice and then stops himself by grasping the railing with his outstretched arms. During this stopping process his center of mass moves 30.0 cm toward the railing. (a) What is the change in the kinetic energy of his center of mass as he stops? (b) What average force must he exert on the railing?

N10. Figure 9N-6 shows a two-ended rocket that is initially stationary on a frictionless floor, with its center at the origin of an x axis. The rocket consists of a central block C (of mass $M = 6.00$ kg) and blocks L and R (each of mass $m = 2.00$ kg) on the left and right sides. Small explosions can shoot either of the side blocks away from block C and along the x axis.

Fig. 9N-6 Problem N10.

Here is the sequence: (1) At time $t = 0$, block L is shot to the left with a speed of 3.00 m/s *relative* to the velocity that the explosion gives the rest of the rocket. (2) Next, at time $t = 0.80$ s, block R is shot to the right with a speed of 3.00 m/s *relative* to the velocity that block C then has. At $t = 2.80$ s, what are (a) the velocity of block C and (b) the position of its center?

N11. An electron (mass $m_1 = 9.11 \times 10^{-31}$ kg) and a proton (mass $m_2 = 1.67 \times 10^{-27}$ kg) attract each other via an electrical force. Suppose that an electron and a proton are released from rest with an initial separation $d = 3.0 \times 10^{-6}$ m. When their separation has decreased to 1.0×10^{-6} m, what is the ratio of (a) the electron's linear momentum to the proton's linear momentum, (b) the electron's speed to the proton's speed, and (c) the electron's kinetic energy to the proton's kinetic energy? (d) As the separation continues to decrease, do the answers to (a) through (c) increase, decrease, or remain the same?

N12. You crouch from a standing position, lowering your center of mass 18 cm in the process. Then you jump vertically. The average force exerted on you by the floor while you jump is three times your weight. What is your upward speed as you pass through your standing position in leaving the floor?

N13. A 2.00 kg particle has the xy coordinates $(-1.20$ m, 0.500 m$)$ and a 4.00 kg particle has the xy coordinates $(0.600$ m, -0.750 m$)$. Both lie on a horizontal plane. At what xy coordinates must you place a 3.00 kg particle such that the center of mass of the three-particle system has the coordinates $(-0.500$ m, -0.700 m$)$?

N14. At time $t = 0$, a ball is struck at ground level and sent over level ground. Figure 9N-7 gives the magnitude p of the ball's momentum versus time t during the flight. At what initial angle is the ball launched?

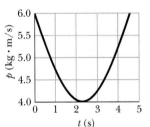

Fig. 9N-7 Problem N14.

N15. Two identical coins are initially held at height $h = 11.0$ m. Coin 1 is dropped at time $t = 0$ and then lands on a muddy field where it sticks. Coin 2 is dropped at $t = 0.500$ s and then lands on the field. What is the acceleration $\vec{a}_{com}$ of the center of mass (com) of the two-coin system (a) from $t = 0$ to $t = 0.500$ s, (b) from $t = 0.500$ s to time t_1 when coin 1 hits and sticks, and (c) from t_1 to time t_2 when coin 2 hits and sticks? What is the speed of the center of mass when t is (d) 0.250 s, (e) 0.750 s, and (f) 1.75 s?

N16. In Figure 9N-8, two particles are launched from the origin of the coordinate system at time $t = 0$. Particle 1 of mass $m_1 = 5.00$ g is shot directly along an x axis on a frictionless floor, where it moves with a constant speed of 10.0 m/s. Particle 2 of mass $m_2 = 3.00$ g is shot with a velocity of magnitude 20.0 m/s, at an upward angle such that it always stays directly above particle 1 during its flight. (a) What is the maximum height H_{max} reached by

the center of mass of the two-particle system? What are the magnitude and direction of (b) the velocity and (c) the acceleration of the center of mass when the center of mass reaches H_{max}?

Fig. 9N-8 Problem N16.

N17. A 2.65 kg stationary package explodes into three parts that then slide across a frictionless floor. The package had been at the origin of a coordinate system. Part 1 has mass $m_1 = 0.500$ kg and velocity $(10.0\hat{i} + 12.0\hat{j})$ m/s. Part 2 has mass $m_2 = 0.750$ kg, a speed of 14.0 m/s, and travels at an angle $110°$ counterclockwise from the positive direction of the x axis. (a) What is the speed of part 3? (b) In what direction does it travel?

N18. At time $t = 0$, force $\vec{F}_1 = (-4.00\hat{i} + 5.00\hat{j})$ N acts on an initially stationary particle of mass 2.00×10^{-3} kg and force $\vec{F}_2 = (2.00\hat{i} - 4.00\hat{j})$ N acts on an initially stationary particle of mass 4.00×10^{-3} kg. From time $t = 0$ to $t = 2.00$ ms, what are (a) the magnitude and (b) the angle of the displacement of the center of mass of the two-particle system? (c) What is the kinetic energy of the center of mass at $t = 2.00$ ms?

N19. A 500.0 kg module is attached to a 400.0 kg shuttle craft, which moves at 1000 m/s relative to the stationary main spaceship. Then a small explosion sends the module backward with speed 100.0 m/s relative to the new speed of the shuttle craft. As measured by someone on the main spaceship, by what fraction did the kinetic energy of the module and shuttle craft increase because of the explosion?

N20. A 0.20 kg hockey puck is sliding on a frictionless ice surface with a velocity of 10 m/s toward the east just before making contact with a hockey stick. What change in momentum (magnitude and direction) does the puck undergo while in contact with the stick if just afterward the velocity of the puck is (a) 20 m/s toward the east, (b) 5.0 m/s toward the east, and (c) 10 m/s toward the west?

N21. A child who is standing in a 95 kg flat-bottom boat is initially 6.0 m from shore. The child starts to walk along the boat toward the shore. When the child has walked 2.5 m relative to the boat, he is 4.1 m from the shore. Assume there is no friction between the boat and the water. What is the child's mass?

N22. What are (a) the x coordinate and (b) the y coordinate of the center of mass for the uniform plate shown in Fig. 9N-9?

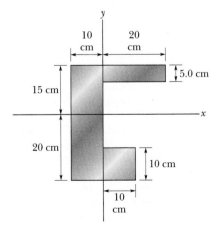

Fig. 9N-9 Problem N22.

10 Collisions

Ronald McNair, a physicist and one of the astronauts killed in the explosion of the *Challenger* space shuttle, held a black belt in karate. Here he breaks several concrete slabs with one blow. In such karate demonstrations, a pine board or a concrete "patio block" is typically used. When struck, the board or block bends, storing energy like a stretched spring does, until a critical energy is reached. Then the object breaks. The energy necessary to break the board is about three times that for the block.

Why then is the board considerably easier to break?

The answer is in this chapter.

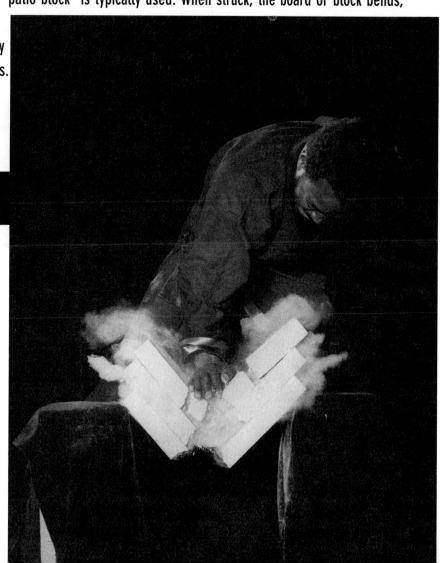

10-1 What Is a Collision?

In everyday language, a *collision* occurs when objects crash into each other. Although we will refine that definition, it conveys the meaning well enough and covers common collisions, such as those between billiard balls, a hammer and a nail, and—too commonly—automobiles. Figure 10-1*a* shows the lasting result of a collision (an impressive crash) that occurred about 20,000 years ago. Collisions range from the microscopic scale of subatomic particles (Fig. 10-1*b*) to the astronomic scale of colliding stars and colliding galaxies. Even when they occur on a human scale, they are often too brief to be visible, although they involve significant distortion of the *colliding bodies* (Fig. 10-1*c*).

We shall use the following more formal definition of collision:

> A **collision** is an isolated event in which two or more bodies (the colliding bodies) exert relatively strong forces on each other for a relatively short time.

We must be able to distinguish times that are *before, during,* and *after* a collision, as suggested in Fig. 10-2. That figure shows a system of two colliding bodies and indicates that the forces the bodies exert on each other are internal to the system.

Note that our formal definition of collision does not require the "crash" of our informal definition. When a space probe swings around a large planet to pick up speed (a *slingshot* encounter), that too is a collision. The probe and planet do not actually "touch," but a collision does not require contact, and a collision force does not have to be a force involving contact; it can just as easily be a gravitational force, as in this case.

(a)

(b)

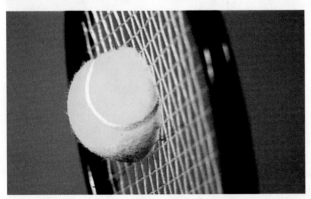

(c)

Fig. 10-1 Collisions range widely in scale. (*a*) Meteor Crater in Arizona is about 1200 m wide and 200 m deep. (*b*) An alpha particle coming in from the left (whose trail is colored yellow in this false-color photograph) bounces off a nitrogen nucleus that had been stationary and that now moves toward the bottom right (red trail). (*c*) In a tennis match, the ball is in contact with the racquet for about 4 ms in each collision (for a cumulative time of only 1 s per set).

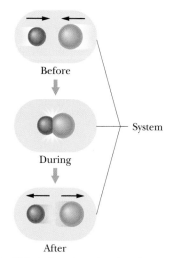

Fig. 10-2 A flowchart showing the system in which a collision occurs.

Many physicists today spend their time playing what we can call "the collision game." A principal goal of this game is to find out as much as possible about the forces that act during a collision, from knowledge of the state of the particles before and after the collision. Virtually all our understanding of the subatomic world—electrons, protons, neutrons, muons, quarks, and the like—comes from experiments involving collisions. The rules of the game are the laws of conservation of momentum and of energy.

10-2 Impulse and Linear Momentum

Single Collision

Figure 10-3 shows the third-law force pair, $\vec{F}(t)$ and $-\vec{F}(t)$, that acts during a simple head-on collision between two particle-like bodies of different masses. These forces will change the linear momentum of both bodies; the amount of the change will depend not only on the average values of the forces, but also on the time Δt during which they act. To see this quantitatively, let us apply Newton's second law in the form $\vec{F} = d\vec{p}/dt$ to, say, body R on the right in Fig. 10-3. We have

$$d\vec{p} = \vec{F}(t)\, dt, \tag{10-1}$$

in which $\vec{F}(t)$ is a time-varying force with magnitude given by the curve in Fig. 10-4a. Let us integrate Eq. 10-1 over the interval Δt—that is, from an initial time t_i (just before the collision) to a final time t_f (just after the collision). We obtain

$$\int_{\vec{p}_i}^{\vec{p}_f} d\vec{p} = \int_{t_i}^{t_f} \vec{F}(t)\, dt. \tag{10-2}$$

Fig. 10-3 Two particle-like bodies L and R collide with each other. During the collision, body L exerts force $\vec{F}(t)$ on body R, and body R exerts force $-\vec{F}(t)$ on body L. Forces $\vec{F}(t)$ and $-\vec{F}(t)$ are a third-law force pair. Their magnitudes vary with time during the collision, but at any given instant those magnitudes are equal.

The left side of this equation is $\vec{p}_f - \vec{p}_i$, the change in linear momentum of body R. The right side, which is a measure of both the strength and the duration of the collision force, is called the **impulse** $\vec{J}$ of the collision. Thus,

$$\vec{J} = \int_{t_i}^{t_f} \vec{F}(t)\, dt \qquad \text{(impulse defined)}. \tag{10-3}$$

Equation 10-3 tells us that the magnitude of the impulse is equal to the area under the $F(t)$ curve of Fig. 10-4a. Because $F(t)$ on body R and $-F(t)$ on body L are third-law force pairs, their impulses have the same magnitudes but opposite directions.

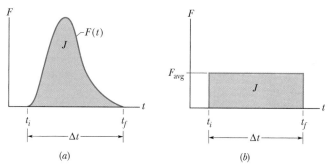

(a)   (b)

Fig. 10-4 (a) The curve shows the magnitude of the time-varying force $F(t)$ that acts on body R during the collision of Fig. 10-3. The area under the curve is equal to the magnitude of the impulse $\vec{J}$ on body R in the collision. (b) The height of the rectangle represents the average force F_{avg} acting on body R over the time interval Δt. The area within the rectangle is equal to the area under the curve in (a) and thus is also equal to the magnitude of the impulse $\vec{J}$ in the collision.

From Eqs. 10-2 and 10-3 we see that the change in the linear momentum of each body in a collision is equal to the impulse that acts on that body:

$$\vec{p}_f - \vec{p}_i = \Delta\vec{p} = \vec{J} \qquad \text{(impulse–linear momentum theorem).} \qquad (10\text{-}4)$$

Equation 10-4 is called the **impulse–linear momentum theorem;** it tells us that impulse and linear momentum are both vectors and have the same units and dimensions. Equation 10-4 can also be written in component form as

$$p_{fx} - p_{ix} = \Delta p_x = J_x, \qquad (10\text{-}5)$$

$$p_{fy} - p_{iy} = \Delta p_y = J_y, \qquad (10\text{-}6)$$

and

$$p_{fz} - p_{iz} = \Delta p_z = J_z. \qquad (10\text{-}7)$$

If F_{avg} is the average magnitude of the force in Fig. 10-4a, we can write the magnitude of the impulse as

$$J = F_{avg}\,\Delta t, \qquad (10\text{-}8)$$

where Δt is the duration of the collision. The value of F_{avg} must be such that the area within the rectangle of Fig. 10-4b is equal to the area under the actual $F(t)$ curve of Fig. 10-4a.

✔**CHECKPOINT 1:**   A paratrooper whose chute fails to open lands in snow; he is hurt slightly. Had he landed on bare ground, the stopping time would have been 10 times shorter and the collision lethal. Does the presence of the snow increase, decrease, or leave unchanged the values of (a) the paratrooper's change in momentum, (b) the impulse stopping the paratrooper, and (c) the force stopping the paratrooper?

Series of Collisions

Now let's consider the force on a body when it undergoes a series of identical, repeated collisions. For example, as a prank, we might adjust one of those machines that fires tennis balls to fire them at a rapid rate directly at a wall. Each collision would produce a force on the wall, but that is not the force we are seeking. We want the average force F_{avg} on the wall during the bombardment—that is, the average force during a large number of collisions.

In Fig. 10-5, a steady stream of projectile bodies, with identical mass m and linear momenta $m\vec{v}$, moves along an x axis and collides with a target body that is fixed in place. Let n be the number of projectiles that collide in a time interval Δt. Because the motion is along only the x axis, we can use the components of the momenta along that axis. Thus, each projectile has initial momentum mv and undergoes a change Δp in linear momentum because of the collision. The total change in linear momentum for n projectiles during interval Δt is $n\,\Delta p$. The resulting impulse J on the target during Δt is along the x axis and has the same magnitude of $n\,\Delta p$ but is in the opposite direction. We can write this relation in component form as

$$J = -n\,\Delta p, \qquad (10\text{-}9)$$

where the minus sign indicates that J and Δp have opposite directions.

By rearranging Eq. 10-8 and substituting Eq. 10-9, we find the average force F_{avg} acting on the target during the collisions:

$$F_{avg} = \frac{J}{\Delta t} = -\frac{n}{\Delta t}\Delta p = -\frac{n}{\Delta t}m\,\Delta v. \qquad (10\text{-}10)$$

This equation gives us F_{avg} in terms of $n/\Delta t$, the rate at which the projectiles collide with the target, and Δv, the change in the velocity of those projectiles.

Fig. 10-5 A steady stream of projectiles, with identical linear momenta, collides with a target, which is fixed in place. The average force F_{avg} on the target is to the right and has a magnitude that depends on the rate at which the projectiles collide or, equivalently, the rate at which mass collides.

If the projectiles stop upon impact, then in Eq. 10-10 we can substitute, for Δv,

$$\Delta v = v_f - v_i = 0 - v = -v, \tag{10-11}$$

where $v_i \, (= v)$ and $v_f \, (= 0)$ are the velocities before and after the collision, respectively. If, instead, the projectiles bounce (rebound) directly backward from the target with no change in speed, then $v_f = -v$ and we can substitute

$$\Delta v = v_f - v_i = -v - v = -2v. \tag{10-12}$$

In time interval Δt, an amount of mass $\Delta m = nm$ collides with the target. With this result, we can rewrite Eq. 10-10 as

$$F_{avg} = -\frac{\Delta m}{\Delta t} \Delta v. \tag{10-13}$$

This equation gives the average force F_{avg} in terms of $\Delta m / \Delta t$, the rate at which mass collides with the target. Here again we can substitute for Δv from Eq. 10-11 or 10-12 depending on what the projectiles do.

✓CHECKPOINT 2:   The figure shows an overhead view of a ball bouncing from a vertical wall without any change in its speed. Consider the change $\Delta \vec{p}$ in the ball's linear momentum. (a) Is Δp_x positive, negative, or zero? (b) Is Δp_y positive, negative, or zero? (c) What is the direction of $\Delta \vec{p}$?

Sample Problem 10-1

A pitched 140 g baseball, in horizontal flight with a speed v_i of 39.0 m/s, is struck by a bat. After leaving the bat, the ball travels in the opposite direction with speed v_f, also 39.0 m/s.

(a) What impulse J acts on the ball while it is in contact with the bat during the collision?

SOLUTION: The Key Idea here is that we can calculate the impulse from the change it produces in the ball's linear momentum, using Eq. 10-4 for one-dimensional motion. Let us choose the direction in which the ball is initially moving to be the negative direction. From Eq. 10-4 we have

$$\begin{aligned} J &= p_f - p_i = mv_f - mv_i \\ &= (0.140 \text{ kg})(39.0 \text{ m/s}) - (0.140 \text{ kg})(-39.0 \text{ m/s}) \\ &= 10.9 \text{ kg} \cdot \text{m/s}. \quad \text{(Answer)} \end{aligned}$$

With our sign convention, the initial velocity of the ball is negative and the final velocity is positive. The impulse turns out to be positive, which tells us that the direction of the impulse vector acting on the ball is the direction in which the bat is swinging.

(b) The impact time Δt for the baseball–bat collision is 1.20 ms. What average force acts on the baseball?

SOLUTION: The Key Idea here is that the average force of the collision is the ratio of the impulse J to the duration Δt of the collision (see Eq. 10-8). Thus,

$$F_{avg} = \frac{J}{\Delta t} = \frac{10.9 \text{ kg} \cdot \text{m/s}}{0.00120 \text{ s}}$$

$$= 9080 \text{ N}. \quad \text{(Answer)}$$

Note that this is the *average* force; the *maximum* force is larger. The sign of the average force on the ball from the bat is positive, which means that the direction of the force vector is the same as that of the impulse vector.

In defining a collision, we assumed that no net external force acts on the colliding bodies. That is not true in this case, because the gravitational force always acts on the ball, whether the ball is in flight or in contact with the bat. However, this force, with a magnitude of $mg = 1.37$ N, is negligible compared to the average force exerted by the bat, which has a magnitude of 9080 N. We are quite safe in treating the collision as "isolated."

(c) Now suppose the collision is not head-on, and the ball leaves the bat with a speed v_f of 45.0 m/s at an upward angle of 30.0° (Fig. 10-6). What now is the impulse on the ball?

SOLUTION: The Key Idea here is that now the collision is two-dimensional because the ball's outward path is not along the same axis as its incoming path. Thus, we must use vectors to find the impulse

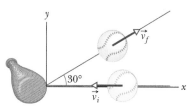

Fig. 10-6 Sample Problem 10-1. A bat collides with a pitched baseball, sending the ball off at an angle of 30° up from the horizontal.

$\vec{J}$. From Eq. 10-4, we can write

$$\vec{J} = \Delta\vec{p} = \vec{p}_f - \vec{p}_i = m\vec{v}_f - m\vec{v}_i.$$

Thus, $$\vec{J} = m(\vec{v}_f - \vec{v}_i). \qquad (10\text{-}14)$$

We can evaluate the right side of this equation directly on a vector-capable calculator, since we know that the mass m is 0.140 kg, the final velocity $\vec{v}_f$ is 45.0 m/s at 30.0°, and the initial velocity $\vec{v}_i$ is 39.0 m/s at 180°.

Instead, we can evaluate Eq. 10-14 in component form. To do so, we first place an xy coordinate system as shown in Fig. 10-6. Then along the x axis we have

$$J_x = p_{fx} - p_{ix} = m(v_{fx} - v_{ix})$$
$$= (0.140 \text{ kg})[(45.0 \text{ m/s})(\cos 30.0°) - (-39.0 \text{ m/s})]$$
$$= 10.92 \text{ kg} \cdot \text{m/s}.$$

Along the y axis,

$$J_y = p_{fy} - p_{iy} = m(v_{fy} - v_{iy})$$
$$= (0.140 \text{ kg})[(45.0 \text{ m/s})(\sin 30.0°) - 0]$$
$$= 3.150 \text{ kg} \cdot \text{m/s}.$$

The impulse is then

$$\vec{J} = (10.9\hat{i} + 3.15\hat{j}) \text{ kg} \cdot \text{m/s}, \qquad \text{(Answer)}$$

and the magnitude and direction of $\vec{J}$ are

$$J = \sqrt{J_x^2 + J_y^2} = 11.4 \text{ kg} \cdot \text{m/s}$$

and

$$\theta = \tan^{-1}\frac{J_y}{J_x} = 16°. \qquad \text{(Answer)}$$

10-3 Momentum and Kinetic Energy in Collisions

Consider a system of two colliding bodies. If there is to be a collision, then at least one of the bodies must be moving, so the system has a certain kinetic energy and a certain linear momentum before the collision. During the collision, the kinetic energy and linear momentum of each body are changed by the impulse from the other body. For the rest of this chapter we shall discuss these changes—and also the changes in the kinetic energy and linear momentum of the system as a whole—without knowing the details of the impulses that determine the changes. The discussion will be limited to collisions in systems that are **closed** (no mass enters or leaves them) and **isolated** (no net external forces act on the bodies within the system).

Kinetic Energy

If the total kinetic energy of the system of two colliding bodies is unchanged by the collision, then the kinetic energy of the system is *conserved* (it is the same before and after the collision). Such a collision is called an **elastic collision.** In everyday collisions of common bodies, such as two cars or a ball and a bat, some energy is always transferred from kinetic energy to other forms of energy, such as thermal energy or energy of sound. Thus, the kinetic energy of the system is *not* conserved. Such a collision is called an **inelastic collision.**

However, in some situations, we can *approximate* a collision of common bodies as elastic. Suppose that you drop a Superball onto a hard floor. If the collision between the ball and floor (or Earth) were elastic, the ball would lose no kinetic energy because of the collision and would rebound to its original height. However, the actual rebound height is somewhat short, showing that at least some kinetic energy is lost in the collision and thus that the collision is somewhat inelastic. Still, we might choose to neglect that small loss of kinetic energy to approximate the collision as elastic.

A dropped golf ball will lose more of its kinetic energy and will rebound to only 60% of its original height. This collision is noticeably inelastic and cannot be approximated as elastic. If you drop a ball of wet putty onto the floor, it sticks to the floor and does not rebound at all. Because the putty sticks, this collision is called a **completely inelastic collision.** Figure 10-7 shows a more dramatic example of a completely inelastic collision. In such collisions, the bodies always stick together and lose kinetic energy.

Fig. 10-7 Two cars after an almost head-on, almost completely inelastic collision.

Linear Momentum

Regardless of the details of the impulses in a collision, and regardless of what happens to the total kinetic energy of the system, the total linear momentum $\vec{P}$ of a closed, isolated system *cannot* change. The reason is that $\vec{P}$ can be changed only by external forces (from outside the system), but the forces in the collision are internal forces (inside the system). Thus, we have this important rule:

> ▶ In a closed, isolated system containing a collision, the linear momentum of each colliding body may change but the total linear momentum $\vec{P}$ of the system cannot change, whether the collision is elastic or inelastic.

This is actually another statement of the **law of conservation of linear momentum** that we first discussed in Section 9-6. In the next two sections we apply this law to some specific collisions, first inelastic and then elastic.

10-4 Inelastic Collisions in One Dimension

One-Dimensional Collision

Fig. 10-8 Bodies 1 and 2 move along an *x* axis, before and after they have an inelastic collision.

Figure 10-8 shows two bodies just before and just after they have a *one-dimensional collision* (meaning that the motions before and after the collision are along a single axis). The velocities before the collision (subscript *i*) and after the collision (subscript *f*) are indicated. The two bodies form our system, which is closed and isolated. We can write the law of conservation of linear momentum for this two-body system as

$$\left(\begin{array}{c} \text{total momentum } \vec{P}_i \\ \text{before the collision} \end{array} \right) = \left(\begin{array}{c} \text{total momentum } \vec{P}_f \\ \text{after the collision} \end{array} \right),$$

which we can symbolize as

$$\vec{p}_{1i} + \vec{p}_{2i} = \vec{p}_{1f} + \vec{p}_{2f} \qquad \text{(conservation of linear momentum).} \qquad (10\text{-}15)$$

Because the motion is one-dimensional, we can drop the overhead arrows for vectors and use only components along the axis. Thus, from $p = mv$, we can rewrite Eq. 10-15 as

$$m_1 v_{1i} + m_2 v_{2i} = m_1 v_{1f} + m_2 v_{2f}. \qquad (10\text{-}16)$$

If we know values for, say, the masses, the initial velocities, and one of the final velocities, we can find the other final velocity with Eq. 10-16.

Completely Inelastic Collision

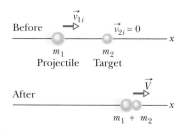

Fig. 10-9 A completely inelastic collision between two bodies. Before the collision, the body with mass m_2 is at rest and the body with mass m_1 moves directly toward it. After the collision, the stuck-together bodies move with the same velocity $\vec{V}$.

Figure 10-9 shows two bodies before and after they have a completely inelastic collision (meaning they stick together). The body with mass m_2 happens to be initially at rest ($v_{2i} = 0$). We can refer to that body as the *target*, and the incoming body as the *projectile*. After the collision, the stuck-together bodies move with velocity V. For this situation, we can rewrite Eq. 10-16 as

$$m_1 v_{1i} = (m_1 + m_2)V \qquad (10\text{-}17)$$

or

$$V = \frac{m_1}{m_1 + m_2} v_{1i}. \qquad (10\text{-}18)$$

If we know values for, say, the masses and the initial velocity v_{1i} of the projectile, we can find the final velocity V with Eq. 10-18. Note that V must be less than v_{1i}, because the mass ratio $m_1/(m_1 + m_2)$ must be less than unity.

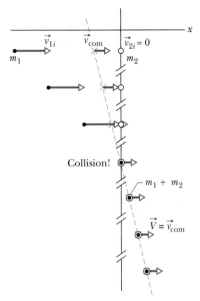

Fig. 10-10 Some freeze-frames of the two-body system in Fig. 10-9, which undergoes a completely inelastic collision. The system's center of mass is shown in each freeze-frame. The velocity $\vec{v}_{com}$ of the center of mass is unaffected by the collision. Because the bodies stick together after the collision, their common velocity $\vec{V}$ must be equal to $\vec{v}_{com}$.

Velocity of Center of Mass

In a closed, isolated system, the velocity $\vec{v}_{com}$ of the center of mass of the system cannot be changed by a collision because, with the system isolated, there is no net external force to change it. To get an expression for $\vec{v}_{com}$, let us return to the two-body system and one-dimensional collision of Fig. 10-8. From Eq. 9-25 ($\vec{P} = M\vec{v}_{com}$), we can relate $\vec{v}_{com}$ to the total linear momentum $\vec{P}$ of that two-body system by writing

$$\vec{P} = M\vec{v}_{com} = (m_1 + m_2)\vec{v}_{com}. \qquad (10\text{-}19)$$

The total linear momentum $\vec{P}$ is conserved during the collision; so it is given by either side of Eq. 10-15. Let us use the left side to write

$$\vec{P} = \vec{p}_{1i} + \vec{p}_{2i}. \qquad (10\text{-}20)$$

Substituting this for $\vec{P}$ in Eq. 10-19 and solving for $\vec{v}_{com}$ give us

$$\vec{v}_{com} = \frac{\vec{P}}{m_1 + m_2} = \frac{\vec{p}_{1i} + \vec{p}_{2i}}{m_1 + m_2}. \qquad (10\text{-}21)$$

The right side of this equation is a constant, and $\vec{v}_{com}$ has that same constant value before and after the collision.

For example, Fig. 10-10 shows, in a series of freeze-frames, the motion of the center of mass for the completely inelastic collision of Fig. 10-9. Body 2 is the target, and its initial linear momentum in Eq. 10-21 is $\vec{p}_{2i} = m_2\vec{v}_{2i} = 0$. Body 1 is the projectile, and its initial linear momentum in Eq. 10-21 is $\vec{p}_{1i} = m_1\vec{v}_{1i}$. Note that as the series of freeze-frames progresses to and then beyond the collision, the center of mass moves at a constant velocity to the right. After the collision, the common final speed V of the bodies is equal to $\vec{v}_{com}$ because then the center of mass travels with the stuck-together bodies.

✔**CHECKPOINT 3:**   Body 1 and body 2 are in a completely inelastic one-dimensional collision. What is their final momentum if their initial momenta are, respectively, (a) 10 kg · m/s and 0; (b) 10 kg · m/s and 4 kg · m/s; (c) 10 kg · m/s and −4 kg · m/s?

Sample Problem 10-2

The *ballistic pendulum* was used to measure the speeds of bullets before electronic timing devices were developed. The version shown in Fig. 10-11 consists of a large block of wood of mass $M = 5.4$ kg, hanging from two long cords. A bullet of mass $m = 9.5$ g is fired into the block, coming quickly to rest. The *block + bullet* then swing upward, their center of mass rising a vertical distance $h = 6.3$ cm before the pendulum comes momentarily to rest at the end of its arc. What is the speed of the bullet just prior to the collision?

SOLUTION: We can see that the bullet's speed v must determine the rise height h. However, a Key Idea is that we cannot use the conservation of mechanical energy to relate these two quantities because surely energy is transferred from mechanical energy to other forms (such as thermal energy and energy to break apart the wood) as the bullet penetrates the block. Another Key Idea helps—we can split this complicated motion into two steps that we can separately analyze: (1) the bullet–block collision and (2) the bullet–block rise, during which mechanical energy *is* conserved.

Fig. 10-11 Sample Problem 10-2. A ballistic pendulum, used to measure the speeds of bullets.

Step 1. Because the collision within the bullet–block system is so brief, we can make two important assumptions: (1) During the collision, the gravitational force on the block and the force on the block from the cords are still balanced. Thus, during the collision, the net external impulse on the bullet–block system is zero. Therefore, the system is isolated and its total linear momentum is conserved. (2) The collision is one-dimensional in the sense that the direction of the bullet and block *just after the collision* is in the bullet's original direction of motion.

Because the collision is one-dimensional, the block is initially at rest, and the bullet sticks in the block, we use Eq. 10-18 to express the conservation of linear momentum. Replacing the symbols there with the corresponding symbols here, we have

$$V = \frac{m}{m + M}\, v. \tag{10-22}$$

Step 2. As the bullet and block now swing up together, the mechanical energy of the bullet–block–Earth system is conserved. (This mechanical energy is not changed by the force of the cords on the block, because that force is always directed perpendicular to the block's direction of travel.) Let's take the block's initial level as our reference level of zero gravitational potential energy. Then conservation of mechanical energy means that the system's kinetic energy at the start of the swing must equal its gravitational potential energy at the highest point of the swing. Because the speed of the bullet and block at the start of the swing is the speed V immediately after the collision, we may write this conservation as

$$\tfrac{1}{2}(m + M)V^2 = (m + M)gh.$$

Substituting for V from Eq. 10-22 leads to

$$v = \frac{m + M}{m}\sqrt{2gh}$$

$$= \left(\frac{0.0095 \text{ kg} + 5.4 \text{ kg}}{0.0095 \text{ kg}}\right)\sqrt{(2)(9.8 \text{ m/s}^2)(0.063 \text{ m})}$$

$$= 630 \text{ m/s}. \tag{Answer}$$

The ballistic pendulum is a kind of "transformer," exchanging the high speed of a light object (the bullet) for the low—and thus more easily measurable—speed of a massive object (the block).

Sample Problem 10-3

A karate expert strikes downward with his fist (of mass $m_1 = 0.70$ kg), breaking a 0.14 kg board (Fig. 10-12a). He then does the same to a 3.2 kg concrete block. The spring constants k for bending are 4.1×10^4 N/m for the board and 2.6×10^6 N/m for the block.

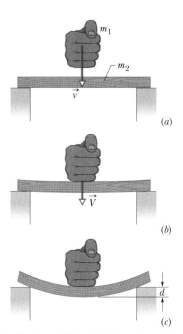

Fig. 10-12 Sample Problem 10-3. (*a*) A karate expert strikes at a flat object with speed v. (*b*) Fist and object undergo a completely inelastic collision, and bending begins. The *fist + object* then have speed V. (*c*) The object breaks when its center has been deflected by an amount d.

Breaking occurs at a deflection d of 16 mm for the board and 1.1 mm for the block (Fig. 10-12c). (The data are taken from "The Physics of Karate," by S. R. Wilk, R. E. McNair, and M. S. Feld, *American Journal of Physics,* September 1983.)

(a) Just before the object (board or block) breaks, what is the energy stored in it?

SOLUTION: The Key Idea here is that we can treat the bending as the compression of a spring for which Hooke's law applies. The stored potential energy is then, from Eq. 8-11, $U = \tfrac{1}{2}kd^2$. For the board,

$$U = \tfrac{1}{2}(4.1 \times 10^4 \text{ N/m})(0.016 \text{ m})^2$$

$$= 5.248 \text{ J} \approx 5.2 \text{ J}. \tag{Answer}$$

For the block,

$$U = \tfrac{1}{2}(2.6 \times 10^6 \text{ N/m})(0.0011 \text{ m})^2$$

$$= 1.573 \text{ J} \approx 1.6 \text{ J}. \tag{Answer}$$

(b) What is the lowest fist speed v_{fist} required to break the object (board or block)? Assume the following: The collisions are completely inelastic collisions of only the fist and the object. Bending begins just after the collision. Mechanical energy is conserved from the beginning of the bending until just before the object breaks. The speed of the fist and object is negligible at that point.

SOLUTION: The Key Idea here is that we can split up this complicated motion into three steps that we can separately analyze:

1. The completely inelastic one-dimensional collision of the fist and the object transfers energy to the kinetic energy of the fist–object system.

2. That energy is then transferred to the potential energy U stored by the bending.

3. The object breaks when U reaches the value calculated in part (a).

In step 1, we can use Eq. 10-18 to relate the fist speed v_{fist} just before the collision to the fist–object speed V_{fo} just after the collision and as the bending begins. With the notation we are using here, Eq. 10-18 becomes

$$V_{\text{fo}} = \frac{m_1}{m_1 + m_2} v_{\text{fist}}. \qquad (10\text{-}23)$$

In step 2, the mechanical energy of the fist–object system is conserved during the bending (up to the break). (Because the downward deflection of the object is small, the change in the gravitational potential energies of fist and object during the deflection are small enough to be neglected.) We can write the conservation of mechanical energy during the bending as

$$\begin{pmatrix} \text{kinetic energy at} \\ \text{start of bending} \end{pmatrix} = \begin{pmatrix} \text{potential energy of bending} \\ \text{just before the break} \end{pmatrix},$$

or

$$\tfrac{1}{2}(m_1 + m_2)V_{\text{fo}}^2 = U. \qquad (10\text{-}24)$$

Substituting for V_{fo} from Eq. 10-23 and solving for v_{fist}, we find

$$v_{\text{fist}} = \frac{1}{m_1}\sqrt{2U(m_1 + m_2)}. \qquad (10\text{-}25)$$

For step 3, we substitute the proper mass value and the breaking value U as obtained in part (a). We find that for the board,

$$v_{\text{fist}} = 4.2 \text{ m/s} \qquad \text{(Answer)}$$

and for the block,

$$v_{\text{fist}} = 5.0 \text{ m/s.} \qquad \text{(Answer)}$$

Thus, from the answer to (a) we see that breaking a board requires the greater energy. However, from the answers to (b) we see why a board is easier to break: The required fist speed is less. The reason lies in Eq. 10-23. If we decrease the target mass in the collision, we increase the speed V_{fo} given to the object. Thus, we also increase the fraction of the fist's energy that is transferred to the object. (One reason why breaking a pencil in the arrangement of Fig. 10-12 would be easy is because of the pencil's small mass.)

10-5 Elastic Collisions in One Dimension

Stationary Target

As we discussed in Section 10-3, everyday collisions are inelastic but we can approximate some of them as being elastic; that is, we can approximate that the total kinetic energy of the colliding bodies is conserved and is not transferred to other forms of energy:

$$\begin{pmatrix} \text{total kinetic energy} \\ \text{before the collision} \end{pmatrix} = \begin{pmatrix} \text{total kinetic energy} \\ \text{after the collision} \end{pmatrix}.$$

This does not mean that the kinetic energy of each colliding body cannot change. Rather, it means this:

▶ In an elastic collision, the kinetic energy of each colliding body may change, but the total kinetic energy of the system does not change.

For example, the collision of a cue ball with an object ball in a game of pool can be approximated as being an elastic collision. If the collision is head-on (the cue ball heads directly toward the object ball), the kinetic energy of the cue ball can be transferred almost entirely to the object ball. (Still, the fact that the collision makes a sound means that at least a little of the kinetic energy is transferred to the energy of the sound.)

Figure 10-13 shows two bodies before and after they have a one-dimensional collision, like a head-on collision between pool balls. A projectile body of mass m_1 and initial velocity v_{1i} moves toward a target body of mass m_2 that is initially at rest ($v_{2i} = 0$). Let's assume that this two-body system is closed and isolated. Then the net linear momentum of the system is conserved, and from Eq. 10-15 we can write that conservation as

$$m_1 v_{1i} = m_1 v_{1f} + m_2 v_{2f} \qquad \text{(linear momentum).} \qquad (10\text{-}26)$$

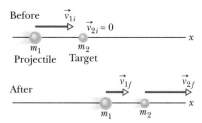

Fig. 10-13 Body 1 moves along an x axis before having an elastic collision with body 2, which is initially at rest. Both bodies move along that axis after the collision.

If the collision is also elastic, then the total kinetic energy is conserved and we can write that conservation as

$$\tfrac{1}{2}m_1v_{1i}^2 = \tfrac{1}{2}m_1v_{1f}^2 + \tfrac{1}{2}m_2v_{2f}^2 \qquad \text{(kinetic energy)}. \qquad (10\text{-}27)$$

In each of these equations, the subscript i identifies the initial velocities and the subscript f the final velocities of the bodies. If we know the masses of the bodies and if we also know v_{1i}, the initial velocity of body 1, the only unknown quantities are v_{1f} and v_{2f}, the final velocities of the two bodies. With two equations at our disposal, we should be able to find these two unknowns.

To do so, we rewrite Eq. 10-26 as

$$m_1(v_{1i} - v_{1f}) = m_2v_{2f} \qquad (10\text{-}28)$$

and Eq. 10-27 as*

$$m_1(v_{1i} - v_{1f})(v_{1i} + v_{1f}) = m_2v_{2f}^2. \qquad (10\text{-}29)$$

After dividing Eq. 10-29 by Eq. 10-28 and doing some more algebra, we obtain

$$v_{1f} = \frac{m_1 - m_2}{m_1 + m_2}\, v_{1i} \qquad (10\text{-}30)$$

and

$$v_{2f} = \frac{2m_1}{m_1 + m_2}\, v_{1i}. \qquad (10\text{-}31)$$

We note from Eq. 10-31 that v_{2f} is always positive (the target body with mass m_2 always moves forward). From Eq. 10-30 we see that v_{1f} may be of either sign (the projectile body with mass m_1 moves forward if $m_1 > m_2$ but rebounds if $m_1 < m_2$).

Let us look at a few special situations.

1. **Equal masses**   If $m_1 = m_2$, Eqs. 10-30 and 10-31 reduce to

$$v_{1f} = 0 \quad \text{and} \quad v_{2f} = v_{1i},$$

which we might call a pool player's result. It predicts that after a head-on collision of bodies with equal masses, body 1 (initially moving) stops dead in its tracks and body 2 (initially at rest) takes off with the initial speed of body 1. In head-on collisions, bodies of equal mass simply exchange velocities. This is true even if the target particle (body 2) is not initially at rest.

2. **A massive target**   In Fig. 10-13, a massive target means that $m_2 \gg m_1$. For example, we might fire a golf ball at a cannonball. Equations 10-30 and 10-31 then reduce to

$$v_{1f} \approx -v_{1i} \quad \text{and} \quad v_{2f} \approx \left(\frac{2m_1}{m_2}\right)v_{1i}. \qquad (10\text{-}32)$$

This tells us that body 1 (the golf ball) simply bounces back along its incoming path, its speed essentially unchanged. Body 2 (the cannonball) moves forward at a low speed, because the quantity in parentheses in Eq. 10-32 is much less than unity. All this is what we should expect.

3. **A massive projectile**   This is the opposite case; that is, $m_1 \gg m_2$. This time, we fire a cannonball at a golf ball. Equations 10-30 and 10-31 reduce to

$$v_{1f} \approx v_{1i} \quad \text{and} \quad v_{2f} \approx 2v_{1i}. \qquad (10\text{-}33)$$

*In this step, we use the identity $a^2 - b^2 = (a - b)(a + b)$. It reduces the amount of algebra needed to solve the simultaneous equations, Eqs. 10-28 and 10-29.

Equation 10-33 tells us that body 1 (the cannonball) simply keeps on going, scarcely slowed by the collision. Body 2 (the golf ball) charges ahead at twice the speed of the cannonball.

You may wonder: "Why twice the speed?" As a starting point in thinking about the matter, recall the collision described by Eq. 10-32, in which the velocity of the incident light body (the golf ball) changed from $+v$ to $-v$, a velocity *change* of $2v$. The same *change* in velocity (but now from zero to $2v$) occurs in this example also.

Moving Target

Fig. 10-14 Two bodies headed for a one-dimensional elastic collision.

Now that we have examined the elastic collision of a projectile and a stationary target, let us examine the situation in which both bodies are moving before they undergo an elastic collision.

For the situation of Fig. 10-14, the conservation of linear momentum is written as

$$m_1 v_{1i} + m_2 v_{2i} = m_1 v_{1f} + m_2 v_{2f}, \qquad (10\text{-}34)$$

and the conservation of kinetic energy is written as

$$\tfrac{1}{2} m_1 v_{1i}^2 + \tfrac{1}{2} m_2 v_{2i}^2 = \tfrac{1}{2} m_1 v_{1f}^2 + \tfrac{1}{2} m_2 v_{2f}^2. \qquad (10\text{-}35)$$

To solve these simultaneous equations for v_{1f} and v_{2f}, we first rewrite Eq. 10-34 as

$$m_1(v_{1i} - v_{1f}) = -m_2(v_{2i} - v_{2f}), \qquad (10\text{-}36)$$

and Eq. 10-35 as

$$m_1(v_{1i} - v_{1f})(v_{1i} + v_{1f}) = -m_2(v_{2i} - v_{2f})(v_{2i} + v_{2f}). \qquad (10\text{-}37)$$

After dividing Eq. 10-37 by Eq. 10-36 and doing some more algebra, we obtain

$$v_{1f} = \frac{m_1 - m_2}{m_1 + m_2} v_{1i} + \frac{2m_2}{m_1 + m_2} v_{2i} \qquad (10\text{-}38)$$

and

$$v_{2f} = \frac{2m_1}{m_1 + m_2} v_{1i} + \frac{m_2 - m_1}{m_1 + m_2} v_{2i}. \qquad (10\text{-}39)$$

Note that the assignment of subscripts 1 and 2 to the bodies is arbitrary. If we exchange those subscripts in Fig. 10-14 and in Eqs. 10-38 and 10-39, we end up with the same set of equations. Note also that if we set $v_{2i} = 0$, body 2 becomes a stationary target as in Fig. 10-13, and Eqs. 10-38 and 10-39 reduce to Eqs. 10-30 and 10-31, respectively.

✔**CHECKPOINT 4:** What is the final linear momentum of the target in Fig. 10-13 if the initial linear momentum of the projectile is 6 kg · m/s and the final linear momentum of the projectile is (a) 2 kg · m/s and (b) −2 kg · m/s? (c) What is the final kinetic energy of the target if the initial and final kinetic energies of the projectile are, respectively, 5 J and 2 J?

Sample Problem 10-4

Two metal spheres, suspended by vertical cords, initially just touch, as shown in Fig. 10-15. Sphere 1, with mass $m_1 = 30$ g, is pulled to the left to height $h_1 = 8.0$ cm, and then released from rest. After swinging down, it undergoes an elastic collision with sphere 2, whose mass $m_2 = 75$ g. What is the velocity v_{1f} of sphere 1 just after the collision?

SOLUTION: A first Key Idea is that we can split this complicated motion into two steps that we can separately analyze: (1) the descent of sphere 1 and (2) the two-sphere collision.

Step 1. The Key Idea here is that as sphere 1 swings down, the mechanical energy of the sphere–Earth system is conserved.

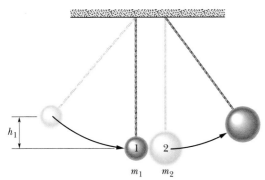

Fig. 10-15 Sample Problem 10-4. Two metal spheres suspended by cords just touch when they are at rest. Sphere 1, with mass m_1, is pulled to the left to height h_1 and then released.

(The mechanical energy is not changed by the force of the cord on sphere 1 because that force is always directed perpendicular to the sphere's direction of travel.) Let's take the lowest level as our reference level of zero gravitational potential energy. Then the kinetic energy of sphere 1 at the lowest level must equal the gravitational potential energy of the sys-

tem when sphere 1 is at the initial height. Thus,

$$\tfrac{1}{2}m_1v_{1i}^2 = m_1gh_1,$$

which we solve for the speed v_{1i} of sphere 1 just before the collision:

$$v_{1i} = \sqrt{2gh_1} = \sqrt{(2)(9.8 \text{ m/s}^2)(0.080 \text{ m})} = 1.252 \text{ m/s}.$$

Step 2. Here we can make two assumptions in addition to the assumption that the collision is elastic. First, we can assume that the collision is one-dimensional because the motions of the spheres are approximately horizontal from just before the collision to just after it. Second, because the collision is so brief, we can assume that the two-sphere system is closed and isolated. This gives the Key Idea that the total linear momentum of the system is conserved. Thus, we can use Eq. 10-30 to find the velocity of sphere 1 just after the collision:

$$v_{1f} = \frac{m_1 - m_2}{m_1 + m_2} v_{1i} = \frac{0.030 \text{ kg} - 0.075 \text{ kg}}{0.030 \text{ kg} + 0.075 \text{ kg}} (1.252 \text{ m/s})$$

$$= -0.537 \text{ m/s} \approx -0.54 \text{ m/s}. \qquad \text{(Answer)}$$

The minus sign tells us that sphere 1 moves to the left just after the collision.

10-6 Collisions in Two Dimensions

When two bodies collide, the impulses of one on the other determine the directions in which they then travel. In particular, when the collision is not head-on, the bodies do not end up traveling along their initial axis. For such two-dimensional collisions in a closed, isolated system, the total linear momentum must still be conserved:

$$\vec{P}_{1i} + \vec{P}_{2i} = \vec{P}_{1f} + \vec{P}_{2f}. \qquad (10\text{-}40)$$

If the collision is also elastic (a special case), then the total kinetic energy is also conserved:

$$K_{1i} + K_{2i} = K_{1f} + K_{2f}. \qquad (10\text{-}41)$$

Equation 10-40 is often more useful for analyzing a two-dimensional collision if we write it in terms of components on an xy coordinate system. For example, Fig. 10-16 shows a *glancing collision* (it is not head-on) between a projectile body and a target body initially at rest. The impulses between the bodies have sent the bodies off at angles θ_1 and θ_2 to the x axis, along which the projectile initially traveled. In this situation we would rewrite Eq. 10-40 for components along the x axis as

$$m_1v_{1i} = m_1v_{1f} \cos \theta_1 + m_2v_{2f} \cos \theta_2, \qquad (10\text{-}42)$$

and along the y axis as

$$0 = -m_1v_{1f} \sin \theta_1 + m_2v_{2f} \sin \theta_2. \qquad (10\text{-}43)$$

We can also write Eq. 10-41 (for the special case of an elastic collision) in terms of speeds:

$$\tfrac{1}{2}m_1v_{1i}^2 = \tfrac{1}{2}m_1v_{1f}^2 + \tfrac{1}{2}m_2v_{2f}^2 \qquad \text{(kinetic energy).} \qquad (10\text{-}44)$$

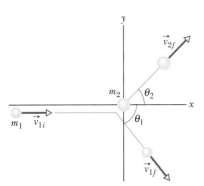

Fig. 10-16 An elastic collision between two bodies in which the collision is not head-on. The body with mass m_2 (the target) is initially at rest.

Equations 10-42 to 10-44 contain seven variables: two masses, m_1 and m_2; three speeds, v_{1i}, v_{1f}, and v_{2f}; and two angles, θ_1 and θ_2. If we know any four of these quantities, we can solve the three equations for the remaining three quantities.

✔**CHECKPOINT 5:** In Fig. 10-16, suppose that the projectile has an initial momentum of 6 kg · m/s, a final x component of momentum of 4 kg · m/s, and a final y component of momentum of -3 kg · m/s. For the target, what then are (a) the final x component of momentum and (b) the final y component of momentum?

Sample Problem 10-5

Two skaters collide and embrace, in a completely inelastic collision. Thus, they stick together after impact, as suggested by Fig. 10-17, where the origin is placed at the point of collision. Alfred, whose mass m_A is 83 kg, is originally moving east with speed $v_A = 6.2$ km/h. Barbara, whose mass m_B is 55 kg, is originally moving north with speed $v_B = 7.8$ km/h.

(a) What is the velocity $\vec{V}$ of the couple after they collide?

SOLUTION: One Key Idea here is the assumption that the two skaters form a closed, isolated system; that is, we assume no *net* external force acts on them. In particular, we neglect any frictional force on their skates from the ice. With that assumption, we can apply the conservation of the total linear momentum $\vec{P}$ to the system by writing $\vec{P}_i = \vec{P}_f$ as

$$m_A\vec{v}_A + m_B\vec{v}_B = (m_A + m_B)\vec{V}. \qquad (10\text{-}45)$$

Solving for $\vec{V}$ gives us

$$\vec{V} = \frac{m_A\vec{v}_A + m_B\vec{v}_B}{m_A + m_B}.$$

We can solve this directly on a vector-capable calculator by substituting given data for the symbols on the right side. We can also solve it by applying a second Key Idea (one we have used before) and then some algebra: The idea is that the total linear momentum of the system is conserved separately for components along the x axis and y axis shown in Fig. 10-17. Writing Eq. 10-45 in component form for the x axis yields

$$m_A v_A + m_B(0) = (m_A + m_B)V \cos\theta, \qquad (10\text{-}46)$$

and for the y axis

$$m_A(0) + m_B v_B = (m_A + m_B)V \sin\theta. \qquad (10\text{-}47)$$

We cannot solve either of these equations separately because they both contain two unknowns (V and θ), but we can solve them simultaneously by dividing Eq. 10-47 by Eq. 10-46. We get

$$\tan\theta = \frac{m_B v_B}{m_A v_A} = \frac{(55 \text{ kg})(7.8 \text{ km/h})}{(83 \text{ kg})(6.2 \text{ km/h})} = 0.834.$$

Thus,

$$\theta = \tan^{-1} 0.834 = 39.8° \approx 40°. \qquad \text{(Answer)}$$

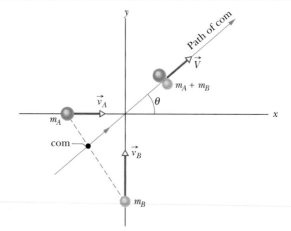

Fig. 10-17 Sample Problem 10-5. Two skaters, Alfred (A) and Barbara (B), represented by spheres in this simplified overhead view, have a completely inelastic collision. Afterward, they move off together at angle θ, with speed V. The path of their center of mass is shown. The position of the center of mass for the indicated positions of the skaters before the collision is also shown.

From Eq. 10-47, with $m_A + m_B = 138$ kg, we then have

$$V = \frac{m_B v_B}{(m_A + m_B)\sin\theta} = \frac{(55 \text{ kg})(7.8 \text{ km/h})}{(138 \text{ kg})(\sin 39.8°)}$$
$$= 4.86 \text{ km/h} \approx 4.9 \text{ km/h}. \qquad \text{(Answer)}$$

(b) What is the velocity $\vec{v}_{\text{com}}$ of the center of mass of the two skaters before the collision and after the collision?

SOLUTION: For the situation after the collision, we use the Key Idea that because the skaters are stuck together, their center of mass must travel with them, as shown in Fig. 10-17. Thus, the velocity $\vec{v}_{\text{com}}$ of their center of mass is equal to $\vec{V}$, as calculated in (a).

To find $\vec{v}_{\text{com}}$ before the collision, we use another Key Idea: The $\vec{v}_{\text{com}}$ of a system can be changed only by a net external force, not an internal force. However, here we assumed that the skaters form an isolated system (no net external force acts on the system). Therefore, $\vec{v}_{\text{com}}$ cannot change because of the collision (which produces only internal forces). Thus, before and after the collision, we have

$$\vec{v}_{\text{com}} = \vec{V}. \qquad \text{(Answer)}$$

Collisions In a **collision,** two bodies exert strong forces on each other for a relatively short time. These forces are internal to the two-body system and are significantly larger than any external force during the collision.

Impulse and Linear Momentum Applying Newton's second law in momentum form to a particle-like body involved in a collision leads to the **impulse–linear momentum theorem:**

$$\vec{p}_f - \vec{p}_i = \Delta\vec{p} = \vec{J},\qquad(10\text{-}4)$$

where $\vec{p}_f - \vec{p}_i = \Delta\vec{p}$ is the change in the body's linear momentum, and $\vec{J}$ is the **impulse** due to the force $\vec{F}(t)$ exerted on the body by the other body in the collision:

$$\vec{J} = \int_{t_i}^{t_f} \vec{F}(t)\,dt.\qquad(10\text{-}3)$$

If F_{avg} is the average magnitude of $\vec{F}(t)$ during the collision and Δt is the duration of the collision, then for one-dimensional motion

$$J = F_{avg}\,\Delta t.\qquad(10\text{-}8)$$

When a steady stream of bodies, each with mass m and speed v, collides with a body whose position is fixed, the average force on the fixed body is

$$F_{avg} = -\frac{n}{\Delta t}\Delta p = -\frac{n}{\Delta t}m\,\Delta v,\qquad(10\text{-}10)$$

where $n/\Delta t$ is the rate at which the bodies collide with the fixed body, and Δv is the change in velocity of each colliding body. This average force can also be written as

$$F_{avg} = -\frac{\Delta m}{\Delta t}\Delta v,\qquad(10\text{-}13)$$

where $\Delta m/\Delta t$ is the rate at which mass collides with the fixed body. In Eqs. 10-10 and 10-13, $\Delta v = -v$ if the bodies stop upon impact, or $\Delta v = -2v$ if they bounce directly backward with no change in their speed.

Inelastic Collision—One Dimension In an *inelastic collision* of two bodies, the kinetic energy of the two-body system is not conserved. If the system is closed and isolated, then the total linear momentum of the system *must* be conserved, which we can write in vector form as

$$\vec{p}_{1i} + \vec{p}_{2i} = \vec{p}_{1f} + \vec{p}_{2f},\qquad(10\text{-}15)$$

where subscripts i and f refer to values just before and just after the collision, respectively.

If the motion of the bodies is along a single axis, the collision is one-dimensional and we can write Eq. 10-15 in terms of velocity components along that axis:

$$m_1 v_{1i} + m_2 v_{2i} = m_1 v_{1f} + m_2 v_{2f}.\qquad(10\text{-}16)$$

If the bodies stick together, the collision is a *completely inelastic collision* and the bodies have the same final velocity V (because they *are* stuck together).

Motion of the Center of Mass The center of mass of a closed, isolated system of two colliding bodies is not affected by the collision. In particular, the velocity $\vec{v}_{com}$ of the center of mass cannot be changed by the collision and is related to the constant total momentum $\vec{P}$ of the system by

$$\vec{v}_{com} = \frac{\vec{P}}{m_1 + m_2} = \frac{\vec{p}_{1i} + \vec{p}_{2i}}{m_1 + m_2}.\qquad(10\text{-}21)$$

Elastic Collisions—One Dimension An *elastic collision* is a special type of collision in which the kinetic energy of the system of colliding bodies is conserved. Some collisions in the everyday world can be approximated as being elastic collisions. If the system is closed and isolated, its linear momentum is also conserved. For a one-dimensional collision in which body 2 is a target and body 1 is an incoming projectile, conservation of kinetic energy and linear momentum yield the following expressions for the velocities immediately after the collision:

$$v_{1f} = \frac{m_1 - m_2}{m_1 + m_2}v_{1i}\qquad(10\text{-}30)$$

and

$$v_{2f} = \frac{2m_1}{m_1 + m_2}v_{1i}.\qquad(10\text{-}31)$$

If both bodies are moving prior to the collision, their velocities immediately after the collision are given by

$$v_{1f} = \frac{m_1 - m_2}{m_1 + m_2}v_{1i} + \frac{2m_2}{m_1 + m_2}v_{2i}\qquad(10\text{-}38)$$

and

$$v_{2f} = \frac{2m_1}{m_1 + m_2}v_{1i} + \frac{m_2 - m_1}{m_1 + m_2}v_{2i}.\qquad(10\text{-}39)$$

Note the symmetry of subscripts 1 and 2 in Eqs. 10-38 and 10-39.

Collisions in Two Dimensions If two bodies collide and their motion is not along a single axis (the collision is not head-on), then the collision is two-dimensional. If the two-body system is closed and isolated, then the law of conservation of momentum applies to the collision and can be written as

$$\vec{P}_{1i} + \vec{P}_{2i} = \vec{P}_{1f} + \vec{P}_{2f},\qquad(10\text{-}40)$$

In component form, the law gives two equations that describe the collision (one equation for each of the two dimensions). If the collision is also elastic (a special case), then the conservation of kinetic energy during the collision gives a third equation:

$$K_{1i} + K_{2i} = K_{1f} + K_{2f}.\qquad(10\text{-}41)$$

QUESTIONS

1. Figure 10-18 shows three graphs of force magnitude versus time for a body involved in a collision. Rank the graphs according to the magnitude of the impulse on the body, greatest first.

Fig. 10-18 Question 1.

2. Two objects that are moving along an xy plane on a frictionless floor collide. Assume that they form a closed, isolated system. The following table gives some of the momentum components (in kilogram-meters per second) before and after the collision. What are the missing values?

Situation	Object	Before p_x	Before p_y	After p_x	After p_y
1	A	3	4	7	2
	B	2	2		
2	C	−4	5	3	
	D		−2	4	2
3	E	−6			3
	F	6	2	−4	−3

3. The following table gives, for three situations, the masses (in kilograms) and velocities (in meters per second) of the two bodies in Fig. 10-14. For which situations is the center of mass of the two-body system stationary?

Situation	m_1	v_1	m_2	v_2
a	2	3	4	−3
b	6	2	3	−4
c	4	3	4	−3

4. Figure 10-19 shows four graphs of position versus time for two bodies and their center of mass. The two bodies form a closed, isolated system and undergo a completely inelastic, one-dimensional collision along an x axis. In graph 1, are (a) the two bodies and (b) the center of mass moving in the positive or negative direction of the x axis? (c) Which graphs correspond to a physically impossible situation? Explain.

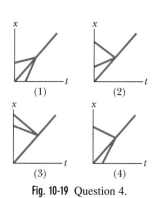

Fig. 10-19 Question 4.

5. In Fig. 10-20, blocks A and B have linear momenta with directions as shown and with magnitudes of 9 kg · m/s and 4 kg · m/s, respectively. (a) What is the direction of motion of the center of mass of the two-block system over the frictionless floor? (b) If the blocks stick together during their collision, in what direction do they move? (c) If, instead, block A ends up moving to the left, is the magnitude of its momentum then smaller than, more than, or the same as that of block B?

Fig. 10-20 Question 5.

6. A projectile body moving in the positive direction of an x axis on a frictionless floor runs into an initially stationary target body (as in Fig. 10-13) in a one-dimensional collision. Assume the particles form a closed, isolated system. Nine choices for a graph of the momenta of the bodies versus time (before and after the collision) are given in Fig. 10-21. Determine which choices represent physically impossible situations and explain why.

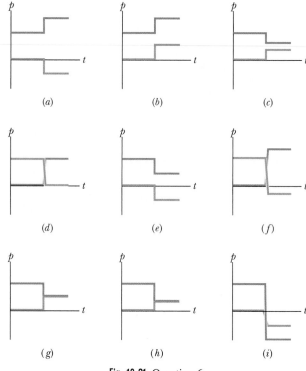

Fig. 10-21 Question 6.

7. Two bodies have undergone an elastic one-dimensional collision along an x axis. Figure 10-22 is a graph of position versus time for those bodies and for their center of mass. (a) Were both bodies initially moving, or was one initially stationary? Which line segment corresponds to the motion of the center of mass (b) before the collision and (c) after the col-

Fig. 10-22 Question 7.

lision? (d) Is the mass of the body that was moving faster before the collision greater than, less than, or equal to that of the other body?

8. Drop, in succession, a baseball and a basketball from about shoulder height above a hard floor, and note how high each rebounds. Then align the baseball above the basketball (with a small separation as in Fig. 10-23a) and drop them simultaneously. (Be prepared to duck, and guard your face.) (a) Is the rebound height of the basketball now higher or lower than before (Fig. 10-23b)? (b) Is the rebound height of the baseball less than or greater than the sum of the individual baseball and basketball rebound heights? (See also Problem 45.)

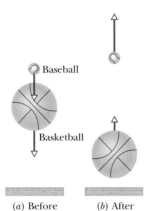

(a) Before   (b) After

Fig. 10-23 Question 8 and Problem 45.

9. A projectile hockey puck A, with initial momentum 5 kg · m/s along an x axis, collides with an initially stationary hockey puck B. The pucks slide over frictionless ice and are shown in an overhead view in Fig. 10-24. Also shown are three general choices for the path taken by puck A after the collision. Which choice is appropriate if the momentum of puck B after the collision has an x component of (a) 5 kg · m/s, (b) more than 5 kg · m/s, and (c) less than 5 kg · m/s? Can that x component be (d) 1 kg · m/s or (e) −1 kg · m/s?

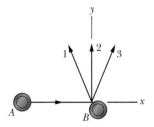

Fig. 10-24 Question 9.

10. Two bodies that form a closed, isolated system undergo a collision along a frictionless floor. Which of the three choices in Fig. 10-25 best represents the paths of those bodies and the path of their center of mass as seen from overhead?

(a)   (b)   (c)

Fig. 10-25 Question 10.

EXERCISES & PROBLEMS

SEC. 10-2   Impulse and Linear Momentum

1E. A cue stick strikes a stationary pool ball, with an average force of 50 N over a time of 10 ms. If the ball has mass 0.20 kg, what speed does it have just after impact?   ssm

2E. The National Transportation Safety Board is testing the crashworthiness of a new car. The 2300 kg vehicle, moving at 15 m/s, is allowed to collide with a bridge abutment, which stops it in 0.56 s. What is the magnitude of the average force that acts on the car during the impact?

3E. A 150 g baseball pitched at a speed of 40 m/s is hit straight back to the pitcher at a speed of 60 m/s. What is the magnitude of the average force on the ball from the bat if the bat is in contact with the ball for 5.0 ms?

4E. Until he was in his seventies, Henri LaMothe excited audiences by belly-flopping from a height of 12 m into 30 cm of water (Fig. 10-26). Assuming that he stops just as he reaches the bottom of the water and estimating his mass, find the magnitudes of (a) the average force and (b) the average impulse on him from the water.

5E. A force that averages 1200 N is applied to a 0.40 kg steel ball moving at 14 m/s in a collision lasting 27 ms. If the force is in a direction opposite the initial velocity of the ball, find the final speed and direction of the ball.   ssm

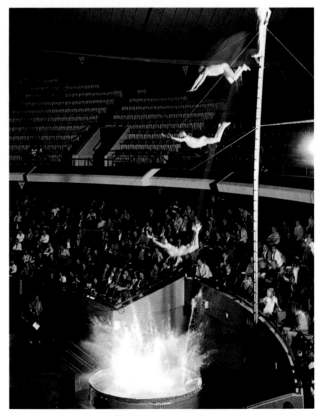

Fig. 10-26 Exercise 4.

6E. In February 1955, a paratrooper fell 370 m from an airplane without being able to open his chute but happened to land in snow, suffering only minor injuries. Assume that his speed at impact was 56 m/s (terminal speed), that his mass (including gear) was 85 kg, and that the magnitude of the force on him from the snow was at the survivable limit of 1.2×10^5 N. What are (a) the minimum depth of snow that would have stopped him safely and (b) the magnitude of the impulse on him from the snow?

7E. A 1.2 kg ball drops vertically onto a floor, hitting with a speed of 25 m/s. It rebounds with an initial speed of 10 m/s. (a) What impulse acts on the ball during the contact? (b) If the ball is in contact with the floor for 0.020 s, what is the magnitude of the average force on the floor from the ball?

8P. It is well known that bullets and other missiles fired at Superman simply bounce off his chest (Fig. 10-27). Suppose that a gangster sprays Superman's chest with 3 g bullets at the rate of 100 bullets/min, and the speed of each bullet is 500 m/s. Suppose too that the bullets rebound straight back with no change in speed. What is the magnitude of the average force on Superman's chest from the stream of bullets?

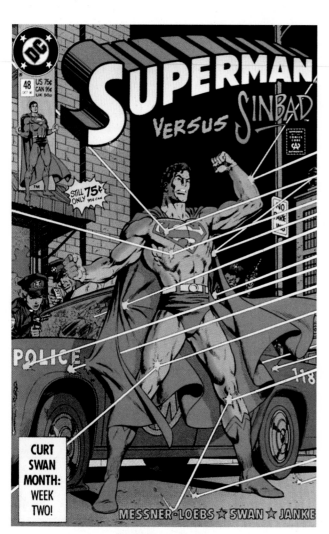

Fig. 10-27 Problem 8.

9P. A 1400 kg car moving at 5.3 m/s is initially traveling north in the positive y direction. After completing a 90° right-hand turn to the positive x direction in 4.6 s, the inattentive operator drives into a tree, which stops the car in 350 ms. In unit-vector notation, what is the impulse on the car (a) due to the turn and (b) due to the collision? What is the magnitude of the average force that acts on the car (c) during the turn and (d) during the collision? (e) What is the angle between the average force in (c) and the positive x direction? ssm www

10P. A 0.30 kg softball has a velocity of 12 m/s at an angle of 35° below the horizontal just before making contact with a bat. The ball leaves the bat 2.0 ms later with a vertical velocity of magnitude 10 m/s as shown in Fig. 10-28. What is the magnitude of the average force of the bat on the ball during the ball–bat contact?

Fig. 10-28 Problem 10.

11P. The magnitude of an unbalanced force on a 10 kg object increases at a constant rate from zero to 50 N in 4.0 s, causing the initially stationary object to move. What is the object's speed at the end of the 4.0 s? ssm

12P. During a violent thunderstorm, hail of diameter 1.0 cm falls directly downward at a speed of 25 m/s. There are estimated to be 120 hailstones per cubic meter of air. (a) What is the mass of each hailstone (density = 0.92 g/cm³)? (b) Assuming that the hail does not bounce, find the magnitude of the average force on a flat roof measuring 10 m × 20 m due to the impact of the hail. (*Hint:* During impact, the force on a hailstone from the roof is approximately equal to the net force on the hailstone, because the gravitational force on it is small.)

13P. A pellet gun fires ten 2.0 g pellets per second with a speed of 500 m/s. The pellets are stopped by a rigid wall. What are (a) the momentum of each pellet, (b) the kinetic energy of each pellet, and (c) the magnitude of the average force on the wall from the stream of pellets? (d) If each pellet is in contact with the wall for 0.6 ms, what is the magnitude of the average force on the wall from each pellet during contact? (e) Why is this average force so different from the average force calculated in (c)? ssm

14P. Figure 10-29 shows an approximate plot of force magnitude versus time during the collision of a 58 g Superball with a wall. The initial velocity of the ball is 34 m/s perpendicular to the wall; it rebounds directly back with approximately the same speed, also perpendicular to the wall. What is F_{max}, the maximum magnitude of the force on the ball from the wall during the collision?

Fig. 10-29 Problem 14.

15P. A spacecraft is separated into two parts by detonating the explosive bolts that hold them together. The masses of the parts are 1200 kg and 1800 kg; the magnitude of the impulse on each part from the bolts is 300 N · s. With what relative speed do the two parts separate because of the detonation? ssm

16P. A ball having a mass of 150 g strikes a wall with a speed of 5.2 m/s and rebounds with only 50% of its initial kinetic energy. (a) What is the speed of the ball immediately after rebounding? (b) What is the magnitude of the impulse on the wall from the ball? (c) If the ball was in contact with the wall for 7.6 ms, what was the magnitude of the average force on the ball from the wall during this time interval?

17P. In the overhead view of Fig. 10-30, a 300 g ball with a speed v of 6.0 m/s strikes a wall at an angle θ of 30° and then rebounds with the same speed and angle. It is in contact with the wall for 10 ms. (a) What is the impulse on the ball from the wall? (b) What is the average force on the wall from the ball?

Fig. 10-30 Problem 17.

18P. A 2500 kg unmanned space probe is moving in a straight line at a constant speed of 300 m/s. Control rockets on the space probe execute a burn in which a thrust of 3000 N acts for 65.0 s. (a) What is the change in the magnitude of the probe's linear momentum if the thrust is backward, forward, or directly sideways? (b) What is the change in kinetic energy under the same three conditions? Assume that the mass of the ejected burn products is negligible compared to the mass of the space probe.

19P. A soccer player kicks a soccer ball of mass 0.45 kg that is initially at rest. The player's foot is in contact with the ball for 3.0×10^{-3} s, and the force of the kick is given by

$$F(t) = [(6.0 \times 10^6)t - (2.0 \times 10^9)t^2] \text{ N},$$

for $0 \le t \le 3.0 \times 10^{-3}$ s, where t is in seconds. Find the magnitudes of the following: (a) the impulse on the ball due to the kick, (b) the average force on the ball from the player's foot during the period of contact, (c) the maximum force on the ball from the player's foot during the period of contact, and (d) the ball's speed immediately after it loses contact with the player's foot. ssm

SEC. 10-4 Inelastic Collisions in One Dimension

20E. A 5.20 g bullet moving at 672 m/s strikes a 700 g wooden block at rest on a frictionless surface. The bullet emerges, traveling in the same direction with its speed reduced to 428 m/s. (a) What is the resulting speed of the block? (b) What is the speed of the bullet–block center of mass?

21E. A 6.0 kg box sled is coasting across frictionless ice at a speed of 9.0 m/s when a 12 kg package is dropped into it from above. What is the new speed of the sled? ssm

22E. A bullet of mass 10 g strikes a ballistic pendulum of mass 2.0 kg. The center of mass of the pendulum rises a vertical distance

of 12 cm. Assuming that the bullet remains embedded in the pendulum, calculate the bullet's initial speed.

23E. Meteor Crater in Arizona (Fig. 10-1a) is thought to have been formed by the impact of a meteor with Earth some 20,000 years ago. The mass of the meteor is estimated at 5×10^{10} kg, and its speed at 7200 m/s. What speed would such a meteor give Earth in a head-on collision? ssm

24E. A bullet of mass 4.5 g is fired horizontally into a 2.4 kg wooden block at rest on a horizontal surface. The coefficient of kinetic friction between block and surface is 0.20. The bullet stops in the block, which slides straight ahead for 1.8 m (without rotation). (a) What is the speed of the block immediately after the bullet stops relative to it? (b) At what speed is the bullet fired?

25P. Two cars A and B slide on an icy road as they attempt to stop at a traffic light. The mass of A is 1100 kg, and the mass of B is 1400 kg. The coefficient of kinetic friction between the locked wheels of either car and the road is 0.13. Car A succeeds in stopping at the light, but car B cannot stop and rear-ends car A. After the collision, A stops 8.2 m ahead of its position at impact, and B 6.1 m ahead; see Fig. 10-31. Both drivers had their brakes locked throughout the incident. From the distance each car moved after the collision, find the speed of (a) car A and (b) car B immediately after impact. (c) Use conservation of linear momentum to find the speed at which car B struck car A. On what grounds can the use of linear momentum conservation be criticized here?

Fig. 10-31 Problem 25.

26P. In Fig. 10-32a, a 3.50 g bullet is fired horizontally at two blocks at rest on a frictionless tabletop. The bullet passes through the first block, with mass 1.20 kg, and embeds itself in the second, with mass 1.80 kg. Speeds of 0.630 m/s and 1.40 m/s, respectively, are thereby given to the blocks (Fig. 10-32b). Neglecting the mass

Fig. 10-32 Problem 26.

removed from the first block by the bullet, find (a) the speed of the bullet immediately after it emerges from the first block and (b) the bullet's original speed.

27P. A box is put on a scale that is marked in units of mass and adjusted to read zero when the box is empty. A stream of marbles is then poured into the box from a height h above its bottom at a rate of R (marbles per second). Each marble has mass m. (a) If the collisions between the marbles and the box are completely inelastic, find the scale reading at time t after the marbles begin to fill the box. (b) Determine a numerical answer when $R = 100$ s^{-1}, $h = 7.60$ m, $m = 4.50$ g, and $t = 10.0$ s. ssm

28P. A 5.0 kg block with a speed of 3.0 m/s collides with a 10 kg block that has a speed of 2.0 m/s in the same direction. After the collision, the 10 kg block is observed to be traveling in the original direction with a speed of 2.5 m/s. (a) What is the velocity of the 5.0 kg block immediately after the collision? (b) By how much does the total kinetic energy of the system of two blocks change because of the collision? (c) Suppose, instead, that the 10 kg block ends up with a speed of 4.0 m/s. What then is the change in the total kinetic energy? (d) Account for the result you obtained in (c). ilw

29P. A railroad freight car of mass 3.18×10^4 kg collides with a stationary caboose car. They couple together, and 27.0% of the initial kinetic energy is transferred to thermal energy, sound, vibrations, and so on. Find the mass of the caboose. ssm www

30P. A 10 g bullet moving directly upward at 1000 m/s strikes and passes through the center of mass of a 5.0 kg block initially at rest (Fig. 10-33). The bullet emerges from the block moving directly upward at 400 m/s. To what maximum height does the block then rise above its initial position?

Fig. 10-33 Problem 30.

31P. In Fig. 10-34, a ball of mass m is shot with speed v_i into the barrel of a spring gun of mass M initially at rest on a frictionless surface. The ball sticks in the barrel at the point of maximum compression of the spring. Assume that the increase in thermal energy due to friction between the ball and the barrel is negligible. (a) What is the speed of the spring gun after the ball stops in the barrel? (b) What fraction of the initial kinetic energy of the ball is stored in the spring? ssm

Fig. 10-34 Problem 31.

32P. A 4.0 kg physics book and a 6.0 kg calculus book, connected by a spring, are stationary on a horizontal frictionless surface. The spring constant is 8000 N/m. The books are pushed together, compressing the spring, and then they are released from rest. When the spring has returned to its unstretched length, the speed of the calculus book is 4.0 m/s. How much energy is stored in the spring at the instant the books are released?

33P. A block of mass $m_1 = 2.0$ kg slides along a frictionless table with a speed of 10 m/s. Directly in front of it, and moving in the same direction, is a block of mass $m_2 = 5.0$ kg moving at 3.0 m/s.

A massless spring with spring constant $k = 1120$ N/m is attached to the near side of m_2, as shown in Fig. 10-35. When the blocks collide, what is the maximum compression of the spring? (*Hint:* At the moment of maximum compression of the spring, the two blocks move as one. Find the velocity by noting that the collision is completely inelastic at this point.) ilw

Fig. 10-35 Problem 33.

34P. A 1.0 kg block at rest on a horizontal frictionless surface is connected to an unstretched spring ($k = 200$ N/m) whose other end is fixed (Fig. 10-36). A 2.0 kg block moving at 4.0 m/s collides with the 1.0 kg block. If the two blocks stick together after the one-dimensional collision, what maximum compression of the spring occurs when the blocks momentarily stop?

Fig. 10-36 Problem 34.

SEC. 10-5   Elastic Collisions in One Dimension

35E. The blocks in Fig. 10-37 slide without friction. (a) What is the velocity $\vec{v}$ of the 1.6 kg block after the collision? (b) Is the collision elastic? (c) Suppose the initial velocity of the 2.4 kg block is the reverse of what is shown. Can the velocity $\vec{v}$ of the 1.6 kg block after the collision be in the direction shown? ssm

Fig. 10-37 Exercise 35.

36E. An electron undergoes a one-dimensional elastic collision with an initially stationary hydrogen atom. What percentage of the electron's initial kinetic energy is transferred to kinetic energy of the hydrogen atom? (The mass of the hydrogen atom is 1840 times the mass of the electron.)

37E. A cart with mass 340 g moving on a frictionless linear air track at an initial speed of 1.2 m/s undergoes an elastic collision with an initially stationary cart of unknown mass. After the collision, the first cart continues in its original direction at 0.66 m/s. (a) What is the mass of the second cart? (b) What is its speed after impact? (c) What is the speed of the two-cart center of mass? ssm

38E. Spacecraft *Voyager 2* (of mass m and speed v relative to the Sun) approaches the planet Jupiter (of mass M and speed V_J relative to the Sun) as shown in Fig. 10-38. The spacecraft rounds the planet and departs in the opposite direction. What is its speed, relative to the Sun, after this slingshot encounter, which can be analyzed as a collision? Assume $v = 12$ km/s and $V_J = 13$ km/s (the orbital speed of Jupiter). The mass of Jupiter is very much greater than the mass of the spacecraft ($M \gg m$).

Fig. 10-38 Exercise 38.

39E. An alpha particle (mass 4 u) experiences an elastic head-on collision with a gold nucleus (mass 197 u) that is originally at rest. (The symbol u represents the atomic mass unit.) What percentage of its original kinetic energy does the alpha particle lose? ilw

40P. A steel ball of mass 0.500 kg is fastened to a cord that is 70.0 cm long and fixed at the far end. The ball is then released when the cord is horizontal (Fig. 10-39). At the bottom of its path, the ball strikes a 2.50 kg steel block initially at rest on a frictionless surface. The collision is elastic. Find (a) the speed of the ball and (b) the speed of the block, both just after the collision.

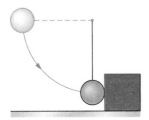

Fig. 10-39 Problem 40.

41P. A body of mass 2.0 kg makes an elastic collision with another body at rest and continues to move in the original direction but with one-fourth of its original speed. (a) What is the mass of the other body? (b) What is the speed of the two-body center of mass if the initial speed of the 2.0 kg body was 4.0 m/s? ssm www

42P. In the two-sphere arrangement of Sample Problem 10-4, assume that sphere 1 has a mass of 50 g and an initial height of 9.0 cm and that sphere 2 has a mass of 85 g. After the collision, what height is reached by (a) sphere 1 and (b) sphere 2? After the next (elastic) collision, what height is reached by (c) sphere 1 and (d) sphere 2? (*Hint:* Do not use rounded-off values.)

43P. Two titanium spheres approach each other head-on with the same speed and collide elastically. After the collision, one of the spheres, whose mass is 300 g, remains at rest. (a) What is the mass of the other sphere? (b) What is the speed of the two-sphere center of mass if the initial speed of each sphere is 2.0 m/s? ssm

44P. In Fig. 10-40, block 1 of mass m_1 is at rest on a long frictionless table that is up against a wall. Block 2 of mass m_2 is placed between block 1 and the wall and sent sliding to the left, toward block 1, with constant speed v_{2i}. Assuming that all collisions are elastic, find the value of m_2 (in terms of m_1) for which both blocks move with the same velocity after block 2 has collided once with block 1 and once with the wall. Assume the wall to have infinite mass.

Fig. 10-40 Problem 44.

45P. A small ball of mass m is aligned above a larger ball of mass M (with a slight separation, as in Fig. 10-23*a*), and the two are dropped simultaneously from height h. (Assume the radius of each ball is negligible compared to h.) (a) If the larger ball rebounds elastically from the floor and then the small ball rebounds elastically from the larger ball, what ratio m/M results in the larger ball stopping upon its collision with the small ball? (The answer is approximately the mass ratio of a baseball to a basketball, as in Question 8.) (b) What height does the small ball then reach? ssm

SEC. 10-6 Collisions in Two Dimensions

46E. Two 2.0 kg bodies, A and B, collide. The velocities before the collision are $\vec{v}_A = 15\hat{i} + 30\hat{j}$ and $\vec{v}_B = -10\hat{i} + 5.0\hat{j}$. After the collision, $\vec{v}'_A = -5.0\hat{i} + 20\hat{j}$. All speeds are given in meters per second. (a) What is the final velocity of B? (b) How much kinetic energy is gained or lost in the collision?

47E. An alpha particle collides with an oxygen nucleus that is initially at rest. The alpha particle is scattered at an angle of 64.0° from its initial direction of motion, and the oxygen nucleus recoils at an angle of 51.0° on the opposite side of that initial direction. The final speed of the nucleus is 1.20×10^5 m/s. Find (a) the final speed and (b) the initial speed of the alpha particle. (In atomic mass units, the mass of an alpha particle is 4.0 u, and the mass of an oxygen nucleus is 16 u.) ilw

48E. A proton with a speed of 500 m/s collides elastically with another proton initially at rest. The projectile and target protons then move along perpendicular paths, with the projectile path at 60° from the original direction. After the collision, what are the speeds of (a) the target proton and (b) the projectile proton?

49E. In a game of pool, the cue ball strikes another ball of the same mass and initially at rest. After the collision, the cue ball moves at 3.50 m/s along a line making an angle of 22.0° with its original direction of motion, and the second ball has a speed of 2.00 m/s. Find (a) the angle between the direction of motion of the second ball and the original direction of motion of the cue ball and (b) the original speed of the cue ball. (c) Is kinetic energy (of the centers of mass, don't consider the rotation) conserved? ssm

50P. Two balls A and B, having different but unknown masses, collide. Initially, A is at rest and B has speed v. After the collision, B has speed $v/2$ and moves perpendicularly to its original motion. (a) Find the direction in which ball A moves after the collision. (b) Show that you cannot determine the speed of A from the information given.

51P. After a completely inelastic collision, two objects of the same mass and same initial speed are found to move away together at $\frac{1}{2}$ their initial speed. Find the angle between the initial velocities of the objects.   ssm

52P. A billiard ball moving at a speed of 2.2 m/s strikes an identical stationary ball a glancing blow. After the collision, one ball is found to be moving at a speed of 1.1 m/s in a direction making a 60° angle with the original line of motion. (a) Find the velocity of the other ball. (b) Can the collision be inelastic, given these data?

53P. In Fig. 10-41, ball 1 with an initial speed of 10 m/s collides elastically with stationary balls 2 and 3, whose centers are on a line perpendicular to the initial velocity of ball 1 and that are initially in contact with each other. The three balls are identical. Ball 1 is aimed directly at the contact point, and all motion is frictionless. After the collision, what are the velocities of (a) ball 2, (b) ball 3, and (c) ball 1? (*Hint:* With friction absent, each impulse is directed along the line connecting the centers of the colliding balls, normal to the colliding surfaces.)   ssm

Fig. 10-41 Problem 53.

54P. Two 30 kg children, each with a speed of 4.0 m/s, are sliding on a frictionless frozen pond when they collide and stick together because they have Velcro straps on their jackets. The two children then collide and stick to a 75 kg man who was sliding at 2.0 m/s. After this collision, the three-person composite is stationary. What is the angle between the initial velocity vectors of the two children?

55P. Show that if a neutron is scattered through 90° in an elastic collision with an initially stationary deuteron, the neutron loses $\frac{2}{3}$ of its initial kinetic energy to the deuteron. (In atomic mass units, the mass of a neutron is 1.0 u and the mass of a deuteron is 2.0 u.)   ssm

Additional Problems

56. Basilisk lizards can run across the top of a water surface (Fig. 10-42). With each step, a lizard first slaps its foot against the water and then pushes it down into the water rapidly enough to form an air cavity around the top of the foot. To avoid having to pull the foot back up against water drag in order to complete the step, the lizard withdraws the foot before water can flow into the air cavity. During this full action of slap, downward push, and withdrawal, the average upward impulse on the lizard must match the downward impulse due to the gravitational force if the lizard is not to sink. Suppose that the mass of a basilisk lizard is 90.0 g, the mass of each foot is 3.00 g, the speed of a foot as it slaps the water is 1.50 m/s, and the time for a single step is 0.600 s. (a) What is the magnitude of the impulse on the lizard during the slap? (Assume this impulse is directly upward.) (b) During the 0.600 s duration of a step, what is the downward impulse on the lizard due to the gravitational force? (c) Which action, the slap or the push, provides the primary support for the lizard, or are they approximately equal in their support?

Fig. 10-42 Problem 56. Basilisk lizard running across water.

57. *Tyrannosaurus rex* may have known from experience not to run particularly fast because of the danger of tripping, in which case their short forearms would have been no help in cushioning the fall. Suppose a *T. Rex* of mass m trips while walking, toppling over, with its center of mass falling freely a distance of 1.5 m. Then its center of mass descends an additional 0.30 m owing to compression of its body and the ground. (a) In multiples of the dinosaur's weight, what is the approximate magnitude of the average vertical force on the dinosaur during its collision with the ground (during the descent of 0.30 m)? Now assume that the dinosaur is running at a speed of 19 m/s (fast) when it trips, falls to the ground, and then slides to a stop with a coefficient of kinetic friction of 0.6. Assume also that the average vertical force during the collision and sliding is that in (a). What, approximately, are (b) the magnitude of the average total force on it from the ground (again in multiples of its weight) and (c) the sliding distance? The force magnitudes of (a) and (b) strongly suggest that the collision would injure the torso of the dinosaur. The head, which would fall farther, would suffer even greater injury.

NEW PROBLEMS

N1. In Fig. 10N-1, block 1 of mass m_1 slides from rest along a frictionless ramp from a height of 2.50 m and then collides with stationary block 2, which has mass $m_2 = 2.00m_1$. After the collision, block 2 slides into a region where the coefficient of kinetic friction is 0.500 and comes to a stop in distance d within that region. What is the value of distance d if the collision is (a) elastic and (b) completely inelastic?

Fig. 10N-1 Problem N1.

N2. In Fig. 10N-2, puck 1 of mass $m_1 = 0.20$ kg is sent sliding across a frictionless lab bench, to undergo a one-dimensional elastic collision with stationary puck 2. Puck 2 then slides off the bench and lands a distance d from the base of the bench. Puck 1 rebounds from the collision and slides off the opposite edge of the bench, landing a distance $2d$ from the base of the bench. What is the mass of puck 2? (*Hint:* Be careful with signs.)

Fig. 10N-2 Problem N2.

N3. A 0.25 kg puck is initially stationary on an ice surface with negligible friction. At time $t = 0$, a horizontal force begins to move the puck. The force is given by $\vec{F} = (12.0 - 3.00t^2)\hat{i}$, with $\vec{F}$ in newtons and t in seconds, and it acts until its magnitude is zero. (a) What is the magnitude of the impulse on the puck from the force between $t = 0.500$ s and $t = 1.25$ s? (b) What is the change in momentum of the puck between $t = 0$ and when the force magnitude is zero?

N4. *Speed amplifier*. In Fig. 10N-3, block 1 of mass m_1 slides along an x axis on a frictionless floor with a speed of $v_{1i} = 4.00$ m/s. Then it undergoes a one-dimensional elastic collision with stationary block 2 of mass $m_2 = 0.500m_1$. Next, block 2 undergoes a one-dimensional elastic collision with stationary block 3 of mass $m_3 = 0.500m_2$. (a) What then is the speed of block 3? Are (b) the speed, (c) the kinetic energy, and (d) the momentum of block 3 greater than, less than, or the same as the initial values for block 1?

Fig. 10N-3 Problem N4.

N5. For the two-collision sequence of Problem N4, Figure 10N-4a shows the velocity of block 1 plotted versus time t, where the velocity is given in terms of the initial velocity v_{1i}. The times for the first collision (t_1) and the second collision (t_2) are indicated. (a) On the same graph, plot the velocities of blocks 2 and 3, all in terms of v_{1i}. Figure 10N-4b shows a plot of the kinetic energy of

Fig. 10N-4 Problem N5.

block 1 versus time, where kinetic energy is given in terms of the initial kinetic energy $K_{1i} = \frac{1}{2}m_1v_{1i}^2$. (b) On the same graph, plot the kinetic energies of blocks 2 and 3, all in terms of K_{1i}. After the second collision, what percentage of the total kinetic energy do (c) block 1, (d) block 2, and (e) block 3 have?

N6. *Speed deamplifier*. In Fig. 10N-5, block 1 of mass m_1 slides along an x axis on a frictionless floor with a speed of 4.00 m/s. Then it undergoes a one-dimensional elastic collision with stationary block 2 of mass $m_2 = 2.00m_1$. Next, block 2 undergoes a one-dimensional elastic collision with stationary block 3 of mass $m_3 = 2.00m_2$. (a) What then is the speed of block 3? Are (b) the speed, (c) the kinetic energy, and (d) the momentum of block 3 greater than, less than, or the same as the initial values for block 1?

Fig. 10N-5 Problem N6.

N7. For the two-collision sequence of Problem N6, Figure 10N-6a shows a plot of the velocity of block 1 versus time t, where the velocity is given in terms of the initial velocity v_{1i}. The times for the first collision (t_1) and the second collision (t_2) are indicated. (a) On the same graph, plot the velocities of blocks 2 and 3, all in terms of v_{1i}. Figure 10N-6b shows a plot of the kinetic energy of block 1 versus time, where kinetic energy is given in terms of the initial kinetic energy $K_{1i} = \frac{1}{2}m_1v_{1i}^2$. (b) On the same graph, plot the kinetic energies of blocks 2 and 3, all in terms of K_{1i}. After the second collision, what percentage of the total kinetic energy do (c) block 1, (d) block 2, and (e) block 3 have?

Fig. 10N-6 Problem N7.

N8. In Fig. 10N-7, particle 1 of mass $m_1 = 0.30$ kg slides rightward along an x axis on a frictionless floor with a speed of 2.0 m/s. When it reaches $x = 0$, it undergoes a one-dimensional elastic collision with stationary particle 2 of mass $m_2 = 0.40$ kg. When particle 2 then reaches a wall at $x = 70$ cm, it bounces from it with no loss of speed. At what position on the x axis does particle 2 then collide with particle 1?

Fig. 10N-7 Problem N8.

N9. In a railroad accident, a boxcar weighing 200 kN and traveling at 3.00 m/s on horizontal track slams into a stationary caboose weighing 400 kN. The collision connects the caboose to the boxcar. How much energy is transferred from kinetic energy to other forms of energy in the collision?

N10. A 2.00 kg "particle" traveling with velocity $\vec{v} = (4.0 \text{ m/s})\hat{i}$ collides with a 4.00 kg "particle" traveling with velocity $\vec{v} = (2.0 \text{ m/s})\hat{j}$. The collision connects the two particles. What then is their velocity in (a) unit-vector notation and (b) magnitude-angle notation?

N11. Block 1, with mass m_1 and speed 4.0 m/s, slides along an x axis on a frictionless floor and then undergoes a one-dimensional elastic collision with stationary block 2, with mass $m_2 = 0.40m_1$. The two blocks then slide into a region where the coefficient of kinetic friction is 0.50; there they stop. How far into that region do (a) block 1 and (b) block 2 slide?

N12. A collision occurs between a 2.00 kg particle traveling with velocity $\vec{v} = (-4.00 \text{ m/s})\hat{i} + (-5.00 \text{ m/s})\hat{j}$ and a 4.00 kg particle traveling with velocity $\vec{v} = (6.00 \text{ m/s})\hat{i} + (-2.00 \text{ m/s})\hat{j}$. The collision connects the two particles. What then is their velocity in (a) unit-vector notation and (b) magnitude-angle notation?

N13. Particle 1 of mass 200 g and speed 3.00 m/s undergoes a one-dimensional collision with stationary particle 2 of mass 400 g. What is the magnitude of the impulse on particle 1 if the collision is (a) elastic and (b) completely inelastic?

N14. A 0.550 kg ball falls directly down onto concrete, hitting it with a speed of 12.0 m/s and rebounding directly upward with a speed of 3.00 m/s. Extend a y axis upward. In unit-vector notation, what are (a) the change in the ball's momentum, (b) the impulse on the ball, and (c) the impulse on the concrete?

N15. A 5.0 kg toy race car can move along an x axis. Figure 10N-8 gives the x component F_x of a force that acts on the car, which begins at rest at time $t = 0$. In unit-vector notation, what are the momenta of the car at (a) $t = 4.0$ s and (b) $t = 7.0$ s, and (c) what is the velocity at $t = 9.0$ s?

Fig. 10N-8 Problem N15.

N16. *Two average forces.* Snowballs, each with a mass of 0.250 kg, are shot perpendicularly into a wall at a speed of 4.00 m/s and at a steady rate. They each stick to the wall. Figure 10N-9 is a graph of the magnitude F of the force on the wall as a function of time t for two of the snowball impacts: impacts occur every 50 ms, last 10 ms, and produce an isosceles triangle on the graph. During each impact, what are the magnitudes of (a) the impulse and (b) the average force on the wall? (c) During a time interval of many impacts, what is the magnitude of the average force on the wall?

Fig. 10N-9 Problem N16.

N17. In Fig. 10N-10, block 1 slides along an x axis on a frictionless floor with a speed of 0.75 m/s. When it reaches stationary block 2, the two blocks undergo an elastic collision.

Fig. 10N-10 Problem N17.

The following table gives the mass and length of the (uniform) blocks and also the locations of their centers at time $t = 0$. Where is the center of mass of the two-block system located (a) at $t = 0$, (b) when the two blocks first touch, and (c) at $t = 4.0$ s?

Block	Mass (kg)	Length (cm)	Center at $t = 0$
1	0.25	5.0	$x = -1.50$ m
2	0.50	6.0	$x = 0$

N18. An object of mass m undergoes a one-dimensional impulse $\vec{J}$, with the speed changing from v to u without any change in the direction of travel. Show that the work is $\frac{1}{2}J(u + v)$.

N19. A ball of mass m and speed v strikes a wall perpendicularly and rebounds in the opposite direction with the same speed. (a) If the duration of the collision is Δt, what is the magnitude of the average force on the wall from the ball? (b) Evaluate this magnitude for $m = 140$ g, $v = 7.80$ m/s, and $\Delta t = 3.80$ ms.

N20. A 0.15 kg ball hits a wall with a velocity of $(5.00 \text{ m/s})\hat{i} + (6.50 \text{ m/s})\hat{j} + (4.00 \text{ m/s})\hat{k}$. It rebounds from the wall with a velocity of $(2.00 \text{ m/s})\hat{i} + (3.50 \text{ m/s})\hat{j} + (-3.20 \text{ m/s})\hat{k}$. What are (a) the change in the ball's momentum, (b) the impulse on the ball, and (c) the impulse on the wall?

N21. Particle 1 with mass 3.0 kg and velocity $(5.0 \text{ m/s})\hat{i}$ undergoes a one-dimensional elastic collision with particle 2 with mass 2.0 kg and velocity $(-6.0 \text{ m/s})\hat{i}$. After the collision, what are the velocities of (a) particle 1 and (b) particle 2?

N22. Figure 10N-11 shows an overhead view of two particles sliding at constant velocity over a frictionless surface. The particles have the same mass and the same initial speed $v = 4.00$ m/s, and they collide where their paths intersect. An x axis is arranged to bisect the angle between their incoming paths, so that $\theta = 40.0°$. The region to the right of the collision is divided into four lettered sections by the x axis and four numbered dashed lines. In what region or along what line do the particles travel if the collision is (a) completely inelastic, (b) elastic, and (c) partially inelastic? What are their final speeds if the collision is (d) completely inelastic and (e) elastic?

Fig. 10N-11 Problem N22.

N23. A completely inelastic collision occurs between two balls of wet putty that move directly toward each other along a vertical axis. Just before the collision, one ball of mass 3.0 kg is moving upward at 20 m/s, and the other ball, of mass 2.0 kg, is moving downward at 12 m/s. How high do the combined two balls of putty rise above the collision point? (Neglect air drag.)

N24. Block 1 of mass m_1 slides along a frictionless floor and into a one-dimensional elastic collision with stationary block 2 of mass $m_2 = 3m_1$. Prior to the collision, the center of mass of the two-block system had a speed of 3.00 m/s. Afterwards, what are the speeds of (a) the center of mass and (b) block 2?

11 Rotation

In judo, a weaker and smaller fighter who understands physics can defeat a stronger and larger fighter who does not. This fact is demonstrated by the basic "hip throw," in which a fighter rotates the fighter's opponent around his hip and—if the throw is successful—onto the mat. Without the proper use of physics, the throw requires considerable strength and can easily fail.

What is the advantage offered by physics?

The answer is in this chapter.

(a)

(b)

Fig. 11-1 Figure skater Michelle Kwan in motion of (*a*) pure translation in a fixed direction and (*b*) pure rotation about a vertical axis.

11-1 Translation and Rotation

The graceful movement of figure skaters can be used to illustrate two kinds of pure, or unmixed, motion. Figure 11-1*a* shows a skater gliding across the ice in a straight line with constant speed. Her motion is one of pure **translation.** Figure 11-1*b* shows her spinning at a constant rate about a vertical axis, in a motion of pure **rotation.**

Translation is motion along a straight line, which has been our focus up to now. Rotation is the motion of wheels, gears, motors, planets, the hands of clocks, the rotors of jet engines, and the blades of helicopters. It is our focus in this chapter.

11-2 The Rotational Variables

We wish to examine the rotation of a rigid body about a fixed axis. A **rigid body** is a body that can rotate with all its parts locked together and without any change in its shape. A **fixed axis** means that the rotation occurs about an axis that does not move. Thus, we shall not examine an object like the Sun, because the parts of the Sun (a ball of gas) are not locked together. We also shall not examine an object like a bowling ball rolling along a bowling alley, because the ball rotates about an axis that moves (the ball's motion is a mixture of rotation and translation).

Figure 11-2 shows a rigid body of arbitrary shape in rotation about a fixed axis, called the **axis of rotation** or the **rotation axis.** Every point of the body moves in a circle whose center lies on the axis of rotation, and every point moves through the same angle during a particular time interval. In pure translation, every point of the body moves in a straight line, and every point moves through the same *linear distance* during a particular time interval. (Comparisons between angular and linear motion will appear throughout this chapter.)

We deal now—one at a time—with the angular equivalents of the linear quantities position, displacement, velocity, and acceleration.

Angular Position

Figure 11-2 shows a *reference line*, fixed in the body, perpendicular to the rotation axis, and rotating with the body. The **angular position** of this line is the angle of the line relative to a fixed direction, which we take as the **zero angular position.** In Fig. 11-3, the angular position θ is measured relative to the positive direction of the x axis. From geometry, we know that θ is given by

$$\theta = \frac{s}{r} \qquad \text{(radian measure).} \qquad (11\text{-}1)$$

Here s is the length of arc (or the arc distance) along a circle and between the x axis (the zero angular position) and the reference line; r is the radius of that circle.

An angle defined in this way is measured in **radians** (rad) rather than in revolutions (rev) or degrees. The radian, being the ratio of two lengths, is a pure number and thus has no dimension. Because the circumference of a circle of radius r is $2\pi r$, there are 2π radians in a complete circle:

$$1 \text{ rev} = 360° = \frac{2\pi r}{r} = 2\pi \text{ rad,} \qquad (11\text{-}2)$$

and thus

$$1 \text{ rad} = 57.3° = 0.159 \text{ rev.} \qquad (11\text{-}3)$$

We do *not* reset θ to zero with each complete rotation of the reference line about the rotation axis. If the reference line completes two revolutions from the zero angular position, then the angular position θ of the line is $\theta = 4\pi$ rad.

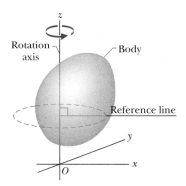

Fig. 11-2 A rigid body of arbitrary shape in pure rotation about the *z* axis of a coordinate system. The position of the *reference line* with respect to the rigid body is arbitrary, but it is perpendicular to the rotation axis. It is fixed in the body and rotates with the body.

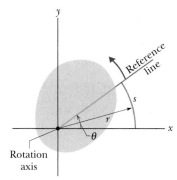

Fig. 11-3 The rotating rigid body of Fig. 11-2 in cross section, viewed from above. The plane of the cross section is perpendicular to the rotation axis, which now extends out of the page, toward you. In this position of the body, the reference line makes an angle θ with the *x* axis.

For pure translational motion along the *x* direction, we can know all there is to know about a moving body if we are given $x(t)$, its position as a function of time. Similarly, for pure rotation, we can know all there is to know about a rotating body if we are given $\theta(t)$, the angular position of the body's reference line as a function of time.

Angular Displacement

If the body of Fig. 11-3 rotates about the rotation axis as in Fig. 11-4, changing the angular position of the reference line from θ_1 to θ_2, the body undergoes an **angular displacement** $\Delta\theta$ given by

$$\Delta\theta = \theta_2 - \theta_1. \tag{11-4}$$

This definition of angular displacement holds not only for the rigid body as a whole but also for *every particle within that body* because the particles are all locked together.

If a body is in translational motion along an *x* axis, its displacement Δx is either positive or negative, depending on whether the body is moving in the positive or negative direction of the axis. Similarly, the angular displacement $\Delta\theta$ of a rotating body is either positive or negative, according to the following rule:

▶ An angular displacement in the counterclockwise direction is positive, and one in the clockwise direction is negative.

The phrase "*clocks are negative*" can help you remember this rule.

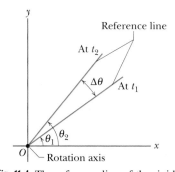

Fig. 11-4 The reference line of the rigid body of Figs. 11-2 and 11-3 is at angular position θ_1 at time t_1 and at angular position θ_2 at a later time t_2. The quantity $\Delta\theta\ (= \theta_2 - \theta_1)$ is the angular displacement that occurs during the interval $\Delta t\ (= t_2 - t_1)$. The body itself is not shown.

Angular Velocity

Suppose (see Fig. 11-4) that our rotating body is at angular position θ_1 at time t_1 and at angular position θ_2 at time t_2. We define the **average angular velocity** of the body in the time interval Δt from t_1 to t_2 to be

$$\omega_{\text{avg}} = \frac{\theta_2 - \theta_1}{t_2 - t_1} = \frac{\Delta\theta}{\Delta t}, \tag{11-5}$$

in which $\Delta\theta$ is the angular displacement that occurs during Δt (ω is the lowercase Greek letter omega).

The **(instantaneous) angular velocity** ω, with which we shall be most concerned, is the limit of the ratio in Eq. 11-5 as Δt approaches zero. Thus,

$$\omega = \lim_{\Delta t \to 0} \frac{\Delta\theta}{\Delta t} = \frac{d\theta}{dt}. \tag{11-6}$$

If we know $\theta(t)$, we can find the angular velocity ω by differentiation.

Equations 11-5 and 11-6 hold not only for the rotating rigid body as a whole but also for *every particle of that body* because the particles are all locked together. The unit of angular velocity is commonly the radian per second (rad/s) or the revolution per second (rev/s). Another measure of angular velocity was used during at least the first three decades of rock: Music was produced by vinyl (phonograph) records that were played on turntables at "$33\frac{1}{3}$ rpm" or "45 rpm," meaning at $33\frac{1}{3}$ rev/min or 45 rev/min.

If a particle moves in translation along an x axis, its linear velocity v is either positive or negative, depending on whether the particle is moving in the positive or negative direction of the axis. Similarly, the angular velocity ω of a rotating rigid body is either positive or negative, depending on whether the body is rotating counterclockwise (positive) or clockwise (negative). ("Clocks are negative" still works.) The magnitude of an angular velocity is called the **angular speed,** which is also represented with ω.

Angular Acceleration

If the angular velocity of a rotating body is not constant, then the body has an angular acceleration. Let ω_2 and ω_1 be its angular velocities at times t_2 and t_1, respectively. The **average angular acceleration** of the rotating body in the interval from t_1 to t_2 is defined as

$$\alpha_{\text{avg}} = \frac{\omega_2 - \omega_1}{t_2 - t_1} = \frac{\Delta\omega}{\Delta t}, \tag{11-7}$$

in which $\Delta\omega$ is the change in the angular velocity that occurs during the time interval Δt. The **(instantaneous) angular acceleration** α, with which we shall be most concerned, is the limit of this quantity as Δt approaches zero. Thus,

$$\alpha = \lim_{\Delta t \to 0} \frac{\Delta\omega}{\Delta t} = \frac{d\omega}{dt}. \tag{11-8}$$

Equations 11-7 and 11-8 hold not only for the rotating rigid body as a whole but also for *every particle of that body*. The unit of angular acceleration is commonly the radian per second-squared (rad/s^2) or the revolution per second-squared (rev/s^2).

Sample Problem 11-1

The disk in Fig. 11-5a is rotating about its central axis like a merry-go-round. The angular position $\theta(t)$ of a reference line on the disk is given by

$$\theta = -1.00 - 0.600t + 0.250t^2, \tag{11-9}$$

with t in seconds, θ in radians, and the zero angular position as indicated in the figure.

(a) Graph the angular position of the disk versus time from $t = -3.0$ s to $t = 6.0$ s. Sketch the disk and its angular position reference line at $t = -2.0$ s, 0 s, and 4.0 s, and when the curve crosses the t axis.

SOLUTION: The Key Idea here is that the angular position of the disk is the angular position $\theta(t)$ of its reference line, which is given by

Eq. 11-9 as a function of time. So we graph Eq. 11-9; the result is shown in Fig. 11-5b.

To sketch the disk and its reference line at a particular time, we need to determine θ for that time. To do so, we substitute the time into Eq. 11-9. For $t = -2.0$ s, we get

$$\theta = -1.00 - (0.600)(-2.0) + (0.250)(-2.0)^2$$

$$= 1.2 \text{ rad} = 1.2 \text{ rad} \frac{360°}{2\pi \text{ rad}} = 69°.$$

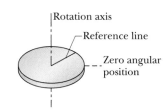

Rotation axis

Reference line

Zero angular position

(a)

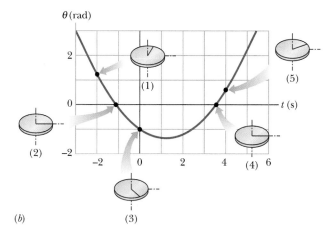

θ (rad)

(1)

(5)

(2)

t (s)

(4) 6

(b)

(3)

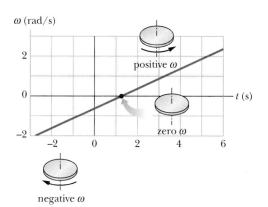

ω (rad/s)

positive ω

zero ω

t (s)

negative ω

(c)

Fig. 11-5 Sample Problem 11-1. (a) A rotating disk. (b) A plot of the disk's angular position $\theta(t)$. Five sketches indicate the angular position of the reference line on the disk for five points on the curve. (c) A plot of the disk's angular velocity $\omega(t)$. Positive values of ω correspond to counterclockwise rotation, and negative values to clockwise rotation.

This means that at $t = -2.0$ s the reference line on the disk is rotated counterclockwise from the zero angular position by 1.2 rad or 69° (counterclockwise because θ is positive). Sketch 1 in Fig. 11-5b shows this angular position of the reference line.

Similarly, for $t = 0$, we find $\theta = -1.00$ rad $= -57°$, which means that the reference line is rotated clockwise from the zero angular position by 1.0 rad or 57°, as shown in sketch 3. For $t = 4.0$ s, we find $\theta = 0.60$ rad $= 34°$ (sketch 5). Drawing sketches for when the curve crosses the t axis is easy, because then $\theta = 0$ and the reference line is momentarily aligned with the zero angular position (sketches 2 and 4).

(b) At what time t_{min} does $\theta(t)$ reach the minimum value shown in Fig. 11-5b? What is that minimum value?

SOLUTION: The Key Idea here is that to find the extreme value (here the minimum) of a function, we take the first derivative of the function and set the result to zero. The first derivative of $\theta(t)$ is

$$\frac{d\theta}{dt} = -0.600 + 0.500t. \tag{11-10}$$

Setting this to zero and solving for t give us the time at which $\theta(t)$ is minimum:

$$t_{min} = 1.20 \text{ s}. \tag{Answer}$$

To get the minimum value of θ, we next substitute t_{min} into Eq. 11-9, finding

$$\theta = -1.36 \text{ rad} \approx -77.9°. \tag{Answer}$$

This *minimum* of $\theta(t)$ (the bottom of the curve in Fig. 11-5b) corresponds to the *maximum clockwise* rotation of the disk from the zero angular position, somewhat more than is shown in sketch 3.

(c) Graph the angular velocity ω of the disk versus time from $t = -3.0$ s to $t = 6.0$ s. Sketch the disk and indicate the direction of turning and the sign of ω at $t = -2.0$ s and 4.0 s, and also at t_{min}.

SOLUTION: The Key Idea here is that, from Eq. 11-6, the angular velocity ω is equal to $d\theta/dt$ as given in Eq. 11-10. So, we have

$$\omega = -0.600 + 0.500t. \tag{11-11}$$

The graph of this function $\omega(t)$ is shown in Fig. 11-5c.

To sketch the disk at $t = -2.0$ s, we substitute that value into Eq. 11-11, obtaining

$$\omega = -1.6 \text{ rad/s}. \tag{Answer}$$

The minus sign tells us that at $t = -2.0$ s, the disk is turning clockwise, as suggested by the lowest sketch in Fig. 11-5c.

Substituting $t = 4.0$ s into Eq. 11-11 gives us

$$\omega = 1.4 \text{ rad/s}. \tag{Answer}$$

The implied plus sign tells us that at $t = 4.0$ s, the disk is turning counterclockwise (the highest sketch in Fig. 11-5c).

For t_{min}, we already know that $d\theta/dt = 0$. So, we must also have $\omega = 0$. That is, the disk momentarily stops when the reference line reaches the minimum value of θ in Fig. 11-5b, as suggested by the center sketch in Fig. 11-5c.

(d) Use the results in parts (a) through (c) to describe the motion of the disk from $t = -3.0$ s to $t = 6.0$ s.

SOLUTION: When we first observe the disk at $t = -3.0$ s, it has a positive angular position and is turning clockwise but slowing. It stops at angular position $\theta = -1.36$ rad and then begins to turn counterclockwise, with its angular position eventually becoming positive again.

✔**CHECKPOINT 1:**   A disk can rotate about its central axis like the one in Fig. 11-5a. Which of the following pairs of values for its initial and final angular positions, respectively, give a negative angular displacement: (a) -3 rad, $+5$ rad, (b) -3 rad, -7 rad, (c) 7 rad, -3 rad?

11-3   Are Angular Quantities Vectors?

We can describe the position, velocity, and acceleration of a single particle by means of vectors. If the particle is confined to a straight line, however, we do not really need vector notation. Such a particle has only two directions available to it, and we can indicate these directions with plus and minus signs.

In the same way, a rigid body rotating about a fixed axis can rotate only clockwise or counterclockwise as seen along the axis, and again we can select between the two directions by means of plus and minus signs. The question arises: "Can we treat the angular displacement, velocity, and acceleration of a rotating body as vectors?" The answer is a qualified "yes" (see the caution below, in connection with angular displacements).

Consider the angular velocity. Figure 11-6a shows a vinyl record rotating on a turntable. The record has a constant angular speed ω ($= 33\frac{1}{3}$ rev/min) in the clockwise direction. We can represent its angular velocity as a vector $\vec{\omega}$ pointing along the axis of rotation, as in Fig. 11-6b. Here's how: We choose the length of this vector according to some convenient scale, for example, with 1 cm corresponding to 10 rev/min. Then we establish a direction for the vector $\vec{\omega}$ by using a **right-hand rule,** as Fig. 11-6c shows: Curl your right hand about the rotating record, your fingers pointing *in the direction of rotation.* Your extended thumb will then point in the direction of the angular velocity vector. If the record were to rotate in the opposite sense, the right-hand rule would tell you that the angular velocity vector then points in the opposite direction.

It is not easy to get used to representing angular quantities as vectors. We instinctively expect that something should be moving *along* the direction of a vector. That is not the case here. Instead, something (the rigid body) is rotating *around* the direction of the vector. In the world of pure rotation, a vector defines an axis of rotation, not a direction in which something moves. Nonetheless, the vector also defines the motion. Furthermore, it obeys all the rules for vector manipulation discussed in Chapter 3. The angular acceleration $\vec{\alpha}$ is another vector, and it too obeys those rules.

Fig. 11-6 (*a*) A record rotating about a vertical axis that coincides with the axis of the spindle. (*b*) The angular velocity of the rotating record can be represented by the vector $\vec{\omega}$, lying along the axis and pointing down, as shown. (*c*) We establish the direction of the angular velocity vector as downward by using the right-hand rule. When the fingers of the right hand curl around the record and point the way it is moving, the extended thumb points in the direction of $\vec{\omega}$.

(*a*)   (*b*)   (*c*)

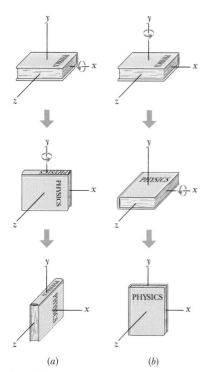

Fig. 11-7 (a) From its initial position, at the top, the book is given two successive 90° rotations, first about the (horizontal) x axis and then about the (vertical) y axis. (b) The book is given the same rotations, but in the reverse order.

In this chapter we consider only rotations that are about a fixed axis. For such situations, we need not consider vectors—we can represent angular velocity with ω and angular acceleration with α, and we can indicate direction with an implied plus sign for counterclockwise or an explicit minus sign for clockwise.

Now for the caution: Angular *displacements* (unless they are very small) *cannot* be treated as vectors. Why not? We can certainly give them both magnitude and direction, as we did for the angular velocity vector in Fig. 11-6. However, to be represented as a vector, a quantity must *also* obey the rules of vector addition, one of which says that if you add two vectors, the order in which you add them does not matter. Angular displacements fail this test.

Figure 11-7 gives an example. An initially horizontal book is given two 90° angular displacements, first in the order of Fig. 11-7a and then in the order of Fig. 11-7b. Although the two angular displacements are identical, their order is not, and the book ends up with different orientations. Thus, the addition of the two angular displacements depends on their order and they cannot be vectors.

11-4 Rotation with Constant Angular Acceleration

In pure translation, motion with a *constant linear acceleration* (for example, that of a falling body) is an important special case. In Table 2-1, we displayed a series of equations that hold for such motion.

In pure rotation, the case of *constant angular acceleration* is also important, and a parallel set of equations holds for this case also. We shall not derive them here, but simply write them from the corresponding linear equations, substituting equivalent angular quantities for the linear ones. This is done in Table 11-1, which lists both sets of equations (Eqs. 2-11 and 2-15 to 2-18; 11-12 to 11-16).

Recall that Eqs. 2-11 and 2-15 are basic equations for constant linear acceleration—the other equations in the Linear list can be derived from them. Similarly, Eqs. 11-12 and 11-13 are the basic equations for constant angular acceleration, and the other equations in the Angular list can be derived from them. To solve a simple problem involving constant angular acceleration, you can usually use an equation from the Angular list (*if* you have the list). Choose an equation for which the only unknown variable will be the variable requested in the problem. A better plan is to remember only Eqs. 11-12 and 11-13, and then solve them as simultaneous equations whenever needed. An example is given in Sample Problem 11-3.

✔CHECKPOINT 2: In four situations, a rotating body has angular position $\theta(t)$ given by (a) $\theta = 3t - 4$, (b) $\theta = -5t^3 + 4t^2 + 6$, (c) $\theta = 2/t^2 - 4/t$, and (d) $\theta = 5t^2 - 3$. To which situations do the angular equations of Table 11-1 apply?

TABLE 11-1 Equations of Motion for Constant Linear Acceleration and for Constant Angular Acceleration

Equation Number	Linear Equation	Missing Variable		Angular Equation	Equation Number
(2-11)	$v = v_0 + at$	$x - x_0$	$\theta - \theta_0$	$\omega = \omega_0 + \alpha t$	(11-12)
(2-15)	$x - x_0 = v_0 t + \frac{1}{2}at^2$	v	ω	$\theta - \theta_0 = \omega_0 t + \frac{1}{2}\alpha t^2$	(11-13)
(2-16)	$v^2 = v_0^2 + 2a(x - x_0)$	t	t	$\omega^2 = \omega_0^2 + 2\alpha(\theta - \theta_0)$	(11-14)
(2-17)	$x - x_0 = \frac{1}{2}(v_0 + v)t$	a	α	$\theta - \theta_0 = \frac{1}{2}(\omega_0 + \omega)t$	(11-15)
(2-18)	$x - x_0 = vt - \frac{1}{2}at^2$	v_0	ω_0	$\theta - \theta_0 = \omega t - \frac{1}{2}\alpha t^2$	(11-16)

Sample Problem 11-2

A grindstone (Fig. 11-8) rotates at constant angular acceleration $\alpha = 0.35$ rad/s^2. At time $t = 0$, it has an angular velocity of $\omega_0 = -4.6$ rad/s and a reference line on it is horizontal, at the angular position $\theta_0 = 0$.

(a) At what time after $t = 0$ is the reference line at the angular position $\theta = 5.0$ rev?

SOLUTION: The Key Idea here is that the angular acceleration is constant, so we can use the rotation equations of Table 11-1. We choose Eq. 11-13,

$$\theta - \theta_0 = \omega_0 t + \tfrac{1}{2}\alpha t^2,$$

because the only unknown variable it contains is the desired time t. Substituting known values and setting $\theta_0 = 0$ and $\theta = 5.0$ rev $= 10\pi$ rad give us

$$10\pi \text{ rad} = (-4.6 \text{ rad/s})t + \tfrac{1}{2}(0.35 \text{ rad/s}^2)t^2.$$

(We converted 5.0 rev to 10π rad to keep the units consistent.) Solving this quadratic equation for t, we find

$$t = 32 \text{ s.} \qquad \text{(Answer)}$$

(b) Describe the grindstone's rotation between $t = 0$ and $t = 32$ s.

SOLUTION: The wheel is initially rotating in the negative (clockwise) direction with angular velocity $\omega_0 = -4.6$ rad/s, but its angular acceleration α is positive. This initial opposition of the signs of angular velocity and angular acceleration means that the wheel

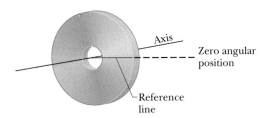

Fig. 11-8 Sample Problem 11-2. A grindstone. At $t = 0$ the reference line (which we imagine to be marked on the stone) is horizontal.

slows in its rotation in the negative direction, stops, and then reverses to rotate in the positive direction. After the reference line comes back through its initial orientation of $\theta = 0$, the wheel turns an additional 5.0 rev by time $t = 32$ s.

(c) At what time t does the grindstone momentarily stop?

SOLUTION: We again go to the table of equations for constant angular acceleration, and again we need an equation that contains only the desired unknown variable t. However, now we use another Key Idea. The equation must also contain the variable ω, so that we can set it to 0 and then solve for the corresponding time t. We choose Eq. 11-12, which yields

$$t = \frac{\omega - \omega_0}{\alpha} = \frac{0 - (-4.6 \text{ rad/s})}{0.35 \text{ rad/s}^2} = 13 \text{ s.} \quad \text{(Answer)}$$

Sample Problem 11-3

While you are operating a Rotor (the rotating cylindrical ride discussed in Sample Problem 6-8), you spot a passenger in acute distress and decrease the angular speed of the cylinder from 3.40 rad/s to 2.00 rad/s in 20.0 rev, at constant angular acceleration. (The passenger is obviously more of a "translation person" than a "rotation person.")

(a) What is the constant angular acceleration during this decrease in angular speed?

SOLUTION: Assume that the rotation is in the counterclockwise direction, and let the acceleration begin at time $t = 0$, at angular position θ_0. The Key Idea here is that, because the angular acceleration is constant, we can relate the cylinder's angular acceleration to its angular velocity and angular displacement via the basic equations for constant angular acceleration (Eqs. 11-12 and 11-13). The initial angular velocity is $\omega_0 = 3.40$ rad/s, the angular displacement is $\theta - \theta_0 = 20.0$ rev, and the angular velocity at the end of that displacement is $\omega = 2.00$ rad/s. But we do not know the angular acceleration α and time t, which are in both basic equations.

To eliminate the unknown t, we use Eq. 11-12 to write

$$t = \frac{\omega - \omega_0}{\alpha},$$

which we then substitute into Eq. 11-13 to write

$$\theta - \theta_0 = \omega_0 \left(\frac{\omega - \omega_0}{\alpha} \right) + \tfrac{1}{2}\alpha \left(\frac{\omega - \omega_0}{\alpha} \right)^2.$$

Solving for α, substituting known data, and converting 20 rev to 125.7 rad, we find

$$\alpha = \frac{\omega^2 - \omega_0^2}{2(\theta - \theta_0)} = \frac{(2.00 \text{ rad/s})^2 - (3.40 \text{ rad/s})^2}{2(125.7 \text{ rad})}$$

$$= -0.0301 \text{ rad/s}^2. \qquad \text{(Answer)}$$

(b) How much time did the speed decrease take?

SOLUTION: Now that we know α, we can use Eq. 11-12 to solve for t:

$$t = \frac{\omega - \omega_0}{\alpha} = \frac{2.00 \text{ rad/s} - 3.40 \text{ rad/s}}{-0.0301 \text{ rad/s}^2}$$

$$= 46.5 \text{ s.} \qquad \text{(Answer)}$$

This sample problem is actually an angular version of Sample Problem 2-5. The data and symbols differ, but the techniques of solution are identical.

11-5 Relating the Linear and Angular Variables

In Section 4-7, we discussed uniform circular motion, in which a particle travels at constant linear speed v along a circle and around an axis of rotation. When a rigid body, such as a merry-go-round, rotates around an axis, each particle in the body moves in its own circle around that axis. Since the body is rigid, all the particles make one revolution in the same amount of time; that is, they all have the same angular speed ω.

However, the farther a particle is from the axis, the greater the circumference of its circle is, and so the faster its linear speed v must be. You can notice this on a merry-go-round. You turn with the same angular speed ω regardless of your distance from the center, but your linear speed v increases noticeably if you move to the outside edge of the merry-go-round.

We often need to relate the linear variables s, v, and a for a particular point in a rotating body to the angular variables θ, ω, and α for that body. The two sets of variables are related by r, the *perpendicular distance* of the point from the rotation axis. This perpendicular distance is the distance between the point and the rotation axis, measured along a perpendicular to the axis. It is also the radius r of the circle traveled by the point around the axis of rotation.

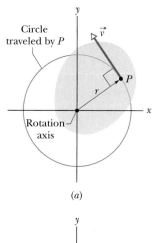

Circle traveled by P

Rotation axis

(a)

Rotation axis

(b)

Fig. 11-9 The rotating rigid body of Fig. 11-2, shown in cross section viewed from above. Every point of the body (such as P) moves in a circle around the rotation axis. (*a*) The linear velocity $\vec{v}$ of every point is tangent to the circle in which the point moves. (*b*) The linear acceleration $\vec{a}$ of the point has (in general) two components: a tangential component a_t and a radial component a_r.

The Position

If a reference line on a rigid body rotates through an angle θ, a point within the body at a position r from the rotation axis moves a distance s along a circular arc, where s is given by Eq. 11-1:

$$s = \theta r \qquad \text{(radian measure).} \qquad (11\text{-}17)$$

This is the first of our linear–angular relations. *Caution:* The angle θ here must be measured in radians because Eq. 11-17 is itself the definition of angular measure in radians.

The Speed

Differentiating Eq. 11-17 with respect to time — with r held constant — leads to

$$\frac{ds}{dt} = \frac{d\theta}{dt} r.$$

However, ds/dt is the linear speed (the magnitude of the linear velocity) of the point in question, and $d\theta/dt$ is the angular speed ω of the rotating body. So

$$v = \omega r \qquad \text{(radian measure).} \qquad (11\text{-}18)$$

Caution: The angular speed ω must be expressed in radian measure.

Equation 11-18 tells us that since all points within the rigid body have the same angular speed ω, points with greater radius r have greater linear speed v. Figure 11-9a reminds us that the linear velocity is always tangent to the circular path of the point in question.

If the angular speed ω of the rigid body is constant, then Eq. 11-18 tells us that the linear speed v of any point within it is also constant. Thus, each point within the body undergoes uniform circular motion. The period of revolution T for the motion of each point and for the rigid body itself is given by Eq. 4-33:

$$T = \frac{2\pi r}{v}.$$ (11-19)

This equation tells us that the time for one revolution is the distance $2\pi r$ traveled in one revolution divided by the speed at which that distance is traveled. Substituting for v from Eq. 11-18 and canceling r, we find also that

$$T = \frac{2\pi}{\omega} \quad \text{(radian measure)}.$$ (11-20)

This equivalent equation says that the time for one revolution is the angular distance 2π rad traveled in one revolution divided by the angular speed (or rate) at which that angle is traveled.

The Acceleration

Differentiating Eq. 11-18 with respect to time—again with r held constant—leads to

$$\frac{dv}{dt} = \frac{d\omega}{dt} r.$$ (11-21)

Here we run up against a complication. In Eq. 11-21, dv/dt represents only the part of the linear acceleration that is responsible for changes in the *magnitude* v of the linear velocity $\vec{v}$. Like $\vec{v}$, that part of the linear acceleration is tangent to the path of the point in question. We call it the *tangential component* a_t of the linear acceleration of the point, and we write

$$a_t = \alpha r \quad \text{(radian measure)},$$ (11-22)

where $\alpha = d\omega/dt$. *Caution:* The angular acceleration α in Eq. 11-22 must be expressed in radian measure.

In addition, as Eq. 4-32 tells us, a particle (or point) moving in a circular path has a *radial component* of linear acceleration, $a_r = v^2/r$ (directed radially inward), that is responsible for changes in the *direction* of the linear velocity $\vec{v}$. By substituting for v from Eq. 11-18, we can write this component as

$$a_r = \frac{v^2}{r} = \omega^2 r \quad \text{(radian measure)}.$$ (11-23)

Thus, as Fig. 11-9b shows, the linear acceleration of a point on a rotating rigid body has, in general, two components. The radially inward component a_r (given by Eq. 11-23) is present whenever the angular velocity of the body is not zero. The tangential component a_t (given by Eq. 11-22) is present whenever the angular acceleration is not zero.

✔**CHECKPOINT 3:** A cockroach rides the rim of a rotating merry-go-round. If the angular speed of this system (*merry-go-round* + *cockroach*) is constant, does the cockroach have (a) radial acceleration and (b) tangential acceleration? If the angular speed is decreasing, does the cockroach have (c) radial acceleration and (d) tangential acceleration?

Sample Problem 11-4

Figure 11-10 shows a centrifuge used to accustom astronaut trainees to high accelerations. The radius r of the circle traveled by an astronaut is 15 m.

(a) At what constant angular speed must the centrifuge rotate if the astronaut is to have a linear acceleration of magnitude $11g$?

Fig. 11-10 Sample Problem 11-4. A centrifuge is used to accustom astronauts to the large acceleration experienced during a liftoff.

we have

$$\omega = \sqrt{\frac{a_r}{r}} = \sqrt{\frac{(11)(9.8 \text{ m/s}^2)}{15 \text{ m}}}$$

$$= 2.68 \text{ rad/s} \approx 26 \text{ rev/min}. \qquad \text{(Answer)}$$

(b) What is the tangential acceleration of the astronaut if the centrifuge accelerates at a constant rate from rest to the angular speed of (a) in 120 s?

SOLUTION: The Key Idea here is that the tangential acceleration a_t, which is the linear acceleration along the circular path, is related to the angular acceleration α by Eq. 11-22 ($a_t = \alpha r$). Also, because the angular acceleration is constant, we can use Eq. 11-12 ($\omega = \omega_0 + \alpha t$) from Table 11-1 to find α from the given angular speeds. Putting these two equations together, we find

$$a_t = \alpha r = \frac{\omega - \omega_0}{t} r$$

$$= \frac{2.68 \text{ rad/s} - 0}{120 \text{ s}} (15 \text{ m}) = 0.34 \text{ m/s}^2$$

$$= 0.034g. \qquad \text{(Answer)}$$

SOLUTION: The Key Idea is this: Because the angular speed is constant, the angular acceleration α ($= d\omega/dt$) is zero and so is the tangential component of the linear acceleration ($a_t = \alpha r$). This leaves only the radial component. From Eq. 11-23 ($a_r = \omega^2 r$), with $a_r = 11g$,

Although the final radial acceleration $a_r = 11g$ is large (and alarming), the astronaut's tangential acceleration a_t during the speed-up is not.

PROBLEM-SOLVING TACTICS

Tactic 1: *Units for Angular Variables*
In Eq. 11-1 ($\theta = s/r$), we began the use of radian measure for all angular variables whenever we are using equations that contain both angular and linear variables. Thus, we must express angular displacements in radians, angular velocities in rad/s and rad/min, and angular accelerations in rad/s² and rad/min². Equations 11-17, 11-18, 11-20, 11-22, and 11-23 are marked to emphasize this. The only exceptions to this rule are equations that involve *only* angular variables, such as the angular equations listed in Table 11-1. Here

you are free to use any unit you wish for the angular variables; that is, you may use radians, degrees, or revolutions, as long as you use them consistently.

In equations where radian measure must be used, you need not keep track of the unit "radian" (rad) algebraically, as you must do for other units. You can add or delete it at will, to suit the context. In Sample Problem 11-4a the unit was added to the answer; in Sample Problem 11-4b it was omitted from the answer.

11-6 Kinetic Energy of Rotation

The rapidly rotating blade of a table saw certainly has kinetic energy due to that rotation. How can we express the energy? We cannot apply the familiar formula $K = \frac{1}{2}mv^2$ to the saw as a whole because that would only give us the kinetic energy of the saw's center of mass, which is zero.

Instead, we shall treat the table saw (and any other rotating rigid body) as a collection of particles with different speeds. We can then add up the kinetic energies of all the particles to find the kinetic energy of the body as a whole. In this way we obtain, for the kinetic energy of a rotating body,

$$K = \tfrac{1}{2}m_1v_1^2 + \tfrac{1}{2}m_2v_2^2 + \tfrac{1}{2}m_3v_3^2 + \cdots$$

$$= \sum \tfrac{1}{2}m_i v_i^2, \qquad (11\text{-}24)$$

in which m_i is the mass of the ith particle and v_i is its speed. The sum is taken over all the particles in the body.

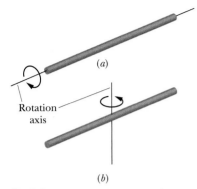

Fig. 11-11 A long rod is much easier to rotate about (*a*) its central (longitudinal) axis than about (*b*) an axis through its center and perpendicular to its length because the mass is distributed closer to the rotation axis in (*a*) than in (*b*).

The problem with Eq. 11-24 is that v_i is not the same for all particles. We solve this problem by substituting for v from Eq. 11-18 ($v = \omega r$), so that we have

$$K = \sum \tfrac{1}{2} m_i (\omega r_i)^2 = \tfrac{1}{2}\left(\sum m_i r_i^2\right)\omega^2, \tag{11-25}$$

in which ω *is* the same for all particles.

The quantity in parentheses on the right side of Eq. 11-25 tells us how the mass of the rotating body is distributed about its axis of rotation. We call that quantity the **rotational inertia** (or **moment of inertia**) *I* of the body with respect to the axis of rotation. It is a constant for a particular rigid body and a particular rotation axis. (That axis must always be specified if the value of *I* is to be meaningful.)

We may now write

$$I = \sum m_i r_i^2 \qquad \text{(rotational inertia)} \tag{11-26}$$

and substitute into Eq. 11-25, obtaining

$$K = \tfrac{1}{2} I \omega^2 \qquad \text{(radian measure)} \tag{11-27}$$

as the expression we seek. Because we have used the relation $v = \omega r$ in deriving Eq. 11-27, ω must be expressed in radian measure. The SI unit for *I* is the kilogram–square meter (kg · m²).

Equation 11-27, which gives the kinetic energy of a rigid body in pure rotation, is the angular equivalent of the formula $K = \tfrac{1}{2} M v_{com}^2$, which gives the kinetic energy of a rigid body in pure translation. In both formulas there is a factor of $\tfrac{1}{2}$. Where mass *M* appears in one equation, *I* (which involves both mass and its distribution) appears in the other. Finally, each equation contains as a factor the square of a speed—translational or rotational as appropriate. The kinetic energies of translation and of rotation are not different kinds of energy. They are both kinetic energy, expressed in ways that are appropriate to the motion at hand.

We noted previously that the rotational inertia of a rotating body involves not only its mass but also how that mass is distributed. Here is an example that you can literally feel. Rotate a long, fairly heavy rod (a pole, a length of lumber, or something similar), first around its central (longitudinal) axis (Fig. 11-11*a*) and then around an axis perpendicular to the rod and through the center (Fig. 11-11*b*). Both rotations involve the very same mass, but the first rotation is much easier than the second. The reason is that the mass is distributed much closer to the rotation axis in the first rotation. As a result, the rotational inertia of the rod is much smaller in Fig. 11-11*a* than in Fig. 11-11*b*. In general, smaller rotational inertia means easier rotation.

✔**CHECKPOINT 4:**   The figure shows three small spheres that rotate about a vertical axis. The perpendicular distance between the axis and the center of each sphere is given. Rank the three spheres according to their rotational inertia about that axis, greatest first.

11-7 Calculating the Rotational Inertia

If a rigid body consists of a few particles, we can calculate its rotational inertia about a given rotation axis with Eq. 11-26 ($I = \sum m_i r_i^2$), that is, we can find the product mr^2 for each particle and then sum the products. (Recall that *r* is the perpendicular distance a particle is from the given rotation axis.)

TABLE 11-2   Some Rotational Inertias

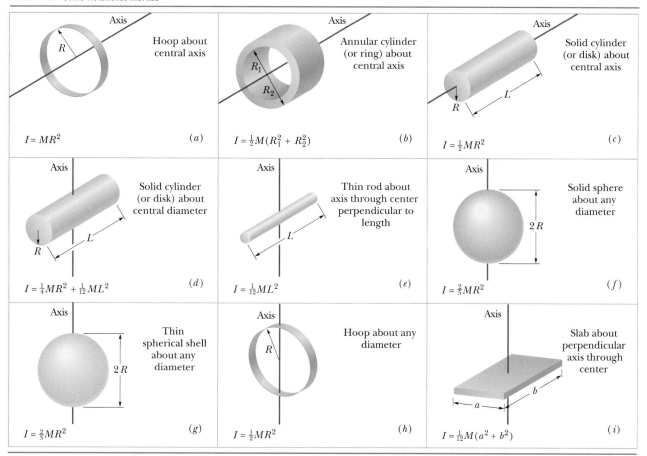

If a rigid body consists of a great many adjacent particles (it is *continuous,* like a Frisbee), using Eq. 11-26 would require a computer. Thus, instead, we replace the sum in Eq. 11-26 with an integral and define the rotational inertia of the body as

$$I = \int r^2 \, dm \qquad \text{(rotational inertia, continuous body).} \qquad (11\text{-}28)$$

Table 11-2 gives the results of such integration for nine common body shapes and the indicated axes of rotation.

Parallel-Axis Theorem

Suppose we want to find the rotational inertia I of a body of mass M about a given axis. In principle, we can always find I with the integration of Eq. 11-28. However, there is a shortcut if we happen to already know the rotational inertia I_{com} of the body about a *parallel* axis that extends through the body's center of mass. Let h be the perpendicular distance between the given axis and the axis through the center of mass (remember these two axes must be parallel). Then the rotational inertia I about the given axis is

$$I = I_{com} + Mh^2 \qquad \text{(parallel-axis theorem).} \qquad (11\text{-}29)$$

This equation is known as the **parallel-axis theorem.** We shall now prove it and then put it to use in Checkpoint 5 and Sample Problem 11-5.

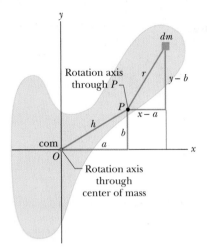

Fig. 11-12 A rigid body in cross section, with its center of mass at O. The parallel-axis theorem (Eq. 11-29) relates the rotational inertia of the body about an axis through O to that about a parallel axis through a point such as P, a distance h from the body's center of mass. Both axes are perpendicular to the plane of the figure.

Proof of the Parallel-Axis Theorem

Let O be the center of mass of the arbitrarily shaped body shown in cross section in Fig. 11-12. Place the origin of the coordinates at O. Consider an axis through O perpendicular to the plane of the figure, and another axis through point P parallel to the first axis. Let the x and y coordinates of P be a and b.

Let dm be a mass element with the general coordinates x and y. The rotational inertia of the body about the axis through P is then, from Eq. 11-28,

$$I = \int r^2 \, dm = \int [(x - a)^2 + (y - b)^2] \, dm,$$

which we can rearrange as

$$I = \int (x^2 + y^2) \, dm - 2a \int x \, dm - 2b \int y \, dm + \int (a^2 + b^2) \, dm. \quad (11\text{-}30)$$

From the definition of the center of mass (Eq. 9-9), the middle two integrals of Eq. 11-30 give the coordinates of the center of mass (multiplied by a constant) and thus must each be zero. Because $x^2 + y^2$ is equal to R^2, where R is the distance from O to dm, the first integral is simply I_{com}, the rotational inertia of the body about an axis through its center of mass. Inspection of Fig. 11-12 shows that the last term in Eq. 11-30 is Mh^2, where M is the body's total mass. Thus, Eq. 11-30 reduces to Eq. 11-29, which is the relation that we set out to prove.

✓CHECKPOINT 5:   The figure shows a booklike object (one side is longer than the other) and four choices of rotation axes, all perpendicular to the face of the object. Rank the choices according to the rotational inertia of the object about the axis, greatest first.

(1)   (2)   (3) (4)

Sample Problem 11-5

Figure 11-13a shows a rigid body consisting of two particles of mass m connected by a rod of length L and negligible mass.

(a) What is the rotational inertia I_{com} of this body about an axis through its center of mass, perpendicular to the rod as shown?

SOLUTION: The **Key Idea** is that because we have only two particles with mass, we can find the body's rotational inertia I_{com} by using Eq. 11-26 rather than by integration. For the two particles, each at perpendicular distance $\frac{1}{2}L$ from the rotation axis, we have

$$I = \sum m_i r_i^2 = (m)(\tfrac{1}{2}L)^2 + (m)(\tfrac{1}{2}L)^2$$

$$= \tfrac{1}{2}mL^2. \qquad \text{(Answer)}$$

(b) What is the rotational inertia I of the body about an axis through the left end of the rod and parallel to the first axis (Fig. 11-13b)?

SOLUTION: This situation is simple enough that we can find I using either of two **Key Ideas.** The first is identical to the one we used in (a). The only difference here is that the perpendicular distance r_i is zero for the particle on the left, and L for the particle on the right. Now Eq. 11-26 gives us

$$I = m(0)^2 + mL^2 = mL^2. \qquad \text{(Answer)}$$

(a)

Fig. 11-13 Sample Problem 11-5. A rigid body of two particles of mass m, which are joined by a rod of negligible mass.

(b)

The second **Key Idea** provides a more powerful technique: Because we already know I_{com} about an axis through the center of mass and because the axis here is parallel to that "com axis," we can apply the parallel-axis theorem (Eq. 11-29). We find

$$I = I_{com} + Mh^2 = \tfrac{1}{2}mL^2 + (2m)(\tfrac{1}{2}L)^2$$

$$= mL^2. \qquad \text{(Answer)}$$

Sample Problem 11-6

Large machine components that undergo prolonged, high-speed rotation are first examined for the possibility of failure in a *spin test system*. In this system, a component is *spun up* (brought up to high speed) while inside a cylindrical arrangement of lead bricks and containment liner, all within a steel shell that is closed by a lid clamped into place. If the rotation causes the component to shatter, the soft lead bricks are supposed to catch the pieces so that the failure can then be analyzed.

In early 1985, Test Devices, Inc. (www.testdevices.com) was spin testing a sample of a solid steel rotor (a disk) of mass $M = 272$ kg and radius $R = 38.0$ cm. When the sample reached an angular speed ω of 14 000 rev/min, the test engineers heard a dull thump from the test system, which was located one floor down and one room over from them. Investigating, they found that lead bricks had been thrown out in the hallway leading to the test room, a door to the room had been hurled into the adjacent parking lot, one lead brick had shot from the test site through the wall of a neighbor's kitchen, the structural beams of the test building had been damaged, the concrete floor beneath the spin chamber had been shoved downward by about 0.5 cm, and the 900 kg lid had been blown upward through the ceiling and had then crashed back onto the test equipment (Fig. 11-14). The exploding pieces had not penetrated the room of the test engineers only by luck.

How much energy was released in the explosion of the rotor?

SOLUTION: The Key Idea here is that this released energy was equal to the rotational kinetic energy K of the rotor just as it reached the angular speed of 14 000 rev/min. We can find K with Eq. 11-27 ($K = \frac{1}{2}I\omega^2$), but first we need an expression for the rotational inertia I. Because the rotor was a disk that rotated like a merry-go-round, I is given by the expression in Table 11-2c ($I = \frac{1}{2}MR^2$).

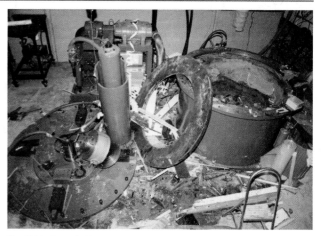

Fig. 11-14 Sample Problem 11-6. Some of the destruction caused by the explosion of a rapidly rotating steel disk.

Thus, we have

$$I = \tfrac{1}{2}MR^2 = \tfrac{1}{2}(272 \text{ kg})(0.38 \text{ m})^2 = 19.64 \text{ kg} \cdot \text{m}^2.$$

The angular speed of the rotor was

$$\omega = (14\ 000 \text{ rev/min})(2\pi \text{ rad/rev})\left(\frac{1 \text{ min}}{60 \text{ s}}\right)$$
$$= 1.466 \times 10^3 \text{ rad/s}.$$

Now we can use Eq. 11-27 to write

$$K = \tfrac{1}{2}I\omega^2 = \tfrac{1}{2}(19.64 \text{ kg} \cdot \text{m}^2)(1.466 \times 10^3 \text{ rad/s})^2$$
$$= 2.1 \times 10^7 \text{ J}. \qquad \text{(Answer)}$$

Being near this explosion was like being near an exploding bomb.

11-8 Torque

A doorknob is located as far as possible from the door's hinge line for a good reason. If you want to open a heavy door, you must certainly apply a force; that alone, however, is not enough. Where you apply that force and in what direction you push are also important. If you apply your force nearer to the hinge line than the knob, or at any angle other than 90° to the plane of the door, you must use a greater force to move the door than if you apply the force at the knob and perpendicular to the door's plane.

Figure 11-15a shows a cross section of a body that is free to rotate about an axis passing through O and perpendicular to the cross section. A force $\vec{F}$ is applied at point P, whose position relative to O is defined by a position vector $\vec{r}$. The directions of vectors $\vec{F}$ and $\vec{r}$ make an angle ϕ with each other. (For simplicity, we consider only forces that have no component parallel to the rotation axis; thus, $\vec{F}$ is in the plane of the page.)

To determine how $\vec{F}$ results in a rotation of the body around the rotation axis, we resolve $\vec{F}$ into two components (Fig. 11-15b). One component, called the *radial component F_r*, points along $\vec{r}$. This component does not cause rotation, because it acts along a line that extends through O. (If you pull on a door parallel to the plane of the door, you do not rotate the door.) The other component of $\vec{F}$, called the

(a)

(b)

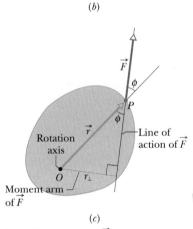

(c)

Fig. 11-15 (a) A force $\vec{F}$ acts at point P on a rigid body that is free to rotate about an axis through O; the axis is perpendicular to the plane of the cross section shown here. (b) The torque due to this force is $(r)(F \sin \phi)$. We can also write it as rF_t, where F_t is the tangential component of $\vec{F}$. (c) The torque can also be written as $r_\perp F$, where $r_\perp$ is the moment arm of $\vec{F}$.

tangential component F_t, is perpendicular to $\vec{r}$ and has magnitude $F_t = F \sin \phi$. This component *does* cause rotation. (If you pull on a door perpendicular to its plane, you can rotate the door.)

The ability of $\vec{F}$ to rotate the body depends not only on the magnitude of its tangential component F_t, but also on just how far from O the force is applied. To include both these factors, we define a quantity called **torque** τ as the product of the two factors and write it as

$$\tau = (r)(F \sin \phi). \tag{11-31}$$

Two equivalent ways of computing the torque are

$$\tau = (r)(F \sin \phi) = rF_t \tag{11-32}$$

and

$$\tau = (r \sin \phi)(F) = r_\perp F, \tag{11-33}$$

where $r_\perp$ is the perpendicular distance between the rotation axis at O and an extended line running through the vector $\vec{F}$ (Fig. 11-15c). This extended line is called the **line of action** of $\vec{F}$, and $r_\perp$ is called the **moment arm** of $\vec{F}$. Figure 11-15b shows that we can describe r, the magnitude of $\vec{r}$, as being the moment arm of the force component F_t.

Torque, which comes from the Latin word meaning "to twist," may be loosely identified as the turning or twisting action of the force $\vec{F}$. When you apply a force to an object—such as a screwdriver or torque wrench—with the purpose of turning that object, you are applying a torque. The SI unit of torque is the newton-meter (N · m). *Caution:* The newton-meter is also the unit of work. Torque and work, however, are quite different quantities and must not be confused. Work is often expressed in joules (1 J = 1 N · m), but torque never is.

In the next chapter we shall discuss torque in a general way as being a vector quantity. Here, however, because we consider only rotation around a single axis, we do not need vector notation. Instead, a torque has either a positive or negative value depending on the direction of rotation it would give a body initially at rest: If the body would rotate counterclockwise, the torque is positive. If the object would rotate clockwise, the torque is negative. (The phrase "clocks are negative" from Section 11-2 still works.)

Torques obey the superposition principle that we discussed in Chapter 5 for forces: When several torques act on a body, the **net torque** (or **resultant torque**) is the sum of the individual torques. The symbol for net torque is τ_{net}.

✔**CHECKPOINT 6:**   The figure shows an overhead view of a meter stick that can pivot about the dot at the position marked 20 (for 20 cm). All five horizontal forces on the stick have the same magnitude. Rank those forces according to the magnitude of the torque that they produce, greatest first.

11-9 Newton's Second Law for Rotation

A torque can cause rotation of a rigid body, as when you use a torque to rotate a door. Here we want to relate the net torque τ_{net} on a rigid body to the angular acceleration α it causes about a rotation axis. We do so by analogy with Newton's second law ($F_{net} = ma$) for the acceleration a of a body of mass m due to a net force F_{net} along a coordinate axis. We replace F_{net} with τ_{net}, m with I, and a with α, writing

$$\tau_{net} = I\alpha \qquad \text{(Newton's second law for rotation)}, \tag{11-34}$$

where α must be in radian measure.

Proof of Equation 11-34

We prove Eq. 11-34 by first considering the simple situation shown in Fig. 11-16. The rigid body there consists of a particle of mass m on one end of a massless rod of length r. The rod can move only by rotating about its other end, around a rotation axis (an axle) that is perpendicular to the plane of the page. Thus, the particle can move only in a circular path that has the rotation axis at its center.

A force $\vec{F}$ acts on the particle. However, because the particle can move only along the circular path, only the tangential component F_t of the force (the component that is tangent to the circular path) can accelerate the particle along the path. We can relate F_t to the particle's tangential acceleration a_t along the path with Newton's second law, writing

$$F_t = ma_t.$$

The torque acting on the particle is, from Eq. 11-32,

$$\tau = F_t r = ma_t r.$$

From Eq. 11-22 ($a_t = \alpha r$) we can write this as

$$\tau = m(\alpha r)r = (mr^2)\alpha. \tag{11-35}$$

The quantity in parentheses on the right side of Eq. 11-35 is the rotational inertia of the particle about the rotation axis (see Eq. 11-26). Thus, Eq. 11-35 reduces to

$$\tau = I\alpha \quad \text{(radian measure).} \tag{11-36}$$

For the situation in which more than one force is applied to the particle, we can generalize Eq. 11-36 as

$$\tau_{\text{net}} = I\alpha \quad \text{(radian measure),} \tag{11-37}$$

which we set out to prove. We can extend this equation to any rigid body rotating about a fixed axis, because any such body can always be analyzed as an assembly of single particles.

✔CHECKPOINT 7: The figure shows an overhead view of a meter stick that can pivot about the point indicated, which is to the left of the stick's midpoint. Two horizontal forces, $\vec{F}_1$ and $\vec{F}_2$, are applied to the stick. Only $\vec{F}_1$ is shown. Force $\vec{F}_2$ is perpendicular to the stick and is applied at the right end. If the stick is not to turn, (a) what should be the direction of $\vec{F}_2$, and (b) should F_2 be greater than, less than, or equal to F_1?

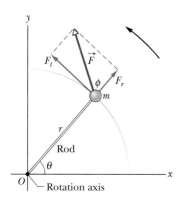

Fig. 11-16 A simple rigid body, free to rotate about an axis through O, consists of a particle of mass m fastened to the end of a rod of length r and negligible mass. An applied force $\vec{F}$ causes the body to rotate.

Sample Problem 11-7

Figure 11-17a shows a uniform disk, with mass $M = 2.5$ kg and radius $R = 20$ cm, mounted on a fixed horizontal axle. A block with mass $m = 1.2$ kg hangs from a massless cord that is wrapped around the rim of the disk. Find the acceleration of the falling block, the angular acceleration of the disk, and the tension in the cord. The cord does not slip, and there is no friction at the axle.

SOLUTION: One Key Idea here is that, taking the block as a system, we can relate its acceleration a to the forces acting on it with Newton's second law ($\vec{F}_{\text{net}} = m\vec{a}$). Those forces are shown in the block's free-body diagram in Fig. 11-17b: The force from the cord is $\vec{T}$ and the gravitational force is $\vec{F}_g$, of magnitude mg. We can

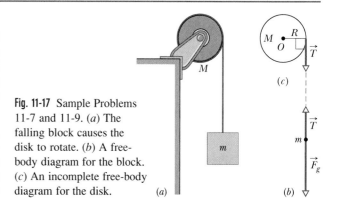

Fig. 11-17 Sample Problems 11-7 and 11-9. (a) The falling block causes the disk to rotate. (b) A free-body diagram for the block. (c) An incomplete free-body diagram for the disk.

now write Newton's second law for components along a vertical y axis ($F_{net,y} = ma_y$) as

$$T - mg = ma. \qquad (11\text{-}38)$$

However, we cannot solve this equation for a because it also contains the unknown T.

Previously, when we got stuck on the y axis, we would switch to the x axis. Here, we switch to the rotation of the disk and use this Key Idea: Taking the disk as a system, we can relate its angular acceleration α to the torque acting on it with Newton's second law for rotation ($\tau_{net} = I\alpha$). To calculate the torques and the rotational inertia I, we take the rotation axis to be perpendicular to the disk and through its center, at point O in Fig. 11-17c.

The torques are then given by Eq. 11-32 ($\tau = rF_t$). The gravitational force on the disk and the force on the disk from the axle both act at the center of the disk and thus at distance $r = 0$, so their torques are zero. The force $\vec{T}$ on the disk due to the cord acts at distance $r = R$ and is tangent to the rim of the disk. Therefore, its torque is $-RT$, negative because the torque rotates the disk clockwise from rest. From Table 11-2c, the rotational inertia I of the disk is $\frac{1}{2}MR^2$. Thus we can write $\tau_{net} = I\alpha$ as

$$-RT = \frac{1}{2}MR^2\alpha. \qquad (11\text{-}39)$$

This equation seems useless because it has two unknowns, α and T, neither of which is the desired a. However, mustering physics courage, we can make it useful with a third Key Idea: Because the cord does not slip, the linear acceleration a of the block and the (tangential) linear acceleration a_t of the rim of the disk are equal. Then, by Eq. 11-22 ($a_t = \alpha r$) we see that here $\alpha = a/R$. Substituting this in Eq. 11-39 yields

$$T = -\frac{1}{2}Ma. \qquad (11\text{-}40)$$

Now combining Eqs. 11-38 and 11-40 leads to

$$a = -g\frac{2m}{M + 2m} = -(9.8 \text{ m/s}^2)\frac{(2)(1.2 \text{ kg})}{2.5 \text{ kg} + (2)(1.2 \text{ kg})}$$

$$= -4.8 \text{ m/s}^2. \qquad \text{(Answer)}$$

We then use Eq. 11-40 to find T:

$$T = -\frac{1}{2}Ma = -\frac{1}{2}(2.5 \text{ kg})(-4.8 \text{ m/s}^2)$$

$$= 6.0 \text{ N}. \qquad \text{(Answer)}$$

As we should expect, the acceleration of the falling block is less than g, and the tension in the cord (= 6.0 N) is less than the gravitational force on the hanging block (= mg = 11.8 N). We see also that the acceleration of the block and the tension depend on the mass of the disk but not on its radius. As a check, we note that the formulas derived above predict $a = -g$ and $T = 0$ for the case of a massless disk ($M = 0$). This is what we would expect; the block simply falls as a free body, trailing the string behind it.

From Eq. 11-22, the angular acceleration of the disk is

$$\alpha = \frac{a}{R} = \frac{-4.8 \text{ m/s}^2}{0.20 \text{ m}} = -24 \text{ rad/s}^2. \qquad \text{(Answer)}$$

Sample Problem 11-8

To throw an 80 kg opponent with a basic judo hip throw, you intend to pull his uniform with a force $\vec{F}$ and a moment arm $d_1 = 0.30$ m from a pivot point (rotation axis) on your right hip (Fig. 11-18). You wish to rotate him about the pivot point with an angular acceleration α of -6.0 rad/s^2—that is, with an angular acceleration that is *clockwise* in the figure. Assume that his rotational inertia I relative to the pivot point is 15 kg $\cdot$ m^2.

(a) What must the magnitude of $\vec{F}$ be if, before you throw him, you bend your opponent forward to bring his center of mass to your hip (Fig. 11-18a)?

SOLUTION: One Key Idea here is that we can relate your pull $\vec{F}$ on him to the given angular acceleration α via Newton's second law for rotation ($\tau_{net} = I\alpha$). As his feet leave the floor, we can assume that only three forces act on him: your pull $\vec{F}$, a force $\vec{N}$ on him from you at the pivot point (this force is not indicated in Fig. 11-18), and the gravitational force $\vec{F}_g$. To use $\tau_{net} = I\alpha$, we need the corresponding three torques, each about the pivot point.

From Eq. 11-33 ($\tau = r_\perp F$), the torque due to your pull $\vec{F}$ is equal to $-d_1 F$, where d_1 is the moment arm $r_\perp$ and the sign indicates the clockwise rotation this torque tends to cause. The torque due to $\vec{N}$ is zero, because $\vec{N}$ acts at the pivot point and thus has moment arm $r_\perp = 0$.

To evaluate the torque due to $\vec{F}_g$, we need a Key Idea from Chapter 9: We can assume that $\vec{F}_g$ acts at your opponent's center of mass. With the center of mass at the pivot point, $\vec{F}_g$ has moment

Fig. 11-18 Sample Problem 11-8. A judo hip throw (a) correctly executed and (b) incorrectly executed.

arm $r_\perp = 0$ and thus the torque due to $\vec{F}_g$ is zero. Thus, the only torque on your opponent is due to your pull $\vec{F}$, and we can write $\tau_{net} = I\alpha$ as

$$-d_1 F = I\alpha.$$

We then find

$$F = \frac{-I\alpha}{d_1} = \frac{-(15 \text{ kg} \cdot \text{m}^2)(-6.0 \text{ rad/s}^2)}{0.30 \text{ m}}$$

$$= 300 \text{ N}. \qquad \text{(Answer)}$$

(b) What must the magnitude of $\vec{F}$ be if your opponent remains upright before you throw him, so that $\vec{F}_g$ has a moment arm $d_2 = 0.12$ m from the pivot point (Fig. 11-18b)?

SOLUTION: The Key Ideas we need here are similar to those in (a) with one exception: Because the moment arm for $\vec{F}_g$ is no longer zero, the torque due to $\vec{F}_g$ is now equal to d_2mg, and is positive because the torque attempts counterclockwise rotation. Now we write $\tau_{net} = I\alpha$ as

$$-d_1F + d_2mg = I\alpha,$$

which gives

$$F = -\frac{I\alpha}{d_1} + \frac{d_2mg}{d_1}.$$

From (a), we know that the first term on the right is equal to 300 N. Substituting this and the given data, we have

$$F = 300 \text{ N} + \frac{(0.12 \text{ m})(80 \text{ kg})(9.8 \text{ m/s}^2)}{0.30 \text{ m}}$$

$$= 613.6 \text{ N} \approx 610 \text{ N}. \qquad \text{(Answer)}$$

The results indicate that you will have to pull much harder if you do not initially bend your opponent to bring his center of mass to your hip. A good judo fighter knows this lesson from physics. (An analysis of the physics of judo and aikido is given in "The Amateur Scientist" by J. Walker, *Scientific American,* July 1980, Vol. 243, pp. 150–161.)

11-10 Work and Rotational Kinetic Energy

As we discussed in Chapter 7, when a force F causes a rigid body of mass m to accelerate along a coordinate axis, it does work W on the body. Thus, the body's kinetic energy ($K = \frac{1}{2}mv^2$) can change. Suppose it is the only energy of the body that changes. Then we relate the change ΔK in kinetic energy to the work W with the work–kinetic energy theorem (Eq. 7-10), writing

$$\Delta K = K_f - K_i = \tfrac{1}{2}mv_f^2 - \tfrac{1}{2}mv_i^2 = W \qquad \text{(work–kinetic energy theorem).} \quad (11\text{-}41)$$

For motion confined to an x axis, we can calculate the work with Eq. 7-32,

$$W = \int_{x_i}^{x_f} F \, dx \qquad \text{(work, one-dimensional motion).} \quad (11\text{-}42)$$

This reduces to $W = Fd$ when F is constant and the body's displacement is d. The rate at which the work is done is the power, which we can find with Eqs. 7-43 and 7-48,

$$P = \frac{dW}{dt} = Fv \qquad \text{(power, one-dimensional motion).} \quad (11\text{-}43)$$

Now let us consider a rotational situation that is similar. When a torque accelerates a rigid body in rotation about a fixed axis, it does work W on the body. Therefore, the body's rotational kinetic energy ($K = \frac{1}{2}I\omega^2$) can change. Suppose that it is the only energy of the body that changes. Then we can still relate the change ΔK in kinetic energy to the work W with the work–kinetic energy theorem, except now the kinetic energy is a rotational kinetic energy:

$$\Delta K = K_f - K_i = \tfrac{1}{2}I\omega_f^2 - \tfrac{1}{2}I\omega_i^2 = W \qquad \text{(work–kinetic energy theorem).} \quad (11\text{-}44)$$

Here, I is the rotational inertia of the body about the fixed axis and ω_i and ω_f are the angular speeds of the body before and after the work is done, respectively.

Also, we can calculate the work with a rotational equivalent of Eq. 11-42,

$$W = \int_{\theta_i}^{\theta_f} \tau \, d\theta \qquad \text{(work, rotation about fixed axis),} \quad (11\text{-}45)$$

where τ is the torque doing the work W, and θ_i and θ_f are the body's angular positions before and after the work is done, respectively. When τ is constant, Eq. 11-45 reduces to

$$W = \tau(\theta_f - \theta_i) \qquad \text{(work, constant torque).} \quad (11\text{-}46)$$

The rate at which the work is done is the power, which we can find with the rotational equivalent of Eq. 11-43,

$$P = \frac{dW}{dt} = \tau\omega \qquad \text{(power, rotation about fixed axis)}. \qquad (11\text{-}47)$$

Table 11-3 summarizes the equations that apply to the rotation of a rigid body about a fixed axis and the corresponding equations for translational motion.

Proof of Eqs. 11-44 through 11-47

Let us again consider the situation of Fig. 11-16, in which force $\vec{F}$ rotates a rigid body consisting of a single particle of mass m fastened to the end of a massless rod. During the rotation, force $\vec{F}$ does work on the body. Let us assume that the only energy of the body that is changed by $\vec{F}$ is the kinetic energy. Then we can apply the work–kinetic energy theorem of Eq. 11-41:

$$\Delta K = K_f - K_i = W. \qquad (11\text{-}48)$$

Using $K = \frac{1}{2}mv^2$ and Eq. 11-18 ($v = \omega r$), we can rewrite Eq. 11-48 as

$$\Delta K = \tfrac{1}{2}mr^2\omega_f^2 - \tfrac{1}{2}mr^2\omega_i^2 = W. \qquad (11\text{-}49)$$

From Eq. 11-26, the rotational inertia for this one-particle body is $I = mr^2$. Substituting this into Eq. 11-49 yields

$$\Delta K = \tfrac{1}{2}I\omega_f^2 - \tfrac{1}{2}I\omega_i^2 = W,$$

which is Eq. 11-44. We derived it for a rigid body with one particle, but it holds for any rigid body rotated about a fixed axis.

We next relate the work W done on the body in Fig. 11-16 to the torque τ on the body due to force $\vec{F}$. When the particle moves a distance ds along its circular path, only the tangential component F_t of the force accelerates the particle along the path. Therefore, only F_t does work on the particle. We write that work dW as $F_t\, ds$. However, we can replace ds with $r\, d\theta$, where $d\theta$ is the angle through which the particle moves. Thus we have

$$dW = F_t r\, d\theta. \qquad (11\text{-}50)$$

From Eq. 11-32, we see that the product $F_t r$ is equal to the torque τ, so we can rewrite Eq. 11-50 as

$$dW = \tau\, d\theta. \qquad (11\text{-}51)$$

TABLE 11-3   **Some Corresponding Relations for Translational and Rotational Motion**

Pure Translation (Fixed Direction)		Pure Rotation (Fixed Axis)	
Position	x	Angular position	θ
Velocity	$v = dx/dy$	Angular velocity	$\omega = d\theta/dt$
Acceleration	$a = dv/dt$	Angular acceleration	$\alpha = d\omega/dt$
Mass	m	Rotational inertia	I
Newton's second law	$F_{\text{net}} = ma$	Newton's second law	$\tau_{\text{net}} = I\alpha$
Work	$W = \int F\, dx$	Work	$W = \int \tau\, d\theta$
Kinetic energy	$K = \tfrac{1}{2}mv^2$	Kinetic energy	$K = \tfrac{1}{2}I\omega^2$
Power (constant force)	$P = Fv$	Power (constant torque)	$P = \tau\omega$
Work–kinetic energy theorem	$W = \Delta K$	Work–kinetic energy theorem	$W = \Delta K$

The work done during a finite angular displacement from θ_i to θ_f is then

$$W = \int_{\theta_i}^{\theta_f} \tau \, d\theta,$$

which is Eq. 11-45. It holds for any rigid body rotating about a fixed axis. Equation 11-46 comes directly from Eq. 11-45.

We can find the power P for rotational motion from Eq. 11-51:

$$P = \frac{dW}{dt} = \tau \frac{d\theta}{dt} = \tau\omega,$$

which is Eq. 11-47.

Sample Problem 11-9

Let the disk in Sample Problem 11-7 and Fig. 11-17 start from rest at time $t = 0$. What is its rotational kinetic energy K at $t = 2.5$ s?

SOLUTION: We can find K with Eq. 11-27 ($K = \frac{1}{2}I\omega^2$). We already know that I is equal to $\frac{1}{2}MR^2$, but we do not yet know ω at $t = 2.5$ s. A Key Idea, though, is that the angular acceleration α has the constant value of -24 rad/s^2, so we can apply the equations for constant angular acceleration in Table 11-1. Because we want ω and know α and ω_0 ($= 0$), we use Eq. 11-12:

$$\omega = \omega_0 + \alpha t = 0 + \alpha t = \alpha t.$$

Substituting $\omega = \alpha t$ and $I = \frac{1}{2}MR^2$ into Eq. 11-27, we find

$$K = \tfrac{1}{2}I\omega^2 = \tfrac{1}{2}(\tfrac{1}{2}MR^2)(\alpha t)^2 = \tfrac{1}{4}M(R\alpha t)^2$$
$$= \tfrac{1}{4}(2.5 \text{ kg})[(0.20 \text{ m})(-24 \text{ rad/s}^2)(2.5 \text{ s})]^2$$
$$= 90 \text{ J.} \qquad \text{(Answer)}$$

We can also get this answer with a different Key Idea: We can find the disk's kinetic energy from the work done on the disk. First, we relate the *change* in the kinetic energy of the disk to the net work W done on the disk, using the work–kinetic energy theorem of Eq. 11-44 ($K_f - K_i = W$). With K substituted for K_f and 0 for

K_i, we get

$$K = K_i + W = 0 + W = W. \qquad (11\text{-}52)$$

Next we want to find the work W. We can relate W to the torques acting on the disk with Eq. 11-45 or 11-46. The only torque causing angular acceleration and doing work is the torque due to force $\vec{T}$ on the disk from the cord. From Sample Problem 11-7, this torque is equal to $-TR$. Another Key Idea is that because α is constant, this torque also must be constant. Thus, we can use Eq. 11-46 to write

$$W = \tau(\theta_f - \theta_i) = -TR(\theta_f - \theta_i). \qquad (11\text{-}53)$$

We need one more Key Idea: Because α is constant, we can use Eq. 11-13 to find $\theta_f - \theta_i$. With $\omega_i = 0$, we have

$$\theta_f - \theta_i = \omega_i t + \tfrac{1}{2}\alpha t^2 = 0 + \tfrac{1}{2}\alpha t^2 = \tfrac{1}{2}\alpha t^2.$$

Now we substitute this into Eq. 11-53 and then substitute the result into Eq. 11-52. With $T = 6.0$ N and $\alpha = -24$ rad/s^2 (from Sample Problem 11-7), we have

$$K = W = -TR(\theta_f - \theta_i) = -TR(\tfrac{1}{2}\alpha t^2) = -\tfrac{1}{2}TR\alpha t^2$$
$$= -\tfrac{1}{2}(6.0 \text{ N})(0.20 \text{ m})(-24 \text{ rad/s}^2)(2.5 \text{ s})^2$$
$$= 90 \text{ J.} \qquad \text{(Answer)}$$

Sample Problem 11-10

A rigid sculpture consists of a thin hoop (of mass m and radius $R = 0.15$ m) and a thin radial rod (of mass m and length $L = 2.0R$), arranged as shown in Fig. 11-19. The sculpture can pivot around a horizontal axis in the plane of the hoop, passing through its center.

(a) In terms of m and R, what is the sculpture's rotational inertia I about the rotation axis?

SOLUTION: A Key Idea here is that we can separately find the rotational inertias of the hoop and the rod and then add the results to get the sculpture's total rotational inertia I. From Table 11-2h, the hoop has rotational inertia $I_{\text{hoop}} = \frac{1}{2}mR^2$ about its diameter. From Table 11-2e, the rod has rotational inertia $I_{\text{com}} = mL^2/12$ about an axis through its center of mass and parallel to the sculpture's rotation axis. To find its rotational inertia I_{rod} about that rotation axis, we

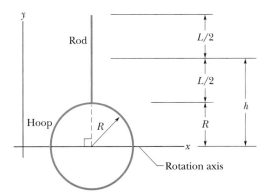

Fig. 11-19 Sample Problem 11-10. A rigid sculpture consisting of a hoop and a rod can rotate around a horizontal axis.

use Eq. 11-29, the parallel-axis theorem:

$$I_{rod} = I_{com} + mh_{com}^2 = \frac{mL^2}{12} + m\left(R + \frac{L}{2}\right)^2$$

$$= 4.33mR^2,$$

where we have used the fact that $L = 2.0R$ and where the perpendicular distance between the rod's center of mass and the rotation axis is $h = R + L/2$. Thus, the rotational inertia I of the sculpture about the rotation axis is

$$I = I_{hoop} + I_{rod} = \tfrac{1}{2}mR^2 + 4.33mR^2$$

$$= 4.83mR^2 \approx 4.8mR^2. \qquad \text{(Answer)}$$

(b) Starting from rest, the sculpture rotates around the rotation axis from the initial upright orientation of Fig. 11-19. What is its angular speed ω about the axis when it is inverted?

SOLUTION: Three Key Ideas are required here:

1. We can relate the sculpture's speed ω to its rotational kinetic energy K with Eq. 11-27 ($K = \tfrac{1}{2}I\omega^2$).

2. We can relate K to the gravitational potential energy U of the sculpture via the conservation of the sculpture's mechanical energy E during the rotation. Thus, during the rotation, E does not change ($\Delta E = 0$) as energy is transferred from U to K.

3. For the gravitational potential energy we can treat the rigid sculpture as a particle located at the center of mass, with the total mass $2m$ concentrated there.

We can write the conservation of mechanical energy ($\Delta E = 0$) as

$$\Delta K + \Delta U = 0. \qquad \text{(11-54)}$$

As the sculpture rotates from its initial position at rest to its inverted

position, when the angular speed is ω, the change ΔK in its kinetic energy is

$$\Delta K = K_f - K_i = \tfrac{1}{2}I\omega^2 - 0 = \tfrac{1}{2}I\omega^2. \qquad \text{(11-55)}$$

From Eq. 8-7 ($\Delta U = mg\,\Delta y$), the corresponding change ΔU in the gravitational potential energy is

$$\Delta U = (2m)g\,\Delta y_{com}, \qquad \text{(11-56)}$$

where $2m$ is the sculpture's total mass, and Δy_{com} is the vertical displacement of its center of mass during the rotation.

To find Δy_{com}, we first find the initial location y_{com} of the center of mass in Fig. 11-19. The hoop (with mass m) is centered at $y = 0$. The rod (with mass m) is centered at $y = R + L/2$. Thus, from Eq. 9-5, the sculpture's center of mass is at

$$y_{com} = \frac{m(0) + m(R + L/2)}{2m} = \frac{0 + m(R + 2R/2)}{2m} = R.$$

When the sculpture is inverted, the center of mass is this same distance R from the rotation axis but *below* it. Therefore, the vertical displacement of the center of mass from the initial position to the inverted position is $\Delta y_{com} = -2R$.

Now let's pull these results together. Substituting Eqs. 11-55 and 11-56 into 11-54 gives us

$$\tfrac{1}{2}I\omega^2 + (2m)g\,\Delta y_{com} = 0.$$

Substituting $I = 4.83mR^2$ from (a) and $\Delta y_{com} = -2R$ from above and solving for ω, we find

$$\omega = \sqrt{\frac{8g}{4.83R}} = \sqrt{\frac{(8)(9.8 \text{ m/s}^2)}{(4.83)(0.15 \text{ m})}}$$

$$= 10 \text{ rad/s}. \qquad \text{(Answer)}$$

REVIEW & SUMMARY

Angular Position   To describe the rotation of a rigid body about a fixed axis, called the **rotation axis,** we assume a **reference line** is fixed in the body, perpendicular to that axis and rotating with the body. We measure the **angular position** θ of this line relative to a fixed direction. When θ is measured in **radians,**

$$\theta = \frac{s}{r} \qquad \text{(radian measure)}, \qquad \text{(11-1)}$$

where s is the arc length of a circular path of radius r and angle θ. Radian measure is related to angle measure in revolutions and degrees by

$$1 \text{ rev} = 360° = 2\pi \text{ rad.} \qquad \text{(11-2)}$$

Angular Displacement   A body that rotates about a rotation axis, changing its angular position from θ_1 to θ_2, undergoes an **angular displacement**

$$\Delta\theta = \theta_2 - \theta_1, \qquad \text{(11-4)}$$

where $\Delta\theta$ is positive for counterclockwise rotation and negative for clockwise rotation.

Angular Velocity and Speed   If a body rotates through an angular displacement $\Delta\theta$ in a time interval Δt, its **average angular velocity** ω_{avg} is

$$\omega_{avg} = \frac{\Delta\theta}{\Delta t}. \qquad \text{(11-5)}$$

The (**instantaneous**) **angular velocity** ω of the body is

$$\omega = \frac{d\theta}{dt}. \qquad \text{(11-6)}$$

Both ω_{avg} and ω are vectors, with directions given by the **right-hand rule** of Fig. 11-6. They are positive for counterclockwise rotation and negative for clockwise rotation. The magnitude of the body's angular velocity is the **angular speed.**

Angular Acceleration   If the angular velocity of a body changes from ω_1 to ω_2 in a time interval $\Delta t = t_2 - t_1$, the **average angular acceleration** α_{avg} of the body is

$$\alpha_{avg} = \frac{\omega_2 - \omega_1}{t_2 - t_1} = \frac{\Delta \omega}{\Delta t}. \tag{11-7}$$

The **(instantaneous) angular acceleration** α of a body is

$$\alpha = \frac{d\omega}{dt}. \tag{11-8}$$

Both α_{avg} and α are vectors.

The Kinematic Equations for Constant Angular Acceleration

Constant angular acceleration (α = constant) is an important special case of rotational motion. The appropriate kinematic equations, given in Table 11-1, are

$$\omega = \omega_0 + \alpha t, \tag{11-12}$$

$$\theta - \theta_0 = \omega_0 t + \tfrac{1}{2}\alpha t^2, \tag{11-13}$$

$$\omega^2 = \omega_0^2 + 2\alpha(\theta - \theta_0), \tag{11-14}$$

$$\theta - \theta_0 = \tfrac{1}{2}(\omega_0 + \omega)t, \tag{11-15}$$

$$\theta - \theta_0 = \omega t - \tfrac{1}{2}\alpha t^2. \tag{11-16}$$

Linear and Angular Variables Related

A point in a rigid rotating body, at a *perpendicular distance* r from the rotation axis, moves in a circle with radius r. If the body rotates through an angle θ, the point moves along an arc with length s given by

$$s = \theta r \qquad \text{(radian measure)}, \tag{11-17}$$

where θ is in radians.

The linear velocity $\vec{v}$ of the point is tangent to the circle; the point's linear speed v is given by

$$v = \omega r \qquad \text{(radian measure)}, \tag{11-18}$$

where ω is the angular speed (in radians per second) of the body.

The linear acceleration $\vec{a}$ of the point has both *tangential* and *radial* components. The tangential component is

$$a_t = \alpha r \qquad \text{(radian measure)}, \tag{11-22}$$

where α is the magnitude of the angular acceleration (in radians per second-squared) of the body. The radial component of $\vec{a}$ is

$$a_r = \frac{v^2}{r} = \omega^2 r \qquad \text{(radian measure)}. \tag{11-23}$$

If the point moves in uniform circular motion, the period T of the motion for the point and the body is

$$T = \frac{2\pi r}{v} = \frac{2\pi}{\omega} \qquad \text{(radian measure)}. \tag{11-19, 11-20}$$

Rotational Kinetic Energy and Rotational Inertia

The kinetic energy K of a rigid body rotating about a fixed axis is given by

$$K = \tfrac{1}{2}I\omega^2 \qquad \text{(radian measure)}, \tag{11-27}$$

in which I is the **rotational inertia** of the body, defined as

$$I = \sum m_i r_i^2 \tag{11-26}$$

for a system of discrete particles and as

$$I = \int r^2\, dm \tag{11-28}$$

for a body with continuously distributed mass. The r and r_i in these expressions represent the perpendicular distance from the axis of rotation to each mass element in the body.

The Parallel-Axis Theorem

The *parallel-axis theorem* relates the rotational inertia I of a body about any axis to that of the same body about a parallel axis through the center of mass:

$$I = I_{com} + Mh^2. \tag{11-29}$$

Here h is the perpendicular distance between the two axes.

Torque

Torque is a turning or twisting action on a body about a rotation axis due to a force $\vec{F}$. If $\vec{F}$ is exerted at a point given by the position vector $\vec{r}$ relative to the axis, then the magnitude of the torque is

$$\tau = rF_t = r_\perp F = rF \sin \phi, \tag{11-32, 11-33, 11-31}$$

where F_t is the component of $\vec{F}$ perpendicular to $\vec{r}$, and ϕ is the angle between $\vec{r}$ and $\vec{F}$. The quantity $r_\perp$ is the perpendicular distance between the rotation axis and an extended line running through the $\vec{F}$ vector. This line is called the **line of action** of $\vec{F}$, and $r_\perp$ is called the **moment arm** of $\vec{F}$. Similarly, r is the moment arm of F_t.

The SI unit of torque is the newton-meter ($N \cdot m$). A torque τ is positive if it tends to rotate a body at rest counterclockwise and negative if it tends to rotate the body in the clockwise direction.

Newton's Second Law in Angular Form

The rotational analog of Newton's second law is

$$\tau_{net} = I\alpha, \tag{11-37}$$

where τ_{net} is the net torque acting on a particle or rigid body, I is the rotational inertia of the particle or body about the rotation axis, and α is the resulting angular acceleration about that axis.

Work and Rotational Kinetic Energy

The equations used for calculating work and power in rotational motion correspond to equations used for translational motion and are

$$W = \int_{\theta_i}^{\theta_f} \tau\, d\theta \tag{11-45}$$

and

$$P = \frac{dW}{dt} = \tau\omega. \tag{11-47}$$

When τ is constant, Eq. 11-45 reduces to

$$W = \tau(\theta_f - \theta_i). \tag{11-46}$$

The form of the work–kinetic energy theorem used for rotating bodies is

$$\Delta K = K_f - K_i = \tfrac{1}{2}I\omega_f^2 - \tfrac{1}{2}I\omega_i^2 = W. \tag{11-44}$$

QUESTIONS

1. Figure 11-20b is a graph of the angular position of the rotating disk of Fig. 11-20a. Is the angular velocity of the disk positive, negative, or zero at (a) $t = 1$ s, (b) $t = 2$ s, and (c) $t = 3$ s? (d) Is the angular acceleration positive or negative?

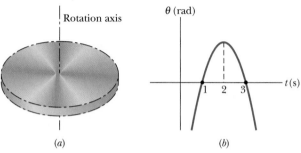

(a) (b)

Fig. 11-20 Question 1.

2. Figure 11-21 is a graph of the angular velocity of the rotating disk of Fig. 11-20a. What are the (a) initial and (b) final directions of rotation? (c) Does the disk momentarily stop? (d) Is the angular acceleration positive or negative? (e) Is the angular acceleration constant or varying?

3. Hold your right arm downward, palm toward your thigh. Keeping your wrist rigid, (1) lift the arm until it is horizontal and forward, (2) move it horizontally until it is pointed toward the right, and (3) then bring it down to your side. Your palm faces forward. If you start over, but reverse the steps, why does your palm *not* face forward.

Fig. 11-21 Question 2.

4. For which of the following expressions for $\omega(t)$ of a rotating object do the angular equations of Table 11-1 apply? (a) $\omega = 3$; (b) $\omega = 4t^2 + 2t - 6$; (c) $\omega = 3t - 4$; (d) $\omega = 5t^2 - 3$, all with ω in radians per second and t in seconds.

5. Figure 11-22 is a graph of the angular velocity versus time for the rotating disk of Fig. 11-20a. For a point on the rim of the disk, rank the four instants a, b, c, and d according to the magnitude of (a) the tangential acceleration and (b) the radial acceleration, greatest first.

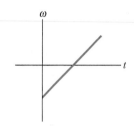

Fig. 11-22 Question 5.

6. The overhead view of Fig. 11-23 is a snapshot of a disk turning counterclockwise like a merry-go-round. The angular speed ω of the disk is decreasing (the disk is turning counterclockwise slower and slower). The figure shows the

Fig. 11-23 Question 6.

position of a cockroach that rides the rim of the disk. At the instant of the snapshot, what are the directions of (a) the cockroach's radial acceleration and (b) its tangential acceleration?

7. Figure 11-24 shows an assembly of three small spheres of the same mass that are attached to a massless rod with the indicated spacings. Consider the rotational inertia I of the assembly about each sphere, in turn. Then rank the spheres according to the rotational inertia about them, greatest first.

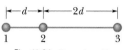

Fig. 11-24 Question 7.

8. Figure 11-25a shows an overhead view of a horizontal bar that can pivot about the point indicated. Two horizontal forces act on the bar, but the bar is stationary. If the angle between the bar and force $\vec{F}_2$ is now decreased from the initial 90° and the bar is still not to turn, should the magnitude of $\vec{F}_2$ be made larger, made smaller, or left the same?

(a) (b)

Fig. 11-25 Questions 8 and 9.

9. Figure 11-25b shows an overhead view of a horizontal bar that is rotated about the pivot point by two horizontal forces, $\vec{F}_1$ and $\vec{F}_2$, at opposite ends of the bar. The direction of $\vec{F}_2$ is at angle ϕ to the bar. Rank the following values of ϕ according to the magnitude of the angular acceleration of the bar, greatest first: 90°, 70°, and 110°.

10. In the overhead view of Fig. 11-26, five forces of the same magnitude act on a merry-go-round for the strange; it is a square that can rotate about point P at midlength along one of the edges. Rank the forces acting on it according to the magnitude of the torque they create about point P, greatest first.

Fig. 11-26 Question 10.

11. In Fig. 11-27, two forces $\vec{F}_1$ and $\vec{F}_2$ act on a disk that turns about its center like a merry-go-round. The forces maintain the indicated angles during the rotation, which is counterclockwise and at a constant rate. However, we are to decrease the angle θ of $\vec{F}_1$ without changing the magnitude of $\vec{F}_1$. (a) To keep the angular speed constant, should we increase, decrease, or maintain

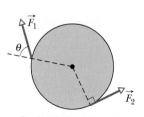

Fig. 11-27 Question 11.

the magnitude of $\vec{F}_2$? Do forces (b) $\vec{F}_1$ and (c) $\vec{F}_2$ tend to rotate the disk clockwise or counterclockwise?

12. Figure 11-28 gives the angular velocity ω versus time t for a merry-go-round being turned by a varying applied force. The magnitudes of the curve's slopes in time intervals 1, 3, 4, and 6 are equal. (a) During which of the time intervals is energy transferred *from* the merry-go-round by the applied force? (b) Rank the time intervals according to the work done on the merry-go-round by the applied force during the interval, most positive work first, most negative work last. (c) Rank the intervals according to the rate at which the applied force transfers energy, greatest rate of transfer *to* the merry-go-round first, greatest rate of transfer *from* it last.

Fig. 11-28 Question 12.

EXERCISES & PROBLEMS

ssm Solution is in the Student Solutions Manual.
www Solution is available on the World Wide Web at:
            http://www.wiley.com/college/hrw
ilw Solution is available on the Interactive LearningWare.

SEC. 11-2   The Rotational Variables

1E. During a time interval t the flywheel of a generator turns through the angle $\theta = at + bt^3 - ct^4$, where a, b, and c are constants. Write expressions for the wheel's (a) angular velocity and (b) angular acceleration.

2E. What is the angular speed of (a) the second hand, (b) the minute hand, and (c) the hour hand of a smoothly running analog watch? Answer in radians per second.

3E. Our Sun is 2.3×10^4 ly (light-years) from the center of our Milky Way galaxy and is moving in a circle around that center at a speed of 250 km/s. (a) How long does it take the Sun to make one revolution about the galactic center? (b) How many revolutions has the Sun completed since it was formed about 4.5×10^9 years ago? ssm

4E. The angular position of a point on the rim of a rotating wheel is given by $\theta = 4.0t - 3.0t^2 + t^3$, where θ is in radians and t is in seconds. What are the angular velocities at (a) $t = 2.0$ s and (b) $t = 4.0$ s? (c) What is the average angular acceleration for the time interval that begins at $t = 2.0$ s and ends at $t = 4.0$ s? What are the instantaneous angular accelerations at (d) the beginning and (e) the end of this time interval?

5E. The angular position of a point on a rotating wheel is given by $\theta = 2 + 4t^2 + 2t^3$, where θ is in radians and t is in seconds. At $t = 0$, what are (a) the point's angular position and (b) its angular velocity? (c) What is its angular velocity at $t = 4.0$ s? (d) Calculate its angular acceleration at $t = 2.0$ s. (e) Is its angular acceleration constant? ssm

6P. The wheel in Fig. 11-29 has eight equally spaced spokes and a radius of 30 cm. It is mounted on a fixed axle and is spinning at

Fig. 11-29 Problem 6.

2.5 rev/s. You want to shoot a 20-cm-long arrow parallel to this axle and through the wheel without hitting any of the spokes. Assume that the arrow and the spokes are very thin. (a) What minimum speed must the arrow have? (b) Does it matter where between the axle and rim of the wheel you aim? If so, what is the best location?

7P. A diver makes 2.5 revolutions on the way from a 10-m-high platform to the water. Assuming zero initial vertical velocity, find the diver's average angular velocity during a dive. ilw

SEC. 11-4   Rotation with Constant Angular Acceleration

8E. The angular speed of an automobile engine is increased at a constant rate from 1200 rev/min to 3000 rev/min in 12 s. (a) What is its angular acceleration in revolutions per minute-squared? (b) How many revolutions does the engine make during this 12 s interval?

9E. A record turntable rotating at $33\frac{1}{3}$ rev/min slows down and stops in 30 s after the motor is turned off. (a) Find its (constant) angular acceleration in revolutions per minute-squared. (b) How many revolutions does it make in this time? ssm

10E. A disk, initially rotating at 120 rad/s, is slowed down with a constant angular acceleration of magnitude 4.0 rad/s². (a) How much time does the disk take to stop? (b) Through what angle does the disk rotate during that time?

11E. A heavy flywheel rotating on its central axis is slowing down because of friction in its bearings. At the end of the first minute of slowing, its angular speed is 0.90 of its initial angular speed of 250 rev/min. Assuming a constant angular acceleration, find its angular speed at the end of the second minute.

12E. Starting from rest, a disk rotates about its central axis with constant angular acceleration. In 5.0 s, it rotates 25 rad. During that time, what are the magnitudes of (a) the angular acceleration and (b) the average angular velocity? (c) What is the instantaneous angular velocity of the disk at the end of the 5.0 s? (d) With the angular acceleration unchanged, through what additional angle will the disk turn during the next 5.0 s?

13P. A wheel has a constant angular acceleration of 3.0 rad/s². During a certain 4.0 s interval, it turns through an angle of 120 rad.

Assuming that the wheel starts from rest, how long is it in motion at the start of this 4.0 s interval? **ssm** **www**

14P. A wheel, starting from rest, rotates with a constant angular acceleration of 2.00 rad/s². During a certain 3.00 s interval, it turns through 90.0 rad. (a) How long is the wheel turning before the start of the 3.00 s interval? (b) What is the angular velocity of the wheel at the start of the 3.00 s interval?

15P. At $t = 0$, a flywheel has an angular velocity of 4.7 rad/s, an angular acceleration of -0.25 rad/s², and a reference line at $\theta_0 = 0$. (a) Through what maximum angle θ_{max} will the reference line turn in the positive direction? At what times t will the reference line be at (b) $\theta = \frac{1}{2}\theta_{max}$ and (c) $\theta = -10.5$ rad (consider both positive and negative values of t)? (d) Graph θ versus t, and indicate the answers to (a), (b), and (c) on the graph.

16P. A disk rotates about its central axis starting from rest and accelerates with constant angular acceleration. At one time it is rotating at 10 rev/s; 60 revolutions later, its angular speed is 15 rev/s. Calculate (a) the angular acceleration, (b) the time required to complete the 60 revolutions, (c) the time required to reach the 10 rev/s angular speed, and (d) the number of revolutions from rest until the time the disk reaches the 10 rev/s angular speed.

17P. A flywheel turns through 40 rev as it slows from an angular speed of 1.5 rad/s to a stop. (a) Assuming a constant angular acceleration, find the time for it to come to rest. (b) What is its angular acceleration? (c) How much time is required for it to complete the first 20 of the 40 revolutions? **ilw**

18P. A wheel rotating about a fixed axis through its center has a constant angular acceleration of 4.0 rad/s². In a certain 4.0 s interval the wheel turns through an angle of 80 rad. (a) What is the angular velocity of the wheel at the start of the 4.0 s interval? (b) Assuming that the wheel starts from rest, how long is it in motion at the start of the 4.0 s interval?

SEC. 11-5 Relating the Linear and Angular Variables

19E. What is the linear acceleration of a point on the rim of a 30-cm-diameter record rotating at a constant angular speed of $33\frac{1}{3}$ rev/min? **ssm**

20E. A vinyl record on a turntable rotates at $33\frac{1}{3}$ rev/min. (a) What is its angular speed in radians per second? What is the linear speed of a point on the record at the needle when the needle is (b) 15 cm and (c) 7.4 cm from the turntable axis?

21E. What is the angular speed of a car traveling at 50 km/h and rounding a circular turn of radius 110 m?

22E. A flywheel with a diameter of 1.20 m is rotating at an angular speed of 200 rev/min. (a) What is the angular speed of the flywheel in radians per second? (b) What is the linear speed of a point on the rim of the flywheel? (c) What constant angular acceleration (in revolutions per minute-squared) will increase the wheel's angular speed to 1000 rev/min in 60 s? (d) How many revolutions does the wheel make during that 60 s?

23E. An astronaut is being tested in a centrifuge. The centrifuge has a radius of 10 m and, in starting, rotates according to $\theta = 0.30t^2$, where t is in seconds and θ is in radians. When $t = 5.0$ s, what are the magnitudes of the astronaut's (a) angular velocity,

(b) linear velocity, (c) tangential acceleration, and (d) radial acceleration?

24E. What are the magnitudes of (a) the angular velocity, (b) the radial acceleration, and (c) the tangential acceleration of a spaceship taking a circular turn of radius 3220 km at a speed of 29 000 km/h?

25P. An early method of measuring the speed of light makes use of a rotating slotted wheel. A beam of light passes through a slot at the outside edge of the wheel, as in Fig. 11-30, travels to a distant mirror, and returns to the wheel just in time to pass through the next slot in the wheel. One such slotted wheel has a radius of 5.0 cm and 500 slots at its edge. Measurements taken when the mirror is $L = 500$ m from the wheel indicate a speed of light of 3.0×10^5 km/s. (a) What is the (constant) angular speed of the wheel? (b) What is the linear speed of a point on the edge of the wheel? **ssm**

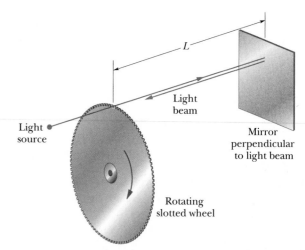

Fig. 11-30 Problem 25.

26P. The flywheel of a steam engine runs with a constant angular velocity of 150 rev/min. When steam is shut off, the friction of the bearings and of the air stops the wheel in 2.2 h. (a) What is the constant angular acceleration, in revolutions per minute-squared, of the wheel during the slowdown? (b) How many rotations does the wheel make before stopping? (c) At the instant the flywheel is turning at 75 rev/min, what is the tangential component of the linear acceleration of a flywheel particle that is 50 cm from the axis of rotation? (d) What is the magnitude of the net linear acceleration of the particle in (c)?

27P. (a) What is the angular speed ω about the polar axis of a point on Earth's surface at a latitude of 40° N? (Earth rotates about that axis.) (b) What is the linear speed v of the point? What are (c) ω and (d) v for a point at the equator? **ssm**

28P. A gyroscope flywheel of radius 2.83 cm is accelerated from rest at 14.2 rad/s² until its angular speed is 2760 rev/min. (a) What is the tangential acceleration of a point on the rim of the flywheel during this spin-up process? (b) What is the radial acceleration of this point when the flywheel is spinning at full speed? (c) Through what distance does a point on the rim move during the spin-up?

29P. In Fig. 11-31, wheel A of radius $r_A = 10$ cm is coupled by belt B to wheel C of radius $r_C = 25$ cm. The angular speed of wheel A is increased from rest at a constant rate of 1.6 rad/s². Find the time for wheel C to reach a rotational speed of 100 rev/min, assuming the belt does not slip. (*Hint:* If the belt does not slip, the linear speeds at the rims of the two wheels must be equal.) ssm www

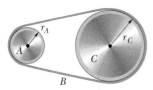

Fig. 11-31 Problem 29.

30P. An object rotates about a fixed axis, and the angular position of a reference line on the object is given by $\theta = 0.40e^{2t}$, where θ is in radians and t is in seconds. Consider a point on the object that is 4.0 cm from the axis of rotation. At $t = 0$, what are the magnitudes of the point's (a) tangential component of acceleration and (b) radial component of acceleration?

31P. A pulsar is a rapidly rotating neutron star that emits a radio beam like a lighthouse emits a light beam. We receive a radio pulse for each rotation of the star. The period T of rotation is found by measuring the time between pulses. The pulsar in the Crab nebula (Fig. 11-32) has a period of rotation of $T = 0.033$ s that is increasing at the rate of 1.26×10^{-5} s/y. (a) What is the pulsar's angular acceleration? (b) If its angular acceleration is constant, how many years from now will the pulsar stop rotating? (c) The pulsar originated in a supernova explosion seen in the year 1054. What was the initial T for the pulsar? (Assume constant angular acceleration since the pulsar originated.) ssm

32P. A record turntable is rotating at $33\frac{1}{3}$ rev/min. A watermelon seed is on the turntable 6.0 cm from the axis of rotation. (a) Calculate the acceleration of the seed, assuming that it does not slip. (b) What is the minimum value of the coefficient of static friction between the seed and the turntable if the seed is not to slip? (c) Suppose that the turntable achieves its angular speed by starting from rest and undergoing a constant angular acceleration for 0.25 s. Calculate the minimum coefficient of static friction required for the seed not to slip during the acceleration period.

SEC. 11-6 Kinetic Energy of Rotation

33E. Calculate the rotational inertia of a wheel that has a kinetic energy of 24 400 J when rotating at 602 rev/min. ssm

34P. The oxygen molecule O_2 has a mass of 5.30×10^{-26} kg and a rotational inertia of 1.94×10^{-46} kg·m² about an axis through the center of the line joining the atoms and perpendicular to that line. Suppose the center of mass of an O_2 molecule in a gas has a translational speed of 500 m/s and the molecule has a rotational kinetic energy that is $\frac{2}{3}$ of the translational kinetic energy of its center of mass. What then is the molecule's angular speed about the center of mass?

SEC. 11-7 Calculating the Rotational Inertia

35E. Two uniform solid cylinders, each rotating about its central (longitudinal) axis, have the same mass of 1.25 kg and rotate with the same angular speed of 235 rad/s, but they differ in radius. What is the rotational kinetic energy of (a) the smaller cylinder, of radius 0.25 m, and (b) the larger cylinder, of radius 0.75 m? ssm

36E. A communications satellite is a solid cylinder with mass 1210 kg, diameter 1.21 m, and length 1.75 m. Prior to launching from the shuttle cargo bay, it is set spinning at 1.52 rev/s about the cylinder axis (Fig. 11-33). Calculate the satellite's (a) rotational inertia about the rotation axis and (b) rotational kinetic energy.

Fig. 11-33 Exercise 36.

37E. In Fig. 11-34, two particles, each with mass m, are fastened to each other, and to a rotation axis at O, by two thin rods, each with length d and mass M. The combination rotates around the rotation axis with angular velocity ω. In terms of these symbols, and measured about O, what are the combination's (a) rotational inertia and (b) kinetic energy?

Fig. 11-32 Problem 31. The Crab nebula resulted from a star whose explosion was seen in 1054. In addition to the gaseous debris seen here, the explosion left a spinning neutron star at its center. The star has a diameter of only 30 km.

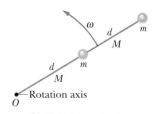

Fig. 11-34 Exercise 37.

38E. Each of the three helicopter rotor blades shown in Fig. 11-35 is 5.20 m long and has a mass of 240 kg. The rotor is rotating at 350 rev/min. (a) What is the rotational inertia of the rotor assembly about the axis of rotation? (Each blade can be considered to be a thin rod rotated about one end.) (b) What is the total kinetic energy of rotation?

Fig. 11-35 Exercise 38.

39E. Calculate the rotational inertia of a meter stick, with mass 0.56 kg, about an axis perpendicular to the stick and located at the 20 cm mark. (Treat the stick as a thin rod.) ssm

40P. Four identical particles of mass 0.50 kg each are placed at the vertices of a 2.0 m × 2.0 m square and held there by four massless rods, which form the sides of the square. What is the rotational inertia of this rigid body about an axis that (a) passes through the midpoints of opposite sides and lies in the plane of the square, (b) passes through the midpoint of one of the sides and is perpendicular to the plane of the square, and (c) lies in the plane of the square and passes through two diagonally opposite particles?

Fig. 11-36 Problem 41.

41P. The uniform solid block in Fig. 11-36 has mass M and edge lengths a, b, and c. Calculate its rotational inertia about an axis through one corner and perpendicular to the large faces. ssm

42P. The masses and coordinates of four particles are as follows: 50 g, $x = 2.0$ cm, $y = 2.0$ cm; 25 g, $x = 0$, $y = 4.0$ cm; 25 g, $x = -3.0$ cm, $y = -3.0$ cm; 30 g, $x = -2.0$ cm, $y = 4.0$ cm. What are the rotational inertias of this collection about the (a) x, (b) y, and (c) z axes? (d) Suppose the answers to (a) and (b) are A and B, respectively. Then what is the answer to (c) in terms of A and B?

43P. (a) Show that the rotational inertia of a solid cylinder of mass M and radius R about its central axis is equal to the rotational inertia of a thin hoop of mass M and radius $R/\sqrt{2}$ about its central axis. (b) Show that the rotational inertia I of any given body of mass M about any given axis is equal to the rotational inertia of an *equivalent hoop* about that axis, if the hoop has the same mass M and a radius k given by

$$k = \sqrt{\frac{I}{M}}.$$

The radius k of the equivalent hoop is called the *radius of gyration* of the given body. ssm

44P. Delivery trucks that operate by making use of energy stored in a rotating flywheel have been used in Europe. The trucks are charged by using an electric motor to get the flywheel up to its top speed of 200π rad/s. One such flywheel is a solid, uniform cylinder with a mass of 500 kg and a radius of 1.0 m. (a) What is the kinetic energy of the flywheel after charging? (b) If the truck operates with an average power requirement of 8.0 kW, for how many minutes can it operate between chargings?

SEC. 11-8 Torque

45E. A small ball of mass 0.75 kg is attached to one end of a 1.25-m-long massless rod, and the other end of the rod is hung from a pivot. When the resulting pendulum is 30° from the vertical, what is the magnitude of the torque about the pivot? ssm

46E. The length of a bicycle pedal arm is 0.152 m, and a downward force of 111 N is applied to the pedal by the rider's foot. What is the magnitude of the torque about the pedal arm's pivot point when the arm makes an angle of (a) 30°, (b) 90°, and (c) 180° with the vertical?

47P. The body in Fig. 11-37 is pivoted at O, and two forces act on it as shown. (a) Find an expression for the net torque on the body about the pivot. (b) If $r_1 =$ 1.30 m, $r_2 = 2.15$ m, $F_1 = 4.20$ N, $F_2 = 4.90$ N, $\theta_1 = 75.0°$, and $\theta_2 = 60.0°$, what is the net torque about the pivot? ssm ilw

Fig. 11-37 Problem 47.

48P. The body in Fig. 11-38 is pivoted at O. Three forces act on it in the directions shown: $F_A = 10$ N at point A, 8.0 m from O; $F_B = 16$ N at point B, 4.0 m from O; and $F_C = 19$ N at point C, 3.0 m from O. What is the net torque about O?

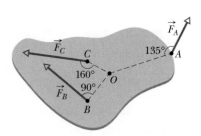

Fig. 11-38 Problem 48.

SEC. 11-9 Newton's Second Law for Rotation

49E. During the launch from a board, a diver's angular speed about her center of mass changes from zero to 6.20 rad/s in 220 ms. Her rotational inertia about her center of mass is 12.0 kg·m². During the launch, what are the magnitudes of (a) her average angular acceleration and (b) the average external torque on her from the board? ssm ilw

50E. A torque of 32.0 N·m on a certain wheel causes an angular acceleration of 25.0 rad/s². What is the wheel's rotational inertia?

51E. A thin spherical shell has a radius of 1.90 m. An applied torque of 960 N·m gives the shell an angular acceleration of 6.20 rad/s² about an axis through the center of the shell. What are (a) the rotational inertia of the shell about that axis and (b) the mass of the shell? ssm

52E. In Fig. 11-39, a cylinder having a mass of 2.0 kg can rotate about its central axis through point O. Forces are applied as shown: $F_1 = 6.0$ N, $F_2 = 4.0$ N, $F_3 = 2.0$ N, and $F_4 = 5.0$ N. Also, $R_1 =$ 5.0 cm and $R_2 = 12$ cm. Find the magnitude and direction of the angular acceleration of the cylinder. (During the rotation, the forces maintain their same angles relative to the cylinder.)

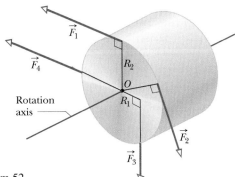

Fig. 11-39 Problem 52.

53P. Figure 11-40 shows the massive shield door at a neutron test facility at Lawrence Livermore Laboratory; this is the world's heaviest hinged door. The door has a mass of 44,000 kg, a rotational inertia about a vertical axis through its huge hinges of 8.7×10^4 kg·m^2, and a (front) face width of 2.4 m. Neglecting friction, what steady force, applied at its outer edge and perpendicular to the plane of the door, can move it from rest through an angle of 90° in 30 s?

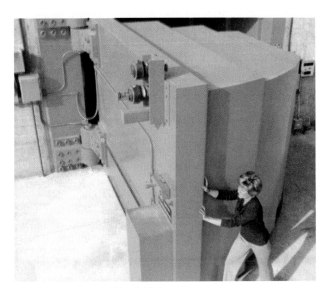

Fig. 11-40 Problem 53.

54P. A wheel of radius 0.20 m is mounted on a frictionless horizontal axis. The rotational inertia of the wheel about the axis is 0.050 kg·m^2. A massless cord wrapped around the wheel is attached to a 2.0 kg block that slides on a horizontal frictionless surface. If a horizontal force of magnitude $P = 3.0$ N is applied to the block as shown in Fig. 11-41, what is the magnitude of the angular acceleration of the wheel? Assume the string does not slip on the wheel.

Fig. 11-41 Problem 54.

55P. In Fig. 11-42, one block has mass $M = 500$ g, the other has mass $m = 460$ g, and the pulley, which is mounted in horizontal frictionless bearings, has a radius of 5.00 cm. When released from rest, the heavier block falls 75.0 cm in 5.00 s (without the cord slipping on the pulley). (a) What is the magnitude of the blocks' acceleration? What is the tension in the part of the cord that supports (b) the heavier block and (c) the lighter block? (d) What is the magnitude of the pulley's angular acceleration? (e) What is its rotational inertia? ssm www

Fig. 11-42 Problem 55.

56P. A pulley, with a rotational inertia of 1.0×10^{-3} kg·m^2 about its axle and a radius of 10 cm, is acted on by a force applied tangentially at its rim. The force magnitude varies in time as $F = 0.50t + 0.30t^2$, with F in newtons and t in seconds. The pulley is initially at rest. At $t = 3.0$ s what are (a) its angular acceleration and (b) its angular speed?

57P. Figure 11-43 shows two blocks, each of mass m, suspended from the ends of a rigid massless rod of length $L_1 + L_2$, with $L_1 = 20$ cm and $L_2 = 80$ cm. The rod is held horizontally on the fulcrum and then released. What are the magnitudes of the initial accelerations of (a) the block closer to the fulcrum and (b) the other block?

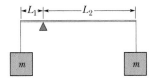

Fig. 11-43 Problem 57.

SEC. 11-10 Work and Rotational Kinetic Energy

58E. (a) If $R = 12$ cm, $M = 400$ g, and $m = 50$ g in Fig. 11-17, find the speed of the block after it has descended 50 cm starting from rest. Solve the problem using energy conservation principles. (b) Repeat (a) with $R = 5.0$ cm.

59E. An automobile crankshaft transfers energy from the engine to the axle at the rate of 100 hp (= 74.6 kW) when rotating at a speed of 1800 rev/min. What torque (in newton-meters) does the crankshaft deliver?

60E. A 32.0 kg wheel, essentially a thin hoop with radius 1.20 m, is rotating at 280 rev/min. It must be brought to a stop in 15.0 s. (a) How much work must be done to stop it? (b) What is the required average power?

61P. A thin rod of length L and mass m is suspended freely from one end. It is pulled to one side and then allowed to swing like a pendulum, passing through its lowest position with angular speed ω. In terms of these symbols and g, and neglecting friction and air resistance, find (a) the rod's kinetic energy at its lowest position and (b) how far above that position the center of mass rises.

62P. Calculate (a) the torque, (b) the energy, and (c) the average power required to accelerate Earth in 1 day from rest to its present angular speed about its axis.

63P. A meter stick is held vertically with one end on the floor and is then allowed to fall. Find the speed of the other end when it hits the floor, assuming that the end on the floor does not slip. (*Hint:* Consider the stick to be a thin rod and use the conservation of energy principle.) ssm ilw

64P. A uniform cylinder of radius 10 cm and mass 20 kg is mounted so as to rotate freely about a horizontal axis that is parallel to and 5.0 cm from the central longitudinal axis of the cylinder. (a) What is the rotational inertia of the cylinder about the axis of rotation? (b) If the cylinder is released from rest with its central longitudinal axis at the same height as the axis about which the cylinder rotates, what is the angular speed of the cylinder as it passes through its lowest position?

65P. A rigid body is made of three identical thin rods, each with length L, fastened together in the form of a letter **H** (Fig. 11-44). The body is free to rotate about a horizontal axis that runs along the length of one of the legs of the **H**.

Fig. 11-44   Problem 65.

The body is allowed to fall from rest from a position in which the plane of the **H** is horizontal. What is the angular speed of the body when the plane of the **H** is vertical?   ssm   www

66P. A uniform spherical shell of mass M and radius R rotates about a vertical axis on frictionless bearings (Fig. 11-45). A massless cord passes around the equator of the shell, over a pulley of rotational inertia I and radius r, and is attached to a small object of mass m. There is no friction on the pulley's axle; the cord does not slip on the pulley. What is the speed of the object after it falls a distance h from rest? Use energy considerations.

Fig. 11-45   Problem 66.

67P. A tall, cylinder-shaped chimney falls over when its base is ruptured. Treat the chimney as a thin rod of length H, and let θ be the angle the chimney makes with the vertical. In terms of these symbols and g, express the following: (a) the angular speed of the chimney, (b) the radial acceleration of the chimney's top, and (c) the tangential acceleration of the top. (*Hint:* Use energy considerations, not a torque. In part (c), recall that $\alpha = d\omega/dt$.) (d) At what angle θ does the tangential acceleration equal g?   ssm

Additional Problems

68. At 7:14 A.M. on June 30, 1908, a huge explosion occurred above remote central Siberia, at latitude 61° N and longitude 102° E; the fireball thus created was the brightest flash seen by anyone before nuclear weapons. The *Tunguska Event,* which according to one chance witness "covered an enormous part of the sky," was probably the explosion of a *stony asteroid* about 140 m wide. (a) Considering only Earth's rotation, determine how much later the asteroid would have had to arrive to put the explosion above Helsinki at longitude 25° E. This would have obliterated the city. (b) If the asteroid had, instead, been a *metallic asteroid,* it

could have reached Earth's surface. How much later would such an asteroid have had to arrive to put the impact in the Atlantic Ocean at longitude 20° W? (The resulting tsunamis would have wiped out coastal civilization on both sides of the Atlantic.)

69. The method in which the massive lintels (top stones) were lifted to the top of the upright stones at Stonehenge has long been debated. One possible method was tested in a small Czech town. A concrete block of mass 5124 kg was pulled up along two oak beams whose top surfaces had been debarked and then lubricated with fat (Fig. 11-46). The beams were 10 m long, and each extended from the ground to the top of one of the two upright pillars onto which the lintel was to be raised. The pillars were 3.9 m high; the coefficient of static friction between the block and the beams was 0.22. The pull on the block was via ropes that wrapped around the block and also around the top ends of two spruce logs of length 4.5 m. A riding platform was strung at the opposite end of each log. When enough workers sat or stood on a riding platform, the corresponding spruce log would pivot about the top of its upright pillar and pull one side of the block a short distance up a beam. For each log, the ropes were approximately perpendicular to the log; the distance between the pivot point and the point where the ropes wrapped around the log was 0.70 m. Assuming that each worker had a mass of 85 kg, find the least number of workers needed on the two platforms so that the block begins to move up along the beams. (About half this number could actually move the block by moving first one side of the block and then the other side.)

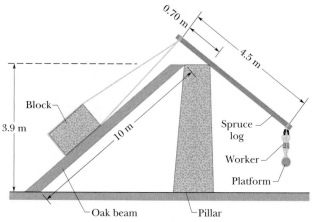

Fig. 11-46   Problem 69.

70. Cheetahs in full run have been reported to have an astounding speed of 114 km/h (about 71 mi/h) by observers moving along with the cheetahs in Jeep-like vehicles. Imagine trying to measure a cheetah's speed by keeping your Jeep abreast of the animal while also glancing at your speedometer, which is registering 114 km/h. You keep the Jeep at a constant 8.0 m from the cheetah, but the noise of the Jeep causes the cheetah to continuously veer away from the Jeep, along a circular path of radius 92 m. Thus, you travel along a circular path of radius 100 m. (a) What is the angular speed of you and the cheetah around the circular paths? (b) What is the linear speed of the cheetah along its path? (If you did not account for the circular motion, you would conclude erroneously that the cheetah's speed is 114 km/h, and that type of error was apparently made in the published reports.)

NEW PROBLEMS

N1. Figure 11N-1a shows a disk that can rotate about an axis that is perpendicular to its face and at a radial distance h from the center of the disk. Figure 11N-1b gives the rotational inertia I of the disk about the axis as a function of that distance h, from the center out to the edge of the disk. What is the mass of the disk?

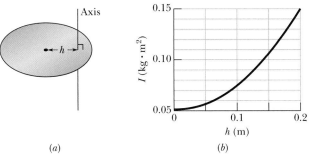

(a)   (b)

Fig. 11N-1 Problem N1.

N2. Figure 11N-2 shows a flat construction of two circular rings that have a common center and are held together by three rods of negligible mass. The construction, which is initially at rest, can rotate around the common center (like a merry-go-round) through which another rod of negligible mass extends. The

Fig. 11N-2 Problem N2.

mass, inner radius, and outer radius of the rings are given in the following table. A tangential force of magnitude 12.0 N is applied to the outer edge of the outer ring for 0.300 s. What is the change in the angular speed of the construction during that time interval?

Ring	Mass (kg)	Inner Radius (m)	Outer Radius (m)
1	0.12	0.016	0.045
2	0.24	0.090	0.140

N3. In Fig. 11N-3a, an irregularly shaped plastic plate with uniform thickness and density (mass per unit volume) is to be rotated around an axle that is perpendicular to the plate face and through point O. The rotational inertia of the plate about that axle is measured with the following method. A circular disk of mass 0.500 kg and radius 2.00 cm is glued to the plate, with its center aligned with point O (Fig. 11N-3b). A string is wrapped around the edge of the disk like a string is wrapped around a top. Then the string is pulled for 5.00 s. As a result, the disk and plate are rotated by a constant force of 0.400 N that is applied by the string tangentially to the edge of the disk. The resulting angular speed is 114 rad/s. What is the rotational inertia of the plate about the axle?

(a)

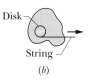

(b)

Fig. 11N-3 Problem N3.

N4. In a judo foot-sweep move, you sweep your opponent's left foot out from under him while pulling on his gi (uniform) toward that side. As a result, your opponent rotates around his right foot and onto the mat. Figure 11N-4 shows a simplified diagram of your opponent as you face him, with his left foot swept out. The rotational axis is through point O. The gravitational force $\vec{F}_g$ on him effectively acts at his center of mass, which is a horizontal distance of $d = 28$ cm from point O. His mass is 70 kg, and his rotational inertia about point O is 65 kg·m². What is the magnitude of his initial angular acceleration about point O if your pull $\vec{F}_a$ on his gi is (a) negligible and (b) horizontal with a magnitude of 300 N and applied at height $h = 1.4$ m?

Fig. 11N-4 Problem N4.

N5. Figure 11N-5 shows three 0.0100 kg particles that have been glued to a rod of length 6.00 cm and negligible mass. The assembly can rotate around a perpendicular axis through point O at the left end. If we remove one particle (that is, 33% of the mass), by what percentage does the rotational inertia of the assembly around the rotation axis decrease when that removed particle is (a) the innermost one and (b) the outermost one?

Fig. 11N-5 Problems N5 and N11.

N6. *Strobing.* A disk that rotates clockwise at angular speed 10π rad/s is illuminated by only a flashing stroboscopic light. The flashes reveal a small black dot on the rim of the disk. In the first flash, the dot appears at the 12:00 position (as on an analog clock face). Where does it appear in the next five flashes if the time between flashes is (a) 0.20 s, (b) 0.050 s, and (c) 40 ms?

N7. A disk, with a radius of 0.25 m, is to be rotated like a merry-go-round through 800 rad, starting from rest, gaining angular speed at the constant rate α_1 through the first 400 rad and then losing angular speed at the constant rate $-\alpha_1$ until it is again at rest. The magnitude of the centripetal acceleration of any portion of the disk is not to exceed 400 m/s². (a) What is the least time required for the rotation? (b) What is the corresponding value of α_1?

N8. A drum rotates around its central axis at an angular velocity of 12.60 rad/s. If the drum then slows at a constant rate of 4.20 rad/s², (a) how much time is required for it to come to rest and (b) through what angle does it rotate as it comes to rest?

N9. Figure 11N-6 shows an arrangement of 15 identical disks that have been glued together in a rod-like shape of length L and (total) mass M. The arrangement can rotate about a perpendicular axis through its central disk at point O. (a) What is the rotational inertia of the arrangement about that axis? (b) If we approximated the arrangement as being a uniform rod of mass M and length L, what percentage error would we make in using the formula in Table 11-2e to calculate the rotational inertia?

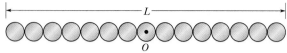

Fig. 11N-6 Problem N9.

N10. A vinyl record is played by rotating the record so that an approximately circular groove in the vinyl slides under a stylus. Bumps in the groove run into the stylus, causing it to oscillate. The equipment converts those oscillations to electrical signals and then to sound. Suppose that a record turns at the rate of $33\frac{1}{3}$ rev/min, the groove being played is at a radius of 10 cm, and the bumps in the groove are uniformly separated by 1.75 mm. At what rate (hits per second) do the bumps hit the stylus?

N11. In Fig. 11N-5, three 0.0100 kg particles have been glued to a rod of length 6.00 cm and negligible mass and can rotate around a perpendicular axis through point O at one end. How much work is required to change the rotational rate (a) from 0 to 20 rad/s, (b) from 20 rad/s to 40 rad/s, and (c) from 40 rad/s to 60 rad/s? (d) What is the slope of a plot of the assembly's kinetic energy (in joules) versus the square of its rotation rate (in radians-squared per second-squared)?

N12. Figure 11N-7 shows a rigid assembly of a thin hoop (of mass m and radius $R = 0.150$ m) and a thin radial rod (of mass m and length $L = 2.00R$). The assembly is upright but if we give it a slight nudge, it will rotate around a horizontal axis in the plane of the rod and hoop, through the lower end

Fig. 11N-7 Problem N12.

of the rod. Assuming that the energy given to the assembly in such a nudge is negligible, what would be the assembly's angular speed about the rotation axis when it passes through the upside-down (inverted) orientation?

N13. A disk rotates at constant angular acceleration, from angular position $\theta_1 = 10.0$ rad to angular position $\theta_2 = 70.0$ rad in 6.00 s. Its angular velocity at θ_2 is 15.0 rad/s. (a) What was its angular velocity at θ_1? (b) What is the angular acceleration? (c) At what angular position was the disk initially at rest? (d) Graph θ versus time t and angular speed ω versus t for the disk, from the beginning of the motion (let $t = 0$ then).

N14. A merry-go-round rotates from rest with an angular acceleration of 1.50 rad/s². How long does it take to rotate through (a) the first 2.00 rev and (b) the next 2.00 rev?

N15. Figure 11N-8 gives the angular speed of a thin rod that rotates around one end. (a) What is the magnitude of the rod's angular acceleration? (b) At $t = 4.0$ s, the rod has a rotational kinetic energy of 1.60 J. What is its kinetic energy at $t = 0$?

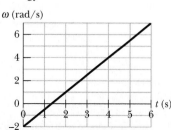

Fig. 11N-8 Problem N15.

N16. Figure 11N-9 shows a uniform disk that can rotate around its center like a merry-go-round. The disk has a radius of 2.00 cm and a mass of 20.0 grams and is initially at rest. Starting at time $t = 0$, two forces are to be applied tangentially to the rim as indicated, so that at time $t = 1.25$ s the disk has an angular velocity of 250 rad/s counterclockwise. Force $\vec{F}_1$ has a magnitude of 0.100 N. What magnitude is required of force $\vec{F}_2$?

Fig. 11N-9 Problem N16.

N17. In Fig. 11N-10, a small disk of radius $r = 2.00$ cm has been glued to the edge of a larger disk of radius $R = 4.00$ cm so that the disks lie in the same plane. The disks can be rotated around a perpendicular axis through point O at the center of the larger disk. The disks both have a uniform density (mass per unit volume) of 1.40×10^3 kg/m³ and a uniform thickness of 5.00 mm. What is the rotational inertia of the two-disk assembly about the rotation axis through O?

Fig. 11N-10 Problem N17.

N18. Five particle-like objects, positioned in the xy plane according to the following table, form a rigidly connected body. What is the rotational inertia of the body about (a) the x axis, (b) the y axis, and (c) the z axis? (d) What is the center of mass of the body?

Object	1	2	3	4	5
Mass (grams)	500	400	300	600	450
x (cm)	15	−13	17	−4.0	−5.0
y (cm)	20	13	−6.0	−7.0	9.0

N19. A golf ball is launched at an angle of 20° to the horizontal, with a speed of 60 m/s and a rotation rate of 90 rad/s. Neglecting air drag, determine the number of revolutions the ball makes by the time it reaches maximum height.

N20. The angular acceleration of a wheel is $\alpha = 6.0t^4 - 4.0t^2$, with α in radians per second-squared and t in seconds. At time $t = 0$, the wheel has an angular velocity of +2.0 rad/s and an angular position of +1.0 rad. Write expressions for (a) the angular velocity and (b) the angular position as functions of time.

N21. A coin of mass M is placed a distance R from the center of a phonograph turntable. The coefficient of static friction is μ_s. The angular speed of the turntable is slowly increased to a value ω_0, at which time the coin slides off. (a) Find ω_0 in terms of the given symbols and g. (b) Sketch the path of sliding.

N22. Figure 11N-11 shows a 2.0 kg uniform rod that is 3.0 m long. The rod is mounted to rotate freely about a horizontal axis perpendicular to the rod that passes through a point 1.0 m from one end of the rod. The rod is released from rest when it is horizontal. (a) What is the rod's maximum angular speed during the resulting rotation? (b) If the rod's mass were increased, would the answer to (a) increase, decrease, or stay the same?

Fig. 11N-11
Problem N22.

12 Rolling, Torque, and Angular Momentum

In 1897, a European "aerialist" made the first triple somersault during the flight from a swinging trapeze to the hands of a partner. For the next 85 years aerialists attempted to complete a *quadruple* somersault, but not until 1982 was it done before an audience: Miguel Vazquez of the Ringling Bros. and Barnum & Bailey Circus rotated his body in four complete circles in midair before his brother Juan caught him. Both were stunned by their success.

Why was the feat so difficult, and what feature of physics made it (finally) possible?

The answer is in this chapter.

Fig. 12-1 A time exposure photograph of a rolling disk. Small lights have been attached to the disk, one at its center and one at its edge. The latter traces out a curve called a *cycloid*.

Fig. 12-2 The center of mass O of a rolling wheel moves a distance s at velocity $\vec{v}_{com}$, while the wheel rotates through angle θ. The point P at which the wheel makes contact with the surface over which the wheel rolls also moves a distance s.

12-1 Rolling

When a bicycle moves along a straight track, the center of each wheel moves forward in pure translation. A point on the rim of the wheel, however, traces out a more complex path, as Fig. 12-1 shows. In what follows, we analyze the motion of a rolling wheel first by viewing it as a combination of pure translation and pure rotation, and then by viewing it as rotation alone.

Rolling as Rotation and Translation Combined

Imagine that you are watching the wheel of a bicycle, which passes you at constant speed while *rolling smoothly* (that is, without sliding) along a street. As shown in Fig. 12-2, the center of mass O of the wheel moves forward at constant speed v_{com}. The point P on the street where the wheel makes contact also moves forward at speed v_{com}, so that it always remains directly below O.

During a time interval t, you see both O and P move forward by a distance s. The bicycle rider sees the wheel rotate through an angle θ about the center of the wheel, with the point of the wheel that was touching the street at the beginning of t moving through arc length s. Equation 11-17 relates the arc length s to the rotation angle θ:

$$s = \theta R, \tag{12-1}$$

where R is the radius of the wheel. The linear speed v_{com} of the center of the wheel (the center of mass of this uniform wheel) is ds/dt. The angular speed ω of the wheel about its center is dθ/dt. Thus, differentiating Eq. 12-1 with respect to time (with R held constant) gives us

$$v_{com} = \omega R \qquad \text{(smooth rolling motion).} \tag{12-2}$$

Figure 12-3 shows that the rolling motion of a wheel is a combination of purely translational and purely rotational motions. Figure 12-3a shows the purely rotational motion (as if the rotation axis through the center were stationary): Every point on the wheel rotates about the center with angular speed ω. (This is the type of motion we considered in Chapter 11.) Every point on the outside edge of the wheel has linear speed v_{com} given by Eq. 12-2. Figure 12-3b shows the purely translational

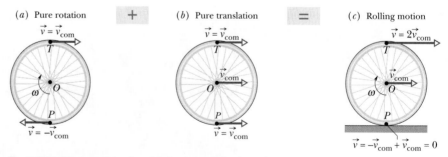

Fig. 12-3 Rolling motion of a wheel as a combination of purely rotational motion and purely translational motion. (a) The purely rotational motion: All points on the wheel move with the same angular speed ω. Points on the outside edge of the wheel all move with the same linear speed $v = v_{com}$. The linear velocities $\vec{v}$ of two such points, at top (T) and bottom (P) of the wheel, are shown. (b) The purely translational motion: All points on the wheel move to the right with the same linear velocity $\vec{v}_{com}$ as the center of the wheel. (c) The rolling motion of the wheel is the combination of (a) and (b).

motion (as if the wheel did not rotate at all): Every point on the wheel moves to the right with speed v_{com}.

The combination of Figs. 12-3a and 12-3b yields the actual rolling motion of the wheel, Fig. 12-3c. Note that in this combination of motions, the portion of the wheel at the bottom (at point P) is stationary and the portion at the top (at point T) is moving at speed $2v_{com}$, faster than any other portion of the wheel. These results are demonstrated in Fig. 12-4, which is a time exposure of a rolling bicycle wheel. You can tell that the wheel is moving faster near its top than near its bottom because the spokes are more blurred at the top than at the bottom.

The motion of any round body rolling smoothly over a surface can be separated into purely rotational and purely translational motions, as in Figs. 12-3a and 12-3b.

Rolling as Pure Rotation

Figure 12-5 suggests another way to look at the rolling motion of a wheel—namely, as pure rotation about an axis that always extends through the point where the wheel contacts the street as the wheel moves. We consider the rolling motion to be pure rotation about an axis passing through point P in Fig. 12-3c and perpendicular to the plane of the figure. The vectors in Fig. 12-5 then represent the instantaneous velocities of points on the rolling wheel.

Question: What angular speed about this new axis will a stationary observer assign to a rolling bicycle wheel?

Answer: The same angular speed ω that the rider assigns to the wheel as she or he observes it in pure rotation about an axis through its center of mass.

To verify this answer, let us use it to calculate the linear speed of the top of the rolling wheel from the point of view of a stationary observer. If we call the wheel's radius R, the top is a distance $2R$ from the axis through P in Fig. 12-5, so the linear speed at the top should be (using Eq. 12-2)

$$v_{top} = (\omega)(2R) = 2(\omega R) = 2v_{com},$$

in exact agreement with Fig. 12-3c. You can similarly verify the linear speeds shown for the portion of the wheel at points O and P in Fig. 12-3c.

Fig. 12-5 Rolling can be viewed as pure rotation, with angular speed ω, about an axis that always extends through P. The vectors show the instantaneous linear velocities of selected points on the rolling wheel. You can obtain the vectors by combining the translational and rotational motions as in Fig. 12-3.

Fig. 12-4 A photograph of a rolling bicycle wheel. The spokes near the top of the wheel are more blurred than those near the bottom of the wheel because they are moving faster, as Fig. 12-3c shows.

✔**CHECKPOINT 1:**   The rear wheel on a clown's bicycle has twice the radius of the front wheel. (a) When the bicycle is moving, is the linear speed at the very top of the rear wheel greater than, less than, or the same as that of the front wheel? (b) Is the angular speed of the rear wheel greater than, less than, or the same as that of the front wheel?

12-2 The Kinetic Energy of Rolling

Let us now calculate the kinetic energy of the rolling wheel as measured by the stationary observer. If we view the rolling as pure rotation about an axis through P in Fig. 12-5, then from Eq. 11-27 we have

$$K = \tfrac{1}{2}I_P\omega^2, \tag{12-3}$$

in which ω is the angular speed of the wheel and I_P is the rotational inertia of the wheel about the axis through P. From the parallel-axis theorem of Eq. 11-29 ($I = I_{com} + Mh^2$), we have

$$I_P = I_{com} + MR^2, \tag{12-4}$$

in which M is the mass of the wheel, I_{com} is its rotational inertia about an axis through its center of mass, and R (the wheel's radius) is the perpendicular distance h. Substituting Eq. 12-4 into Eq. 12-3, we obtain

$$K = \tfrac{1}{2}I_{com}\omega^2 + \tfrac{1}{2}MR^2\omega^2,$$

and using the relation $v_{com} = \omega R$ (Eq. 12-2) yields

$$K = \tfrac{1}{2}I_{com}\omega^2 + \tfrac{1}{2}Mv^2_{com}. \qquad (12\text{-}5)$$

We can interpret the term $\tfrac{1}{2}I_{com}\omega^2$ as the kinetic energy associated with the rotation of the wheel about an axis through its center of mass (Fig. 12-3a), and the term $\tfrac{1}{2}Mv^2_{com}$ as the kinetic energy associated with the translational motion of the wheel's center of mass (Fig. 12-3b). Thus, we have the following rule:

> ▶ A rolling object has two types of kinetic energy: a rotational kinetic energy ($\tfrac{1}{2}I_{com}\omega^2$) due to its rotation about its center of mass and a translational kinetic energy ($\tfrac{1}{2}Mv^2_{com}$) due to translation of its center of mass.

Sample Problem 12-1

A uniform solid cylindrical disk, of mass $M = 1.4$ kg and radius $R = 8.5$ cm, rolls smoothly across a horizontal table at a speed of 15 cm/s. What is its kinetic energy K?

SOLUTION: Equation 12-5 gives the kinetic energy of a rolling object, but we need three Key Ideas to use it.

1. When we speak of the speed of a rolling object, we always mean the speed of the center of mass, so here $v_{com} = 15$ cm/s.

2. Equation 12-5 requires the angular speed ω of the rolling object, which we can relate to v_{com} with Eq. 12-2, writing $\omega = v_{com}/R$.

3. Equation 12-5 also requires the rotational inertia I_{com} of the object about its center of mass. From Table 11-2c, we find that for a solid disk $I_{com} = \tfrac{1}{2}MR^2$.

Now Eq. 12-5 gives us

$$\begin{aligned}K &= \tfrac{1}{2}I_{com}\omega^2 + \tfrac{1}{2}Mv^2_{com}\\ &= (\tfrac{1}{2})(\tfrac{1}{2}MR^2)(v_{com}/R)^2 + \tfrac{1}{2}Mv^2_{com} = \tfrac{3}{4}Mv^2_{com}\\ &= \tfrac{3}{4}(1.4\ \text{kg})(0.15\ \text{m/s})^2\\ &= 0.024\ \text{J} = 24\ \text{mJ}. \qquad\text{(Answer)}\end{aligned}$$

12-3 The Forces of Rolling

Friction and Rolling

If a wheel rolls at constant speed, as in Fig. 12-2, it has no tendency to slide at the point of contact P, and thus no frictional force acts there. However, if a net force acts on the rolling wheel to speed it up or to slow it, then that net force causes acceleration $\vec{a}_{com}$ of the center of mass along the direction of travel. It also causes the wheel to rotate faster or slower, which means it causes an angular acceleration α about the center of mass. These accelerations tend to make the wheel slide at P. Thus, a frictional force must act on the wheel at P to oppose that tendency.

If the wheel *does not* slide, the force is a *static* frictional force $\vec{f}_s$ and the motion is smooth rolling. We can then relate the magnitudes of the linear acceleration $\vec{a}_{com}$ and the angular acceleration α by differentiating Eq. 12-2 with respect to time (with R held constant). On the left side, dv_{com}/dt is a_{com}, and on the right side $d\omega/dt$ is α. So, for smooth rolling we have

$$a_{com} = \alpha R \qquad \text{(smooth rolling motion).} \qquad (12\text{-}6)$$

If the wheel *does* slide when the net force acts on it, the frictional force that acts at P in Fig. 12-2 is a *kinetic* frictional force $\vec{f}_k$. The motion then is not smooth rolling, and Eq. 12-6 does not apply to the motion. In this chapter we discuss only smooth rolling motion.

Fig. 12-6 A wheel rolls horizontally without sliding while accelerating with linear acceleration $\vec{a}_{\text{com}}$. A static frictional force $\vec{f}_s$ acts on the wheel at P, opposing its tendency to slide.

Figure 12-6 shows an example in which a wheel is being made to rotate faster while rolling along a flat surface, as on a bicycle wheel at the start of a race. The faster rotation tends to make the bottom of the wheel slide to the left at point P. A frictional force at P, directed to the right, opposes this tendency to slide. If the wheel does not slide, that frictional force is a static frictional force $\vec{f}_s$ (as shown), the motion is smooth rolling, and Eq. 12-6 applies to the motion. (Without friction, bicycle races would be stationary and very boring.)

If the wheel in Fig. 12-6 were made to rotate slower, as on a slowing bicycle, we would change the figure in two ways: The directions of the center-of-mass acceleration $\vec{a}_{\text{com}}$ and the frictional force $\vec{f}_s$ at point P would now be to the left.

Rolling down a Ramp

Figure 12-7 shows a round uniform body of mass M and radius R rolling smoothly down a ramp at angle θ, along an x axis. We want to find expressions for the body's acceleration $a_{\text{com},x}$ down the ramp. We do this by using Newton's second law in both its linear version ($F_{\text{net}} = Ma$) and its angular version ($\tau_{\text{net}} = I\alpha$).

We start by drawing the forces on the body as shown in Fig. 12-7:

1. The gravitational force $\vec{F}_g$ on the body is directed downward. The tail of the vector is placed at the center of mass of the body. The component along the ramp is $F_g \sin \theta$, which is equal to $Mg \sin \theta$.

2. A normal force $\vec{N}$ is perpendicular to the ramp. It acts at the point of contact P, but in Fig. 12-7 the vector has been shifted along its direction until its tail is at the body's center of mass.

3. A static frictional force $\vec{f}_s$ acts at the point of contact P and is directed up the ramp. (Do you see why? If the body were to slide at P, it would slide *down* the ramp. Thus, the frictional force opposing the sliding must be *up* the ramp.)

We can write Newton's second law for components along the x axis in Fig. 12-7 ($F_{\text{net},x} = ma_x$) as

$$f_s - Mg \sin \theta = Ma_{\text{com},x}. \qquad (12\text{-}7)$$

This equation contains two unknowns, f_s and $a_{\text{com},x}$. (We should *not* assume that f_s is at its maximum value $f_{s,\text{max}}$. All we know is that the value of f_s is just right for the body to roll smoothly down the ramp, without sliding.)

We now wish to apply Newton's second law in angular form to the body's rotation about its center of mass. First, we shall use Eq. 11-33 ($\tau = r_{\perp}F$) to write the torques on the body about that point. The frictional force $\vec{f}_s$ has moment arm R and thus produces a torque Rf_s, which is positive because it tends to rotate the body counterclockwise in Fig. 12-7. Forces $\vec{F}_g$ and $\vec{N}$ have zero moment arms about the center of mass and thus produce zero torques. So we can write the angular form of Newton's second law ($\tau_{\text{net}} = I\alpha$) about an axis through the body's center of mass as

$$Rf_s = I_{\text{com}}\alpha. \qquad (12\text{-}8)$$

This equation contains two unknowns, f_s and α.

Because the body is rolling smoothly, we can use Eq. 12-6 ($a_{\text{com}} = \alpha R$) to relate the unknowns $a_{\text{com},x}$ and α. But we must be cautious because here $a_{\text{com},x}$ is negative (in the negative direction of the x axis) and α is positive (counterclockwise). Thus we substitute $-a_{\text{com},x}/R$ for α in Eq. 12-8. Then, solving for f_s, we obtain

$$f_s = -I_{\text{com}}\frac{a_{\text{com},x}}{R^2}. \qquad (12\text{-}9)$$

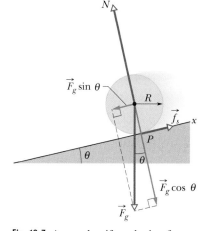

Fig. 12-7 A round uniform body of radius R rolls down a ramp. The forces that act on the body are the gravitational force $\vec{F}_g$, a normal force $\vec{N}$, and a frictional force $\vec{f}_s$ pointing up the ramp. (For clarity, vector $\vec{N}$ has been shifted in the direction it points until its tail is at the center of the body.)

Substituting the right side of Eq. 12-9 for f_s in Eq. 12-7, we then find

$$a_{com,x} = -\frac{g\sin\theta}{1 + I_{com}/MR^2}. \tag{12-10}$$

We can use this equation to find the linear acceleration $a_{com,x}$ of any body rolling along an incline of angle θ with the horizontal.

✓**CHECKPOINT 2:**   Disks A and B are identical and roll across a floor with equal speeds. Then disk A rolls up an incline, reaching a maximum height h, and disk B moves up an incline that is identical except that it is frictionless. Is the maximum height reached by disk B greater than, less than, or equal to h?

Sample Problem 12-2

A uniform ball, of mass $M = 6.00$ kg and radius R, rolls smoothly from rest down a ramp at angle $\theta = 30.0°$ (Fig. 12-7).

(a) The ball descends a vertical height $h = 1.20$ m to reach the bottom of the ramp. What is its speed at the bottom?

SOLUTION: One Key Idea here is that we can relate the ball's speed at the bottom to its kinetic energy K_f there. A second Key Idea is that the mechanical energy E of the ball–Earth system is conserved as the ball rolls down the ramp. The reason is that the only force doing work on the ball is the gravitational force, a conservative force. The normal force on the ball from the ramp does zero work because it is perpendicular to the ball's path. The frictional force on the ball from the ramp does not transfer any energy to thermal energy because the ball does not slide (it *rolls smoothly*).

Therefore, we can write the conservation of mechanical energy ($E^f = E^i$) as

$$K_f + U_f = K_i + U_i, \tag{12-11}$$

where subscripts f and i refer to the final values (at the bottom) and initial values (at rest), respectively. The gravitational potential energy is initially $U_i = Mgh$ (where M is the ball's mass) and finally $U_f = 0$. The kinetic energy is initially $K_i = 0$. For the final kinetic energy K_f, we need a third Key Idea: Because the ball rolls, the kinetic energy involves both translation *and* rotation, so we include them both with the right side of Eq. 12-5. Substituting all these expressions into Eq. 12-11 gives us

$$(\tfrac{1}{2}I_{com}\omega^2 + \tfrac{1}{2}Mv_{com}^2) + 0 = 0 + Mgh, \tag{12-12}$$

where I_{com} is the ball's rotational inertia about an axis through its center of mass, v_{com} is the requested speed at the bottom, and ω is the angular speed there.

Because the ball rolls smoothly, we can use Eq. 12-2 to sub-stitute v_{com}/R for ω to reduce the unknowns in Eq. 12-12. Doing so, substituting $\tfrac{2}{5}MR^2$ for I_{com} (from Table 11-2f), and then solving for v_{com} give us

$$v_{com} = \sqrt{(\tfrac{10}{7})gh} = \sqrt{(\tfrac{10}{7})(9.8\text{ m/s}^2)(1.2\text{ m})}$$

$$= 4.1\text{ m/s.} \qquad\qquad \text{(Answer)}$$

Note that the answer does not depend on the mass M or radius R of the ball.

(b) What are the magnitude and direction of the friction force on the ball as it rolls down the ramp?

SOLUTION: The Key Idea here is that, because the ball rolls smoothly, Eq. 12-9 gives the frictional force on the ball. However, first we need the ball's acceleration $a_{com,x}$ from Eq. 12-10:

$$a_{com,x} = -\frac{g\sin\theta}{1 + I_{com}/MR^2} = -\frac{g\sin\theta}{1 + \tfrac{2}{5}MR^2/MR^2}$$

$$= -\frac{(9.8\text{ m/s}^2)\sin 30.0°}{1 + \tfrac{2}{5}} = -3.50\text{ m/s}^2.$$

Note that we needed neither mass M nor radius R to find $a_{com,x}$. Thus, any size ball with any uniform mass would have this accel-eration down a 30.0° ramp, provided the ball rolls smoothly.

We can now solve Eq. 12-9 as

$$f_s = -I_{com}\frac{a_{com,x}}{R^2} = -\tfrac{2}{5}MR^2\frac{a_{com,x}}{R^2} = -\tfrac{2}{5}Ma_{com,x}$$

$$= -\tfrac{2}{5}(6.00\text{ kg})(-3.50\text{ m/s}^2) = 8.40\text{ N.} \qquad \text{(Answer)}$$

Note that here we needed mass M but not radius R. Thus, a 30.0° ramp would produce a frictional force of magnitude 8.40 N on any size ball with a uniform mass of 6.00 kg, for smooth rolling.

12-4   The Yo-Yo

A yo-yo is a physics lab that you can fit in your pocket. If a yo-yo rolls down its string for a distance h, it loses potential energy in amount mgh but gains kinetic energy in both translational ($\tfrac{1}{2}mv_{com}^2$) and rotational ($\tfrac{1}{2}I_{com}\omega^2$) forms. As it climbs back up, it loses kinetic energy and regains potential energy.

In a modern yo-yo, the string is not tied to the axle but is looped around it. When the yo-yo "hits" the bottom of its string, an upward force on the axle from the string stops the descent. The yo-yo then spins, axle inside loop, with only rotational kinetic energy. The yo-yo keeps spinning ("sleeping") until you "wake it" by jerking on the string, causing the string to catch on the axle and the yo-yo to climb back up. The rotational kinetic energy of the yo-yo at the bottom of its string (and thus the sleeping time) can be considerably increased by throwing the yo-yo downward so that it starts down the string with initial speeds v_{com} and ω instead of rolling down from rest.

To find an expression for the linear acceleration a_{com} of a yo-yo rolling down a string, we could use Newton's second law just as we did for the body rolling down a ramp in Fig. 12-7. The analysis is the same except for the following:

1. Instead of rolling down a ramp at angle θ with the horizontal, the yo-yo rolls down a string at angle $\theta = 90°$ with the horizontal.

2. Instead of rolling on its outer surface at radius R, the yo-yo rolls on an axle of radius R_0 (Fig. 12-8a).

3. Instead of being slowed by frictional force $\vec{f}_s$, the yo-yo is slowed by the force $\vec{T}$ on it from the string (Fig. 12-8b).

The analysis would again lead us to Eq. 12-10. Therefore, let us just change the notation in Eq. 12-10 and set $\theta = 90°$ to write the linear acceleration as

$$a_{com} = -\frac{g}{1 + I_{com}/MR_0^2},$$ (12-13)

where I_{com} is the yo-yo's rotational inertia about its center, and M is its mass. A yo-yo has the same downward acceleration when it is climbing back up the string, because the forces on it are still those shown in Fig. 12-8b.

Fig. 12-8 (a) A yo-yo, shown in cross section. The string, of assumed negligible thickness, is wound around an axle of radius R_0. (b) A free-body diagram for the falling yo-yo. Only the axle is shown.

12-5 Torque Revisited

In Chapter 11 we defined torque τ for a rigid body that can rotate around a fixed axis, with each particle in the body forced to move in a path that is a circle about that axis. We now expand the definition of torque to apply it to an individual particle that moves along any path relative to a fixed *point* (rather than a fixed axis). The path need no longer be a circle, and we must write the torque as a vector $\vec{\tau}$ that may have any direction.

Figure 12-9a shows such a particle at point A in the xy plane. A single force $\vec{F}$ in that plane acts on the particle, and the particle's position relative to the origin O is given by position vector $\vec{r}$. The torque $\vec{\tau}$ acting on the particle relative to the fixed point O is a vector quantity defined as

$$\vec{\tau} = \vec{r} \times \vec{F} \qquad \text{(torque defined).}$$ (12-14)

We can evaluate the vector (or cross) product in this definition of $\vec{\tau}$ by using the rules for such products given in Section 3-7. To find the direction of $\vec{\tau}$, we slide the vector $\vec{F}$ (without changing its direction) until its tail is at the origin O, so that the two vectors in the vector product are tail to tail as in Fig. 12-9b. We then use the right-hand rule for vector products in Fig. 3-20a, sweeping the fingers of the right hand from $\vec{r}$ (the first vector in the product) into $\vec{F}$ (the second vector). The outstretched right thumb then gives the direction of $\vec{\tau}$. In Fig. 12-9b, the direction of $\vec{\tau}$ is in the positive direction of the z axis.

To determine the magnitude of $\vec{\tau}$, we apply the general result of Eq. 3-27

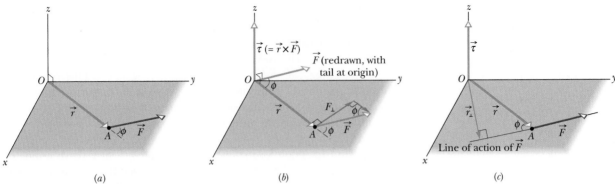

(a)   *(b)*   *(c)*

Fig. 12-9 Defining torque. (*a*) A force $\vec{F}$, lying in the *xy* plane, acts on a particle at point *A*. (*b*) This force produces a torque $\vec{\tau}\,(=\vec{r}\times\vec{F})$ on the particle with respect to the origin *O*. By the right-hand rule for vector (cross) products, the torque vector points in the positive direction of *z*. Its magnitude is given by $rF_\perp$ in (*b*) and by $r_\perp F$ in (*c*).

$(c = ab \sin\phi)$, finding

$$\tau = rF \sin\phi, \tag{12-15}$$

where ϕ is the angle between the directions of $\vec{r}$ and $\vec{F}$ when the vectors are tail to tail. From Fig. 12-9*b*, we see that Eq. 12-15 can be rewritten as

$$\tau = rF_\perp, \tag{12-16}$$

where $F_\perp\,(=F\sin\phi)$ is the component of $\vec{F}$ perpendicular to $\vec{r}$. From Fig. 12-9*c*, we see that Eq. 12-15 can also be rewritten as

$$\tau = r_\perp F, \tag{12-17}$$

where $r_\perp\,(=r\sin\phi)$ is the moment arm of $\vec{F}$ (the perpendicular distance between *O* and the line of action of $\vec{F}$).

Sample Problem 12-3

In Fig. 12-10*a*, three forces, each of magnitude 2.0 N, act on a particle. The particle is in the *xz* plane at point *A* given by position vector $\vec{r}$, where $r = 3.0$ m and $\theta = 30°$. Force $\vec{F}_1$ is parallel to the *x* axis, force $\vec{F}_2$ is parallel to the *z* axis, and force $\vec{F}_3$ is parallel to the *y* axis. What is the torque, about the origin *O*, due to each force?

SOLUTION: The Key Idea here is that, because the three force vectors do not lie in a plane, we cannot evaluate their torques as in Chapter 11. Instead, we must use vector (or cross) products, with magnitudes given by Eq. 12-15 ($\tau = rF\sin\phi$) and directions given by the right-hand rule for vector products.

Because we want the torques with respect to the origin *O*, the vector $\vec{r}$ required for each cross product is the given position vector. To determine the angle ϕ between the direction of $\vec{r}$ and the direction of each force, we shift the force vectors of Fig. 12-10*a*, each in turn, so that their tails are at the origin. Figures 12-10*b, c,*

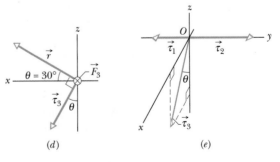

Fig. 12-10 Sample Problem 12-3. (*a*) A particle at point *A* is acted on by three forces, each parallel to a coordinate axis. The angle ϕ (used in finding torque) is shown (*b*) for $\vec{F}_1$ and (*c*) for $\vec{F}_2$. (*d*) Torque $\vec{\tau}_3$ is perpendicular to both $\vec{r}$ and $\vec{F}_3$ (force $\vec{F}_3$ is directed into the plane of the figure). (*e*) The torques (relative to the origin *O*) acting on the particle.

and *d*, which are direct views of the *xz* plane, show the shifted force vectors $\vec{F}_1$, $\vec{F}_2$, and $\vec{F}_3$, respectively. (Note how much easier the angles are to see.) In Fig. 12-10*d*, the angle between the directions of $\vec{r}$ and $\vec{F}_3$ is 90° and the symbol $\otimes$ means $\vec{F}_3$ is directed into the page. If it were directed out of the page, it would be represented with the symbol $\odot$.

Now, applying Eq. 12-15 for each force, we find the magnitudes of the torques to be

$$\tau_1 = rF_1 \sin \phi_1 = (3.0 \text{ m})(2.0 \text{ N})(\sin 150°) = 3.0 \text{ N} \cdot \text{m},$$

$$\tau_2 = rF_2 \sin \phi_2 = (3.0 \text{ m})(2.0 \text{ N})(\sin 120°) = 5.2 \text{ N} \cdot \text{m},$$

and $\quad \tau_3 = rF_3 \sin \phi_3 = (3.0 \text{ m})(2.0 \text{ N})(\sin 90°)$

$$= 6.0 \text{ N} \cdot \text{m}. \qquad \text{(Answer)}$$

To find the directions of these torques, we use the right-hand rule, placing the fingers of the right hand so as to rotate $\vec{r}$ into $\vec{F}$ through the *smaller* of the two angles between their directions. The thumb points in the direction of the torque. Thus $\vec{\tau}_1$ is directed into the page in Fig. 12-10*b*; $\vec{\tau}_2$ is directed out of the page in Fig. 12-10*c*; and $\vec{\tau}_3$ is directed as shown in Fig. 12-10*d*. All three torque vectors are shown in Fig. 12-10*e*.

✔**CHECKPOINT 3:** The position vector $\vec{r}$ of a particle points along the positive direction of the *z* axis. If the torque on the particle is (a) zero, (b) in the negative direction of *x*, and (c) in the negative direction of *y*, in what direction is the force producing the torque?

Tactic 1: *Vector Products and Torques*
Equation 12-15 for torques is our first application of the vector (or cross) product. You might want to review Section 3-7, where the rules for the vector product are given. In that section, Problem-Solving Tactic 5 lists many common errors in finding the direction of a vector product.

Keep in mind that a torque is calculated *with respect to* (or *about*) a point, which must be known if the value of the torque is to be meaningful. Changing the point can change the torque in both magnitude and direction. For example, in Sample Problem 12-3, the torques due to the three forces are calculated about the origin *O*. You can show that the torques due to the same three forces are all zero if they are calculated about point *A* (at the position of the particle), because then *r* = 0 for each force.

12-6 Angular Momentum

Recall that the concept of linear momentum $\vec{p}$ and the principle of conservation of linear momentum are extremely powerful tools. They allow us to predict the outcome of, say, a collision of two cars without knowing the details of the collision. Here we begin a discussion of the angular counterpart of $\vec{p}$. We shall end the chapter by discussing the angular counterpart of the conservation principle.

Figure 12-11 shows a particle of mass *m* with linear momentum $\vec{p}\ (= m\vec{v})$ as it passes through point *A* in the *xy* plane. The **angular momentum** $\vec{\ell}$ of this particle with respect to the origin *O* is a vector quantity defined as

$$\vec{\ell} = \vec{r} \times \vec{p} = m(\vec{r} \times \vec{v}) \qquad \text{(angular momentum defined),} \qquad (12\text{-}18)$$

where $\vec{r}$ is the position vector of the particle with respect to *O*. As the particle moves relative to *O* in the direction of its momentum $\vec{p}\ (= m\vec{v})$, position vector $\vec{r}$ rotates around *O*. Note carefully that to have angular momentum about *O*, the particle does *not* itself have to rotate around *O*. Comparison of Eqs. 12-14 and 12-18 shows that angular momentum bears the same relation to linear momentum that torque does to force. The SI unit of angular momentum is the kilogram-meter-squared per second (kg · m²/s), equivalent to the joule-second (J · s).

To find the direction of the angular momentum vector $\vec{\ell}$ in Fig. 12-11, we slide the vector $\vec{p}$ until its tail is at the origin *O*. Then we use the right-hand rule for

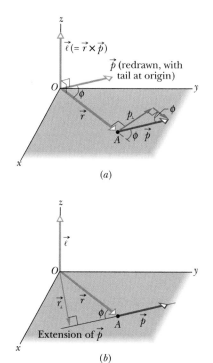

(a)

(b)

Fig. 12-11 Defining angular momentum. A particle passing through point *A* has linear momentum $\vec{p}\ (= m\vec{v})$, with the vector $\vec{p}$ lying in the *xy* plane. The particle has angular momentum $\vec{\ell}\ (= \vec{r} \times \vec{p})$ with respect to the origin *O*. By the right-hand rule, the angular momentum vector points in the positive direction of *z*. (a) The magnitude of $\vec{\ell}$ is given by $\ell = rp_\perp = rmv_\perp$. (b) The magnitude of $\vec{\ell}$ is also given by $\ell = r_\perp p = r_\perp mv$.

vector products, sweeping the fingers from $\vec{r}$ into $\vec{p}$. The outstretched thumb then shows that the direction of $\vec{\ell}$ is in the positive direction of the z axis in Fig. 12-11. This positive direction is consistent with the counterclockwise rotation of the particle's position vector $\vec{r}$ about the z axis, as the particle continues to move. (A negative direction of $\vec{\ell}$ would be consistent with a clockwise rotation of $\vec{r}$ about the z axis.)

To find the magnitude of $\vec{\ell}$, we use the general result of Eq. 3-27, to write

$$\ell = rmv \sin \phi, \tag{12-19}$$

where ϕ is the angle between $\vec{r}$ and $\vec{p}$ when these two vectors are tail to tail. From Fig. 12-11a, we see that Eq. 12-19 can be rewritten as

$$\ell = rp_\perp = rmv_\perp, \tag{12-20}$$

where $p_\perp$ is the component of $\vec{p}$ perpendicular to $\vec{r}$ and $v_\perp$ is the component of $\vec{v}$ perpendicular to $\vec{r}$. From Fig. 12-11b, we see that Eq. 12-19 can also be rewritten as

$$\ell = r_\perp p = r_\perp mv, \tag{12-21}$$

where $r_\perp$ is the perpendicular distance between O and the extension of $\vec{p}$.

Just as is true for torque, angular momentum has meaning only with respect to a specified origin. Moreover, if the particle in Fig. 12-11 did not lie in the xy plane, or if the linear momentum $\vec{p}$ of the particle did not also lie in that plane, the angular momentum $\vec{\ell}$ would not be parallel to the z axis. The direction of the angular momentum vector is always perpendicular to the plane formed by the position and linear momentum vectors $\vec{r}$ and $\vec{p}$.

✔CHECKPOINT 4:   In part a of the figure, particles 1 and 2 move around point O in opposite directions, in circles with radii 2 m and 4 m. In part b, particles 3 and 4 travel in the same direction, along straight lines at perpendicular distances of 4 m and 2 m from point O. Particle 5 moves directly away from O. All five particles have the same mass and the same constant speed.
(a) Rank the particles according to the magnitudes of their angular momentum about point O, greatest first.
(b) Which particles have negative angular momentum about point O?

(a)

(b)

Sample Problem 12-4

Figure 12-12 shows an overhead view of two particles moving at constant momentum along horizontal paths. Particle 1, with momentum magnitude $p_1 = 5.0$ kg · m/s, has position vector $\vec{r}_1$ and will pass 2.0 m from point O. Particle 2, with momentum magnitude $p_2 = 2.0$ kg · m/s, has position vector $\vec{r}_2$ and will pass 4.0 m from point O. What is the net angular momentum $\vec{L}$ about point O of the two-particle system?

SOLUTION: The Key Idea here is that to find $\vec{L}$, we can first find the individual angular momenta $\vec{\ell}_1$ and $\vec{\ell}_2$ and then add them. To evaluate their magnitudes, we can use any one of Eqs. 12-18 through 12-21. However, Eq. 12-21 is easiest, because we are given the perpendicular distances $r_{1\perp}$ (= 2.0 m) and $r_{2\perp}$ (= 4.0 m) and the

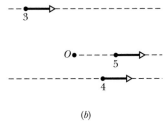

Fig. 12-12 Sample Problem 12-4. Two particles pass near point O.

momentum magnitudes p_1 and p_2. We are not given all the values required for the other equations.

For particle 1, Eq. 12-21 yields

$$\ell_1 = r_{1\perp}p_1 = (2.0 \text{ m})(5.0 \text{ kg} \cdot \text{m/s})$$
$$= 10 \text{ kg} \cdot \text{m}^2/\text{s}.$$

To find the direction of vector $\vec{\ell}_1$, we use Eq. 12-18 and the right-hand rule for vector products. For $\vec{r}_1 \times \vec{p}_1$, the vector product is out of the page, perpendicular to the plane of Fig. 12-12. This is the positive direction, consistent with the counterclockwise rotation of the particle's position vector $\vec{r}_1$ around O as particle 1 moves. Thus, the angular momentum vector for particle 1 is

$$\ell_1 = +10 \text{ kg} \cdot \text{m}^2/\text{s}.$$

Similarly, the magnitude of $\vec{\ell}_2$ is

$$\ell_2 = r_{\perp 2}p_2 = (4.0 \text{ m})(2.0 \text{ kg} \cdot \text{m/s})$$
$$= 8.0 \text{ kg} \cdot \text{m}^2/\text{s},$$

and the vector product $\vec{r}_2 \times \vec{p}_2$ is into the page, which is the negative direction, consistent with the clockwise rotation of $\vec{r}_2$ around O as particle 2 moves. Thus, the angular momentum vector for particle 2 is

$$\ell_2 = -8.0 \text{ kg} \cdot \text{m}^2/\text{s}.$$

The net angular momentum for the two-particle system is then

$$L = \ell_1 + \ell_2 = +10 \text{ kg} \cdot \text{m}^2/\text{s} + (-8.0 \text{ kg} \cdot \text{m}^2/\text{s})$$
$$= +2.0 \text{ kg} \cdot \text{m}^2/\text{s}. \qquad \text{(Answer)}$$

The plus sign means that the system's net angular momentum about point O is out of the page.

12-7 Newton's Second Law In Angular Form

Newton's second law written in the form

$$\vec{F}_{\text{net}} = \frac{d\vec{p}}{dt} \qquad \text{(single particle)} \qquad (12\text{-}22)$$

expresses the close relation between force and linear momentum for a single particle. We have seen enough of the parallelism between linear and angular quantities to be pretty sure that there is also a close relation between torque and angular momentum. Guided by Eq. 12-22, we can even guess that it must be

$$\vec{\tau}_{\text{net}} = \frac{d\vec{\ell}}{dt} \qquad \text{(single particle).} \qquad (12\text{-}23)$$

Equation 12-23 is indeed an angular form of Newton's second law for a single particle:

> The (vector) sum of all the torques acting on a particle is equal to the time rate of change of the angular momentum of that particle.

Equation 12-23 has no meaning unless the torques $\vec{\tau}$ and the angular momentum $\vec{\ell}$ are defined with respect to the same origin.

Proof of Equation 12-23

We start with Eq. 12-18, the definition of the angular momentum of a particle:

$$\vec{\ell} = m(\vec{r} \times \vec{v}),$$

where $\vec{r}$ is the position vector of the particle and $\vec{v}$ is the velocity of the particle. Differentiating* each side with respect to time t yields

$$\frac{d\vec{\ell}}{dt} = m\left(\vec{r} \times \frac{d\vec{v}}{dt} + \frac{d\vec{r}}{dt} \times \vec{v}\right). \qquad (12\text{-}24)$$

However, $d\vec{v}/dt$ is the acceleration $\vec{a}$ of the particle, and $d\vec{r}/dt$ is its velocity $\vec{v}$. Thus, we can rewrite Eq. 12-24 as

$$\frac{d\vec{\ell}}{dt} = m(\vec{r} \times \vec{a} + \vec{v} \times \vec{v}).$$

*In differentiating a vector product, be sure not to change the order of the two quantities (here $\vec{r}$ and $\vec{v}$) that form that product. (See Eq. 3-28.)

Now $\vec{v} \times \vec{v} = 0$ (the vector product of any vector with itself is zero because the angle between the two vectors is necessarily zero). This leads to

$$\frac{d\vec{\ell}}{dt} = m(\vec{r} \times \vec{a}) = \vec{r} \times m\vec{a}.$$

We now use Newton's second law ($\vec{F}_{net} = m\vec{a}$) to replace $m\vec{a}$ with its equal, the vector sum of the forces that act on the particle, obtaining

$$\frac{d\vec{\ell}}{dt} = \vec{r} \times \vec{F}_{net} = \sum(\vec{r} \times \vec{F}). \tag{12-25}$$

Here the symbol Σ indicates that we must sum the vector products $\vec{r} \times \vec{F}$ for all the forces. However, from Eq. 12-14, we know that each one of those vector products is the torque associated with one of the forces. Therefore, Eq. 12-25 tells us that

$$\vec{\tau}_{net} = \frac{d\vec{\ell}}{dt}.$$

This is Eq. 12-23, the relation that we set out to prove.

✔**CHECKPOINT 5:**   The figure shows the position vector $\vec{r}$ of a particle at a certain instant, and four choices for the direction of a force that is to accelerate the particle. All four choices lie in the xy plane. (a) Rank the choices according to the magnitude of the time rate of change ($d\vec{\ell}/dt$) they produce in the angular momentum of the particle about point O, greatest first. (b) Which choice results in a negative rate of change about O?

Sample Problem 12-5

In Fig. 12-13, a penguin of mass m falls from rest at point A, a horizontal distance D from the origin O of an xyz coordinate system. (The positive direction of the z axis is directly outward from the plane of the figure.)

(a) What is the angular momentum $\vec{\ell}$ of the falling penguin about O?

SOLUTION: One Key Idea here is that we can treat the penguin as a particle, and thus its angular momentum $\vec{\ell}$ is given by Eq. 12-18 ($\vec{\ell} = \vec{r} \times \vec{p}$), where $\vec{r}$ is the penguin's position vector (extending from O to the penguin) and $\vec{p}$ is the penguin's linear momentum. The second Key Idea is that the penguin has *angular* momentum about O even though it moves in a straight line, because $\vec{r}$ rotates about O as the penguin falls.

To find the magnitude of $\vec{\ell}$, we can use any one of the scalar equations derived from Eq. 12-18—namely, Eqs. 12-19 through 12-21. However, Eq. 12-21 ($\ell = r_\perp mv$) is easiest because the perpendicular distance $r_\perp$ between O and an extension of vector $\vec{p}$ is the given distance D. A third Key Idea is an old one: The speed of an object that has fallen from rest for a time t is $v = gt$. We can now write Eq. 12-21 in terms of given quantities as

$$\ell = r_\perp mv = Dmgt. \qquad \text{(Answer)}$$

To find the direction of $\vec{\ell}$, we use the right-hand rule for the vector product $\vec{r} \times \vec{p}$ in Eq. 12-18. Mentally shift $\vec{p}$ until its tail

Fig. 12-13 Sample Problem 12-5. A penguin falls vertically from point A. The torque $\vec{\tau}$ and the angular momentum $\vec{\ell}$ of the falling penguin with respect to the origin O are directed into the plane of the figure.

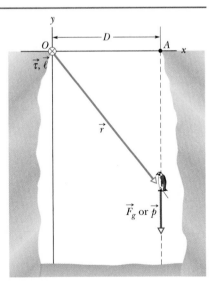

is at the origin, and then use the fingers of your right hand to rotate $\vec{r}$ into $\vec{p}$ through the smaller angle between the two vectors. Your outstretched thumb then points into the plane of the figure, indicating that the product $\vec{r} \times \vec{p}$ and thus also $\vec{\ell}$ are directed into that plane, in the negative direction of the z axis. We represent $\vec{\ell}$ with an encircled cross $\otimes$ at O. The vector $\vec{\ell}$ changes with time in magnitude only; its direction remains unchanged.

(b) About the origin O, what is the torque $\vec{\tau}$ on the penguin due to the gravitational force $\vec{F}_g$?

SOLUTION: One Key Idea here is that the torque is given by Eq. 12-14 ($\vec{\tau} = \vec{r} \times \vec{F}$), where now the force is $\vec{F}_g$. An associated Key Idea is that $\vec{F}_g$ causes a torque on the penguin, even though the penguin moves in a straight line, because $\vec{r}$ rotates about O as the penguin moves.

To find the magnitude of $\vec{\tau}$, we can use any one of the scalar equations derived from Eq. 12-14—namely, Eqs. 12-15 through 12-17. However, Eq. 12-17 ($\tau = r_{\perp}F$) is easiest because the perpendicular distance $r_{\perp}$ between O and the line of action of $\vec{F}_g$ is the given distance D. So, substituting D and using mg for the magnitude of $\vec{F}_g$, we can write Eq. 12-17 as

$$\tau = DF_g = Dmg. \qquad \text{(Answer)}$$

Using the right-hand rule for the vector product $\vec{r} \times \vec{F}$ in Eq. 12-14, we find that the direction of $\vec{\tau}$ is the negative direction of the z axis, the same as $\vec{\ell}$.

The results we obtained in parts (a) and (b) must be consistent with Newton's second law in the angular form of Eq. 12-23 ($\vec{\tau}_{net} = d\vec{\ell}/dt$). To check the magnitudes we got, we write Eq. 12-23 in component form for the z axis and then substitute our result $\ell = Dmgt$. We find

$$\tau = \frac{d\ell}{dt} = \frac{d(Dmgt)}{dt} = Dmg,$$

which is the magnitude we found for $\vec{\tau}$. To check the directions, we note that Eq. 12-23 tells us that $\vec{\tau}$ and $d\vec{\ell}/dt$ must have the same direction. So $\vec{\tau}$ and $\vec{\ell}$ must also have the same direction, which is what we found.

12-8 The Angular Momentum of a System of Particles

Now we turn our attention to the angular momentum of a system of particles with respect to an origin. The total angular momentum $\vec{L}$ of the system is the (vector) sum of the angular momenta $\vec{\ell}$ of the individual particles:

$$\vec{L} = \vec{\ell}_1 + \vec{\ell}_2 + \vec{\ell}_3 + \cdots + \vec{\ell}_n = \sum_{i=1}^{n} \vec{\ell}_i, \qquad (12\text{-}26)$$

in which i ($= 1, 2, 3, \ldots$) labels the particles.

With time, the angular momenta of individual particles may change, either because of interactions within the system (between the individual particles) or because of influences that may act on the system from the outside. We can find the change in $\vec{L}$ as these changes take place by taking the time derivative of Eq. 12-26. Thus,

$$\frac{d\vec{L}}{dt} = \sum_{i=1}^{n} \frac{d\vec{\ell}_i}{dt}. \qquad (12\text{-}27)$$

From Eq. 12-23, we see that $d\vec{\ell}_i/dt$ is equal to the net torque $\vec{\tau}_{net,i}$ on the ith particle. We can rewrite Eq. 12-27 as

$$\frac{d\vec{L}}{dt} = \sum_{i=1}^{n} \vec{\tau}_{net,i}. \qquad (12\text{-}28)$$

That is, the rate of change of the system's angular momentum $\vec{L}$ is equal to the vector sum of the torques on its individual particles. Those torques include *internal torques* (due to forces between the particles) and *external torques* (due to forces on the particles from bodies external to the system). However, the forces between the particles always come in third-law force pairs so their torques sum to zero. Thus, the only torques that can change the total angular momentum $\vec{L}$ of the system are the external torques acting on the system.

Let $\vec{\tau}_{net}$ represent the net external torque, the vector sum of all external torques on all particles in the system. Then we can write Eq. 12-28 as

$$\vec{\tau}_{net} = \frac{d\vec{L}}{dt} \qquad \text{(system of particles).} \qquad (12\text{-}29)$$

This equation is Newton's second law for rotation in angular form, for a system of particles. It says:

> The net external torque $\vec{\tau}_{net}$ acting on a system of particles is equal to the time rate of change of the system's total angular momentum $\vec{L}$.

Equation 12-29 is analogous to $\vec{F}_{net} = d\vec{P}/dt$ (Eq. 9-23) but requires extra caution: Torques and the system's angular momentum must be measured relative to the same origin. If the center of mass of the system is not accelerating relative to an inertial frame, that origin can be any point. However, if the center of mass of the system *is* accelerating, the origin can be only at that center of mass. As an example, consider a wheel as the system of particles. If the wheel is rotating about an axis that is fixed relative to the ground, then the origin for applying Eq. 12-29 can be any point that is stationary relative to the ground. However, if the wheel is rotating about an axis that is accelerating (such as when the wheel rolls down a ramp), then the origin can be only at the wheel's center of mass.

12-9   The Angular Momentum of a Rigid Body Rotating About a Fixed Axis

We next evaluate the angular momentum of a system of particles that form a rigid body that rotates about a fixed axis. Figure 12-14a shows such a body. The fixed axis of rotation is the z axis, and the body rotates about it with constant angular speed ω. We wish to find the angular momentum of the body about that axis.

We can find the angular momentum by summing the z components of the angular momenta of the mass elements in the body. In Fig. 12-14a, a typical mass element, of mass Δm_i, moves around the z axis in a circular path. The position of the mass element is located relative to the origin O by position vector $\vec{r}_i$. The radius of the mass element's circular path is $r_{\perp i}$, the perpendicular distance between the element and the z axis.

The magnitude of the angular momentum $\vec{\ell}_i$ of this mass element, with respect to O, is given by Eq. 12-19:

$$\ell_i = (r_i)(p_i)(\sin 90°) = (r_i)(\Delta m_i v_i),$$

where p_i and v_i are the linear momentum and linear speed of the mass element, and 90° is the angle between $\vec{r}_i$ and $\vec{p}_i$. The angular momentum vector $\vec{\ell}_i$ for the mass element in Fig. 12-14a is shown in Fig. 12-14b; its direction must be perpendicular to those of $\vec{r}_i$ and $\vec{p}_i$.

We are interested in the component of $\vec{\ell}_i$ that is parallel to the rotation axis, here the z axis. That z component is

$$\ell_{iz} = \ell_i \sin \theta = (r_i \sin \theta)(\Delta m_i v_i) = r_{\perp i} \Delta m_i v_i.$$

The z component of the angular momentum for the rotating rigid body as a whole is found by adding up the contributions of all the mass elements that make up the body. Thus, because $v = \omega r_\perp$, we may write

$$L_z = \sum_{i=1}^{n} \ell_{iz} = \sum_{i=1}^{n} \Delta m_i v_i r_{\perp i} = \sum_{i=1}^{n} \Delta m_i (\omega r_{\perp i}) r_{\perp i}$$

$$= \omega \left(\sum_{i=1}^{n} \Delta m_i r_{\perp i}^2 \right). \tag{12-30}$$

We can remove ω from the summation here because it has the same value for all points of the rotating rigid body.

(a)

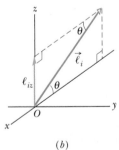

(b)

Fig. 12-14 (a) A rigid body rotates about the z axis with angular speed ω. A mass element of mass Δm_i within the body moves about the z axis in a circle with radius $r_{\perp i}$. The mass element has linear momentum $\vec{p}_i$, and it is located relative to the origin O by position vector $\vec{r}_i$. Here the mass element is shown when $r_{\perp i}$ is parallel to the x axis. (b) The angular momentum $\vec{\ell}_i$, with respect to O, of the mass element in (a). The z component ℓ_{iz} is also shown.

TABLE 12-1   More Corresponding Variables and Relations for Translational and Rotational Motion[a]

Translational		Rotational	
Force	$\vec{F}$	Torque	$\vec{\tau} \, (= \vec{r} \times \vec{F})$
Linear momentum	$\vec{p}$	Angular momentum	$\vec{\ell} \, (= \vec{r} \times \vec{p})$
Linear momentum[b]	$\vec{P} \, (= \Sigma \vec{p}_i)$	Angular momentum[b]	$\vec{L} \, (= \Sigma \vec{\ell}_i)$
Linear momentum[b]	$\vec{P} = M\vec{v}_{com}$	Angular momentum[c]	$L = I\omega$
Newton's second law[b]	$\vec{F}_{net} = \dfrac{d\vec{P}}{dt}$	Newton's second law[b]	$\vec{\tau}_{net} = \dfrac{d\vec{L}}{dt}$
Conservation law[d]	$\vec{P}$ = a constant	Conservation law[d]	$\vec{L}$ = a constant

[a]See also Table 11-3.   [b]For systems of particles, including rigid bodies.
[c]For a rigid body about a fixed axis, with L being the component along that axis.
[d]For a closed, isolated system.

The quantity $\Sigma \, \Delta m_i \, r_{\perp i}^2$ in Eq. 12-30 is the rotational inertia I of the body about the fixed axis (see Eq. 11-26). Thus Eq. 12-30 reduces to

$$L = I\omega \qquad \text{(rigid body, fixed axis).} \qquad (12\text{-}31)$$

We have dropped the subscript z, but you must remember that the angular momentum defined by Eq. 12-31 is the angular momentum about the rotation axis. Also, I in that equation is the rotational inertia about that same axis.

Table 12-1, which supplements Table 11-3, extends our list of corresponding linear and angular relations.

✔CHECKPOINT 6:   In the figure, a disk, a hoop, and a solid sphere are made to spin about fixed central axes (like a top) by means of strings wrapped around them, with the strings producing the same constant tangential force $\vec{F}$ on all three objects. The three objects have the same mass and radius, and they are initially stationary. Rank the objects according to (a) their angular momentum about their central axes and (b) their angular speed, greatest first, when the strings have been pulled for a certain time t.

Disk

Hoop

Sphere

Sample Problem 12-6

George Washington Gale Ferris, Jr., a civil engineering graduate from Rensselaer Polytechnic Institute, built the original Ferris wheel (Fig. 12-15) for the 1893 World's Columbian Exposition in Chicago. The wheel, an astounding engineering construction at the time, carried 36 wooden cars, each holding as many as 60 passengers, around a circle of radius $R = 38$ m. The mass of each car was about 1.1×10^4 kg. The mass of the wheel's structure was about 6.0×10^5 kg, which was mostly in the circular grid from which the cars were suspended. The cars were loaded 6 at a time, and once all 36 cars were full, the wheel made a complete rotation at an angular speed ω_F in about 2 min.

Fig. 12-15 Sample Problem 12-6. The original Ferris wheel, built in 1893 near the University of Chicago, towered over the surrounding buildings.

(a) Estimate the magnitude L of the angular momentum of the wheel and its passengers while the wheel rotated at ω_F.

SOLUTION: The Key Idea here is that we can treat the wheel, cars, and passengers as a rigid object rotating about a fixed axis, at the wheel's axle. Then Eq. 12-31 ($L = I\omega$) gives the magnitude of the angular momentum of that object. We need to find the rotational inertia I of this object and the angular speed ω_F.

To find I, let us start with the loaded cars. Because we can treat them as particles, at distance R from the axis of rotation, we know from Eq. 11-26 that their rotational inertia is $I_{pc} = M_{pc}R^2$, where M_{pc} is their total mass. Let us assume that the 36 cars are each filled with 60 passengers, each of mass 70 kg. Then their total mass is

$$M_{pc} = 36[1.1 \times 10^4 \text{ kg} + 60(70 \text{ kg})] = 5.47 \times 10^5 \text{ kg}$$

and their rotational inertia is

$$I_{pc} = M_{pc}R^2 = (5.47 \times 10^5 \text{ kg})(38 \text{ m})^2 = 7.90 \times 10^8 \text{ kg} \cdot \text{m}^2.$$

Next we consider the structure of the wheel. Let us assume that the rotational inertia of the structure is due mainly to the circular grid suspending the cars. Further, let us assume that the grid forms a hoop of radius R, with a mass M_{hoop} of 3.0×10^5 kg (half the wheel's mass). From Table 11-2a, the rotational inertia of the hoop is

$$I_{hoop} = M_{hoop}R^2 = (3.0 \times 10^5 \text{ kg})(38 \text{ m})^2$$
$$= 4.33 \times 10^8 \text{ kg} \cdot \text{m}^2.$$

The combined rotational inertia I of the cars, passengers, and hoop is then

$$I = I_{pc} + I_{hoop} = 7.90 \times 10^8 \text{ kg} \cdot \text{m}^2 + 4.33 \times 10^8 \text{ kg} \cdot \text{m}^2$$
$$= 1.22 \times 10^9 \text{ kg} \cdot \text{m}^2.$$

To find the rotational speed ω_F, we use Eq. 11-5 ($\omega_{avg} = \Delta\theta/\Delta t$). Here the wheel goes through an angular displacement of $\Delta\theta = 2\pi$ rad in a time period $\Delta t = 2$ min. Thus, we have

$$\omega_F = \frac{2\pi \text{ rad}}{(2 \text{ min})(60 \text{ s/min})} = 0.0524 \text{ rad/s}.$$

Now we can find the magnitude L of the angular momentum with Eq. 12-31:

$$L = I\omega_F = (1.22 \times 10^9 \text{ kg} \cdot \text{m}^2)(0.0524 \text{ rad/s})$$
$$= 6.39 \times 10^7 \text{ kg} \cdot \text{m}^2/\text{s} \approx 6.4 \times 10^7 \text{ kg} \cdot \text{m}^2/\text{s}. \quad \text{(Answer)}$$

(b) Assume that the fully loaded wheel is rotated from rest to ω_F in a time period $\Delta t_1 = 5.0$ s. What is the magnitude τ_{avg} of the average net external torque acting on it during Δt_1?

SOLUTION: The Key Idea here is that the average net external torque is related to the change ΔL in the angular momentum of the loaded wheel by Eq. 12-29 ($\vec{\tau}_{net} = d\vec{L}/dt$). Because the wheel rotates about a fixed axis to reach angular speed ω_F in time period Δt_1, we can rewrite Eq. 12-29 as $\tau_{avg} = \Delta L/\Delta t_1$. The change ΔL is from zero to the answer for part (a). Thus, we have

$$\tau_{avg} = \frac{\Delta L}{\Delta t_1} = \frac{6.39 \times 10^7 \text{ kg} \cdot \text{m}^2/\text{s} - 0}{5.0 \text{ s}}$$
$$\approx 1.3 \times 10^7 \text{ N} \cdot \text{m}. \quad \text{(Answer)}$$

12-10   Conservation of Angular Momentum

So far we have discussed two powerful conservation laws, the conservation of energy and the conservation of linear momentum. Now we meet a third law of this type, involving the conservation of angular momentum. We start from Eq. 12-29 ($\vec{\tau}_{net} = d\vec{L}/dt$), which is Newton's second law in angular form. If no net external torque acts on the system, this equation becomes $d\vec{L}/dt = 0$, or

$$\vec{L} = \text{a constant} \qquad \text{(isolated system).} \qquad (12\text{-}32)$$

This result, called the **law of conservation of angular momentum,** can also be written as

$$\left(\begin{array}{c} \text{net angular momentum} \\ \text{at some initial time } t_i \end{array} \right) = \left(\begin{array}{c} \text{net angular momentum} \\ \text{at some later time } t_f \end{array} \right),$$

or
$$\vec{L}_i = \vec{L}_f \qquad \text{(isolated system).} \qquad (12\text{-}33)$$

Equations 12-32 and 12-33 tell us:

> If the net external torque acting on a system is zero, the angular momentum $\vec{L}$ of the system remains constant, no matter what changes take place within the system.

Equations 12-32 and 12-33 are vector equations; as such, they are equivalent to three component equations corresponding to the conservation of angular momentum in three mutually perpendicular directions. Depending on the torques acting on a system, the angular momentum of the system might be conserved in only one or two directions but not in all directions:

> If the component of the net *external* torque on a system along a certain axis is zero, then the component of the angular momentum of the system along that axis cannot change, no matter what changes take place within the system.

We can apply this law to the isolated body in Fig. 12-14, which rotates around the z axis. Suppose that the initially rigid body somehow redistributes its mass relative to that rotation axis, changing its rotational inertia about that axis. Equations 12-32 and 12-33 state that the angular momentum of the body cannot change. Substituting Eq. 12-31 (for the angular momentum along the rotational axis) into Eq. 12-33, we write this conservation law as

$$I_i\omega_i = I_f\omega_f. \tag{12-34}$$

Here the subscripts refer to the values of the rotational inertia I and angular speed ω before and after the redistribution of mass.

Like the other two conservation laws that we have discussed, Eqs. 12-32 and 12-33 hold beyond the limitations of Newtonian mechanics. They hold for particles whose speeds approach that of light (where the theory of special relativity reigns), and they remain true in the world of subatomic particles (where quantum physics reigns). No exceptions to the law of conservation of angular momentum have ever been found.

We now discuss four examples involving this law.

1. **The spinning volunteer**   Figure 12-16 shows a student seated on a stool that can rotate freely about a vertical axis. The student, who has been set into rotation at a modest initial angular speed ω_i, holds two dumbbells in his outstretched hands. His angular momentum vector $\vec{L}$ lies along the vertical rotation axis, pointing upward.

The instructor now asks the student to pull in his arms; this action reduces his rotational inertia from its initial value I_i to a smaller value I_f because he moves mass closer to the rotation axis. His rate of rotation increases markedly, from ω_i to ω_f. The student can then slow down by extending his arms once more.

No net external torque acts on the system consisting of the student, stool, and dumbbells. Thus, the angular momentum of that system about the rotation

Fig. 12-16 (*a*) The student has a relatively large rotational inertia about the rotation axis and a relatively small angular speed. (*b*) By decreasing his rotational inertia, the student automatically increases his angular speed. The angular momentum $\vec{L}$ of the rotating system remains unchanged.

(*a*)   (*b*)

Fig. 12-17 The diver's angular momentum $\vec{L}$ is constant throughout the dive, being represented by the tail $\otimes$ of an arrow that is perpendicular to the plane of the figure. Note also that her center of mass (see the dots) follows a parabolic path.

axis must remain constant, no matter how the student maneuvers the dumbbells. In Fig. 12-16a, the student's angular speed ω_i is relatively low and his rotational inertia I_i is relatively high. According to Eq. 12-34, his angular speed in Fig. 12-16b must be greater to compensate for the decreased rotational inertia.

2. The springboard diver   Figure 12-17 shows a diver doing a forward one-and-a-half-somersault dive. As you should expect, her center of mass follows a parabolic path. She leaves the springboard with a definite angular momentum $\vec{L}$ about an axis through her center of mass, represented by a vector pointing into the plane of Fig. 12-17, perpendicular to the page. When she is in the air, no net external torque acts on her about her center of mass, so her angular momentum about her center of mass cannot change. By pulling her arms and legs into the closed *tuck position*, she can considerably reduce her rotational inertia about the same axis and thus, according to Eq. 12-34, considerably increase her angular speed. Pulling out of the tuck position (into the *open layout position*) at the end of the dive increases her rotational inertia and thus slows her rotation rate so she can enter the water with little splash. Even in a more complicated dive involving both twisting and somersaulting, the angular momentum of the diver must be conserved, in both magnitude *and* direction, throughout the dive.

3. Spacecraft orientation   Figure 12-18, which represents a spacecraft with a rigidly mounted flywheel, suggests a scheme (albeit crude) for orientation control. The *spacecraft + flywheel* form an isolated system. Therefore, if the system's total angular momentum $\vec{L}$ is zero because neither spacecraft nor flywheel is turning, it must remain zero (as long as the system remains isolated).

To change the orientation of the spacecraft, the flywheel is made to rotate (Fig. 12-18a). The spacecraft will start to rotate in the opposite sense to maintain the system's angular momentum at zero. When the flywheel is then brought to rest, the spacecraft will also stop rotating but will have changed its orientation (Fig. 12-18b). Throughout, the angular momentum of the system *spacecraft + flywheel* never differs from zero.

Interestingly, the spacecraft *Voyager 2*, on its 1986 flyby of the planet Uranus, was set into unwanted rotation by this flywheel effect every time its tape recorder was turned on at high speed. The ground staff at the Jet Propulsion Laboratory had to program the on-board computer to turn on counteracting thruster jets every time the tape recorder was turned on or off.

4. The incredible shrinking star   When the nuclear fire in the core of a star burns low, the star may eventually begin to collapse, building up pressure in its interior. The collapse may go so far as to reduce the radius of the star from something like that of the Sun to the incredibly small value of a few kilometers. The star then becomes a *neutron star*—its material has been compressed to an incredibly dense gas of neutrons.

During this shrinking process, the star is an isolated system and its angular momentum $\vec{L}$ cannot change. Because its rotational inertia is greatly reduced, its angular speed is correspondingly greatly increased, to as much as 600 to 800 revolutions per *second*. For comparison, the Sun, a typical star, rotates at about one revolution per month.

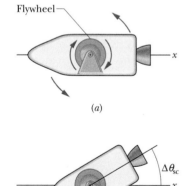

Fig. 12-18 (a) An idealized spacecraft containing a flywheel. If the flywheel is made to rotate clockwise as shown, the spacecraft itself will rotate counter-clockwise. (b) When the flywheel is braked to a stop, the spacecraft will also stop rotating but will be reoriented by the angle $\Delta\theta_{sc}$.

✓**CHECKPOINT 7:**   A rhinoceros beetle rides the rim of a small disk that rotates like a merry-go-round. If the beetle crawls toward the center of the disk, do the following (each relative to the central axis) increase, decrease, or remain the same: (a) the rotational inertia of the beetle–disk system, (b) the angular momentum of the system, and (c) the angular speed of the beetle and disk?

Sample Problem 12-7

Figure 12-19a shows a student, again sitting on a stool that can rotate freely about a vertical axis. The student, initially at rest, is holding a bicycle wheel whose rim is loaded with lead and whose rotational inertia I_{wh} about its central axis is 1.2 kg·m². The wheel is rotating at an angular speed ω_{wh} of 3.9 rev/s; as seen from overhead, the rotation is counterclockwise. The axis of the wheel is vertical, and the angular momentum $\vec{L}_{wh}$ of the wheel points vertically upward. The student now inverts the wheel (Fig. 12-19b) so that, as seen from overhead, it is rotating clockwise. Its angular momentum is then $-\vec{L}_{wh}$. The inversion results in the student, the stool, and the wheel's center rotating together as a composite rigid body about the stool's rotation axis, with rotational inertia $I_b = 6.8$ kg·m². (The fact that the wheel is also rotating about its center does not affect the mass distribution of this composite body; thus, I_b has the same value whether or not the wheel rotates.) With what angular speed ω_b and in what direction does the composite body rotate after the inversion of the wheel?

SOLUTION: The Key Ideas here are these:

1. The angular speed ω_b we seek is related to the final angular momentum $\vec{L}_b$ of the composite body about the stool's rotation axis by Eq. 12-31 ($L = I\omega$).

2. The initial angular speed ω_{wh} of the wheel is related to the angular momentum $\vec{L}_{wh}$ of the wheel's rotation about its center by the same equation.

3. The vector addition of $\vec{L}_b$ and $\vec{L}_{wh}$ gives the total angular momentum $\vec{L}_{tot}$ of the system of student, stool, and wheel.

4. As the wheel is inverted, no net *external* torque acts on that system to change $\vec{L}_{tot}$ about any vertical axis. (Torques due to forces between the student and the wheel as the student inverts the wheel are *internal* to the system.) So, the system's total angular momentum is conserved about any vertical axis.

The conservation of $\vec{L}_{tot}$ is represented with vectors in Fig. 12-19c. We can also write it in terms of components along a vertical axis as

$$L_{b,f} + L_{wh,f} = L_{b,i} + L_{wh,i}, \qquad (12\text{-}35)$$

where i and f refer to the initial state (before inversion of the wheel) and the final state (after inversion). Because inversion of the wheel inverted the angular momentum vector of the wheel's rotation, we substitute $-L_{wh,i}$ for $L_{wh,f}$. Then, if we set $L_{b,i} = 0$ (because the student, the stool, and the wheel's center were initially at rest),

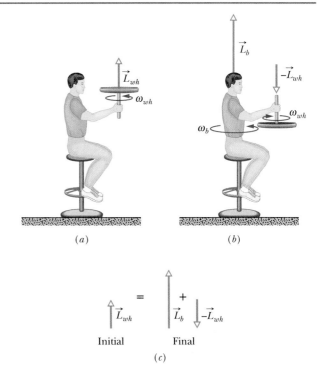

Fig. 12-19 Sample Problem 12-7. (*a*) A student holds a bicycle wheel rotating around the vertical. (*b*) The student inverts the wheel, setting himself into rotation. (*c*) The net angular momentum of the system must remain the same in spite of the inversion.

Eq. 12-35 yields

$$L_{b,f} = 2L_{wh,i}.$$

Using Eq. 12-31, we next substitute $I_b\omega_b$ for $L_{b,f}$ and $I_{wh}\omega_{wh}$ for $L_{wh,i}$ and solve for ω_b, finding

$$\omega_b = \frac{2I_{wh}}{I_b}\,\omega_{wh}$$

$$= \frac{(2)(1.2\ \text{kg}\cdot\text{m}^2)(3.9\ \text{rev/s})}{6.8\ \text{kg}\cdot\text{m}^2} = 1.4\ \text{rev/s}. \quad (\text{Answer})$$

This positive result tells us that the student rotates counterclockwise about the stool axis as seen from overhead. If the student wishes to stop rotating, he has only to invert the wheel once more.

Sample Problem 12-8

During a jump to his partner, an aerialist is to make a quadruple somersault lasting a time $t = 1.87$ s. For the first and last quarter-revolution, he is in the extended orientation shown in Fig. 12-20, with rotational inertia $I_1 = 19.9$ kg·m² around his center of mass (the dot). During the rest of the flight he is in a tight tuck, with rotational inertia $I_2 = 3.93$ kg·m². What must be his angular speed ω_2 around his center of mass during the tuck?

SOLUTION: Obviously he must turn fast enough to complete the 4.0 rev required for a quadruple somersault in the given 1.87 s. To do so, he increases his angular speed to ω_2 by tucking. We can relate ω_2 to his initial angular speed ω_1 with this Key Idea: His angular momentum about his center of mass is conserved throughout the free flight because there is no net external torque about his center of mass to change it. From Eq. 12-34, we can write the

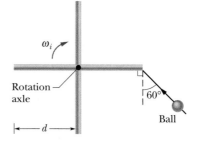

Fig. 12-20 Sample Problem 12-8. An aerialist performing a multiple somersault to a partner.

conservation of angular momentum ($L_1 = L_2$) as

$$I_1\omega_1 = I_2\omega_2$$

or
$$\omega_1 = \frac{I_2}{I_1}\,\omega_2. \qquad (12\text{-}36)$$

A second **Key Idea** is that these angular speeds are related to the angles through which he must rotate and the time available to do so. At the start and at the end, he must rotate in the extended orientation for a total angle of $\theta_1 = 0.500$ rev (two quarter-turns) in a time we shall call t_1. In the tuck, he must rotate through an angle of $\theta_2 = 3.50$ rev in a time t_2. From Eq. 11-5 ($\omega_{\text{avg}} = \Delta\theta/\Delta t$), we can write

$$t_1 = \frac{\theta_1}{\omega_1} \quad \text{and} \quad t_2 = \frac{\theta_2}{\omega_2}.$$

Thus, his total flight time is

$$t = t_1 + t_2 = \frac{\theta_1}{\omega_1} + \frac{\theta_2}{\omega_2}, \qquad (12\text{-}37)$$

which we know to be 1.87 s. Now substituting from Eq. 12-36 for ω_1 yields

$$t = \frac{\theta_1 I_1}{\omega_2 I_2} + \frac{\theta_2}{\omega_2} = \frac{1}{\omega_2}\left(\theta_1\frac{I_1}{I_2} + \theta_2\right).$$

Inserting the known data, we obtain

$$1.87\text{ s} = \frac{1}{\omega_2}\left((0.500\text{ rev})\frac{19.9\text{ kg}\cdot\text{m}^2}{3.93\text{ kg}\cdot\text{m}^2} + 3.50\text{ rev}\right),$$

which gives us

$$\omega_2 = 3.23\text{ rev/s.} \qquad \text{(Answer)}$$

This angular speed is so fast that the aerialist cannot clearly see his surroundings or fine-tune his rotation by adjusting his tuck. The possibility of an aerialist making a four-and-a-half somersault flight, which would require a greater value of ω_2 and thus a smaller I_2 via a tighter tuck, seems very small.

Sample Problem 12-9

(This final sample problem of the chapter is long and challenging, but it is helpful because it pulls together many ideas of Chapters 11 and 12.) In the overhead view of Fig. 12-21, four thin, uniform rods, each of mass M and length $d = 0.50$ m, are rigidly connected to a vertical axle to form a turnstile. The turnstile rotates clockwise about the axle, which is attached to a floor, with initial angular velocity $\omega_i = -2.0$ rad/s. A mud ball of mass $m = \frac{1}{3}M$ and initial speed $v_i = 12$ m/s is thrown along the path shown and sticks to the end of one rod. What is the final angular velocity ω_f of the ball–turnstile system?

SOLUTION: A **Key Idea** here can be stated in a question-and-answer format. The question is this: Does the system have a quantity that is conserved during the collision and that involves angular velocity, so that we can solve for ω_f? To answer, let us check the conservation possibilities:

1. The total kinetic energy K is *not* conserved, because the collision between ball and rod is completely inelastic (the ball

Fig. 12-21 Sample Problem 12-9. An overhead view of four rigidly connected rods rotating freely around a central axle, and the path a mud ball takes to stick onto one of the rods.

sticks). So, some energy must be transferred from kinetic energy to other types of energy (such as thermal energy). For the same reason, total mechanical energy is not conserved.

2. The total linear momentum $\vec{P}$ is also *not* conserved, because during the collision an external force acts on the turnstile at the attachment of the axle to the floor. (This is the force that keeps the turnstile from moving across the floor when it is hit by the mud ball.)

3. The total angular momentum $\vec{L}$ of the system about the axle *is* conserved because there is no net external torque to change $\vec{L}$. (The forces in the collision produce only internal torques; the external force on the turnstile acts at the axle, has zero moment arm, and thus does not produce an external torque.)

We can write the conservation of the system's total angular momentum ($L_f = L_i$) about the axle as

$$L_{\mathrm{ts},f} + L_{\mathrm{ball},f} = L_{\mathrm{ts},i} + L_{\mathrm{ball},i}, \tag{12-38}$$

where ts stands for "turnstile." The final angular velocity ω_f is contained in the terms $L_{\mathrm{ts},f}$ and $L_{\mathrm{ball},f}$ because those final angular momenta depend on how fast the turnstile and ball are rotating. To find ω_f, we consider first the turnstile and then the ball, and then we return to Eq. 12-38.

Turnstile: The Key Idea here is that, because the turnstile is a rotating rigid object, Eq. 12-31 ($L = I\omega$) gives its angular momentum. Thus, we can write its final and initial angular momenta about the axle as

$$L_{\mathrm{ts},f} = I_{\mathrm{ts}}\omega_f \quad \text{and} \quad L_{\mathrm{ts},i} = I_{\mathrm{ts}}\omega_i. \tag{12-39}$$

Because the turnstile consists of four rods, each rotating around an end, the rotational inertia I_{ts} of the turnstile is four times the rotational inertia I_{rod} of each rod about its end. From Table 11-2*e*, we know that the rotational inertia I_{com} of a rod about its center is $\frac{1}{12}Md^2$, where M is its mass and d is its length. To get I_{rod} we use the parallel-axis theorem of Eq. 11-29 ($I = I_{\mathrm{com}} + Mh^2$). Here perpendicular distance h is $d/2$. Thus, we find

$$I_{\mathrm{rod}} = \tfrac{1}{12}Md^2 + M\left(\frac{d}{2}\right)^2 = \tfrac{1}{3}Md^2.$$

With four rods in the turnstile, we then have

$$I_{\mathrm{ts}} = \tfrac{4}{3}Md^2. \tag{12-40}$$

Ball: Before the collision, the ball is like a particle moving along a straight line, as in Fig. 12-11. So, to find the ball's initial angular momentum $L_{\mathrm{ball},i}$ about the axle, we can use any of Eqs. 12-18 through 12-21, but Eq. 12-20 ($\ell = rv_\perp$) is easiest. Here ℓ is $L_{\mathrm{ball},i}$; just before the ball hits, its radial distance r from the axle is d and the component $v_\perp$ of the ball's velocity perpendicular to r is $v_i \cos 60°$.

To give a sign to this angular momentum, we mentally draw a position vector from the turnstile's axle to the ball. As the ball approaches the turnstile, this position vector rotates counterclockwise about the axle, so the ball's angular momentum is a positive quantity. We can now rewrite $\ell = rmv_\perp$ as

$$L_{\mathrm{ball},i} = mdv_i \cos 60°. \tag{12-41}$$

After the collision, the ball is like a particle rotating in a circle of radius d. So, from Eq. 11-26 ($I = \Sigma m_i r_i^2$), we have $I_{\mathrm{ball}} = md^2$ about the axle. Then from Eq. 12-31 ($L = I\omega$), we can write the final angular momentum of the ball about the axle as

$$L_{\mathrm{ball},f} = I_{\mathrm{ball}}\omega_f = md^2\omega_f. \tag{12-42}$$

Return to Eq. 12-38: Substituting from Eqs. 12-39 through 12-42 into Eq.12-38, we have

$$\tfrac{4}{3}Md^2\omega_f + md^2\omega_f = \tfrac{4}{3}Md^2\omega_i + mdv_i \cos 60°.$$

Substituting $M = 3m$ and solving for ω_f, we find

$$\omega_f = \frac{1}{5d}(4d\omega_i + v_i \cos 60°)$$

$$= \frac{1}{5(0.50\ \mathrm{m})}[4(0.50\ \mathrm{m})(-2.0\ \mathrm{rad/s}) + (12\ \mathrm{m/s})(\cos 60°)]$$

$$= 0.80\ \mathrm{rad/s}. \tag{Answer}$$

Thus, the turnstile is now turning counterclockwise.

REVIEW & SUMMARY

Rolling Bodies For a wheel of radius R that is rolling smoothly (no sliding),

$$v_{\mathrm{com}} = \omega R, \tag{12-2}$$

where v_{com} is the linear speed of the wheel's center and ω is the angular speed of the wheel about its center. The wheel may also be viewed as rotating instantaneously about the point P of the "road" that is in contact with the wheel. The angular speed of the wheel about this point is the same as the angular speed of the wheel about its center. The rolling wheel has kinetic energy

$$K = \tfrac{1}{2}I_{\mathrm{com}}\omega^2 + \tfrac{1}{2}Mv_{\mathrm{com}}^2, \tag{12-5}$$

where I_{com} is the rotational moment of the wheel about its center and M is the mass of the wheel. If the wheel is being accelerated

but is still rolling smoothly, the acceleration of the center of mass $\vec{a}_{\mathrm{com}}$ is related to the angular acceleration α about the center with

$$a_{\mathrm{com}} = \alpha R. \tag{12-6}$$

If the wheel rolls smoothly down a ramp of angle θ, its acceleration along an x axis extending up the ramp is

$$a_{\mathrm{com},x} = -\frac{g \sin \theta}{1 + I_{\mathrm{com}}/MR^2}. \tag{12-10}$$

Torque as a Vector In three dimensions, *torque* $\vec{\tau}$ is a vector quantity defined relative to a fixed point (usually an origin); it is

$$\vec{\tau} = \vec{r} \times \vec{F}, \tag{12-14}$$

where $\vec{F}$ is a force applied to a particle and $\vec{r}$ is a position vector locating the particle relative to the fixed point (or origin). The magnitude of $\vec{\tau}$ is given by

$$\tau = rF\sin\phi = rF_\perp = r_\perp F, \qquad \text{(12-15, 12-16, 12-17)}$$

where ϕ is the angle between $\vec{F}$ and $\vec{r}$, $F_\perp$ is the component of $\vec{F}$ perpendicular to $\vec{r}$, and $r_\perp$ is the moment arm of $\vec{F}$. The direction of $\vec{\tau}$ is given by the right-hand rule for cross products.

Angular Momentum of a Particle   The *angular momentum* $\vec{\ell}$ of a particle with linear momentum $\vec{p}$, mass m, and linear velocity $\vec{v}$ is a vector quantity defined relative to a fixed point (usually an origin); it is

$$\vec{\ell} = \vec{r} \times \vec{p} = m(\vec{r} \times \vec{v}). \qquad \text{(12-18)}$$

The magnitude of $\vec{\ell}$ is given by

$$\ell = rmv\sin\phi \qquad \text{(12-19)}$$
$$= rp_\perp = rmv_\perp \qquad \text{(12-20)}$$
$$= r_\perp p = r_\perp mv, \qquad \text{(12-21)}$$

where ϕ is the angle between $\vec{r}$ and $\vec{p}$, $p_\perp$ and $v_\perp$ are the components of $\vec{p}$ and $\vec{v}$ perpendicular to $\vec{r}$, and $r_\perp$ is the perpendicular distance between the fixed point and the extension of $\vec{p}$. The direction of $\vec{\ell}$ is given by the right-hand rule for cross products.

Newton's Second Law in Angular Form   Newton's second law for a particle can be written in angular form as

$$\vec{\tau}_{\text{net}} = \frac{d\vec{\ell}}{dt}, \qquad \text{(12-23)}$$

where $\vec{\tau}_{\text{net}}$ is the net torque acting on the particle, and $\vec{\ell}$ is the angular momentum of the particle.

Angular Momentum of a System of Particles   The angular momentum $\vec{L}$ of a system of particles is the vector sum of the angular momenta of the individual particles:

$$\vec{L} = \vec{\ell}_1 + \vec{\ell}_2 + \cdots + \vec{\ell}_n = \sum_{i=1}^{n} \vec{\ell}_i. \qquad \text{(12-26)}$$

The time rate of change of this angular momentum is equal to the net external torque on the system (the vector sum of the torques due to interactions of the particles of the system with particles external to the system):

$$\vec{\tau}_{\text{net}} = \frac{d\vec{L}}{dt} \qquad \text{(system of particles).} \qquad \text{(12-29)}$$

Angular Momentum of a Rigid Body   For a rigid body rotating about a fixed axis, the component of its angular momentum parallel to the rotation axis is

$$L = I\omega \qquad \text{(rigid body, fixed axis).} \qquad \text{(12-31)}$$

Conservation of Angular Momentum   The angular momentum $\vec{L}$ of a system remains constant if the net external torque acting on the system is zero:

$$\vec{L} = \text{a constant} \qquad \text{(isolated system)} \qquad \text{(12-32)}$$

or $\qquad\qquad \vec{L}_i = \vec{L}_f \qquad \text{(isolated system).} \qquad \text{(12-33)}$

This is the **law of conservation of angular momentum.** It is one of the fundamental conservation laws of nature, having been verified even in situations (involving high-speed particles or subatomic dimensions) in which Newton's laws are not applicable.

QUESTIONS

1. In Fig. 12-22, a block slides down a frictionless ramp and a sphere rolls without sliding down a ramp of the same angle θ. The block and sphere have the same mass, start from rest at point A, and descend through point B. (a) In that descent, is the work done by the gravitational force on the block greater than, less than, or the same as the work done by the gravitational force on the sphere? At B, which object has more (b) translational kinetic energy and (c) speed down the ramp?

2. A cannonball rolls down an incline without sliding. If the roll is now repeated with an incline that is less steep but of the same

height as the first incline, are (a) the ball's time to reach the bottom and (b) its translational kinetic energy at the bottom greater than, less than, or the same as previously?

3. In Fig. 12-23, a woman rolls a cylindrical drum by pushing a board across its top. The drum moves through the distance $L/2$, which is half the board's length. The drum rolls smoothly without sliding or bouncing, and the board does not slide over the drum. (a) What length of board rolls over the top of the drum? (b) How far does the woman walk?

Fig. 12-22 Question 1.

Fig. 12-23 Question 3.

4. The position vector $\vec{r}$ of a particle relative to a certain point has a magnitude of 3 m, and the force $\vec{F}$ on the particle has a magnitude of 4 N. What is the angle between the directions of $\vec{r}$ and $\vec{F}$ if the magnitude of the associated torque equals (a) zero and (b) 12 N·m?

5. Figure 12-24 shows a particle moving at constant velocity $\vec{v}$ and five points with their xy coordinates. Rank the points according to the magnitude of the angular momentum of the particle measured about them, greatest first.

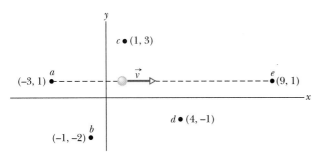

Fig. 12-24 Question 5.

6. (a) In Checkpoint 4, what is the torque on particles 1 and 2 about point O due to the centripetal forces that cause those particles to circle at constant speed? (b) As particles 3, 4, and 5 move from the left to the right of point O, do their individual angular momenta increase, decrease, or stay the same?

7. Figure 12-25 shows three particles of the same mass and the same constant speed moving as indicated by the velocity vectors. Points a, b, c, and d form a square, with point e at the center. Rank the points according to the magnitude of the net angular momentum of the three-particle system when measured about the points, greatest first.

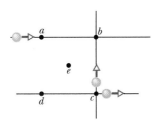

Fig. 12-25 Question 7.

8. A bola, which consists of three heavy balls connected to a common point by identical lengths of sturdy string, is readied for launch by holding one of the balls overhead and rotating the wrist, causing the other two balls to rotate in a horizontal circle about the hand. The bola is then released, and its configuration rapidly changes from that in the overhead view of Fig. 12-26a to that of Fig. 12-26b. Thus, the rotation is initially around axis 1 through the ball that was held. Then it is around axis 2 through the center of mass. Are

Fig. 12-26 Question 8.   (a)   (b)

(a) the angular momentum and (b) the angular speed around axis 2 greater than, less than, or the same as that around axis 1?

9. A rhinoceros beetle rides the rim of a horizontal disk rotating counterclockwise like a merry-go-round. If it then walks along the rim in the direction of the rotation, will the magnitudes of the following increase, decrease, or remain the same: (a) the angular momentum of the beetle–disk system, (b) the angular momentum and angular velocity of the beetle, and (c) the angular momentum and angular velocity of the disk? (d) What are your answers if the beetle walks in the direction opposite the rotation?

10. Figure 12-27 shows an overhead view of a rectangular slab that can spin like a merry-go-round about its center at O. Also shown are seven paths along which wads of bubble gum can be thrown (all with the same speed and mass) to stick onto the stationary slab. (a) Rank the paths according to the angular speed that the slab (and gum) will have after the gum sticks, greatest first. (b) For which paths will the angular momentum of the slab (and gum) about O be negative from the view of Fig. 12-27?

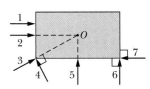

Fig. 12-27 Question 10.

11. In Fig. 12-28, three forces of the same magnitude are applied to a particle at the origin ($\vec{F}_1$ acts directly into the plane of the figure). Rank the forces according to the magnitudes of the torques they create about (a) point P_1, (b) point P_2, and (c) point P_3, greatest first.

Fig. 12-28 Question 11.

12. Figure 12-29 shows two particles A and B at xyz coordinates (1 m, 1 m, 0) and (1 m, 0, 1 m). Acting on each particle are three numbered forces, all of the same magnitude and each directed parallel to an axis. (a) Which of the forces produce a torque about the origin that is directed parallel to y? (b) Rank the forces according to the magnitudes of the torques they produce on the particles about the origin, greatest first.

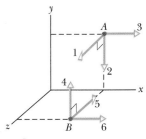

Fig. 12-29 Question 12.

13. Figure 11-24 in Chapter 11 shows an assembly of three small spheres of the same mass that are attached to a massless rod with the indicated spacings. The assembly is to be rotated at 3.0 rad/s about an axis through one of the spheres and perpendicular to the plane of the page. There are, of course, three such choices of the axis. Rank those choices according to (a) the magnitude of the angular momentum the assembly will have about the chosen rotation axis and (b) the rotational kinetic energy the assembly will have, greatest first.

EXERCISES & PROBLEMS

ssm	Solution is in the Student Solutions Manual.
www	Solution is available on the World Wide Web at: http://www.wiley.com/college/hrw
ilw	Solution is available on the Interactive LearningWare.

SEC. 12-1   Rolling

1E. An automobile traveling 80.0 km/h has tires of 75.0 cm diameter. (a) What is the angular speed of the tires about their axles? (b) If the car is brought to a stop uniformly in 30.0 complete turns of the tires (without skidding), what is the magnitude of the angular acceleration of the wheels? (c) How far does the car move during the braking?

2P. Consider a 66-cm-diameter tire on a car traveling at 80 km/h on a level road in the positive direction of an x axis. Relative to a woman in the car, what are (a) the linear velocity $\vec{v}$ and (b) the magnitude a of the linear acceleration of the center of the wheel? What are (c) $\vec{v}$ and (d) a for a point at the top of the tire? What are (e) $\vec{v}$ and (f) a for a point at the bottom of the tire?

Now repeat the questions relative to a hitchhiker sitting near the road: What are (g) $\vec{v}$ at the wheel's center, (h) a at the wheel's center, (i) $\vec{v}$ at the tire top, (j) a at the tire top, (k) $\vec{v}$ at the tire bottom, and (1) a at the tire bottom?

SEC. 12-2   The Kinetic Energy of Rolling

3E. A 140 kg hoop rolls along a horizontal floor so that its center of mass has a speed of 0.150 m/s. How much work must be done on the hoop to stop it?   ssm

4E. A thin-walled pipe rolls along the floor. What is the ratio of its translational kinetic energy to its rotational kinetic energy about an axis parallel to its length and through its center of mass?

5E. A 1000 kg car has four 10 kg wheels. When the car is moving, what fraction of the total kinetic energy of the car is due to rotation of the wheels about their axles? Assume that the wheels have the same rotational inertia as uniform disks of the same mass and size. Why do you not need the radius of the wheels?   ssm   ilw   www

6P. A body of radius R and mass m is rolling smoothly with speed v on a horizontal surface. It then rolls up a hill to a maximum height h. (a) If $h = 3v^2/4g$, what is the body's rotational inertia about the rotational axis through its center of mass? (b) What might the body be?

SEC. 12-3   The Forces of Rolling

7E. A uniform solid sphere rolls down an incline. (a) What must be the incline angle if the linear acceleration of the center of the sphere is to have a magnitude of 0.10g? (b) If a frictionless block were to slide down the incline at that angle, would its acceleration magnitude be more than, less than, or equal to 0.10g? Why?

8P. A constant horizontal force of magnitude 10 N is applied to a wheel of mass 10 kg and radius 0.30 m as shown in Fig. 12-30. The wheel rolls smoothly on the horizontal surface, and the acceleration of its center of mass has magnitude 0.60 m/s². (a) What are the magnitude and direction of the frictional force on the wheel?

Fig. 12-30 Problem 8.

(b) What is the rotational inertia of the wheel about the rotation axis through its center of mass?

9P. A solid ball starts from rest at the upper end of the track shown in Fig. 12-31 and rolls without slipping until it rolls off the right-hand end. If $H = 6.0$ m and $h = 2.0$ m and the track is horizontal at the right-hand end, how far horizontally from point A does the ball land on the floor?   ssm

Fig. 12-31 Problem 9.

10P. A small sphere, with radius r and mass m, rolls without slipping on the inside of a large fixed hemisphere with radius R and a vertical axis of symmetry. It starts at the top from rest. (a) What is its kinetic energy at the bottom? (b) What fraction of its kinetic energy at the bottom is associated with rotation about an axis through its center of mass? (c) Assuming $r \ll R$, find the magnitude of the normal force on the hemisphere from the ball when the ball reaches the bottom.

11P. A solid cylinder of radius 10 cm and mass 12 kg starts from rest and rolls without slipping a distance of 6.0 m down a house roof that is inclined at 30°. (See Fig. 12-32.) (a) What is the angular speed of the cylinder about its center as it leaves the house roof? (b) The roof's edge is 5.0 m high. How far horizontally from the roof's edge does the cylinder hit the level ground?   ilw

Fig. 12-32 Problem 11.

12P. A small solid marble of mass m and radius r will roll without slipping along the loop-the-loop track shown in Fig. 12-33 if it is released from rest somewhere on the straight section of track. (a) From what initial height h above the bottom of the track must the marble be released if it is to be on the verge of leaving the track at the top of the loop? (The radius of the loop-the-loop is R; assume $R \gg r$.) (b) If the marble is released from height $6R$ above the bottom of the track, what is the horizontal component of the force acting on it at point Q?

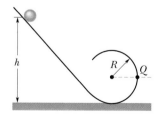

Fig. 12-33 Problem 12.

13P. A hollow sphere of radius 0.15 m, with rotational inertia $I = 0.040$ kg $\cdot$ m^2 about a line through its center of mass, rolls without slipping up a surface inclined at 30° to the horizontal. At a certain initial position, the sphere's total kinetic energy is 20 J. (a) How much of this initial kinetic energy is rotational? (b) What is the speed of the center of mass of the sphere at the initial position? What are (c) the total kinetic energy of the sphere and (d) the speed of its center of mass after it has moved 1.0 m up along the incline from its initial position?

14P. A bowler throws a bowling ball of radius $R = 11$ cm along a lane. The ball slides on the lane, with initial speed $v_{com,0} = 8.5$ m/s and initial angular speed $\omega_0 = 0$. The coefficient of kinetic friction between the ball and the lane is 0.21. The kinetic frictional force $\vec{f}_k$ acting on the ball (Fig. 12-34) causes a linear acceleration of the ball while producing a torque that causes an angular acceleration of the ball. When speed v_{com} has decreased enough and angular speed ω has increased enough, the ball stops sliding and then rolls smoothly. (a) What then is v_{com} in terms of ω? During the sliding, what are the ball's (b) linear acceleration and (c) angular acceleration? (d) How long does the ball slide? (e) How far does the ball slide? (f) What is the speed of the ball when smooth rolling begins?

Fig. 12-34 Problem 14.

SEC. 12-4   The Yo-Yo

15E. A yo-yo has a rotational inertia of 950 g $\cdot$ cm^2 and a mass of 120 g. Its axle radius is 3.2 mm, and its string is 120 cm long. The yo-yo rolls from rest down to the end of the string. (a) What is the magnitude of its linear acceleration? (b) How long does it take to reach the end of the string? As it reaches the end of the string, what are its (c) linear speed, (d) translational kinetic energy, (e) rotational kinetic energy, and (f) angular speed? ssm

16P. Suppose that the yo-yo in Exercise 15, instead of rolling from rest, is thrown so that its initial speed down the string is 1.3 m/s. (a) How long does the yo-yo take to reach the end of the string? As it reaches the end of the string, what are its (b) total kinetic energy, (c) linear speed, (d) translational kinetic energy, (e) angular speed, and (f) rotational kinetic energy?

SEC. 12-5   Torque Revisited

17E. Show that, if $\vec{r}$ and $\vec{F}$ lie in a given plane, the torque $\vec{\tau} = \vec{r} \times \vec{F}$ has no component in that plane.

18E. What are the magnitude and direction of the torque about the origin on a plum located at coordinates $(-2.0$ m, 0, 4.0 m$)$ due to force $\vec{F}$ whose only component is (a) $F_x = 6.0$ N, (b) $F_x = -6.0$ N, (c) $F_z = 6.0$ N, and (d) $F_z = -6.0$ N?

19E. What are the magnitude and direction of the torque about the origin on a particle located at coordinates $(0, -4.0$ m, 3.0 m$)$ due to (a) force $\vec{F}_1$ with components $F_{1x} = 2.0$ N and $F_{1y} = F_{1z} = 0$, and (b) force $\vec{F}_2$ with components $F_{2x} = 0$, $F_{2y} = 2.0$ N, and $F_{2z} = 4.0$ N?

20P. Force $\vec{F} = (2.0$ N$)\hat{i} - (3.0$ N$)\hat{k}$ acts on a pebble with position vector $\vec{r} = (0.50$ m$)\hat{j} - (2.0$ m$)\hat{k}$, relative to the origin. What is

the resulting torque acting on the pebble about (a) the origin and (b) a point with coordinates $(2.0$ m, 0, -3.0 m$)$?

21P. Force $\vec{F} = (-8.0$ N$)\hat{i} + (6.0$ N$)\hat{j}$ acts on a particle with position vector $\vec{r} = (3.0$ m$)\hat{i} + (4.0$ m$)\hat{j}$. What are (a) the torque on the particle about the origin and (b) the angle between the directions of $\vec{r}$ and $\vec{F}$? ssm

22P. What is the torque about the origin on a jar of jalapeño peppers located at coordinates $(3.0$ m, -2.0 m, 4.0 m$)$ due to (a) force $\vec{F}_1 = (3.0$ N$)\hat{i} - (4.0$ N$)\hat{j} + (5.0$ N$)\hat{k}$, (b) force $\vec{F}_2 = (-3.0$ N$)\hat{i} - (4.0$ N$)\hat{j} - (5.0$ N$)\hat{k}$, and (c) the vector sum of $\vec{F}_1$ and $\vec{F}_2$? (d) Repeat part (c) about a point with coordinates $(3.0$ m, 2.0 m, 4.0 m$)$ instead of about the origin.

SEC. 12-6   Angular Momentum

23E. Two objects are moving as shown in Fig. 12-35. What is their total angular momentum about point O? ilw

Fig. 12-35 Exercise 23.

24E. In Fig. 12-36, a particle P with mass 2.0 kg has position vector $\vec{r}$ of magnitude 3.0 m and velocity $\vec{v}$ of magnitude 4.0 m/s. A force $\vec{F}$ of magnitude 2.0 N acts on the particle. All three vectors lie in the xy plane oriented as shown. About the origin, what are (a) the angular momentum of the particle and (b) the torque acting on the particle?

25E. At a certain time, a 0.25 kg object has a position vector $\vec{r} = 2.0\hat{i} - 2.0\hat{k}$, in meters. At that instant, its velocity in meters per second is $\vec{v} = -5.0\hat{i} + 5.0\hat{k}$, and the force in newtons acting on it is $\vec{F} = 4.0\hat{j}$. (a) What is the angular momentum of the object about the origin? (b) What torque acts on it? ssm

Fig. 12-36 Exercise 24.

26P. A 2.0 kg particle-like object moves in a plane with velocity components $v_x = 30$ m/s and $v_y = 60$ m/s as it passes through the point with (x, y) coordinates of $(3.0, -4.0)$ m. Just then, what is its angular momentum relative to (a) the origin and (b) the point $(-2.0, -2.0)$ m?

27P. Two particles, each of mass m and speed v, travel in opposite directions along parallel lines separated by a distance d. (a) In terms of m, v, and d, find an expression for the magnitude L of the angular momentum of the two-particle system around a point midway between the two lines. (b) Does the expression change if the point about which L is calculated is not midway between the lines? (c) Now reverse the direction of travel for one of the particles and repeat (a) and (b). ssm

28P. A 4.0 kg particle moves in an xy plane. At the instant when the particle's position and velocity are $\vec{r} = (2.0\hat{i} + 4.0\hat{j})$ m and $\vec{v} = -4.0\hat{j}$ m/s, the force on the particle is $\vec{F} = -3.0\hat{i}$ N. At this instant, determine (a) the particle's angular momentum about the origin, (b) the particle's angular momentum about the point $x = 0$,

$y = 4.0$ m, (c) the torque acting on the particle about the origin, and (d) the torque acting on the particle about the point $x = 0$, $y = 4.0$ m.

SEC. 12-7   Newton's Second Law in Angular Form

29E. A 3.0 kg particle with velocity $\vec{v} = (5.0 \text{ m/s})\hat{i} - (6.0 \text{ m/s})\hat{j}$ is at $x = 3.0$ m, $y = 8.0$ m. It is pulled by a 7.0 N force in the negative x direction. (a) What is the angular momentum of the particle about the origin? (b) What torque about the origin acts on the particle? (c) At what rate is the angular momentum of the particle changing with time?   ssm   ilw

30E. A particle is acted on by two torques about the origin: $\vec{\tau}_1$ has a magnitude of 2.0 N · m and is directed in the positive direction of the x axis, and $\vec{\tau}_2$ has a magnitude of 4.0 N · m and is directed in the negative direction of the y axis. What are the magnitude and direction of $d\vec{\ell}/dt$, where $\vec{\ell}$ is the angular momentum of the particle about the origin?

31E. What torque about the origin acts on a particle moving in the xy plane, clockwise about the origin, if the particle has the following magnitudes of angular momentum about the origin:
(a) 4.0 kg · m²/s,
(b) $4.0t^2$ kg · m²/s,
(c) $4.0\sqrt{t}$ kg · m²/s,
(d) $4.0/t^2$ kg · m²/s?

32P. At time $t = 0$, a 2.0 kg particle has position vector $\vec{r} = (4.0 \text{ m})\hat{i} - (2.0 \text{ m})\hat{j}$ relative to the origin. Its velocity just then is given by $\vec{v} = (-6.0t^2 \text{ m/s})\hat{i}$. About the origin and for $t > 0$, what are (a) the particle's angular momentum and (b) the torque acting on the particle? (c) Repeat (a) and (b) about a point with coordinates $(-2.0 \text{ m}, -3.0 \text{ m}, 0)$ instead of about the origin.

SEC. 12-9   The Angular Momentum of a Rigid Body Rotating About a Fixed Axis

33E. The angular momentum of a flywheel having a rotational inertia of 0.140 kg · m² about its central axis decreases from 3.00 to 0.800 kg · m²/s in 1.50 s. (a) What is the magnitude of the average torque acting on the flywheel about its central axis during this period? (b) Assuming a constant angular acceleration, through what angle does the flywheel turn? (c) How much work is done on the wheel? (d) What is the average power of the flywheel?   ssm

34E. A sanding disk with rotational inertia 1.2×10^{-3} kg · m² is attached to an electric drill whose motor delivers a torque of 16 N · m. Find (a) the angular momentum of the disk about its central axis and (b) the angular speed of the disk 33 ms after the motor is turned on.

35E. Three particles, each of mass m, are fastened to each other and to a rotation axis at O by three massless strings, each with length d as shown in Fig. 12-37. The combination rotates around the rotational axis with angular velocity ω in such a way that the particles remain in a straight line. In terms of m, d, and ω, and rel-

Fig. 12-37 Exercise 35.

ative to point O, what are (a) the rotational inertia of the combination, (b) the angular momentum of the middle particle, and (c) the total angular momentum of the three particles?   ssm

36P. An impulsive force $F(t)$ acts for a short time Δt on a rotating rigid body with rotational inertia I. Show that

$$\int \tau \, dt = F_{\text{avg}}R \, \Delta t = I(\omega_f - \omega_i),$$

where τ is the torque due to the force, R is the moment arm of the force, F_{avg} is the average value of the force during the time it acts on the body, and ω_i and ω_f are the angular velocities of the body just before and just after the force acts. (The quantity $\int \tau \, dt = F_{\text{avg}} \, \Delta t$ is called the *angular impulse*, in analogy with $F_{\text{avg}} \, \Delta t$, the linear impulse.)

37P*. Two cylinders having radii R_1 and R_2 and rotational inertias I_1 and I_2 about their central axes are supported by axles perpendicular to the plane of Fig. 12-38. The large cylinder is initially rotating clockwise with angular velocity ω_0. The small cylinder is moved to the right until it touches the large cylinder and is caused to rotate by the frictional force between the two. Eventually, slipping ceases, and the two cylinders rotate at constant rates in opposite directions. Find the final angular velocity ω_2 of the small cylinder in terms of I_1, I_2, R_1, R_2, and ω_0. (*Hint:* Neither angular momentum nor kinetic energy is conserved. Apply the angular impulse equation of Problem 36.)   ssm   www

Fig. 12-38 Problem 37.

38P. Figure 12-39 shows a rigid structure consisting of a circular hoop of radius R and mass m, and a square made of four thin bars, each of length R and mass m. The rigid structure rotates at a constant speed about a vertical axis, with a period of rotation of 2.5 s. Assuming $R = 0.50$ m and $m = 2.0$ kg, calculate (a) the structure's rotational inertia about the axis of rotation and (b) its angular momentum about that axis.

Fig. 12-39 Problem 38.

SEC. 12-10   Conservation of Angular Momentum

39E. A man stands on a platform that is rotating (without friction) with an angular speed of 1.2 rev/s; his arms are outstretched and he holds a brick in each hand. The rotational inertia of the system consisting of the man, bricks, and platform about the central axis is 6.0 kg · m². If by moving the bricks the man decreases the rotational inertia of the system to 2.0 kg · m², (a) what is the resulting angular speed of the platform and (b) what is the ratio of the new kinetic energy of the system to the original kinetic energy? (c) What provided the added kinetic energy?   ssm

40E. The rotor of an electric motor has rotational inertia $I_m = 2.0 \times 10^{-3}$ kg · m² about its central axis. The motor is used to change the orientation of the space probe in which it is mounted.

The motor axis is mounted parallel to the axis of the probe, which has rotational inertia $I_p = 12 \text{ kg} \cdot \text{m}^2$ about its axis. Calculate the number of revolutions of the rotor required to turn the probe through 30° about its axis.

41E. A wheel is rotating freely at angular speed 800 rev/min on a shaft whose rotational inertia is negligible. A second wheel, initially at rest and with twice the rotational inertia of the first, is suddenly coupled to the same shaft. (a) What is the angular speed of the resultant combination of the shaft and two wheels? (b) What fraction of the original rotational kinetic energy is lost? ssm ilw

42E. Two disks are mounted on low-friction bearings on the same axle and can be brought together so that they couple and rotate as one unit. (a) The first disk, with rotational inertia $3.3 \text{ kg} \cdot \text{m}^2$ about its central axis, is set spinning at 450 rev/min. The second disk, with rotational inertia $6.6 \text{ kg} \cdot \text{m}^2$ about its central axis, is set spinning at 900 rev/min in the same direction as the first. They then couple together. What is their angular speed after coupling? (b) If instead the second disk is set spinning at 900 rev/min in the direction opposite the first disk's rotation, what is their angular speed and direction of rotation after coupling?

43E. In a playground, there is a small merry-go-round of radius 1.20 m and mass 180 kg. Its radius of gyration (see Problem 43 of Chapter 11) is 91.0 cm. A child of mass 44.0 kg runs at a speed of 3.00 m/s along a path that is tangent to the rim of the initially stationary merry-go-round and then jumps on. Neglect friction between the bearings and the shaft of the merry-go-round. Calculate (a) the rotational inertia of the merry-go-round about its axis of rotation, (b) the magnitude of the angular momentum of the running child about the axis of rotation of the merry-go-round, and (c) the angular speed of the merry-go-round and child after the child has jumped on. ssm

44E. The rotational inertia of a collapsing spinning star changes to $\frac{1}{3}$ its initial value. What is the ratio of the new rotational kinetic energy to the initial rotational kinetic energy?

45P. A track is mounted on a large wheel that is free to turn with negligible friction about a vertical axis (Fig. 12-40). A toy train of mass m is placed on the track and, with the system initially at rest, the electrical power is turned on. The train reaches a steady speed v with respect to the track. What is the angular speed of the wheel if its mass is M and its radius is R? (Treat the wheel as a hoop, and neglect the mass of the spokes and hub.) ssm www

Fig. 12-40 Problem 45.

46P. In Fig. 12-41, two skaters, each of mass 50 kg, approach each other along parallel paths separated by 3.0 m. They have opposite velocities of 1.4 m/s each. One skater carries one end of a long pole with negligible mass, and the other skater grabs the other end

of it as she passes. Assume frictionless ice. (a) Describe quantitatively the motion of the skaters after they have become connected by the pole. (b) What is the kinetic energy of the two-skater system?

Fig. 12-41 Problem 46.

Next, the skaters each pull along the pole so as to reduce their separation to 1.0 m. What then are (c) their angular speed and (d) the kinetic energy of the system? (e) Explain the source of the increased kinetic energy.

47P. A cockroach of mass m runs counterclockwise around the rim of a lazy Susan (a circular dish mounted on a vertical axle) of radius R and rotational inertia I and having frictionless bearings. The cockroach's speed (relative to the ground) is v, whereas the lazy Susan turns clockwise with angular speed ω_0. The cockroach finds a bread crumb on the rim and, of course, stops. (a) What is the angular speed of the lazy Susan after the cockroach stops? (b) Is mechanical energy conserved as they stop?

48P. A girl of mass M stands on the rim of a frictionless merry-go-round of radius R and rotational inertia I that is not moving. She throws a rock of mass m horizontally in a direction that is tangent to the outer edge of the merry-go-round. The speed of the rock, relative to the ground, is v. Afterward, what are (a) the angular speed of the merry-go-round and (b) the linear speed of the girl?

49P. A horizontal vinyl record of mass 0.10 kg and radius 0.10 m rotates freely about a vertical axis through its center with an angular speed of 4.7 rad/s. The rotational inertia of the record about its axis of rotation is $5.0 \times 10^{-4} \text{ kg} \cdot \text{m}^2$. A wad of wet putty of mass 0.020 kg drops vertically onto the record from above and sticks to the edge of the record. What is the angular speed of the record immediately after the putty sticks to it?

50P. A uniform thin rod of length 0.50 m and mass 4.0 kg can rotate in a horizontal plane about a vertical axis through its center. The rod is at rest when a 3.0 g bullet traveling in the horizontal plane of the rod is fired into one end of the rod. As viewed from above, the direction of the bullet's velocity makes an angle of 60° with

Fig. 12-42 Problem 50.

the rod (Fig. 12-42). If the bullet lodges in the rod and the angular velocity of the rod is 10 rad/s immediately after the collision, what is the bullet's speed just before impact?

51P*. Two 2.00 kg balls are attached to the ends of a thin rod of negligible mass, 50.0 cm long. The rod is free to rotate in a vertical plane without friction about a horizontal axis through its center. With the rod initially horizontal (Fig. 12-43), a 50.0 g wad of wet putty drops onto one of the balls, hitting it with a speed of 3.00 m/s and then sticking to it. (a) What is the angular speed of the system just after the putty wad hits? (b) What is the ratio of the kinetic

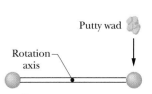

Fig. 12-43 Problem 51.

energy of the entire system after the collision to that of the putty wad just before? (c) Through what angle will the system rotate until it momentarily stops? ssm

52P. A cockroach of mass m lies on the rim of a uniform disk of mass $10.0m$ that can rotate freely about its center like a merry-go-round. Initially the cockroach and disk rotate together with an angular velocity of ω_0. Then the cockroach walks halfway to the center of the disk. (a) What is the change $\Delta\omega$ in the angular velocity of the cockroach–disk system? (b) What is the ratio K/K_0 of the new kinetic energy of the system to its initial kinetic energy? (c) What accounts for the change in the kinetic energy?

53P. If Earth's polar ice caps fully melted and the water returned to the oceans, the oceans would be deeper by about 30 m. What effect would this have on Earth's rotation? Make an estimate of the resulting change in the length of the day. (Concern has been expressed that warming of the atmosphere resulting from industrial pollution could cause the ice caps to melt.)

54P. A horizontal platform in the shape of a circular disk rotates on a frictionless bearing about a vertical axle through the center of the disk. The platform has a mass of 150 kg, a radius of 2.0 m, and a rotational inertia of $300 \text{ kg} \cdot \text{m}^2$ about the axis of rotation. A 60 kg student walks slowly from the rim of the platform toward the center. If the angular speed of the system is 1.5 rad/s when the student starts at the rim, what is the angular speed when she is 0.50 m from the center?

55P. A uniform disk of mass $10m$ and radius $3.0r$ can rotate freely about its fixed center like a merry-go-round. A smaller uniform disk of mass m and radius r lies on top of the larger disk, concentric with it. Initially the two disks rotate together with an angular velocity of 20 rad/s. Then a slight disturbance causes the smaller disk to slide outward across the larger disk, until the outer edge of the smaller disk catches on the outer edge of the larger disk. Afterward, the two disks again rotate together (without further sliding). (a) What then is their angular velocity about the center of the larger disk? (b) What is the ratio K/K_0 of the new kinetic energy of the two-disk system to the system's initial kinetic energy?

56P. A 30 kg child stands on the edge of a stationary merry-go-round of mass 100 kg and radius 2.0 m. The rotational inertia of the merry-go-round about its axis

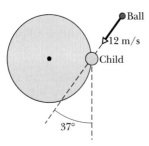
Fig. 12-44 Problem 56.

of rotation is $150 \text{ kg} \cdot \text{m}^2$. The child catches a ball of mass 1.0 kg thrown by a friend. Just before the ball is caught, it has a horizontal velocity of 12 m/s that makes an angle of 37° with a line tangent to the outer edge of the merry-go-round, as shown in the overhead view of Fig. 12-44. What is the angular speed of the merry-go-round just after the ball is caught?

57P. In Fig. 12-45, a 1.0 g bullet is fired into a 0.50 kg block that is mounted on the end of a 0.60 m nonuniform rod of mass 0.50 kg. The block–rod–bullet system then rotates about a fixed axis at point A. The rotational inertia of the rod alone about A is $0.060 \text{ kg} \cdot \text{m}^2$. Assume the block is small enough to treat as a particle on the end of the rod. (a) What is the rotational inertia of the block–rod–bullet system about point A? (b) If the angular speed of the system about A just after the bullet's impact is 4.5 rad/s, what is the speed of the bullet just before the impact?

Fig. 12-45 Problem 57.

58P. In Fig. 12-46, a uniform rod (length = 0.60 m, mass = 1.0 kg) rotates about an axis through one end, with a rotational inertia of $0.12 \text{ kg} \cdot \text{m}^2$. As the rod swings through its lowest position, the end of the rod collides with a small 0.20 kg putty wad that sticks to the end of the rod. If the angular speed of the rod just before the collision is 2.4 rad/s, what is the angular speed of the rod–putty system immediately after the collision?

Fig. 12-46 Problem 58.

59P*. The particle of mass m in Fig. 12-47 slides down the frictionless surface through height h and collides with the uniform vertical rod (of mass M and length d), sticking to it. The rod pivots about point O through the angle θ before momentarily stopping. Find θ.

Fig. 12-47 Problem 59.

NEW PROBLEMS

N1. A wheel rotates clockwise about its central axis with an angular momentum of 600 kg · m²/s. At time $t = 0$, a torque of magnitude 50 N · m is applied to the wheel to reverse the rotation. When does the wheel have an angular speed of zero?

N2. A Texas cockroach first rides at the center of a circular disk that rotates freely like a merry-go-round without external torques. It then walks out to the edge of the disk, at radius R. Figure 12N-1 gives the angular speed ω of the cockroach–disk system during the outward walk. When the cock-roach is on the edge at radius R, what is the ratio of its rotational inertia to that of the disk, both calculated about the rotation axis?

Fig. 12N-1 Problem N2.

N3. At time t, $\vec{r} = 4.0t^2\hat{i} - (2.0t + 6.0t^2)\hat{j}$ gives the position of a 3.0 kg particle relative to the origin of an xy coordinate system ($\vec{r}$ is in meters and t is in seconds). (a) Find an expression for the torque acting on the particle relative to the origin. (b) Is the magnitude of the particle's angular momentum relative to the origin increasing, decreasing, or remaining the same?

N4. *Nonuniform ball.* In Fig. 12N-2, a ball of mass M and radius R smoothly rolls from rest, at a height of $h = 0.36$ m, along a ramp and onto a circular loop of radius 0.48 m. At the bottom of the loop, the magnitude of the normal force on the ball is

Fig. 12N-2 Problem N4.

$2.00Mg$. The ball consists of an outer spherical shell with a certain uniform density (mass per unit volume) that is glued to a central solid sphere with a different uniform density. The rotational inertia of the ball can be expressed in the general form $I = \beta MR^2$, but β is not 0.4 as for a ball with a single uniform density. Determine β.

N5. A 2.5 kg particle that is moving horizontally over a floor with velocity $(-3.00 \text{ m/s})\hat{j}$ undergoes a totally inelastic collision with a 4.00 kg particle that is moving horizontally over the floor with velocity $(4.50 \text{ m/s})\hat{i}$. The collision occurs at xy coordinates $(-0.50 \text{ m}, -0.10 \text{ m})$. After the collision, what are the magnitude and direction of the angular momentum of the stuck-together particles with respect to the origin?

N6. *Nonuniform cylinder.* In Fig. 12N-3, a cylinder of mass M and radius R smoothly rolls from rest along a ramp and onto a final horizontal section. From there it rolls off the ramp and lands on a floor at a horizontal distance of $d = 0.506$ m from the end of the ramp. The initial height of the cylinder is $H = 0.90$ m; the height h of the ramp is 0.10 m. The cylinder consists of an outer cylindrical shell with a certain uniform density (mass per unit volume) that is glued to a central cylinder with a different uniform density. The rotational inertia of the cylinder can be expressed in the general form $I = \beta MR^2$, but β is not 0.5 as for a cylinder with a single uniform density. Determine β.

Fig. 12N-3 Problem N6.

N7. Figure 12N-4 gives the torque τ that acts on an initially stationary disk that can rotate about its center like a merry-go-round. What is the angular momentum of the disk about the rotation axis at times (a) $t = 7.0$ s and (b) $t = 20$ s?

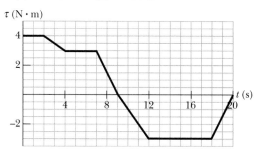

Fig. 12N-4 Problem N7.

N8. In Fig. 12N-5, a small, solid, uniform ball is to be shot from point P so that it rolls smoothly along a horizontal path, up along a ramp and onto a plateau. Then it leaves the plateau horizontally to land on a game board, at a horizontal distance of d from the plateau. The vertical heights are $h_1 = 5.00$ cm and $h_2 = 1.60$ cm. With what speed must the ball be shot at point P for it to land at $d = 6.00$ cm?

Fig. 12N-5 Problem N8.

N9. Figure 12N-6 shows the potential energy $U(x)$ of a solid ball that can roll along an x axis. The ball is uniform, rolls smoothly, and has a mass of 0.400 kg. It is released at $x = 7.0$ m with a mechanical energy of 75 J. (a) If it is initially headed to the left, can it reach $x = 0$? If so, what is its speed there; if not, where is its turning point? (b) If it is initially headed to the right, can it reach $x = 13$ m? If so, what is its speed there; if not, where is its turning point?

Fig. 12N-6 Problem N9.

N10. Figure 12N-7 shows an overhead view of a ring that can rotate about its center like a merry-go-round. Its outer radius R_2 is 0.80 m, its inner radius R_1 is $R_2/2$, its mass M is 8.00 kg, and the mass of the crossbars at its center is negligible. It initially rotates at an angular speed of 8.00 rad/s with a cat of mass $m = M/4$ on its outer edge, at radius R_2. By how much does the cat increase the kinetic energy of the cat–ring system if the cat crawls to the inner edge, at radius R_1?

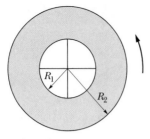

Fig. 12N-7 Problem N10.

N11. (a) In Sample Problem 11-6, when the rotating disk exploded, how much angular momentum, calculated about the rotation axis, was released to the surroundings of the disk? (b) If we assume that (most) of the pieces of the disk were stopped within 0.025 s after the explosion, what was the magnitude of the average torque acting on those pieces, calculated about the rotation axis?

N12. Figure 12N-8 gives the speed v versus time t for an object of mass 0.500 kg and radius 6.00 cm that rolls smoothly down a 30° ramp. What is the rotational inertia of the object?

Fig. 12N-8 Problem N12.

N13. A uniform block of granite in the shape of a book has face dimensions of 20 cm and 15 cm and a thickness of 1.2 cm. The density (mass per unit volume) of granite is 2.64 g/cm³. The block rotates around an axis that is perpendicular to its face and half way between its center and a corner. Its angular momentum about that axis is 0.104 kg · m²/s. What is its rotational kinetic energy about that axis?

N14. At the instant the displacement of a 2.00 kg object relative to the origin is $\vec{d} = (2.00 \text{ m})\hat{i} + (4.00 \text{ m})\hat{j} - (3.00 \text{ m})\hat{k}$, its velocity is $\vec{v} = -(6.00 \text{ m/s})\hat{i} + (3.00 \text{ m/s})\hat{j} + (3.00 \text{ m/s})\hat{k}$, and it is subject to a force $\vec{F} = (6.00 \text{ N})\hat{i} - (8.00 \text{ N})\hat{j} + (4.00 \text{ N})\hat{k}$. Find (a) the acceleration of the object, (b) the angular momentum of the object about the origin, (c) the torque about the origin acting on the object, and (d) the angle between the velocity of the object and the force acting on the object.

N15. A particle moves through an xyz coordinate system while a force acts on it. When the particle has the position vector $\vec{r} = (2.00 \text{ m})\hat{i} - (3.00 \text{ m})\hat{j} + (2.00 \text{ m})\hat{k}$, the force is $\vec{F} = F_x\hat{i} + (7.00 \text{ N})\hat{j} - (6.00 \text{ N})\hat{k}$ and the corresponding torque about the origin is $\vec{\tau} = (4.00 \text{ N} \cdot \text{m})\hat{i} + (2.00 \text{ N} \cdot \text{m})\hat{j} - (1.00 \text{ N} \cdot \text{m})\hat{k}$. Determine F_x.

N16. Figure 12N-9 shows three rotating, uniform disks that are coupled by belts. One belt runs around the rims of disks A and C. Another belt runs around a central hub on disk A and the rim of disk B. The belts move smoothly without slippage on the rims and hub. Disk A has radius R; its hub has radius $R/2$; disk B has radius $R/4$; and disk C has radius $2R$. Disks B and C have the same density (mass per unit volume) and thickness. What is the ratio of the magnitudes of the angular momentum of disk C to that of disk B?

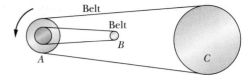

Fig. 12N-9 Problem N16.

N17. If we are given r, p, and ϕ, we can calculate the angular momentum of a particle from Eq. 12N-19. Sometimes, however, we are given the components (x, y, z) of $\vec{r}$ and (v_x, v_y, v_z) of $\vec{v}$ instead. (a) Show that the components of $\vec{\ell}$ along the x, y, and z axes are then given by $\ell_x = m(yv_z - zv_y)$, $\ell_y = m(zv_x - xv_z)$, and $\ell_z = m(xv_y - yv_x)$. (b) Show that if the particle moves only in the xy plane, the angular momentum vector has only a z component.

N18. In Fig. 12N-10, a thin, uniform rod of length $L = 0.800$ m and mass M rotates horizontally about an axis through its center, at an angular velocity 20.0 rad/s. A particle with mass $M/3.00$ is at-

Fig. 12N-10 Problem N18.

tached to one end. The particle is then ejected from the rod by a small explosion. The explosion sends the particle along a path that is perpendicular to the rod at the instant of the explosion, with a speed that is 6.00 m/s greater than the speed of the end of the rod after the explosion. What is the speed of the ejected particle?

N19. A uniform solid ball rolls smoothly along a floor, then up a ramp at 15°. It momentarily stops when it has rolled 1.50 m along the ramp. What was its initial speed?

N20. Figure 12N-11 shows a thin, uniform rod of length 0.600 m and mass M that is rotating horizontally counterclockwise about an axis through its center, at an angular velocity of 80.0 rad/s. A particle of mass $M/3.00$ and speed 40.0 m/s hits and then sticks to the rod. The particle's path is perpendicular to the rod at the instant of the hit, at a distance d from the rod's center. (a) At what value of d are the rod and particle stationary after the hit? (b) In which direction do the rod and particle rotate if d has a greater value?

Fig. 12N-11 Problem N20.

N21. In Fig. 12N-12, a 0.400 kg ball is shot directly upward with an initial speed of 40.0 m/s. What is its angular momentum about point P, at a horizontal distance of 2.00 m from the launch point, when the ball is (a) at its maximum height and (b) half way back to the ground? What is the torque on the ball about point P due to the gravitational force when the ball is (c) at its maximum height and (d) half way back to the ground?

Fig. 12N-12 Problem N21.

APPENDIX A
The International System of Units (SI)*

1. The SI Base Units

Quantity	Name	Symbol	Definition
length	meter	m	". . . the length of the path traveled by light in vacuum in 1/299,792,458 of a second." (1983)
mass	kilogram	kg	". . . this prototype [a certain platinum–iridium cylinder] shall henceforth be considered to be the unit of mass." (1889)
time	second	s	". . . the duration of 9,192,631,770 periods of the radiation corresponding to the transition between the two hyperfine levels of the ground state of the cesium-133 atom." (1967)
electric current	ampere	A	". . . that constant current which, if maintained in two straight parallel conductors of infinite length, of negligible circular cross section, and placed 1 meter apart in vacuum, would produce between these conductors a force equal to 2×10^{-7} newton per meter of length." (1946)
thermodynamic temperature	kelvin	K	". . . the fraction 1/273.16 of the thermodynamic temperature of the triple point of water." (1967)
amount of substance	mole	mol	". . . the amount of substance of a system which contains as many elementary entities as there are atoms in 0.012 kilogram of carbon-12." (1971)
luminous intensity	candela	cd	". . . the luminous intensity, in the perpendicular direction, of a surface of 1/600,000 square meter of a blackbody at the temperature of freezing platinum under a pressure of 101.325 newtons per square meter." (1967)

*Adapted from "The International System of Units (SI)," National Bureau of Standards Special Publication 330, 1972 edition. The definitions above were adopted by the General Conference of Weights and Measures, an international body, on the dates shown. In this book we do not use the candela.

2. Some SI Derived Units

Quantity	Name of Unit	Symbol	
area	square meter	m^2	
volume	cubic meter	m^3	
frequency	hertz	Hz	s^{-1}
mass density (density)	kilogram per cubic meter	kg/m^3	
speed, velocity	meter per second	m/s	
angular velocity	radian per second	rad/s	
acceleration	meter per second per second	m/s^2	
angular acceleration	radian per second per second	rad/s^2	
force	newton	N	$kg \cdot m/s^2$
pressure	pascal	Pa	N/m^2
work, energy, quantity of heat	joule	J	$N \cdot m$
power	watt	W	J/s
quantity of electric charge	coulomb	C	$A \cdot s$
potential difference, electromotive force	volt	V	W/A
electric field strength	volt per meter (or newton per coulomb)	V/m	N/C
electric resistance	ohm	Ω	V/A
capacitance	farad	F	$A \cdot s/V$
magnetic flux	weber	Wb	$V \cdot s$
inductance	henry	H	$V \cdot s/A$
magnetic flux density	tesla	T	Wb/m^2
magnetic field strength	ampere per meter	A/m	
entropy	joule per kelvin	J/K	
specific heat	joule per kilogram kelvin	$J/(kg \cdot K)$	
thermal conductivity	watt per meter kelvin	$W/(m \cdot K)$	
radiant intensity	watt per steradian	W/sr	

3. The SI Supplementary Units

Quantity	Name of Unit	Symbol
plane angle	radian	rad
solid angle	steradian	sr

Some Fundamental Constants of Physics*

Constant	Symbol	Computational Value	Best (1998) Value Value[a]	Best (1998) Value Uncertainty[b]
Speed of light in a vacuum	c	3.00×10^8 m/s	2.997 924 58	exact
Elementary charge	e	1.60×10^{-19} C	1.602 176 462	0.039
Gravitational constant	G	6.67×10^{-11} m³/s²·kg	6.673	1500
Universal gas constant	R	8.31 J/mol·K	8.314 472	1.7
Avogadro constant	N_A	6.02×10^{23} mol^{-1}	6.022 141 99	0.079
Boltzmann constant	k	1.38×10^{-23} J/K	1.380 650 3	1.7
Stefan–Boltzmann constant	σ	5.67×10^{-8} W/m²·K⁴	5.670 400	7.0
Molar volume of ideal gas at STP[d]	V_m	2.27×10^{-2} m³/mol	2.271 098 1	1.7
Permittivity constant	ϵ_0	8.85×10^{-12} F/m	8.854 187 817 62	exact
Permeability constant	μ_0	1.26×10^{-6} H/m	1.256 637 061 43	exact
Planck constant	h	6.63×10^{-34} J·s	6.626 068 76	0.078
Electron mass[c]	m_e	9.11×10^{-31} kg	9.109 381 88	0.079
		5.49×10^{-4} u	5.485 799 110	0.0021
Proton mass[c]	m_p	1.67×10^{-27} kg	1.672 621 58	0.079
		1.0073 u	1.007 276 466 88	1.3×10^{-4}
Ratio of proton mass to electron mass	m_p/m_e	1840	1836.152 667 5	0.0021
Electron charge-to-mass ratio	e/m_e	1.76×10^{11} C/kg	1.758 820 174	0.040
Neutron mass[c]	m_n	1.68×10^{-27} kg	1.674 927 16	0.079
		1.0087 u	1.008 664 915 78	5.4×10^{-4}
Hydrogen atom mass[c]	m_{1H}	1.0078 u	1.007 825 031 6	0.0005
Deuterium atom mass[c]	m_{2H}	2.0141 u	2.014 101 777 9	0.0005
Helium atom mass[c]	m_{4He}	4.0026 u	4.002 603 2	0.067
Muon mass	m_μ	1.88×10^{-28} kg	1.883 531 09	0.084
Electron magnetic moment	μ_e	9.28×10^{-24} J/T	9.284 763 62	0.040
Proton magnetic moment	μ_p	1.41×10^{-26} J/T	1.410 606 663	0.041
Bohr magneton	μ_B	9.27×10^{-24} J/T	9.274 008 99	0.040
Nuclear magneton	μ_N	5.05×10^{-27} J/T	5.050 783 17	0.040
Bohr radius	r_B	5.29×10^{-11} m	5.291 772 083	0.0037
Rydberg constant	R	1.10×10^7 m^{-1}	1.097 373 156 854 8	7.6×10^{-6}
Electron Compton wavelength	λ_C	2.43×10^{-12} m	2.426 310 215	0.0073

[a]Values given in this column should be given the same unit and power of 10 as the computational value.

[b]Parts per million.

[c]Masses given in u are in unified atomic mass units, where 1 u = 1.660 538 73 $\times$ 10^{-27} kg.

[d]STP means standard temperature and pressure: 0°C and 1.0 atm (0.1 MPa).

*The values in this table were selected from the 1998 CODATA recommended values (www.physics.nist.gov).

Some Astronomical Data

Some Distances from Earth

To the Moon*	3.82×10^8 m	To the center of our galaxy	2.2×10^{20} m
To the Sun*	1.50×10^{11} m	To the Andromeda Galaxy	2.1×10^{22} m
To the nearest star (Proxima Centauri)	4.04×10^{16} m	To the edge of the observable universe	$\sim 10^{26}$ m

*Mean distance.

The Sun, Earth, and the Moon

Property	Unit	Sun		Earth	Moon
Mass	kg	1.99×10^{30}		5.98×10^{24}	7.36×10^{22}
Mean radius	m	6.96×10^8		6.37×10^6	1.74×10^6
Mean density	kg/m³	1410		5520	3340
Free-fall acceleration at the surface	m/s²	274		9.81	1.67
Escape velocity	km/s	618		11.2	2.38
Period of rotation[a]	—	37 d at poles[b]	26 d at equator[b]	23 h 56 min	27.3 d
Radiation power[c]	W	3.90×10^{26}			

[a]Measured with respect to the distant stars.
[b]The Sun, a ball of gas, does not rotate as a rigid body.
[c]Just outside Earth's atmosphere solar energy is received, assuming normal incidence, at the rate of 1340 W/m².

Some Properties of the Planets

	Mercury	Venus	Earth	Mars	Jupiter	Saturn	Uranus	Neptune	Pluto
Mean distance from Sun, 10^6 km	57.9	108	150	228	778	1430	2870	4500	5900
Period of revolution, y	0.241	0.615	1.00	1.88	11.9	29.5	84.0	165	248
Period of rotation,[a] d	58.7	−243[b]	0.997	1.03	0.409	0.426	−0.451[b]	0.658	6.39
Orbital speed, km/s	47.9	35.0	29.8	24.1	13.1	9.64	6.81	5.43	4.74
Inclination of axis to orbit	<28°	≈3°	23.4°	25.0°	3.08°	26.7°	97.9°	29.6°	57.5°
Inclination of orbit to Earth's orbit	7.00°	3.39°		1.85°	1.30°	2.49°	0.77°	1.77°	17.2°
Eccentricity of orbit	0.206	0.0068	0.0167	0.0934	0.0485	0.0556	0.0472	0.0086	0.250
Equatorial diameter, km	4880	12100	12800	6790	143000	120000	51800	49500	2300
Mass (Earth = 1)	0.0558	0.815	1.000	0.107	318	95.1	14.5	17.2	0.002
Density (water = 1)	5.60	5.20	5.52	3.95	1.31	0.704	1.21	1.67	2.03
Surface value of g,[c] m/s²	3.78	8.60	9.78	3.72	22.9	9.05	7.77	11.0	0.5
Escape velocity,[c] km/s	4.3	10.3	11.2	5.0	59.5	35.6	21.2	23.6	1.1
Known satellites	0	0	1	2	16 + ring	18 + rings	17 + rings	8 + rings	1

[a]Measured with respect to the distant stars.
[b]Venus and Uranus rotate opposite their orbital motion.
[c]Gravitational acceleration measured at the planet's equator.

APPENDIX D
Conversion Factors

Conversion factors may be read directly from these tables. For example, 1 degree $= 2.778 \times 10^{-3}$ revolutions, so $16.7° = 16.7 \times 2.778 \times 10^{-3}$ rev. The SI units are fully capitalized. Adapted in part from G. Shortley and D. Williams, *Elements of Physics,* 1971, Prentice-Hall, Englewood Cliffs, NJ.

Plane Angle

	°	′	″	RADIAN	rev
1 degree = 1		60	3600	1.745×10^{-2}	2.778×10^{-3}
1 minute = 1.667×10^{-2}		1	60	2.909×10^{-4}	4.630×10^{-5}
1 second = 2.778×10^{-4}		1.667×10^{-2}	1	4.848×10^{-6}	7.716×10^{-7}
1 RADIAN = 57.30		3438	2.063×10^5	1	0.1592
1 revolution = 360		2.16×10^4	1.296×10^6	6.283	1

Solid Angle

1 sphere = 4π steradians = 12.57 steradians

Length

	cm	METER	km	in.	ft	mi
1 centimeter = 1		10^{-2}	10^{-5}	0.3937	3.281×10^{-2}	6.214×10^{-6}
1 METER = 100		1	10^{-3}	39.37	3.281	6.214×10^{-4}
1 kilometer = 10^5		1000	1	3.937×10^4	3281	0.6214
1 inch = 2.540		2.540×10^{-2}	2.540×10^{-5}	1	8.333×10^{-2}	1.578×10^{-5}
1 foot = 30.48		0.3048	3.048×10^{-4}	12	1	1.894×10^{-4}
1 mile = 1.609×10^5		1609	1.609	6.336×10^4	5280	1

1 angström $= 10^{-10}$ m
1 nautical mile = 1852 m
  = 1.151 miles = 6076 ft

1 fermi $= 10^{-15}$ m
1 light-year $= 9.460 \times 10^{12}$ km
1 parsec $= 3.084 \times 10^{13}$ km

1 fathom = 6 ft
1 Bohr radius $= 5.292 \times 10^{-11}$ m
1 yard = 3 ft

1 rod = 16.5 ft
1 mil $= 10^{-3}$ in.
1 nm $= 10^{-9}$ m

Area

	METER^2	cm^2	ft^2	in.^{2}
1 SQUARE METER = 1		10^4	10.76	1550
1 square centimeter = 10^{-4}		1	1.076×10^{-3}	0.1550
1 square foot = 9.290×10^{-2}		929.0	1	144
1 square inch = 6.452×10^{-4}		6.452	6.944×10^{-3}	1

1 square mile $= 2.788 \times 10^7$ ft^2 = 640 acres
1 barn $= 10^{-28}$ m^2

1 acre = 43 560 ft^2
1 hectare $= 10^4$ m^2 = 2.471 acres

Volume

	METER^3	cm^3	L	ft^3	in.^{3}
1 CUBIC METER = 1		10^6	1000	35.31	6.102×10^4
1 cubic centimeter = 10^{-6}		1	1.000×10^{-3}	3.531×10^{-5}	6.102×10^{-2}
1 liter = 1.000×10^{-3}		1000	1	3.531×10^{-2}	61.02
1 cubic foot = 2.832×10^{-2}		2.832×10^4	28.32	1	1728
1 cubic inch = 1.639×10^{-5}		16.39	1.639×10^{-2}	5.787×10^{-4}	1

1 U.S. fluid gallon = 4 U.S. fluid quarts = 8 U.S. pints = 128 U.S. fluid ounces = 231 in.^{3}

1 British imperial gallon = 277.4 in.^{3} = 1.201 U.S. fluid gallons

Mass

Quantities in the colored areas are not mass units but are often used as such. When we write, for example, 1 kg "=" 2.205 lb, this means that a kilogram is a *mass* that *weighs* 2.205 pounds at a location where g has the standard value of 9.80665 m/s^2.

	g	KILOGRAM	slug	u	oz	lb	ton
1 gram = 1		0.001	6.852×10^{-5}	6.022×10^{23}	3.527×10^{-2}	2.205×10^{-3}	1.102×10^{-6}
1 KILOGRAM = 1000		1	6.852×10^{-2}	6.022×10^{26}	35.27	2.205	1.102×10^{-3}
1 slug = 1.459×10^4		14.59	1	8.786×10^{27}	514.8	32.17	1.609×10^{-2}
1 atomic mass unit = 1.661×10^{-24}		1.661×10^{-27}	1.138×10^{-28}	1	5.857×10^{-26}	3.662×10^{-27}	1.830×10^{-30}
1 ounce = 28.35		2.835×10^{-2}	1.943×10^{-3}	1.718×10^{25}	1	6.250×10^{-2}	3.125×10^{-5}
1 pound = 453.6		0.4536	3.108×10^{-2}	2.732×10^{26}	16	1	0.0005
1 ton = 9.072×10^5		907.2	62.16	5.463×10^{29}	3.2×10^4	2000	1

1 metric ton = 1000 kg

Density

Quantities in the colored areas are weight densities and, as such, are dimensionally different from mass densities. See note for mass table.

	slug/ft^3	KILOGRAM/ METER^3	g/cm^3	lb/ft^3	lb/in.^{3}
1 slug per foot^3 = 1		515.4	0.5154	32.17	1.862×10^{-2}
1 KILOGRAM per METER^3 = 1.940×10^{-3}		1	0.001	6.243×10^{-2}	3.613×10^{-5}
1 gram per centimeter^3 = 1.940		1000	1	62.43	3.613×10^{-2}
1 pound per foot^3 = 3.108×10^{-2}		16.02	16.02×10^{-2}	1	5.787×10^{-4}
1 pound per inch^3 = 53.71		2.768×10^4	27.68	1728	1

Time

	y	d	h	min	SECOND
1 year = 1		365.25	8.766×10^3	5.259×10^5	3.156×10^7
1 day = 2.738×10^{-3}		1	24	1440	8.640×10^4
1 hour = 1.141×10^{-4}		4.167×10^{-2}	1	60	3600
1 minute = 1.901×10^{-6}		6.944×10^{-4}	1.667×10^{-2}	1	60
1 SECOND = 3.169×10^{-8}		1.157×10^{-5}	2.778×10^{-4}	1.667×10^{-2}	1

Speed

	ft/s	km/h	METER/SECOND	mi/h	cm/s
1 foot per second = 1	1.097	0.3048	0.6818	30.48
1 kilometer per hour = 0.9113	1	0.2778	0.6214	27.78
1 METER per SECOND = 3.281	3.6	1	2.237	100
1 mile per hour = 1.467	1.609	0.4470	1	44.70
1 centimeter per second = 3.281×10^{-2}	3.6×10^{-2}	0.01	2.237×10^{-2}	1

1 knot = 1 nautical mi/h = 1.688 ft/s   1 mi/min = 88.00 ft/s = 60.00 mi/h

Force

Force units in the colored areas are now little used. To clarify: 1 gram-force (= 1 gf) is the force of gravity that would act on an object whose mass is 1 gram at a location where g has the standard value of 9.80665 m/s^2.

	dyne	NEWTON	lb	pdl	gf	kgf
1 dyne = 1	10^{-5}	2.248×10^{-6}	7.233×10^{-5}	1.020×10^{-3}	1.020×10^{-6}
1 NEWTON = 10^5	1	0.2248	7.233	102.0	0.1020
1 pound = 4.448×10^5	4.448	1	32.17	453.6	0.4536
1 poundal = 1.383×10^4	0.1383	3.108×10^{-2}	1	14.10	1.410×10^2
1 gram-force = 980.7	9.807×10^{-3}	2.205×10^{-3}	7.093×10^{-2}	1	0.001
1 kilogram-force = 9.807×10^5	9.807	2.205	70.93	1000	1

1 ton = 2000 lb

Pressure

	atm	dyne/cm^2	inch of water	cm Hg	PASCAL	lb/in.^{2}	lb/ft^2
1 atmosphere = 1	1.013×10^6	406.8	76	1.013×10^5	14.70	2116
1 dyne per centimeter^2 = 9.869×10^{-7}	1	4.015×10^{-4}	7.501×10^{-5}	0.1	1.405×10^{-5}	2.089×10^{-3}
1 inch of water^a at 4°C = 2.458×10^{-3}	2491	1	0.1868	249.1	3.613×10^{-2}	5.202
1 centimeter of mercury^a at 0°C = 1.316×10^{-2}	1.333×10^4	5.353	1	1333	0.1934	27.85
1 PASCAL = 9.869×10^{-6}	10	4.015×10^{-3}	7.501×10^{-4}	1	1.450×10^{-4}	2.089×10^{-2}
1 pound per inch^2 = 6.805×10^{-2}	6.895×10^4	27.68	5.171	6.895×10^3	1	144
1 pound per foot^2 = 4.725×10^{-4}	478.8	0.1922	3.591×10^{-2}	47.88	6.944×10^{-3}	1

^{a}Where the acceleration of gravity has the standard value of 9.80665 m/s^2.

1 bar = 10^6 dyne/cm^2 = 0.1 MPa   1 millibar = 10^3 dyne/cm^2 = 10^2 Pa   1 torr = 1 mm Hg

Energy, Work, Heat

Quantities in the colored areas are not energy units but are included for convenience. They arise from the relativistic mass–energy equivalence formula $E = mc^2$ and represent the energy released if a kilogram or unified atomic mass unit (u) is completely converted to energy (bottom two rows) or the mass that would be completely converted to one unit of energy (rightmost two columns).

	Btu	erg	ft · lb	hp · h	JOULE	cal	kW · h	eV	MeV	kg	u
1 British thermal unit =	1	1.055×10^{10}	777.9	3.929×10^{-4}	1055	252.0	2.930×10^{-4}	6.585×10^{21}	6.585×10^{15}	1.174×10^{-14}	7.070×10^{12}
1 erg =	9.481×10^{-11}	1	7.376×10^{-8}	3.725×10^{-14}	10^{-7}	2.389×10^{-8}	2.778×10^{-14}	6.242×10^{11}	6.242×10^{5}	1.113×10^{-24}	670.2
1 foot-pound =	1.285×10^{-3}	1.356×10^{7}	1	5.051×10^{-7}	1.356	0.3238	3.766×10^{-7}	8.464×10^{18}	8.464×10^{12}	1.509×10^{-17}	9.037×10^{9}
1 horsepower-hour =	2545	2.685×10^{13}	1.980×10^{6}	1	2.685×10^{6}	6.413×10^{5}	0.7457	1.676×10^{25}	1.676×10^{19}	2.988×10^{-11}	1.799×10^{16}
1 JOULE =	9.481×10^{-4}	10^{7}	0.7376	3.725×10^{-7}	1	0.2389	2.778×10^{-7}	6.242×10^{18}	6.242×10^{12}	1.113×10^{-17}	6.702×10^{9}
1 calorie =	3.969×10^{-3}	4.186×10^{7}	3.088	1.560×10^{-6}	4.186	1	1.163×10^{-6}	2.613×10^{19}	2.613×10^{13}	4.660×10^{-17}	2.806×10^{10}
1 kilowatt-hour =	3413	3.600×10^{13}	2.655×10^{6}	1.341	3.600×10^{6}	8.600×10^{5}	1	2.247×10^{25}	2.247×10^{19}	4.007×10^{-11}	2.413×10^{16}
1 electron-volt =	1.519×10^{-22}	1.602×10^{-12}	1.182×10^{-19}	5.967×10^{-26}	1.602×10^{-19}	3.827×10^{-20}	4.450×10^{-26}	1	10^{-6}	1.783×10^{-36}	1.074×10^{-9}
1 million electron-volts =	1.519×10^{-16}	1.602×10^{-6}	1.182×10^{-13}	5.967×10^{-20}	1.602×10^{-13}	3.827×10^{-14}	4.450×10^{-20}	10^{-6}	1	1.783×10^{-30}	1.074×10^{-3}
1 kilogram =	8.521×10^{13}	8.987×10^{23}	6.629×10^{16}	3.348×10^{10}	8.987×10^{16}	2.146×10^{16}	2.497×10^{10}	5.610×10^{35}	5.610×10^{29}	1	6.022×10^{26}
1 unified atomic mass unit =	1.415×10^{-13}	1.492×10^{-3}	1.101×10^{-10}	5.559×10^{-17}	1.492×10^{-10}	3.564×10^{-11}	4.146×10^{-17}	9.320×10^{8}	932.0	1.661×10^{-27}	1

Power

	Btu/h	ft · lb/s	hp	cal/s	kW	WATT
1 British thermal unit per hour =	1	0.2161	3.929×10^{-4}	6.998×10^{-2}	2.930×10^{-4}	0.2930
1 foot-pound per second =	4.628	1	1.818×10^{-3}	0.3239	1.356×10^{-3}	1.356
1 horsepower =	2545	550	1	178.1	0.7457	745.7
1 calorie per second =	14.29	3.088	5.615×10^{-3}	1	4.186×10^{-3}	4.186
1 kilowatt =	3413	737.6	1.341	238.9	1	1000
1 WATT =	3.413	0.7376	1.341×10^{-3}	0.2389	0.001	1

Magnetic Field

	gauss	TESLA	milligauss
1 gauss =	1	10^{-4}	1000
1 TESLA =	10^{4}	1	10^{7}
1 milligauss =	0.001	10^{-7}	1

1 tesla = 1 weber/meter^2

Magnetic Flux

	maxwell	WEBER
1 maxwell =	1	10^{-8}
1 WEBER =	10^{8}	1

APPENDIX E
Mathematical Formulas

Geometry

Circle of radius r: circumference $= 2\pi r$; area $= \pi r^2$.
Sphere of radius r: area $= 4\pi r^2$; volume $= \frac{4}{3}\pi r^3$.
Right circular cylinder of radius r and height h:
    area $= 2\pi r^2 + 2\pi rh$; volume $= \pi r^2 h$.
Triangle of base a and altitude h: area $= \frac{1}{2}ah$.

Quadratic Formula

If $ax^2 + bx + c = 0$, then $x = \dfrac{-b \pm \sqrt{b^2 - 4ac}}{2a}$.

Trigonometric Functions of Angle θ

$\sin \theta = \dfrac{y}{r}$ $\cos \theta = \dfrac{x}{r}$

$\tan \theta = \dfrac{y}{x}$ $\cot \theta = \dfrac{x}{y}$

$\sec \theta = \dfrac{r}{x}$ $\csc \theta = \dfrac{r}{y}$

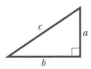

Pythagorean Theorem

In this right triangle,
$$a^2 + b^2 = c^2$$

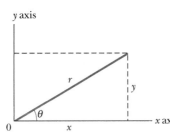

Triangles

Angles are A, B, C
Opposite sides are a, b, c
Angles $A + B + C = 180°$

$\dfrac{\sin A}{a} = \dfrac{\sin B}{b} = \dfrac{\sin C}{c}$

$c^2 = a^2 + b^2 - 2ab \cos C$

Exterior angle $D = A + C$

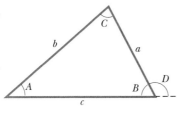

Mathematical Signs and Symbols

$=$ equals
$\approx$ equals approximately
$\sim$ is the order of magnitude of
$\neq$ is not equal to
$\equiv$ is identical to, is defined as
$>$ is greater than ($\gg$ is much greater than)
$<$ is less than ($\ll$ is much less than)
$\geq$ is greater than or equal to (or, is no less than)
$\leq$ is less than or equal to (or, is no more than)
$\pm$ plus or minus
$\propto$ is proportional to
Σ the sum of
x_{avg} the average value of x

Trigonometric Identities

$\sin(90° - \theta) = \cos \theta$

$\cos(90° - \theta) = \sin \theta$

$\sin \theta / \cos \theta = \tan \theta$

$\sin^2 \theta + \cos^2 \theta = 1$

$\sec^2 \theta - \tan^2 \theta = 1$

$\csc^2 \theta - \cot^2 \theta = 1$

$\sin 2\theta = 2 \sin \theta \cos \theta$

$\cos 2\theta = \cos^2 \theta - \sin^2 \theta = 2 \cos^2 \theta - 1 = 1 - 2 \sin^2 \theta$

$\sin(\alpha \pm \beta) = \sin \alpha \cos \beta \pm \cos \alpha \sin \beta$

$\cos(\alpha \pm \beta) = \cos \alpha \cos \beta \mp \sin \alpha \sin \beta$

$\tan(\alpha \pm \beta) = \dfrac{\tan \alpha \pm \tan \beta}{1 \mp \tan \alpha \tan \beta}$

$\sin \alpha \pm \sin \beta = 2 \sin \frac{1}{2}(\alpha \pm \beta) \cos \frac{1}{2}(\alpha \mp \beta)$

$\cos \alpha + \cos \beta = 2 \cos \frac{1}{2}(\alpha + \beta) \cos \frac{1}{2}(\alpha - \beta)$

$\cos \alpha - \cos \beta = -2 \sin \frac{1}{2}(\alpha + \beta) \sin \frac{1}{2}(\alpha - \beta)$

Binomial Theorem

$$(1 + x)^n = 1 + \frac{nx}{1!} + \frac{n(n-1)x^2}{2!} + \cdots \qquad (x^2 < 1)$$

Exponential Expansion

$$e^x = 1 + x + \frac{x^2}{2!} + \frac{x^3}{3!} + \cdots$$

Logarithmic Expansion

$$\ln(1 + x) = x - \tfrac{1}{2}x^2 + \tfrac{1}{3}x^3 - \cdots \qquad (|x| < 1)$$

Trigonometric Expansions
(θ in radians)

$$\sin \theta = \theta - \frac{\theta^3}{3!} + \frac{\theta^5}{5!} - \cdots$$

$$\cos \theta = 1 - \frac{\theta^2}{2!} + \frac{\theta^4}{4!} - \cdots$$

$$\tan \theta = \theta + \frac{\theta^3}{3} + \frac{2\theta^5}{15} + \cdots$$

Cramer's Rule

Two simultaneous equations in unknowns x and y,

$$a_1x + b_1y = c_1 \quad \text{and} \quad a_2x + b_2y = c_2,$$

have the solutions

$$x = \frac{\begin{vmatrix} c_1 & b_1 \\ c_2 & b_2 \end{vmatrix}}{\begin{vmatrix} a_1 & b_1 \\ a_2 & b_2 \end{vmatrix}} = \frac{c_1b_2 - c_2b_1}{a_1b_2 - a_2b_1}$$

and

$$y = \frac{\begin{vmatrix} a_1 & c_1 \\ a_2 & c_2 \end{vmatrix}}{\begin{vmatrix} a_1 & b_1 \\ a_2 & b_2 \end{vmatrix}} = \frac{a_1c_2 - a_2c_1}{a_1b_2 - a_2b_1}.$$

Products of Vectors

Let $\hat{i}$, $\hat{j}$, and $\hat{k}$ be unit vectors in the x, y, and z directions. Then

$$\hat{i} \cdot \hat{i} = \hat{j} \cdot \hat{j} = \hat{k} \cdot \hat{k} = 1, \qquad \hat{i} \cdot \hat{j} = \hat{j} \cdot \hat{k} = \hat{k} \cdot \hat{i} = 0,$$

$$\hat{i} \times \hat{i} = \hat{j} \times \hat{j} = \hat{k} \times \hat{k} = 0,$$

$$\hat{i} \times \hat{j} = \hat{k}, \qquad \hat{j} \times \hat{k} = \hat{i}, \qquad \hat{k} \times \hat{i} = \hat{j}.$$

Any vector $\vec{a}$ with components a_x, a_y, and a_z along the x, y, and z axes can be written as

$$\vec{a} = a_x\hat{i} + a_y\hat{j} + a_z\hat{k}.$$

Let $\vec{a}$, $\vec{b}$, and $\vec{c}$ be arbitrary vectors with magnitudes a, b, and c. Then

$$\vec{a} \times (\vec{b} + \vec{c}) = (\vec{a} \times \vec{b}) + (\vec{a} \times \vec{c})$$

$$(s\vec{a}) \times \vec{b} = \vec{a} \times (s\vec{b}) = s(\vec{a} \times \vec{b}) \quad (s = \text{a scalar}).$$

Let θ be the smaller of the two angles between $\vec{a}$ and $\vec{b}$. Then

$$\vec{a} \cdot \vec{b} = \vec{b} \cdot \vec{a} = a_xb_x + a_yb_y + a_zb_z = ab \cos \theta$$

$$\vec{a} \times \vec{b} = -\vec{b} \times \vec{a} = \begin{vmatrix} \hat{i} & \hat{j} & \hat{k} \\ a_x & a_y & a_z \\ b_x & b_y & b_z \end{vmatrix}$$

$$= \hat{i} \begin{vmatrix} a_y & a_z \\ b_y & b_z \end{vmatrix} - \hat{j} \begin{vmatrix} a_x & a_z \\ b_x & b_z \end{vmatrix} + \hat{k} \begin{vmatrix} a_x & a_y \\ b_x & b_y \end{vmatrix}$$

$$= (a_yb_z - b_ya_z)\hat{i} + (a_zb_x - b_za_x)\hat{j} + (a_xb_y - b_xa_y)\hat{k}$$

$$|\vec{a} \times \vec{b}| = ab \sin \theta$$

$$\vec{a} \cdot (\vec{b} \times \vec{c}) = \vec{b} \cdot (\vec{c} \times \vec{a}) = \vec{c} \cdot (\vec{a} \times \vec{b})$$

$$\vec{a} \times (\vec{b} \times \vec{c}) = (\vec{a} \cdot \vec{c})\vec{b} - (\vec{a} \cdot \vec{b})\vec{c}$$

Derivatives and Integrals

In what follows, the letters u and v stand for any functions of x, and a and m are constants. To each of the indefinite integrals should be added an arbitrary constant of integration. The *Handbook of Chemistry and Physics* (CRC Press Inc.) gives a more extensive tabulation.

1. $\dfrac{dx}{dx} = 1$

2. $\dfrac{d}{dx}(au) = a\dfrac{du}{dx}$

3. $\dfrac{d}{dx}(u + v) = \dfrac{du}{dx} + \dfrac{dv}{dx}$

4. $\dfrac{d}{dx}x^m = mx^{m-1}$

5. $\dfrac{d}{dx}\ln x = \dfrac{1}{x}$

6. $\dfrac{d}{dx}(uv) = u\dfrac{dv}{dx} + v\dfrac{du}{dx}$

7. $\dfrac{d}{dx}e^x = e^x$

8. $\dfrac{d}{dx}\sin x = \cos x$

9. $\dfrac{d}{dx}\cos x = -\sin x$

10. $\dfrac{d}{dx}\tan x = \sec^2 x$

11. $\dfrac{d}{dx}\cot x = -\csc^2 x$

12. $\dfrac{d}{dx}\sec x = \tan x \sec x$

13. $\dfrac{d}{dx}\csc x = -\cot x \csc x$

14. $\dfrac{d}{dx}e^u = e^u\dfrac{du}{dx}$

15. $\dfrac{d}{dx}\sin u = \cos u\dfrac{du}{dx}$

16. $\dfrac{d}{dx}\cos u = -\sin u\dfrac{du}{dx}$

1. $\displaystyle\int dx = x$

2. $\displaystyle\int au\,dx = a\int u\,dx$

3. $\displaystyle\int (u + v)\,dx = \int u\,dx + \int v\,dx$

4. $\displaystyle\int x^m\,dx = \dfrac{x^{m+1}}{m + 1}\quad (m \neq -1)$

5. $\displaystyle\int \dfrac{dx}{x} = \ln |x|$

6. $\displaystyle\int u\dfrac{dv}{dx}\,dx = uv - \int v\dfrac{du}{dx}\,dx$

7. $\displaystyle\int e^x\,dx = e^x$

8. $\displaystyle\int \sin x\,dx = -\cos x$

9. $\displaystyle\int \cos x\,dx = \sin x$

10. $\displaystyle\int \tan x\,dx = \ln |\sec x|$

11. $\displaystyle\int \sin^2 x\,dx = \tfrac{1}{2}x - \tfrac{1}{4}\sin 2x$

12. $\displaystyle\int e^{-ax}\,dx = -\dfrac{1}{a}e^{-ax}$

13. $\displaystyle\int xe^{-ax}\,dx = -\dfrac{1}{a^2}(ax + 1)e^{-ax}$

14. $\displaystyle\int x^2 e^{-ax}\,dx = -\dfrac{1}{a^3}(a^2x^2 + 2ax + 2)e^{-ax}$

15. $\displaystyle\int_0^\infty x^n e^{-ax}\,dx = \dfrac{n!}{a^{n+1}}$

16. $\displaystyle\int_0^\infty x^{2n} e^{-ax^2}\,dx = \dfrac{1 \cdot 3 \cdot 5 \cdots (2n - 1)}{2^{n+1}a^n}\sqrt{\dfrac{\pi}{a}}$

17. $\displaystyle\int \dfrac{dx}{\sqrt{x^2 + a^2}} = \ln(x + \sqrt{x^2 + a^2})$

18. $\displaystyle\int \dfrac{x\,dx}{(x^2 + a^2)^{3/2}} = -\dfrac{1}{(x^2 + a^2)^{1/2}}$

19. $\displaystyle\int \dfrac{dx}{(x^2 + a^2)^{3/2}} = \dfrac{x}{a^2(x^2 + a^2)^{1/2}}$

20. $\displaystyle\int_0^\infty x^{2n+1} e^{-ax^2}\,dx = \dfrac{n!}{2a^{n+1}}\quad (a > 0)$

21. $\displaystyle\int \dfrac{x\,dx}{x + d} = x - d\ln(x + d)$

APPENDIX F
Properties of the Elements

All physical properties are for a pressure of 1 atm unless otherwise specified.

Element	Symbol	Atomic Number Z	Molar Mass, g/mol	Density, g/cm³ at 20°C	Melting Point, °C	Boiling Point, °C	Specific Heat, J/(g·°C) at 25°C
Actinium	Ac	89	(227)	10.06	1323	(3473)	0.092
Aluminum	Al	13	26.9815	2.699	660	2450	0.900
Americium	Am	95	(243)	13.67	1541	—	—
Antimony	Sb	51	121.75	6.691	630.5	1380	0.205
Argon	Ar	18	39.948	1.6626×10^{-3}	−189.4	−185.8	0.523
Arsenic	As	33	74.9216	5.78	817 (28 atm)	613	0.331
Astatine	At	85	(210)	—	(302)	—	—
Barium	Ba	56	137.34	3.594	729	1640	0.205
Berkelium	Bk	97	(247)	14.79	—	—	—
Beryllium	Be	4	9.0122	1.848	1287	2770	1.83
Bismuth	Bi	83	208.980	9.747	271.37	1560	0.122
Bohrium	Bh	107	262.12	—	—	—	—
Boron	B	5	10.811	2.34	2030	—	1.11
Bromine	Br	35	79.909	3.12 (liquid)	−7.2	58	0.293
Cadmium	Cd	48	112.40	8.65	321.03	765	0.226
Calcium	Ca	20	40.08	1.55	838	1440	0.624
Californium	Cf	98	(251)	—	—	—	—
Carbon	C	6	12.01115	2.26	3727	4830	0.691
Cerium	Ce	58	140.12	6.768	804	3470	0.188
Cesium	Cs	55	132.905	1.873	28.40	690	0.243
Chlorine	Cl	17	35.453	3.214×10^{-3} (0°C)	−101	−34.7	0.486
Chromium	Cr	24	51.996	7.19	1857	2665	0.448
Cobalt	Co	27	58.9332	8.85	1495	2900	0.423
Copper	Cu	29	63.54	8.96	1083.40	2595	0.385
Curium	Cm	96	(247)	13.3	—	—	—
Dubnium	Db	105	262.114	—	—	—	—
Dysprosium	Dy	66	162.50	8.55	1409	2330	0.172
Einsteinium	Es	99	(254)	—	—	—	—
Erbium	Er	68	167.26	9.15	1522	2630	0.167
Europium	Eu	63	151.96	5.243	817	1490	0.163
Fermium	Fm	100	(237)	—	—	—	—
Fluorine	F	9	18.9984	1.696×10^{-3} (0°C)	−219.6	−188.2	0.753
Francium	Fr	87	(223)	—	(27)	—	—
Gadolinium	Gd	64	157.25	7.90	1312	2730	0.234
Gallium	Ga	31	69.72	5.907	29.75	2237	0.377

Element	Symbol	Atomic Number Z	Molar Mass, g/mol	Density, g/cm^3 at 20°C	Melting Point, °C	Boiling Point, °C	Specific Heat, J/(g · °C) at 25°C
Germanium	Ge	32	72.59	5.323	937.25	2830	0.322
Gold	Au	79	196.967	19.32	1064.43	2970	0.131
Hafnium	Hf	72	178.49	13.31	2227	5400	0.144
Hassium	Hs	108	(265)	—	—	—	—
Helium	He	2	4.0026	0.1664×10^{-3}	−269.7	−268.9	5.23
Holmium	Ho	67	164.930	8.79	1470	2330	0.165
Hydrogen	H	1	1.00797	0.08375×10^{-3}	−259.19	−252.7	14.4
Indium	In	49	114.82	7.31	156.634	2000	0.233
Iodine	I	53	126.9044	4.93	113.7	183	0.218
Iridium	Ir	77	192.2	22.5	2447	(5300)	0.130
Iron	Fe	26	55.847	7.874	1536.5	3000	0.447
Krypton	Kr	36	83.80	3.488×10^{-3}	−157.37	−152	0.247
Lanthanum	La	57	138.91	6.189	920	3470	0.195
Lawrencium	Lr	103	(257)	—	—	—	—
Lead	Pb	82	207.19	11.35	327.45	1725	0.129
Lithium	Li	3	6.939	0.534	180.55	1300	3.58
Lutetium	Lu	71	174.97	9.849	1663	1930	0.155
Magnesium	Mg	12	24.312	1.738	650	1107	1.03
Manganese	Mn	25	54.9380	7.44	1244	2150	0.481
Meitnerium	Mt	109	(266)	—	—	—	—
Mendelevium	Md	101	(256)	—	—	—	—
Mercury	Hg	80	200.59	13.55	−38.87	357	0.138
Molybdenum	Mo	42	95.94	10.22	2617	5560	0.251
Neodymium	Nd	60	144.24	7.007	1016	3180	0.188
Neon	Ne	10	20.183	0.8387×10^{-3}	−248.597	−246.0	1.03
Neptunium	Np	93	(237)	20.25	637	—	1.26
Nickel	Ni	28	58.71	8.902	1453	2730	0.444
Niobium	Nb	41	92.906	8.57	2468	4927	0.264
Nitrogen	N	7	14.0067	1.1649×10^{-3}	−210	−195.8	1.03
Nobelium	No	102	(255)	—	—	—	—
Osmium	Os	76	190.2	22.59	3027	5500	0.130
Oxygen	O	8	15.9994	1.3318×10^{-3}	−218.80	−183.0	0.913
Palladium	Pd	46	106.4	12.02	1552	3980	0.243
Phosphorus	P	15	30.9738	1.83	44.25	280	0.741
Platinum	Pt	78	195.09	21.45	1769	4530	0.134
Plutonium	Pu	94	(244)	19.8	640	3235	0.130
Polonium	Po	84	(210)	9.32	254	—	—
Potassium	K	19	39.102	0.862	63.20	760	0.758
Praseodymium	Pr	59	140.907	6.773	931	3020	0.197
Promethium	Pm	61	(145)	7.22	(1027)	—	—
Protactinium	Pa	91	(231)	15.37 (estimated)	(1230)	—	—
Radium	Ra	88	(226)	5.0	700	—	—
Radon	Rn	86	(222)	9.96×10^{-3} (0°C)	(−71)	−61.8	0.092
Rhenium	Re	75	186.2	21.02	3180	5900	0.134

Element	Symbol	Atomic Number Z	Molar Mass, g/mol	Density, g/cm³ at 20°C	Melting Point, °C	Boiling Point, °C	Specific Heat, J/(g·°C) at 25°C
Rhodium	Rh	45	102.905	12.41	1963	4500	0.243
Rubidium	Rb	37	85.47	1.532	39.49	688	0.364
Ruthenium	Ru	44	101.107	12.37	2250	4900	0.239
Rutherfordium	Rf	104	261.11	—	—	—	—
Samarium	Sm	62	150.35	7.52	1072	1630	0.197
Scandium	Sc	21	44.956	2.99	1539	2730	0.569
Seaborgium	Sg	106	263.118	—	—	—	—
Selenium	Se	34	78.96	4.79	221	685	0.318
Silicon	Si	14	28.086	2.33	1412	2680	0.712
Silver	Ag	47	107.870	10.49	960.8	2210	0.234
Sodium	Na	11	22.9898	0.9712	97.85	892	1.23
Strontium	Sr	38	87.62	2.54	768	1380	0.737
Sulfur	S	16	32.064	2.07	119.0	444.6	0.707
Tantalum	Ta	73	180.948	16.6	3014	5425	0.138
Technetium	Tc	43	(99)	11.46	2200	—	0.209
Tellurium	Te	52	127.60	6.24	449.5	990	0.201
Terbium	Tb	65	158.924	8.229	1357	2530	0.180
Thallium	Tl	81	204.37	11.85	304	1457	0.130
Thorium	Th	90	(232)	11.72	1755	(3850)	0.117
Thulium	Tm	69	168.934	9.32	1545	1720	0.159
Tin	Sn	50	118.69	7.2984	231.868	2270	0.226
Titanium	Ti	22	47.90	4.54	1670	3260	0.523
Tungsten	W	74	183.85	19.3	3380	5930	0.134
Un-named	Uun	110	(269)	—	—	—	—
Un-named	Uuu	111	(272)	—	—	—	—
Un-named	Uub	112	(264)	—	—	—	—
Un-named	Uut	113	—	—	—	—	—
Un-named	Unq	114	(285)	—	—	—	—
Un-named	Uup	115	—	—	—	—	—
Un-named	Uuh	116	(289)	—	—	—	—
Un-named	Uus	117	—	—	—	—	—
Un-named	Uuo	118	(293)	—	—	—	—
Uranium	U	92	(238)	18.95	1132	3818	0.117
Vanadium	V	23	50.942	6.11	1902	3400	0.490
Xenon	Xe	54	131.30	5.495×10^{-3}	−111.79	−108	0.159
Ytterbium	Yb	70	173.04	6.965	824	1530	0.155
Yttrium	Y	39	88.905	4.469	1526	3030	0.297
Zinc	Zn	30	65.37	7.133	419.58	906	0.389
Zirconium	Zr	40	91.22	6.506	1852	3580	0.276

The values in parentheses in the column of molar masses are the mass numbers of the longest-lived isotopes of those elements that are radioactive. Melting points and boiling points in parentheses are uncertain.

The data for gases are valid only when these are in their usual molecular state, such as H_2, He, O_2, Ne, etc. The specific heats of the gases are the values at constant pressure.

Source: Adapted from J. Emsley, *The Elements,* 3rd ed., 1998, Clarendon Press, Oxford. See also www.webelements.com for the latest values and newest elements.

APPENDIX G
Periodic Table of the Elements

	Metals
	Metalloids
	Nonmetals

Alkali metals
IA

Transition metals

Noble gases
0

IIA

IIIA IVA VA VIA VIIA

1																	2
H																	He

1

3	4												5	6	7	8	9	10
Li	Be												B	C	N	O	F	Ne

2

VIIIB

11	12											13	14	15	16	17	18
Na	Mg											Al	Si	P	S	Cl	Ar

3

IIIB IVB VB VIB VIIB IB IIB

19	20	21	22	23	24	25	26	27	28	29	30	31	32	33	34	35	36
K	Ca	Sc	Ti	V	Cr	Mn	Fe	Co	Ni	Cu	Zn	Ga	Ge	As	Se	Br	Kr

4

37	38	39	40	41	42	43	44	45	46	47	48	49	50	51	52	53	54
Rb	Sr	Y	Zr	Nb	Mo	Tc	Ru	Rh	Pd	Ag	Cd	In	Sn	Sb	Te	I	Xe

5

55	56	57-71	72	73	74	75	76	77	78	79	80	81	82	83	84	85	86
Cs	Ba	*	Hf	Ta	W	Re	Os	Ir	Pt	Au	Hg	Tl	Pb	Bi	Po	At	Rn

6

87	88	89-103	104	105	106	107	108	109	110	111	112	113	114	115	116	117	118
Fr	Ra	†	Rf	Db	Sg	Bh	Hs	Mt									

7

THE HORIZONTAL PERIODS

Inner transition metals

Lanthanide series *

57	58	59	60	61	62	63	64	65	66	67	68	69	70	71
La	Ce	Pr	Nd	Pm	Sm	Eu	Gd	Tb	Dy	Ho	Er	Tm	Yb	Lu

Actinide series †

89	90	91	92	93	94	95	96	97	98	99	100	101	102	103
Ac	Th	Pa	U	Np	Pu	Am	Cm	Bk	Cf	Es	Fm	Md	No	Lr

The names for elements 104 through 109 (Rutherfordium, Dubnium, Seaborgium, Bohrium, Hassium, and Meitnerium, respectively) were adopted by the International Union of Pure and Applied Chemistry (IUPAC) in 1997. Elements 110, 111, 112, 114, 116, and 118 have been discovered but, as of 2000, have not yet been named. See www.webelements.com for the latest information and newest elements.

ANSWERS
to Checkpoints and Odd-Numbered Questions, Exercises, and Problems

CHAPTER 1

EP **1.** (a) 10^9; (b) 10^{-4}; (c) 9.1×10^5 **3.** (a) 160 rods; (b) 40 chains **5.** (a) 4.00×10^4 km; (b) 5.10×10^8 km²; (c) 1.08×10^{12} km³ **7.** 1.9×10^{22} cm³ **9.** 1.1×10^3 acre-feet **11.** (a) 0.98 ft/ns; (b) 0.30 mm/ps **13.** C, D, A, B, E; the important criterion is the constancy of the daily variation, not its magnitude **15.** 0.12 AU/min **17.** 2.1 h **19.** 9.0×10^{49} **21.** (a) 10^3 kg; (b) 158 kg/s **23.** (a) 1.18×10^{-29} m³; (b) 0.282 nm **25.** (a) 60.8 W; (b) 43.3 Z **27.** 89 km **29.** $\approx 1 \times 10^{36}$ **N1.** 3.8 mg/s **N3.** 8×10^2 km **N5.** 6.0×10^{26} **N7.** 700 to 1500 **N9.** (a) 293 U.S. bushels; (b) 3.81×10^3 U.S. bushels **N11.** 9.4×10^{-3} **N13.** 5.95 km **N15.** 1.9×10^5 kg **N17.** 2×10^4 to 4×10^4 **N19.** 10.7

CHAPTER 2

CP **1.** b and c **2.** zero (zero displacement for entire trip) **3.** (check the derivative dx/dt) (a) 1 and 4; (b) 2 and 3 **4.** (see Tactic 5) (a) plus; (b) minus; (c) minus; (d) plus **5.** 1 and 4 ($a = d^2x/dt^2$ must be a constant) **6.** (a) plus (upward displacement on y axis); (b) minus (downward displacement on y axis); (c) $a = -g = -9.8$ m/s² **Q** **1.** (a) all tie; (b) 4, tie of 1 and 2, then 3 **3.** E **5.** a and c **7.** $x = t^2$ and $x = 8(t - 2) + (1.5)(t - 2)^2$ **9.** same **EP** **1.** 414 ms **3.** (a) $+40$ km/h; (b) 40 km/h **5.** (a) 73 km/h; (b) 68 km/h; (c) 70 km/h; (d) 0 **7.** (a) 0, -2, 0, 12 m; (b) $+12$ m; (c) $+7$ m/s **9.** 1.4 m **11.** (a) -6 m/s; (b) negative x direction; (c) 6 m/s; (d) first smaller, then zero, and then larger; (e) yes ($t = 2$ s); (f) no **13.** 100 m **15.** (a) velocity squared; (b) acceleration; (c) m²/s², m/s² **17.** 20 m/s², in the direction opposite to its initial velocity **19.** (a) 80 m/s; (b) 110 m/s; (c) 20 m/s² **21.** (a) m/s², m/s³; (b) 1.0 s; (c) 82 m; (d) -80 m; (e) 0, -12, -36, -72 m/s; (f) -6, -18, -30, -42 m/s² **23.** 0.10 m **25.** (a) 1.6 m/s; (b) 18 m/s **27.** (a) 3.1×10^6 s = 1.2 months; (b) 4.6×10^{13} m **29.** 1.62×10^{15} m/s² **31.** 2.5 s **33.** (a) 3.56 m/s²; (b) 8.43 m/s **35.** (a) 5.00 m/s; (b) 1.67 m/s²; (c) 7.50 m **37.** (a) 0.74 s; (b) -6.2 m/s² **39.** (a) 10.6 m; (b) 41.5 s **41.** (a) 29.4 m; (b) 2.45 s **43.** (a) 31 m/s; (b) 6.4 s **45.** (a) 3.2 s; (b) 1.3 s **47.** (a) 3.70 m/s; (b) 1.74 m/s; (c) 0.154 m **49.** 4.0 m/s **51.** 857 m/s², upward **53.** 1.26×10^3 m/s², upward **55.** 22 cm and 89 cm below the nozzle **57.** 1.5 s **59.** (a) 5.4 s; (b) 41 m/s **61.** (a) 76 m; (b) 4.2 s **63.** (a) 1.23 cm; (b) 4 times, 9 times, 16 times, 25 times **65.** 2.34 m **N1.** (a) 56.6 s; (b) 31.8 m/s **N3.** (a) 0.45 s; (b) 38 m/s; (c) 42 m/s **N5.** 25 km/h **N7.** 13 m **N9.** 2 cm/y **N11.** (a) 2.00 s; (b) 12 cm from left edge of screen; (c) 9.00 cm/s², to the left; (d) to the right; (e) to the left;

(f) 3.46 s **N13.** (a) 0.75 s; (b) 50 m **N15.** (a) 2.0 m/s²; (b) 12 m/s; (c) 45 m **N19.** (a) 30 s; (b) 300 m **N21.** (a) 15.7 m/s; (b) 12.5 m; (c) 82.3 m **N23.** (a) 54 m, 18 m/s, -12 m/s²; (b) 64 m at $t = 4.0$ s; (c) 24 m/s at $t = 2.0$ s; (d) -24 m/s²; (e) 18 m/s **N27.** (a) 1.0 cm/s; (b) 1.6 cm/s, 1.1 cm/s, 0; (c) -0.79 cm/s²; (d) 0, -0.87 cm/s², -1.2 cm/s² **N29.** (a) 18 m/s; (b) 83 m **N31.** (a) $-v_0^2/2a_1$, with a_1 negative; (b) $v_0 t_R - v_0^2/2a_2$, with a_2 negative; (c) $v_0 t_R - 0.5 v_0^2(1/a_2 - 1/a_1)$; (e) 40.0 m; (f) 42.5 m; (h) $a_1 = a_2$; (i) $-0.5 v_0^2(1/a_2 - 1/a_1)$; (k) increase; (l) 10 m; (m) 12.5 m

CHAPTER 3

CP **1.** (a) 7 m ($\vec{a}$ and $\vec{b}$ are in same direction); (b) 1 m ($\vec{a}$ and $\vec{b}$ are in opposite directions) **2.** c, d, f (components must be head-to-tail; $\vec{a}$ must extend from tail of one component to head of the other) **3.** (a) $+$, $+$; (b) $+$, $-$; (c) $+$, $+$ (draw vector from tail of $\vec{d}_1$ to head of $\vec{d}_2$) **4.** (a) 90°; (b) 0° (vectors are parallel—same direction); (c) 180° (vectors are antiparallel—opposite directions) **5.** (a) 0° or 180°; (b) 90° **Q** **1.** $\vec{A}$ and $\vec{B}$ **3.** no, but $\vec{a}$ and $-\vec{b}$ are commutative: $\vec{a} + (-\vec{b}) = (-\vec{b}) + \vec{a}$ **5.** (a) $\vec{a}$ and $\vec{b}$ are parallel; (b) $\vec{b} = 0$; (c) $\vec{a}$ and $\vec{b}$ are perpendicular **7.** all but (e) **9.** (a) 0 (vectors are parallel); (b) 0 (vectors are antiparallel) **EP** **1.** The displacements should be (a) parallel, (b) antiparallel, (c) perpendicular **3.** (a) -2.5 m; (b) -6.9 m **5.** (a) 47.2 m; (b) 122° **7.** (a) 168 cm; (b) 32.5° above the floor **9.** (a) 6.42 m; (b) no; (c) yes; (d) yes; (e) a possible answer: $(4.30 \text{ m})\hat{i} + (3.70 \text{ m})\hat{j} + (3.00 \text{ m})\hat{k}$; (f) 7.96 m **11.** (a) 370 m; (b) 36° north of east; (c) 425 m; (d) the distance **13.** (a) $(-9 \text{ m})\hat{i} + (10 \text{ m})\hat{j}$; (b) 13 m; (c) $+132°$ **15.** (a) 4.2 m; (b) 40° east of north; (c) 8.0 m; (d) 24° north of west **17.** (a) $(3.0 \text{ m})\hat{i} - (2.0 \text{ m})\hat{j} + (5.0 \text{ m})\hat{k}$; (b) $(5.0 \text{ m})\hat{i} - (4.0 \text{ m})\hat{j} - (3.0 \text{ m})\hat{k}$; (c) $(-5.0 \text{ m})\hat{i} + (4.0 \text{ m})\hat{j} + (3.0 \text{ m})\hat{k}$ **19.** (a) 38 m; (b) 320°; (c) 130 m; (d) 1.2°; (e) 62 m; (f) 130° **21.** (a) 1.59 m; (b) 12.1 m; (c) 12.2 m; (d) 82.5° **27.** (a) Put axes along cube edges, with the origin at one corner. Diagonals are $a\hat{i} + a\hat{j} + a\hat{k}$, $a\hat{i} + a\hat{j} - a\hat{k}$, $a\hat{i} - a\hat{j} - a\hat{k}$, $a\hat{i} - a\hat{j} + a\hat{k}$; (b) 54.7°; (c) $\sqrt{3}a$ **29.** (a) 30; (b) 52 **31.** 22° **35.** (b) $a^2b \sin \phi$ **37.** (a) 3.0 m; (b) 0; (c) 3.46 m; (d) 2.00 m; (e) -5.00 m; (f) 8.66 m; (g) -6.67; (h) 4.33 **N1.** 5.0 km, 4.3° south of due west **N3.** 5.39 m at 21.8° left of forward **N5.** 4.1 **N7.** (a) 0; (b) -16; (c) -9 **N9.** (a) -18.8; (b) 26.9, $+z$ direction **N11.** (a) $31\hat{k}$; (b) 8.0; (c) 33; (d) 1.6 **N13.** (a) 0; (b) 0; (c) -1; (d) west; (e) up; (f) west **N15.** (a) $(9 \text{ m})\hat{i} + (6 \text{ m})\hat{j} - (7 \text{ m})\hat{k}$; (b) 123°; (c) -3.2 m; (d) 8.2 m **N17.** (a) 15 m; (b) south; (c) 6.0 m; (d) north **N19.** (a) 103 km; (b) 60.9° north of due

west   **N21.** 0   **N23.** (a) 140°; (b) 90°; (c) 99.1°
N25. (a) −80 m; (b) 110 m; (c) 143 m; (d) +168° (counter-clockwise)

CHAPTER 4

CP   1. (a) $(8\hat{i} - 6\hat{j})$ m; (b) yes, the xy plane (no z component)
2. (draw $\vec{v}$ tangent to path, tail on path) (a) first; (b) third
3. (take second derivative with respect to time) (1) and (3) a_x and a_y are both constant and thus $\vec{a}$ is constant; (2) and (4) a_y is constant but a_x is not, thus $\vec{a}$ is not   **4.** 4 m/s³, −2 m/s, 3 m
5. (a) v_x constant; (b) v_y initially positive, decreases to zero, and then becomes progressively more negative; (c) $a_x = 0$ throughout; (d) $a_y = -g$ throughout   **6.** (a) $-(4 \text{ m/s})\hat{i}$; (b) $-(8 \text{ m/s}^2)\hat{j}$
7. (a) 0, distance not changing; (b) +70 km/h, distance increasing; (c) +80 km/h, distance decreasing   **8.** (a)–(c) increase
Q   1. (a) $(7 \text{ m})\hat{i} + (1 \text{ m})\hat{j} + (-2 \text{ m})\hat{k}$; (b) $(5 \text{ m})\hat{i} + (-3 \text{ m})\hat{j} + (1 \text{ m})\hat{k}$; (c) $(-2 \text{ m})\hat{i}$ **3.** a, b, c **5.** (a) all tie; (b) 1 and 2 tie (the rocket is shot upward), then 3 and 4 tie (it is shot into the ground!)   **7.** (a) 3, 2, 1; (b) 1, 2, 3; (c) all tie; (d) 6, 5, 4
9. (a) less; (b) unanswerable; (c) equal; (d) higher
11. (a) 2; (b) 3; (c) 1; (d) 2; (e) 3; (f) 1   **13.** (a) yes; (b) no; (c) yes   **EP   1.** (a) $(-5.0\hat{i} + 8.0\hat{j})$ m; (b) 9.4 m; (c) 122°; (e) $(8\hat{i} - 8\hat{j})$ m; (f) 11 m; (g) −45°   **3.** (a) $(-7.0\hat{i} + 12\hat{j})$ m; (b) xy plane   **5.** 7.59 km/h, 22.5° east of north   **7.** (a) $(3.00\hat{i} - 8.00\hat{j})$ m/s; (b) $(3.00\hat{i} - 16.0\hat{j})$ m/s; (c) 16.3 m/s; (d) −79.4°
9. (a) $(8t\hat{j} + \hat{k})$ m/s; (b) $8\hat{j}$ m/s²   **11.** (a) $(6.00\hat{i} - 106\hat{j})$ m; (b) $(19.0\hat{i} - 224\hat{j})$ m/s; (c) $(24.0\hat{i} - 336\hat{j})$ m/s²; (d) −85.2° to $+x$ **13.** (a) $(-1.5\hat{j})$ m/s; (b) $(4.5\hat{i} - 2.25\hat{j})$ m   **15.** (a) 45 m; (b) 22 m/s   **17.** (a) 62 ms; (b) 480 m/s   **19.** (a) 0.205 s; (b) 0.205 s; (c) 20.5 cm; (d) 61.5 cm   **21.** (a) 2.00 ns; (b) 2.00 mm; (c) 1.00×10^7 m/s; (d) 2.00×10^6 m/s
23. (a) 16.9 m; (b) 8.21 m; (c) 27.6 m; (d) 7.26 m; (e) 40.2 m; (f) 0   **25.** 4.8 cm   **29.** (a) 11 m; (b) 23 m; (c) 17 m/s; (d) 63° below horizontal   **31.** (a) 24 m/s; (b) 65° above the horizontal
33. (a) 10 s; (b) 897 m   **35.** the third   **37.** (a) 202 m/s; (b) 806 m; (c) 161 m/s; (d) −171 m/s   **39.** (a) yes; (b) 2.56 m
41. between the angles 31° and 63° above the horizontal
43. (a) 7.49 km/s; (b) 8.00 m/s²   **45.** (a) 19 m/s; (b) 35 rev/min; (c) 1.7 s   **47.** (a) 0.034 m/s²; (b) 84 min   **49.** (a) 12 s; (b) 4.1 m/s², down; (c) 4.1 m/s², up   **51.** 160 m/s²   **53.** (a) 13 m/s², eastward; (b) 13 m/s², eastward   **55.** 36 s, no   **57.** 60°
59. 32 m/s   **61.** (a) 38 knots, 1.5° east of north; (b) 4.2 h; (c) 1.5° west of south   **63.** (a) 37° west of north; (b) 62.6 s
N1. (a) 2.7 km; (b) 76° clockwise   **N3.** (a) 73 ft; (b) 7.6°; (c) 1.0 s   **N5.** (a) 2.1 m/s; (b) not accidental, horizontal launch speed is about 20% of world-class sprint speed   **N9.** (a) $y = 7.5 - 4.0t + 0.5t^2$, with y in meters and t in seconds; (b) 3.0 s, 5.0 s; (c) 3.0 s; (d) 21 m; (e) $(-1.85\hat{i} + 1.07\hat{j})$ m/s²
N11. (a) 22 m; (b) 15 s   **N13.** 240 km/h   **N15.** (a) $(1.00 \text{ m})\hat{i} - (2.00 \text{ m})\hat{j} + (1.00 \text{ m})\hat{k}$; (b) 2.45 m; (c) $(2.50 \text{ cm/s})\hat{i} - (5.00 \text{ cm/s})\hat{j} + (2.50 \text{ cm/s})\hat{k}$; (d) insufficient information
N17. 185 km/h, 22° south of west   **N19.** (a) 44 m; (b) 13 m; (c) 8.9 m   **N21.** (a) 0, 0; 2.0 m, 1.4 m; 4.0 m, 2.0 m; 6.0 m, 1.4 m; 8.0 m, 0; (b) 2.0 m/s, 1.1 m/s; 2.0 m/s, 0; 2.0 m/s, −1.1 m/s; (c) 0, −0.87 m/s²; 0, −1.2 m/s²; 0, −0.87 m/s²
N23. (8.82 m, 6.00 m)   **N25.** 143 km/h   **N27.** 8.43 m at −129°
N29. 0.421 m/s at 3.1° west of due north   **N31.** (a) 11 m;

(b) 45 m/s   **N33.** $(3.00 \text{ m/s}^2)\hat{i} + (6.00 \text{ m/s}^2)\hat{j}$ **N35.** (a) 1.5; (b) (36 m, 54 m)   **N37.** 15.8 m/s at 42.6°

CHAPTER 5

CP   1. c, d, and e ($\vec{F}_1$ and $\vec{F}_2$ must be head-to-tail, $\vec{F}_{net}$ must be from tail of one of them to head of the other)   **2.** (a) and (b) 2 N, leftward (acceleration is zero in each situation)   **3.** (a) and (b) 1, 4, 3, 2   **4.** (a) equal; (b) greater (acceleration is upward, thus net force on body must be upward)   **5.** (a) equal; (b) greater; (c) less   **6.** (a) increase; (b) yes; (c) same; (d) yes
7. (a) $F \sin \theta$; (b) increase   **8.** 0 (because now $a = -g$)
Q   1. (a) 5; (b) 7; (c) $(2 \text{ N})\hat{i}$; (d) $-(6 \text{ N})\hat{j}$; (e) fourth; (f) fourth
3. (a) 2 and 4; (b) 2 and 4   **5.** (a) 2, 3, 4; (b) 1, 3, 4; (c) 1, $+y$; 2, $+x$; 3, fourth quadrant; 4, third quadrant   **7.** (a) less; (b) greater   **9.** (a) 20 kg; (b) 18 kg; (c) 10 kg; (d) all tie; (e) 3, 2, 1   **11.** (a) 4 or 5, choose 4; (b) 2; (c) 1; (d) 4 or 5, choose 5; (e) 3; (f) 6; (g) 3 and 6; 1, 2, and 5; (h) 3 and 6; (i) 1, 2, and 5
EP   1. (a) $F_x = 1.88$ N; (b) $F_y = 0.684$ N; (c) $(1.88\hat{i} + 0.684\hat{j})$ N   **3.** 2.9 m/s²   **5.** $(3\hat{i} - 11\hat{j} + 4\hat{k})$ N   **7.** (a) $(-32\hat{i} - 21\hat{j})$ N; (b) 38 N; (c) 213° from $+x$ **9.** (a) 108 N; (b) 108 N; (c) 108 N   **11.** (a) 11 N; (b) 2.2 kg; (c) 0; (d) 2.2 kg   **13.** 16 N
15. (a) 42 N; (b) 72 N; (c) 4.9 m/s²   **17.** (a) 0.02 m/s²; (b) 8×10^4 km; (c) 2×10^3 m/s   **19.** 1.2×10^5 N   **21.** 1.5 mm
23. (a) $(285\hat{i} + 705\hat{j})$ N; (b) $285\hat{i} - 115\hat{j})$ N; (c) 307 N; (d) −22° from $+x$; (e) 3.67 m/s²; (f) −22° from $+x$
25. (a) 0.62 m/s²; (b) 0.13 m/s²; (c) 2.6 m   **27.** (a) 494 N, up; (b) 494 N, down   **29.** (a) 2.2×10^{-3} N; (b) 3.7×10^{-3} N
31. (a) 1.1 N   **33.** 1.8×10^4 N   **35.** (a) 620 N; (b) 580 N
37. (a) 3260 N; (b) 2.7×10^3 kg; (c) 1.2 m/s²   **39.** (a) 180 N; (b) 640 N   **41.** (a) 1.23 N; (b) 2.46 N; (c) 3.69 N; (d) 4.92 N; (e) 6.15 N; (f) 0.25 N   **43.** (a) 0.735 m/s²; (b) downward; (c) 20.8 N   **45.** (a) 1.18 m; (b) 0.674 s; (c) 3.50 m/s   **47.** (a) 4.9 m/s²; (b) 2.0 m/s²; (c) upward; (d) 120 N   **49.** (a) 2.18 m/s²; (b) 116 N; (c) 21.0 m/s²
51. (b) $F/(m + M)$; (c) $MF/(m + M)$; (d) $F(m + 2M)/2(m + M)$
53. $2Ma/(a + g)$ **55.** (a) 31.3 kN; (b) 24.3 kN
N1. (a) 13 597 kg; (b) 4917 L; (c) 6172 kg; (d) 20 075 L; (e) 45%   **N3.** (a) 3.5 N, due west; (b) 2.7 N, 22° west of due south   **N5.** 2.2 kg   **N7.** $(-34\hat{i} - 12\hat{j})$ N   **N9.** 47.4 N
N11. (a) $(1.70 \text{ N})\hat{i} + (3.06 \text{ N})\hat{j}$; (b) $(1.70 \text{ N})\hat{i} + (3.06 \text{ N})\hat{j}$; (c) $(2.02 \text{ N})\hat{i} + (2.71 \text{ N})\hat{j}$ **N13.** (a) $(-6.26\hat{i} - 3.23\hat{j})$ N; (b) 7.0 N, 207° counterclockwise from positive direction of x axis   **N15.** 1.2×10^6 N   **N17.** 2.4 N   **N19.** (a) 4.9×10^5 N; (b) 1.5×10^6 N   **N21.** (a) 245 m/s²; (b) 20.4 kN
N23. 8.0 cm/s²

CHAPTER 6

CP   1. (a) zero (because there is no attempt at sliding); (b) 5 N; (c) no; (d) yes; (e) 8 N   **2.** (a) same (10 N); (b) decreases; (c) decreases (because N decreases)   **3.** greater (from Sample Problem 6-5, v_t depends on $\sqrt{R}$)   **4.** ($\vec{a}$ is directed toward center of circular path) (a) $\vec{a}$ downward, $\vec{N}$ upward; (b) $\vec{a}$ and $\vec{N}$ upward   **5.** (a) same (must still match the gravitational force on the rider); (b) increases ($N = mv^2/R$); (c) increases ($f_{s,max} = \mu_s N$)   **6.** (a) $4R_1$; (b) $4R_1$ **Q   1.** (a) F_1, F_2, F_3; (b) all tie
3. (a) same; (b) increases; (c) increases; (d) no   **5.** (a) decrease; (b) decrease; (c) increase; (d) increase; (e) increase   **7.** (a) the

block's mass m; (b) equal (they are a third-law pair); (c) that on the slab is in the direction of the applied force; that on the block is in the opposite direction; (d) the slab's mass M **9.** 4, 3, then 1, 2, and 5 tie   **EP**   **1.** (a) 200 N; (b) 120 N   **3.** 0.61   **5.** (a) 190 N; (b) 0.56 m/s²   **7.** (a) 0.13 N; (b) 0.12   **9.** (a) no; (b) $(-12\hat{i} + 5\hat{j})$ N   **13.** (a) 300 N; (b) 1.3 m/s²   **15.** (a) 66 N; (b) 2.3 m/s²   **17.** (b) 3.0×10^7 N   **19.** 100 N   **21.** (a) 0; (b) 3.9 m/s² down the incline; (c) 1.0 m/s² down the incline   **23.** (a) 3.5 m/s²; (b) 0.21 N; (c) blocks move independently   **25.** 490 N   **27.** (a) 6.1 m/s², leftward; (b) 0.98 m/s², leftward   **29.** $g(\sin \theta - \sqrt{2}\mu_k \cos \theta)$ **31.** 9.9 s   **33.** 6200 N   **35.** 2.3   **37.** about 48 km/h   **39.** 21 m   **41.** $\sqrt{Mgr/m}$ **43.** (a) light; (b) 778 N; (c) 223 N   **45.** 2.2 km   **47.** (b) 8.74 N; (c) 37.9 N, radially inward; (d) 6.45 m/s   **N1.** (a) no, 222 N; (b) no, 334 N; (c) yes, 311 N; (d) yes, 311 N   **N3.** (a) 7.5 m/s² down the slope; (b) 9.5 m/s² down the slope   **N5.** 36 m   **N7.** (a) $\mu_k mg/(\sin \theta - \mu_k \cos \theta)$; (b) $\theta_0 = \tan^{-1} \mu_s$ **N9.** (a) 27 N; (b) 3.0 m/s²   **N11.** 8.8 N   **N13.** 0.74   **N15.** (a) 3.0 N, up the incline; (b) 3.0 N, up the incline; (c) 1.6 N, up the incline; (d) 4.4 N, up the incline; (e) 1.0 N, down the incline   **N17.** (a) 10 s; (b) 4.9×10^2 N; (c) 1.1×10^3 N   **N19.** (a) 0.40 N; (b) 1.9 s   **N21.** 147 m/s   **N23.** (a) 84.2 N (slightly more than $f_{s,\text{max}}$); (b) 52.8 N; (c) 1.9 m/s²   **N25.** 0.54   **N27.** 118 N   **N29.** (a) 11°; (b) 0.19   **N31.** 0.60   **N33.** (a) does not move, 12 N, down; (b) does not move, 10 N, up; (c) does not move, 26 N, down; (d) moves up, 23 N, down; (e) does not move, 32 N, up; (f) moves down, 23 N, up   **N35.** (a) 1.74 m/s²; (b) 0.33

CHAPTER 7

CP   **1.** (a) decrease; (b) same; (c) negative, zero   **2.** d, c, b, a **3.** (a) same; (b) smaller   **4.** (a) positive; (b) negative; (c) zero   **5.** zero   **Q**   **1.** all tie   **3.** (a) positive; (b) zero; (c) negative; (d) negative; (e) zero; (f) positive   **5.** (a) A, B, C; (b) C, B, A; (c) C, B, A; (d) $A, 2; B, 3; C, 1$ **7.** all tie   **9.** c, d, then a and b tie, then f, e **11.** (a) $2F_1$; (b) $2W_1$ **13.** B, C, A **EP**   **1.** 1.2×10^6 m/s   **3.** (a) 3610 J; (b) 1900 J; (c) 1.1×10^{10} J   **5.** (a) 2.9×10^7 m/s; (b) 2.1×10^{-13} J   **7.** (a) 590 J; (b) 0; (c) 0; (d) 590 J   **9.** (a) 170 N; (b) 340 m; (c) -5.8×10^4 J; (d) 340 N; (e) 170 m; (f) -5.8×10^4 J   **11.** (a) 1.50 J; (b) increases   **13.** 15.3 J   **15.** (a) 98 N; (b) 4.0 cm; (c) 3.9 J; (d) -3.9 J   **17.** (a) 1.2×10^4 J; (b) -1.1×10^4 J; (c) 1100 J; (d) 5.4 m/s   **19.** (a) $-3Mgd/4$; (b) Mgd; (c) $Mgd/4$; (d) $\sqrt{gd/2}$ **21.** (a) -0.043 J; (b) -0.13 J   **23.** (a) 6.6 m/s; (b) 4.7 m   **25.** 800 J   **27.** 0, by both methods   **29.** -6 J   **31.** 490 W   **33.** (a) 0.83 J; (b) 2.5 J; (c) 4.2 J; (d) 5.0 W   **35.** 740 W   **37.** 68 kW   **39.** (a) 1.8×10^5 ft · lb; (b) 0.55 hp   **N1.** (a) 42 J; (b) 30 J; (c) 12 J; (d) 6.48 m/s, positive direction of x axis; (e) 5.48 m/s, positive direction of x axis; (f) 3.46 m/s, positive direction of x axis   **N3.** 0.47 J   **N5.** 6.8 J   **N7.** (a) 6.0 N; (b) -2.5 N; (c) 15 N   **N9.** (a) 101 J; (b) 8.42 W   **N11.** (a) 1.20 J; (b) 1.10 m/s   **N13.** (b) $x = 3.00$ m; (c) 13.50 J; (d) $x = 4.50$ m; (e) $x = 4.50$ m   **N15.** 6.0 J   **N17.** AB: +, BC: 0, CD: −, DE: +   **N19.** 235 kW

CHAPTER 8

CP   **1.** no (consider round trip on the small loop)   **2.** 3, 1, 2 (see Eq. 8-6)   **3.** (a) all tie; (b) all tie   **4.** (a) CD, AB, BC (zero)

(check slope magnitudes); (b) positive direction of x **5.** all tie   **Q**   **1.** (a) 12 J; (b) -2 J   **3.** (a) all tie; (b) all tie   **5.** (a) 4; (b) returns to its starting point and repeats the trip; (c) 1; (d) 1   **7.** (a) fL; (b) 0.50; (c) 1.25; (d) 2.25; (e) b, center; c, right; d, left   **9.** (a) increasing; (b) decreasing; (c) decreasing; (d) constant in AB and BC, decreasing in CD **EP**   **1.** 89 N/cm   **3.** (a) 4.31 mJ; (b) -4.31 mJ; (c) 4.31 mJ; (d) -4.31 mJ; (e) all increase   **5.** (a) mgL; (b) $-mgL$; (c) 0; (d) $-mgL$; (e) mgL; (f) 0; (g) same   **7.** (a) 184 J; (b) -184 J; (c) -184 J   **9.** (a) 2.08 m/s; (b) 2.08 m/s; (c) increase   **11.** (a) $\sqrt{2gL}$; (b) $2\sqrt{gL}$; (c) $\sqrt{2gL}$; (d) all the same   **13.** (a) 260 m; (b) same; (c) decrease   **15.** (a) 21.0 m/s; (b) 21.0 m/s; (c) 21.0 m/s   **17.** (a) 0.98 J; (b) -0.98 J; (c) 3.1 N/cm   **19.** (a) 39.2 J; (b) 39.2 J; (c) 4.00 m   **21.** (a) 35 cm; (b) 1.7 m/s   **23.** (a) 4.8 m/s; (b) 2.4 m/s   **25.** 10 cm   **27.** 1.25 cm   **31.** (a) $2\sqrt{gL}$; (b) $5mg$; (c) 71°   **33.** $mgL/32$ **37.** (a) $1.12(A/B)^{1/6}$; (b) repulsive; (c) attractive   **39.** (a) 5.6 J; (b) 3.5 J   **41.** (a) 30.1 J; (b) 30.1 J; (c) 0.22   **43.** (a) -2900 J; (b) 390 J; (c) 210 N   **45.** 11 kJ   **47.** 20 ft · lb   **49.** (a) 1.5 MJ; (b) 0.51 MJ; (c) 1.0 MJ; (d) 63 m/s   **51.** (a) 67 J; (b) 67 J; (c) 46 cm   **53.** (a) 31.0 J; (b) 5.35 m/s; (c) conservative   **55.** (a) 44 m/s; (b) 0.036   **57.** (a) -0.90 J; (b) 0.46 J; (c) 1.0 m/s   **59.** 1.2 m   **63.** in the center of the flat part   **65.** (a) 216 J; (b) 1180 N; (c) 432 J; (d) motor also supplies thermal energy to crate and belt   **67.** (b) $\rho(L - x)/2$; (c) $v = v_0[2(\rho L + m_f)/(\rho L + 2m_f - \rho x)]^{0.5}$; (e) 35 m/s   **N1.** on the ramp at the right, at a height of $0.75d$ from the lower plateau   **N3.** (a) 12 m/s; (b) 11 cm   **N5.** (a) -0.80 J; (b) -0.80 J; (c) $+1.1$ J   **N7.** 181 W   **N9.** (a) no, $v = 0.95$ m/s; (b) yes, $x = 11.0$ m   **N11.** (a) 109 J; (b) 60.3 J; (c) 68.2 J; (d) 41.0 J   **N13.** (a) 7.0 J; (b) 22 J   **N15.** (a) 3500 J; (b) 3500 J   **N17.** (a) 39.6 cm; (b) 3.64 cm   **N19.** (a) 2.35×10^3 J; (b) 352 J   **N21.** 56 m/s   **N23.** (a) 2.7 J; (b) 1.8 J; (c) 0.39 m   **N25.** 54%   **N27.** (a) $U(x) = -Gm_1m_2/x$; (b) $Gm_1m_2d/x_1(x_1 + d)$ **N29.** yes, 3.5 m/s

CHAPTER 9

CP   **1.** (a) origin; (b) fourth quadrant; (c) on y axis below origin; (d) origin; (e) third quadrant; (f) origin   **2.** (a) to (c) at the center of mass, still at the origin (their forces are internal to the system and cannot move the center of mass)   **3.** (Consider slopes and Eq. 9-23.) (a) 1, 3, and then 2 and 4 tie (zero force); (b) 3   **4.** (No net external force; $\vec{P}$ conserved.) (a) 0; (b) no; (c) $-x$ **5.** (a) 500 km/h; (b) 2600 km/h; (c) 1600 km/h   **6.** (a) yes; (b) no (because of net force along y)   **Q**   **1.** (a) to (d) at the origin   **3.** (a) at the center of the sled; (b) $L/4$, to the right; (c) not at all (no net external force); (d) $L/4$, to the left; (e) L; (f) $L/2$; (g) $L/2$ **5.** (a) ac, cd, and bc; (b) bc; (c) bd and ad **7.** c, d, and then a and b tie   **9.** b, c, a   **EP**   **1.** (a) 4600 km; (b) $0.73R_e$ **3.** (a) 1.1 m; (b) 1.3 m; (c) shifts toward topmost particle   **5.** (a) -0.25 m; (b) 0   **7.** 6.8×10^{-12} m from the nitrogen atom, along axis of symmetry   **9.** (a) $H/2$; (b) $H/2$; (c) descends to lowest point and then ascends to $H/2$; (d) $\dfrac{HM}{m}\left(\sqrt{1 + \dfrac{m}{M}} - 1\right)$ **11.** 72 km/h   **13.** (a) 28 cm; (b) 2.3 m/s   **15.** 53 m   **17.** (a) halfway between the containers; (b) 26 mm toward the heavier container; (c) down; (d) $-1.6 \times$

10^{-2} m/s^2 **19.** 4.2 m   **21.** 24 km/h   **23.** (a) 7.5×10^4 J;
(b) 3.8×10^4 kg·m/s; (c) 38° south of east   **25.** (a) $(-4.0 \times 10^4 \hat{\imath})$ kg·m/s; (b) west; (c) 0   **27.** 3.0 mm/s, away from
the stone   **29.** increases by 4.4 m/s   **31.** 4400 km/h
33. (a) 7290 m/s; (b) 8200 m/s; (c) 1.271×10^{10} J;
(d) 1.275×10^{10} J   **35.** (a) 1.4×10^{-22} kg·m/s; (b) 150°;
(c) 120°; (d) 1.6×10^{-19} J   **37.** (a) 1010 m/s, 9.48° clockwise
from the $+x$ direction; (b) 3.23 MJ   **39.** 14 m/s, 135° from
the other pieces   **41.** 108 m/s   **43.** (a) 1.57×10^6 N;
(b) 1.35×10^5 kg; (c) 2.08 km/s   **45.** 2.2×10^{-3}
47. (a) 46 N; (b) none   **49.** (a) 0.2 to 0.3 MJ; (b) same amount
51. (a) 8.8 m/s; (b) 2600 J; (c) 1.6 kW   **53.** 24 W
55. (a) 860 N; (b) 2.4 m/s   **57.** (a) 2.1×10^6 kg;
(b) $\sqrt{100 + 1.5t}$ m/s; (c) $(1.5 \times 10^6)/\sqrt{100 + 1.5t}$ N;
(d) 6.7 km   **59.** 0.5 cm/s downward (the bubbles rise but the
layers descend)   **N1.** 3.4 kg   **N3.** (a) 0, 0; (b) 0
N5. (a) 40 m/s, m_1 approaching zero; (b) 60 m/s, m_1 approaching M **N9.** (a) -495 J; (b) 1.65 kN   **N11.** (a) -1; (b) 1830;
(c) 1830; (d) same   **N13.** $(-1.50$ m, -1.43 m)
N15. (a) $-4.9\hat{\jmath}$; (b) $-9.8\hat{\jmath}$, with a large, brief, positive acceleration when first coin hits; (c) $-4.9\hat{\jmath}$, with a large, brief, positive
acceleration when second coin hits; (d) 1.23 m/s; (e) 4.90 m/s;
(f) 6.13 m/s   **N17.** (a) 11.4 m/s; (b) 95.1° clockwise from $+x$
N19. 2.5×10^{-3} **N21.** 30 kg

CHAPTER 10

CP   1. (a) unchanged; (b) unchanged (see Eq. 10-4); (c) decrease
(see Eq. 10-8)   **2.** (a) zero; (b) positive (initial p_y down y; final
p_y up y); (c) positive direction of y **3.** (a) 10 kg·m/s;
(b) 14 kg·m/s; (c) 6 kg·m/s   **4.** (a) 4 kg·m/s; (b) 8 kg·m/s;
(c) 3 J   **5.** (a) 2 kg·m/s (conserve momentum along x);
(b) 3 kg·m/s (conserve momentum along y)   **Q   1.** all tie
3. b and c   **5.** (a) rightward; (b) rightward; (c) smaller
7. (a) one was stationary; (b) 2; (c) 5; (d) equal (pool player's
result)   **9.** (a) 2; (b) 1; (c) 3; (d) yes; (e) no   **EP   1.** 2.5 m/s
3. 3000 N   **5.** 67 m/s, in opposite direction   **7.** (a) 42 N·s;
(b) 2100 N   **9.** (a) $(7.4 \times 10^3 \hat{\imath} - 7.4 \times 10^3 \hat{\jmath})$ N·s;
(b) $(-7.4 \times 10^3 \hat{\imath})$ N·s; (c) 2.3×10^3 N; (d) 2.1×10^4 N;
(e) $-45°$ **11.** 10 m/s   **13.** (a) 1.0 kg·m/s; (b) 250 J;
(c) 10 N; (d) 1700 N; (e) answer for (c) includes time between
pellet collisions   **15.** 41.7 cm/s   **17.** (a) 1.8 N·s, upward in
figure; (b) 180 N, downward in figure   **19.** (a) 9.0 kg·m/s;
(b) 3000 N; (c) 4500 N; (d) 20 m/s   **21.** 3.0 m/s
23. ≈ 2 mm/y   **25.** (a) 4.6 m/s; (b) 3.9 m/s; (c) 7.5 m/s
27. (a) $mR(\sqrt{2gh} + gt)$; (b) 5.06 kg   **29.** 1.18×10^4 kg
31. (a) $mv_i/(m + M)$; (b) $M/(m + M)$ **33.** 25 cm
35. (a) 1.9 m/s, to the right; (b) yes; (c) no, total kinetic energy
would have increased   **37.** (a) 99 g; (b) 1.9 m/s; (c) 0.93 m/s
39. 7.8%   **41.** (a) 1.2 kg; (b) 2.5 m/s   **43.** (a) 100 g;
(b) 1.0 m/s   **45.** (a) 1/3; (b) 4h **47.** (a) 4.15×10^5 m/s;
(b) 4.84×10^5 m/s   **49.** (a) 41°; (b) 4.76 m/s; (c) no
51. 120°   **53.** (a) 6.9 m/s, 30° to $+x$ direction; (b) 6.9 m/s,
$-30°$ to $+x$ direction; (c) 2.0 m/s, $-x$ direction   **57.** (a) 5mg;
(b) 7mg; (c) 5 m   **N1.** (a) 2.22 m; (b) 0.556 m
N3. (a) 7.17 N·s; (b) 16.0 kg·m/s   **N5.** (a) 11%; (d) 10%;
(e) 79%   **N7.** (c) 11%; (d) 10%; (e) 79%   **N9.** 61.2 kJ
N11. (a) 30 cm; (b) 3.3 m   **N13.** (a) 0.800 kg·m/s;

(b) 0.400 kg·m/s   **N15.** (a) $(30$ kg·m/s$)\hat{\imath}$; (b) $(38$ kg·m/s$)\hat{\imath}$;
(c) (6.0 m/s)$\hat{\imath}$ **N17.** (a) -0.50 m; (b) -1.8 cm; (c) 0.50 m
N19. (a) $2mv/\Delta t$; (b) 575 N   **N21.** (a) $(-3.8$ m/s$)\hat{\imath}$; (b) $(7.2$ m/s$)\hat{\imath}$
N23. 2.6 m

CHAPTER 11

CP   1. (b) and (c)   **2.** (a) and (d) ($\alpha = d^2\theta/dt^2$ must be a constant)   **3.** (a) yes; (b) no; (c) yes; (d) yes   **4.** all tie   **5.** 1, 2, 4,
3 (see Eq. 11-29)   **6.** (see Eq. 11-32) 1 and 3 tie, 4, then 2 and
5 tie (zero)   **7.** (a) downward in the figure ($\tau_{net} = 0$); (b) less
(consider moment arms)   **Q   1.** (a) positive; (b) zero; (c) negative; (d) negative   **3.** finite angular displacements are not commutative   **5.** (a) c, a, then b and d tie; (b) b, then a and c tie,
then d **7.** 3, 1, 2   **9.** 90°, then 70° and 110° tie   **11.** (a) decrease; (b) clockwise; (c) counterclockwise   **EP   1.** (a) $a +$
$3bt^2 - 4ct^3$; (b) $6bt - 12ct^2$ **3.** (a) 5.5×10^{15} s; (b) 26
5. (a) 2 rad; (b) 0; (c) 130 rad/s; (d) 32 rad/s^2; (e) no   **7.** 11 rad/s
9. (a) -67 rev/min^2; (b) 8.3 rev   **11.** 200 rev/min   **13.** 8.0 s
15. (a) 44 rad; (b) 5.5 s, 32 s; (c) -2.1 s, 40 s   **17.** (a) 340 s;
(b) -4.5×10^{-3} rad/s^2; (c) 98 s   **19.** 1.8 m/s^2, toward the center
21. 0.13 rad/s   **23.** (a) 3.0 rad/s; (b) 30 m/s; (c) 6.0 m/s^2;
(d) 90 m/s^2 **25.** (a) 3.8×10^3 rad/s; (b) 190 m/s
27. (a) 7.3×10^{-5} rad/s; (b) 350 m/s; (c) 7.3×10^{-5} rad/s;
(d) 460 m/s   **29.** 16 s   **31.** (a) -2.3×10^{-9} rad/s^2; (b) 2600 y;
(c) 24 ms   **33.** 12.3 kg·m^2 **35.** (a) 1100 J; (b) 9700 J
37. (a) $5md^2 + \frac{8}{3}Md^2$; (b) $(\frac{5}{2}m + \frac{4}{3}M)d^2\omega^2$ **39.** 0.097 kg·m^2
41. $\frac{1}{3}M(a^2 + b^2)$ **45.** 4.6 N·m   **47.** (a) $r_1F_1 \sin \theta_1 -$
$r_2F_2 \sin \theta_2$; (b) -3.8 N·m   **49.** (a) 28.2 rad/s^2; (b) 338 N·m
51. (a) 155 kg·m^2; (b) 64.4 kg   **53.** 130 N   **55.** (a) 6.00 cm/s^2;
(b) 4.87 N; (c) 4.54 N; (d) 1.20 rad/s^2; (e) 0.0138 kg·m^2
57. (a) 1.73 m/s^2; (b) 6.92 m/s^2 **59.** 396 N·m
61. (a) $mL^2\omega^2/6$; (b) $L^2\omega^2/6g$ **63.** 5.42 m/s   **65.** $\sqrt{9g/4L}$
67. (a) $[(3g/H)(1 - \cos \theta)]^{0.5}$; (b) $3g(1 - \cos \theta)$; (c) $\frac{3}{2}g \sin \theta$;
(d) 41.8°   **69.** 17   **N1.** 2.5 kg   **N3.** 2.51×10^{-4} kg·m^2
N5. (a) 7.1%; (b) 64%   **N7.** (a) 40 s; (b) 2.0 rad/s^2
N9. (a) $0.083519ML^2 \approx 0.084ML^2$; (b) low by (only) 0.22%
N11. (a) 11.2 mJ; (b) 33.6 mJ; (c) 56.0 mJ; (d) 2.8 $\times$
10^{-5} J·s^2/rad^2 **N13.** (a) 5.0 rad/s; (b) 1.67 rad/s^2; (c) 2.5 rad
N15. (a) 1.5 rad/s^2; (b) 0.40 J   **N17.** 6.16×10^{-5} kg·m^2
N19. 30   **N21.** (a) $\omega_0 = (\mu_s g/R)^{0.5}$

CHAPTER 12

CP   1. (a) same; (b) less   **2.** less (consider the transfer of energy from rotational kinetic energy to gravitational potential energy)   **3.** (draw the vectors, use right-hand rule) (a) $\pm z$; (b) $+y$;
(c) $-x$ **4.** (see Eq. 12-21) (a) 1 and 3 tie, then 2 and 4 tie, then
5 (zero); (b) 2 and 3   **5.** (see Eqs. 12-23 and 12-16) (a) 3, 1;
then 2 and 4 tie (zero); (b) 3   **6.** (a) all tie (same τ, same t, thus
same ΔL); (b) sphere, disk, hoop (reverse order of I)   **7.** (a) decreases; (b) same ($\tau_{net} = 0$, so L conserved) (c) increases
Q   1. (a) same; (b) block; (c) block   **3.** (a) 0.5L; (b) L
5. b, then c and d tie, then a and e tie (zero)   **7.** a, then b and c
tie, then e, d (zero)   **9.** (a) same; (b) increase; (c) decrease;
(d) same, decrease, increase   **11.** (a) 1, 2, 3 (zero); (b) 1 and 2
tie, then 3; (c) 1 and 3 tie, then 2   **13.** (a) 3, 1, 2; (b) 3, 1, 2
EP   1. (a) 59.3 rad/s; (b) 9.31 rad/s^2; (c) 70.7 m   **3.** -3.15 J
5. 1/50   **7.** (a) 8.0°; (b) more   **9.** 4.8 m   **11.** (a) 63 rad/s;

(b) 4.0 m   **13.** (a) 8.0 J; (b) 3.0 m/s; (c) 6.9 J; (d) 1.8 m/s
15. (a) 13 cm/s²; (b) 4.4 s; (c) 55 cm/s; (d) 1.8×10^{-2} J;
(e) 1.4 J; (f) 27 rev/s   **19.** (a) 10 N · m, parallel to yz plane,
at 53° to $+y$; (b) 22 N · m, $-x$ **21.** (a) $50\hat{k}$ N · m; (b) 90°
23. 9.8 kg · m²/s   **25.** (a) 0; (b) $(8.0\hat{i} + 8.0\hat{k})$ N · m
27. (a) mvd; (b) no; (c) 0, yes   **29.** (a) $-170\hat{k}$ kg · m²/s;
(b) $+56\hat{k}$ N · m; (c) $+56\hat{k}$ kg · m²/s²   **31.** (a) 0; (b) $8t$ N · m,
in $-z$ direction; (c) $2/\sqrt{t}$ N · m, $-z$; (d) $8/t^3$ N · m, $+z$
33. (a) -1.47 N · m; (b) 20.4 rad; (c) -29.9 J; (d) 19.9 W
35. (a) $14md^2$; (b) $4md^2\omega$; (c) $14md^2\omega$
37. $\omega_0 R_1 R_2 I_1/(I_1 R_2^2 + I_2 R_1^2)$ **39.** (a) 3.6 rev/s; (b) 3.0; (c) in moving the bricks in, the forces on them from the man transferred
energy from internal energy of the man to kinetic energy
41. (a) 267 rev/min; (b) 2/3   **43.** (a) 149 kg · m²;
(b) 158 kg · m²/s; (c) 0.746 rad/s   **45.** $\dfrac{m}{M + m}\left(\dfrac{v}{R}\right)$

47. (a) $(mRv - I\omega_0)/(I + mR^2)$; (b) no, energy transferred to internal energy of cockroach   **49.** 3.4 rad/s   **51.** (a) 0.148 rad/s;
(b) 0.0123; (c) 181°   **53.** the day would be longer by about 0.8 s
55. (a) 18 rad/s; (b) 0.92   **57.** (a) 0.24 kg · m²; (b) 1800 m/s
59. $\theta = \cos^{-1}\left[1 - \dfrac{6m^2h}{d(2m + M)(3m + M)}\right]$ **N1.** 12 s
N3. (a) $\vec{\tau} = (48t)\hat{k}$, with $\vec{\tau}$ in newton-meters and t in seconds;
(b) increasing   **N5.** $(5.55$ kg·m²/s$)\hat{k}$ **N7.** (a) 24 kg·m²/s;
(b) 1.5 kg·m²/s   **N9.** (a) no, $x = 2.0$ m; (b) yes, 7.3 m/s
N11. (a) 2.9×10^4 kg·m²/s; (b) 1.2×10^6 N·m   **N13.** 0.624 J
N15. -5.00 N   **N19.** 2.33 m/s   **N21.** (a) 0; (b) -22.6 kg·m²/s;
(c) -7.84 N·m; (d) -7.84 N·m

PHOTO CREDITS

INDEX

Figures are noted by page numbers in *italics;* tables are indicated by t following the page number.